Nanocarriers: Drug Delivery System

Nirmal Shah
Editor

Nanocarriers: Drug Delivery System

An Evidence Based Approach

Springer

Editor
Nirmal Shah
Department of Pharmacy
Sumandeep Vidyapeeth Deemed to be
University
Piparia, Vadodara, Gujarat, India

ISBN 978-981-33-4496-9 ISBN 978-981-33-4497-6 (eBook)
https://doi.org/10.1007/978-981-33-4497-6

This Springer imprint is published by the registered company Springer Nature Singapore Pte Ltd.
The registered company address is: 152 Beach Road, #21-01/04 Gateway East, Singapore 189721, Singapore

Foreword

Pharmacy is an evergreen and ever-expanding field of applied science where new and novel concepts arise daily as certainly as the Sun. In the ocean of pharmaceutical world, nanotechnology and nanocarriers have emerged as a giant Iceland of good hopes. For pharmacy students, researchers, and educators to become successful in their carrier, a foundation of nanotechnology knowledge is one of the essential elements for developing an adequate knowledge, critical thinking, and problem-solving skills.

In this digital era, it is very much easy to find information about any topic from digital resources, but it is very much difficult to get right information in right format. But the books provide an easy and ready material of the fundamental concept of the subject to students, researchers, and educators which in turn is an essential element of teaching-learning system. There are hundreds of books available for each topic by different authors, but selection of book with perfectly amassed information and dedicated authors is an important step towards the journey of success.

The increasing emphasis and focus on nanotechnology and novel drug delivery systems in pharmacy curriculum make it challenging for students with limited time available to learn and internalize concepts in this area of science. This book brings those memories of my early years as a pharmacy student and the evolution of the pharmacy education and gives me the confidence that it will definitely match the needs of current as well as future learners.

As a pharmacy educator and pharmaceutical scientist, I am pleased to testify and endorse this book to the pharmacy educators and learners as a novel and innovative addition to the available literature. One of the key advantages of this book is that it focuses its approach on a student-friendly manner that incorporates the appropriate illustrative diagrams and study guides as well as self-assessments to enable students to enhance their skills as self-learners.

I can assure you for the content of this book as the main author and editor-in-chief, Dr. Nirmal Shah, is an experienced educator and dedicated researcher in the field of pharmacy. I know all the authors and coauthors of this book since last many years, and I have full confidence for their knowledge, experience, and opinion.

These authors have carefully integrated the key aspect of nanotechnology and nanocarriers in the design and organization of the contents in this book. Each chapter of this book includes three distinct areas—the fundamental concepts, detailed formulation approach of different nanocarriers, and recent developments in the area of novel drug delivery systems. It offers the opportunity for readers to learn the principles, properties, applications, and advancements of nanocarriers in a logical stepwise manner. The research approach and advances included in each chapter are unique features of this book that would be beneficial to all the readers, students of UG, PG, and Ph.D.; faculty members; and other persons working in the field of nanocarriers.

Although there are several books with the similar title available on the market, to my understanding, this is the first textbook of its kind focusing on the integration of majority of nanocarriers. It is my hope and expectation that this book will provide an effective learning experience and referenced resource for all pharmacy students, research scholars, and faculties.

Jigar Ramanlal Vyas

Sigma Institute of Pharmacy
Vadodara, Gujarat, India
27 July 2020

Preface

Health science has an important role in sustaining individuals' life pleasure. In the twenty-first century, with the remarkable developments in industry and in technology, the usual cycle of nature has been distracted. This has led to massive global warming changes and the generation of more critical and life threatening diseases. The pharmaceutical sciences with its advanced technologies have gained the status of considerable and trustful profession amongst all other professional streams. This book provides valuable and unrivalled information of novel delivery system—"**The Nanocarriers**," which have tremendous therapeutical outputs compared to available conventional dosage forms. These nanocarriers have modified pharmacokinetic and pharmacodynamic behavior after its administration. The nanosized carriers have a marvelous ability in curing acute to chronic medical conditions with excellent therapeutic responses. They have the ability to enhance bioavailability with protection and stabilization of more sensitive molecules such as proteins and peptides. Nanocarriers have the capability of reducing side effects and offer means for active targeting. They play a remarkable role in the enhancement of therapeutic efficacy of phytopharmaceuticals. Therefore, such novel nanocarriers are the need of the globe for the betterment of society as far as the treatment regimen is concerned.

The objective of this book is to provide basic to advanced information on different novel carriers, which are under scale-up study for extensive commercialization. The book also includes recent discoveries and latest patents of such nanocarriers along with their broad applications in pharmaceuticals as well as in cosmeceuticals.

This book has been written with the virtue of providing thorough knowledge to all readers of pharmaceutical sciences as well as nanotechnology sciences. The latest information accessible from this book may be beneficial to students, research scholars, fellows, and new research scientists for the implementation of this knowledge in further advanced research which may be useful commercially as well as socially too.

The editor is heartily thankful to all contributing authors of the book who have furnished thorough information about existing and newly discovered nanocarriers. All authors have spent an imperative time and strength in collecting fundamental information about this novel dosage form. I am extending my greetings to all contributing authors for their fruitful attempt in making this book a success.

I expect that this book will furnish all the required information and justify the need of all readers for their study and further research.

Vadodara, Gujarat, India
27 July 2020

Nirmal Shah

Contents

Editor and Contributors

About the Editor

Nirmal Shah holds an M. Pharm Degree in New Drug Delivery System from the Maharaja Sayajirao University of Baroda, Vadodara. He holds a Doctor of Philosophy Degree in Pharmaceutical Sciences from Sumandeep Vidyapeeth, an Institution Deemed to be University, Vadodara. Presently, he is working as an Professor at the Department of Pharmacy, Sumandeep Vidyapeeth, an Institution Deemed to be University, Vadodara. He has more than 13.5 years of academic, research, and pharmaceutical industry experience. So far, he has published more than 65 scientific papers in reputed national and international journals. He is an editorial member and reviewer of many prominent journals. His area of research includes pharmaceutics, development of novel dosage forms, optimization of pharmaceutical nanoformulations, and other aspects of pharmaceutical sciences.

Contributors

Chintan Aundhia Department of Pharmacy, Sumandeep Vidyapeeth Deemed to be University, Vadodara, Gujarat, India

Kalyani Barve Shobhaben Pratapbhai Patel School of Pharmacy & Technology Management SVKMs NMIMS, Mumbai, Maharashtra, India

Vishakha S. Bhave Exxat LLC, Warren, NJ, USA
Department of Pharmaceutical Sciences, Philadelphia College of Osteopathic Medicine School of Pharmacy, Suwanee, GA, USA

Rahi M. Brahmbhatt Department of Neuroscience, University of Illinois at Chicago, Chicago, IL, USA

Pallavi M. Chaudhari Department of Pharmaceutics, Dr. D. Y. Patil College of Pharmacy, Pune, Maharashtra, India

Jagruti L. Desai Ramanbhai Patel College of Pharmacy, Charotar University of Science and Technology, Changa, Gujarat, India

Jaimini Gandhi Maliba Pharmacy College, Uka Tarsadia University, Surat, Gujarat, India

Praful Giradkar Formulation and Development, Alembic Research Centre, Vadodara, Gujarat, India

Dipti Gohil Department of Pharmacy, Sumandeep Vidyapeeth Deemed to be University, Vadodara, Gujarat, India

Shweta Gupta Ideal College of Pharmacy and Research, University of Mumbai, Mumbai, Maharashtra, India

Rinkal Jariwala Maliba Pharmacy College, Uka Tarsadia University, Surat, Gujarat, India

Yashwini Kansara Maliba Pharmacy College, Uka Tarsadia University, Surat, Gujarat, India

Swinal Kapadiya Maliba Pharmacy College, Uka Tarsadia University, Surat, Gujarat, India

Sunil Kardani Department of Pharmacy, Sumandeep Vidyapeeth Deemed to be University, Vadodara, Gujarat, India

Rajesh A. Maheshwari Department of Pharmacy, Sumandeep Vidyapeeth Deemed to be University, Vadodara, Gujarat, India

Hiral M. Manani Department of Pharmaceutical Sciences, Saurashtra University, Rajkot, Gujarat, India

Vicky Mody Department of Pharmaceutical Sciences, Philadelphia College of Osteopathic Medicine School of Pharmacy, Suwanee, GA, USA

Chandramani B. More Department of Oral Medicine and Maxillofacial Radiology, K. M. Shah Dental College and Hospital, Sumandeep Vidyapeeth Deemed to be University, Vadodara, Gujarat, India

Tosha Pandya L.J. Institute of Pharmacy, Ahmedabad, Gujarat, India

Ghanshyam Parmar Department of Pharmacy, Sumandeep Vidyapeeth Deemed to be University, Vadodara, Gujarat, India

Ashwini Patel Babaria Institute of Pharmacy, Babaria Institute of Technology, Vadodara, Gujarat, India

Brijesh Patel Faculty of Pharmacy, The Maharaja Sayajirao University of Baroda, Vadodara, Gujarat, India

Deepa Patel Parul Institute of Pharmacy and Research, Faculty of Pharmacy, Parul University, Vadodara, Gujarat, India

Dhrumi Patel Shobhaben Pratapbhai Patel School of Pharmacy and Technology Management SVKMs NMIMS, Mumbai, Maharashtra, India

Priya Patel Department of Pharmaceutical Sciences, Saurashtra University, Rajkot, Gujarat, India

Snehal Patel Department of Pharmacy, Sumandeep Vidyapeeth Deemed to be University, Vadodara, Gujarat, India

Drashti Pathak Parul Institute of Pharmacy and Research, Faculty of Pharmacy, Parul University, Vadodara, Gujarat, India

Naman R. Rao Harvard Medical School, Harvard University, Boston, MA, USA
Henry M. Goldman School of Dental Medicine, Boston, MA, USA

Jose Ruiz Department of Pharmaceutical Sciences, Philadelphia College of Osteopathic Medicine School of Pharmacy, Suwanee, GA, USA

Vidya P. Sabale Dadasaheb Balpande College of Pharmacy, Nagpur, Maharashtra, India

Piyushkumar K. Sadhu Department of Pharmacy, Sumandeep Vidyapeeth Deemed to be University, Vadodara, Gujarat, India

Girish U. Sailor Bhagwan Mahavir College of Pharmacy, Surat, Gujarat, India

Ashim Kumar Sen Department of Pharmacy, Sumandeep Vidyapeeth Deemed to be University, Vadodara, Gujarat, India

Dhanya B. Sen Department of Pharmacy, Sumandeep Vidyapeeth Deemed to be University, Vadodara, Gujarat, India

Dhiren P. Shah Department of Pharmaceutics, Shree Naranjibhai Lalbhai Patel College of Pharmacy, Umrakh, Bardoli, Surat, Gujarat, India

Pranav Shah Maliba Pharmacy College, Uka Tarsadia University, Surat, Gujarat, India

Noah Tedla Department of Pharmaceutical Sciences, Philadelphia College of Osteopathic Medicine School of Pharmacy, Suwanee, GA, USA

Hetal P. Thakkar Department of Pharmacy, Faculty of Pharmacy, The Maharaja Sayajirao University of Baroda, Vadodara, Gujarat, India

Theivasanthi Thirugnanasambandan International Research Centre, Kalasalingam Academy of Research and Education (Deemed University), Krishnankoil, Tamil Nadu, India

Devendra Vaishnav Department of Pharmaceutical Sciences, Saurashtra University, Rajkot, Gujarat, India

Sarika Wairkar Shobhaben Pratapbhai Patel School of Pharmacy and Technology Management SVKMs NMIMS, Mumbai, Maharashtra, India

Xinyu Wang Department of Pharmaceutical Sciences, Philadelphia College of Osteopathic Medicine School of Pharmacy, Suwanee, GA, USA

Aarti S. Zanwar Department of Pharmacy, Sumandeep Vidyapeeth Deemed to be University, Vadodara, Gujarat, India

Part I

Targeted Drug Delivery

Fundamentals of Nanocarriers and Drug Targeting

1

Dhiren P. Shah

Abstract

In the most recent decades, advancement of novel methodologies for the development of nano-formulations (nanocarriers) as the proficient vehicle of drug particles has offered an unlimited scope of applications in pharmaceuticals. Delivery of drugs through conventional systems has various issues associated with poor specificity, drug resistance and high toxicity. These hurdles may reduce the efficacy of numerous medications. Nanocarriers, inferable from their large surface area-to-volume ratio, can alter fundamental properties and bioactivity of medicines. Nanocarriers are colloidal medication systems having submicron molecules commonly <500 nm. Nanocarriers have been widely examined within the previous number of decades. In the present day, smart nanostructured frameworks may be comprehensively separated into inorganic and organic nanocarriers. There are a few characteristics of nanocarriers like better stability and solubility, site specificity, targeting and modified release of medications which can be fused into drug delivery systems. This is useful for the betterment of therapeutic behaviour of various drugs.

Keywords

Nanocarriers · Drug targeting · Site-specific delivery · Bioavailability

D. P. Shah (✉)
Department of Pharmaceutics, Shree Naranjibhai Lalbhai Patel College of Pharmacy, Umrakh, Bardoli, Surat, Gujarat, India

N. Shah (ed.), *Nanocarriers: Drug Delivery System*,
https://doi.org/10.1007/978-981-33-4497-6_1

1.1 Introduction

In the most recent decades, advancement of novel methodologies for the development of nano-formulations (nanocarriers) as the proficient vehicle of drug particles has offered a vast scope of applications in pharmaceuticals and biotechnology (Neubert 2011; Mishra et al. 2010). A drug can be delivered to the target with smart nanostructured product with less dosage frequency and in a controlled (spatial/transient) way to moderate the symptoms experienced with conventional systems. Specifically, they permit settling the principal basic issues experienced with traditional pharmaceutical treatments such as the rapid clearance, nonspecific distribution, low bioavailability and uncontrollable release (How et al. 2013; Sun et al. 2014; Wong et al. 2007).

The general impact is an exceptional improvement of ongoing approaches; above all else nanocarriers' activity is related with various undesirable side effects that lessen their productive use in nanomedicine. This highlights some fundamental issues in the structure, design and building of nanocarrier frameworks for biotechnology applications, emerging from the complex environment and multiform interaction set-up inside the particular biological media (Wang et al. 2012; Muller et al. 2000; Malam et al. 2009).

Nanocarriers are colloidal medication systems having submicron molecule commonly <500 nm. Nanocarriers have been widely examined in the previous couple of decades. They have demonstrated extraordinary performance in the field of medicines. Nanocarriers, inferable from their large surface area-to-volume ratio, can change fundamental properties and bioactivity of medications. Improved stability and solubility, site specificity and modified release of medicaments are few of the characteristics that nanocarriers can fuse into drug delivery systems.

Traditional delivery systems of numerous drugs have basic issues like high toxicity, poor drug resistance and medication opposition enlistment, which delicately decline the remedial proficiency of many drug systems. The nanostructured models have empowered compelling that enables effective delivery of active (including anticancer) drugs into the diseased tissues. Nanocarrier-based stages have enabled successful effective delivery of medications by exploiting the pathophysiology of microenvironment. Thus, by the way, the therapeutic results for disease conditions may essentially increase.

In the present day, smart nanostructured frameworks can be comprehensively separated into inorganic and organic nanocarriers, while their properties (physiochemical) can be tuned by modifying their creations (inorganic, organic, or half and half), dimensions (large or small sizes), shapes (rod, sphere) and properties of surface (functional group PEGylation, surface charge, covering procedures or connection of focusing on moieties). Meanwhile various nanocarrier-based stages are endorsed for the treatment of different diseases (including tumours), and numerous others are in various phases of clinical trials and clinical preliminaries (Uner and Yener 2007; Mehnert and Mader 2001). In the accompanying areas, we will talk about the principal highlights of the various kinds of nanocarriers.

Nanocarriers can offer numerous favourable circumstances over free drugs. They:

1. Protect the medication from untimely degradation.
2. Prevent drugs from rashly connecting with the natural condition.
3. Enhance incorporation of the medications into a chosen tissue (for instance, strong tumour).
4. Have pharmacokinetic and medication tissue appropriation profile control.
5. Enhance intracellular infiltration.

For quick and viable clinical interpretation, the nanocarrier ought to be produced using a biocompatible material that is very much described and effectively functionalized, which shows large differential take-up productivity in the objective cells or tissues over ordinary ones (or tissue). They should be dissolvable/colloidal under watery state for expanded adequacy. They have an all-incorporating surrounding half-life, a low pace of accumulation and a long time span of usability.

1.1.1 Advantages of Nanocarriers Over Other Novel Drug Delivery Systems

1. The acceptable size of nanocarriers to be presently administered by means of intravenous route is distinct from that of colloidal system which might occlude inside blood capillaries and needle.
2. They can simply pass through the sinusoidal spaces inside the bone marrow and spleen as compared to other systems, owing to its smaller size than microspheres and other formulations.
3. Nanocarriers contain elevated loading capability, owing to superior surface area.
4. Nanocarriers enhance the stability of drug/proteins against enzymatic degradation.
5. Nanocarriers are harmless as well as successful in site-specific and targeted drug delivery systems.
6. The targeting of drugs can be enhanced by attaching monoclonal antibodies with nanocarriers for specificity.
7. The solubility of poorly water-soluble drugs is drastically increased.
8. Due to better bioavailability of nanocarriers, there is decrease in the fluctuation of therapeutic ranges.

1.1.2 Disadvantages of Nanocarriers

1. It is likely to have high level of aggregation inside biological systems because of high surface area and surface free energy.
2. It is promptly scavenged via reticuloendothelial system ensuing in low biological half-life.

3. Sometimes it causes toxicity due to the presence of minute residual amount of organic solvent.
4. It has extreme immunogenicity or foreignness.

1.2 Different Types of Nanocarrier Systems

Nanostructured carriers have demonstrated their significance at more noteworthy degree because of low lethality, high biocompatibility, improved bioavailability, protection from degradation in GIT, high drug-loading efficiency and so on. Both lipophilic and hydrophilic medications can be loaded in lipid nanocarriers. Nanocarriers are categorized based on their compositions.

1.2.1 Organic Nanocarriers

1.2.1.1 Liposomes

Liposomes have achieved an extraordinary consideration during the foremost recent couple of decades in biomedicine, particularly for antitumour drug delivery medications (Patel et al. 2014). Liposomes have many advantages over conventional systems. These are anticipation of early degradation of encapsulated medicaments, advanced delivery of medications, formulation of low-cost medications, protection of the drug from ecological elements, improved performance of the dosage form and effective treatment with diminished foundational lethality (Lee et al. 2015). They are encapsulated with polymers, for example, polyethylene glycol, to display delayed half-life in blood flow. Furthermore, liposome drugs have unique pharmacokinetic properties as compared to free drugs available in solution form (Malam et al. 2009). The structure of liposomes involves circular vesicles having a fluid centre encased by lipid bilayers. They consume single or numerous bilayers (Fig. 1.1a). Those containing one bilayer film are named as unilamellar vesicles or enormous unilamellar vesicles dependent on their sizes. On the off chance that more than one bilayer is available, they are called multilamellar vesicles. Liposomes vary with respect to the technique for designing size, surface charge and composition (Torchilin 2005). Liposomes are ordinarily utilized as model cells or transporters for various bioactive specialists including medications, immunizations and nutraceuticals (Lee et al. 2015).

Liposomes loaded with active drug arrangements control medication conveyance inside plasma, accordingly prompting improved pharmacokinetics and biodistribution of medications. For instance, doxorubicin-loaded PEGylated liposome diminishes volume distribution of doxorubicin from 1000 to 2.8 L/m^2 in plasma when contrasted with free tranquillisers in arrangement. It prompts a higher medication focus inside tumour yet lessens sedation and fixation in typical tissues (Torchilin 2005).

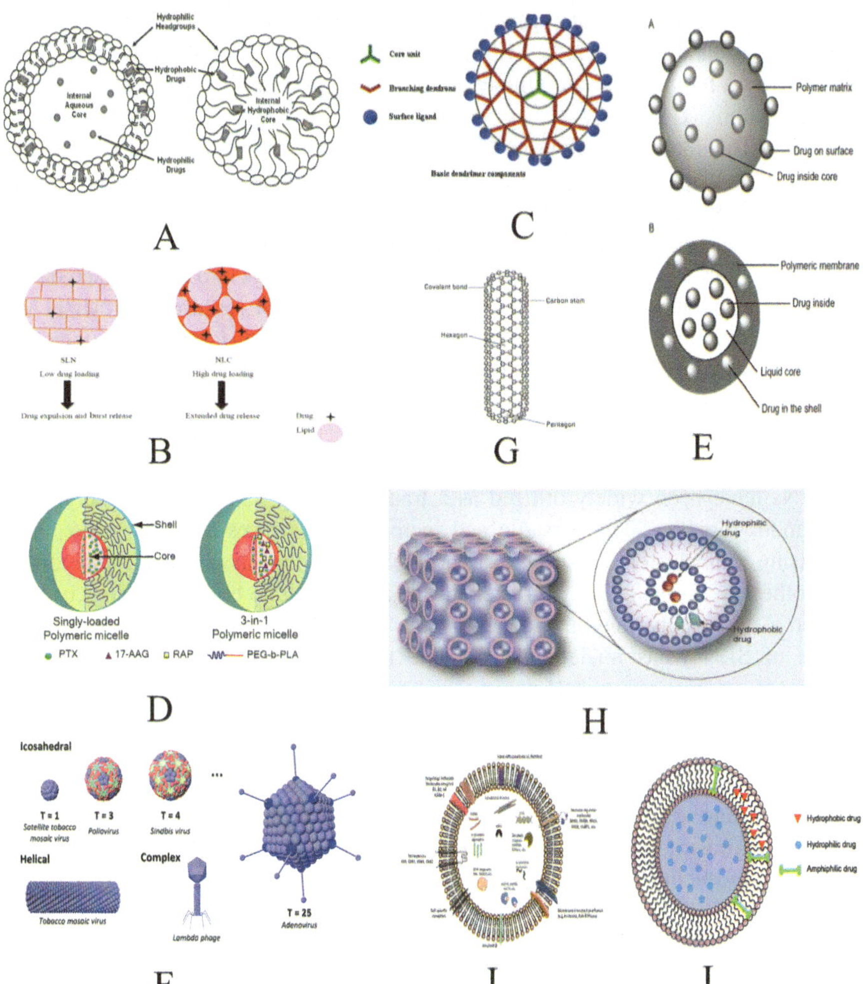

Fig. 1.1 Schematic diagram depicting the structure of various nanocarriers which can accommodate both hydrophilic and hydrophobic drugs either in the internal aqueous core or in the phospholipid bilayer, respectively. (**a**) Liposome. (**b**) Solid lipid nanoparticle. (**c**) Basic dendrimer components. (**d**) Polymeric micelles. (**e**) Polymeric nanoparticles. (**f**) Virus-based nanoparticles. (**g**) Carbon nanotube. (**h**) Cubosomes. (**i**) Exosomes. (**j**) Niosomes (Khan 2010)

1.2.1.2 Solid Lipid Nanoparticles (SLNs)

In the mid-1990s, SLNs were created within the size range of 50–1000 nm. SLNs are formulated by mixing melted solid lipid(s) in water together with emulsifier(s) to stabilize the final product. The foremost generally utilized strategies for formulating SLNs are high-weight homogenization and smaller scale emulsification. SLNs provide an extremely lipophilic lipid network for medications to be distributed or disintegrated. A large variety of solid lipids include mono-, di- and triglycerides;

steroids; free unsaturated fats; waxes; and free greasy alcohols which are utilized for the formulation/arrangement of SLNs. In the formulation of SLNs lipids which are solid at room temperature are utilized rather than liquid lipids (oils), particularly in nanoemulsions (Uner 2007).

SLNs offer more attractions in areas like controlled or modified drug delivery system. This might be due to good stability, absence of toxicity, high drug-loading capacity of inadequate hydrophilic drugs, improved bioavailability and simplicity just as affordability of enormous scale formation (Mehnert and Mader 2001; Zeb et al. 2017). Different models are available to associate drugs into SLNs based on the following characteristics: dependency on the structure of SLNs (drug, surfactant and lipid) and manufacturing conditions (cold or hot homogenization), homogeneous drug dispersal in the network of lipid (strong arrangement/homogeneous grid) of SLNs (Fig. 1.1b), fusion into the shell encompassing the lipid centre (sedation-enhanced shell model) or fusion into the centre encompassed by a lipid shell (tranquillity-improved centre model) (Muller et al. 2002).

SLNs have been widely utilized as a transporter for various chemotherapeutic antitumour moieties. The versatile and incredible properties of SLNs make them promising nanocarriers to survive or possibly limit the disadvantages of normal chemotherapy. SLNs give a characteristic stage to integrate lipophilic drugs used in cancer. Additionally, ongoing development (for example, polymer–lipid mixture nature of SLNs) has made it conceivable to consolidate ionic and hydrophilic anticancer medications too. However, fast end of SLNs from the bloodstream by reticular endothelial system (RES), exemplification of ionic and hydrophilic anti-cancer medications and maintaining of the rate and degree of medication discharge from SLNs are the real hindrances seen with SLNs that limit them from becoming convincing nanocarriers in anticancer drug delivery. SLN systems have been used for the administration of various anticancer drugs like docetaxel (Qureshi et al. 2017), doxorubicin (Battaglia et al. 2014), paclitaxel (Yuan et al. 2008), methotrexate (Kakkar et al. 2015) and 5-fluorouracil (Patel et al. 2014).

1.2.1.3 Dendrimers

Dendrimer systems have their various unique characteristics like multivalency, actual number of atomic weight, circular shapes, expanded number of branching and, importantly, monodispersed macromolecules with a typical nanosize (Basu et al. 2004). It is shown that an average dendrimer atom contains exceedingly expanded layers comprising emphasizing units and various active terminals gathered with initiator centre. Due to this, such compositional arrangement offers an uncommon control over the surface effectiveness, dendrimer size, scattering length and shape.

The exceedingly scattered structure of dendrimers realizes lumbering outside get-togethers. The medications focusing on ligands can be joined to change the surface functionalities so as to acquire explicit goals, which for the most part embroil exact contact at cell dividers and organically dynamic places. The use of dendrimers has been comprehensively broad in fields of biomedicine, which includes immunology; antibodies; antibacterial, antiviral and anticancer medication transportation to

Table 1.1 Various drug conjugations with dendrimers

Name of drug	Process	References
Doxorubicin	Photochemical internalization technology	Lai et al. (2007)
Cisplatin	Platinate conjugation	Malik et al. (1999)
5-FU	Time-sequenced propagation technique	Zhuo et al. (1999)
Doxorubicin	PEGylated based	Lee et al. (2006)
5-FU	PEGylated polyamidoamine	Bhadra et al. (2003)

particular sites; and MRI (Stiriba et al. 2002). An expressive structure of dendrimers made out of centre, halfway layers and outside bunches is showed in Fig. 1.1c. Table 1.1 shows various drug conjugations with dendrimers.

1.2.1.4 Polymeric Micelles (PMs)

Polymeric micelles are 10–100 nm colloidal particles. It is framed by the self-gathering of engineered amphiphilic di- or tri-square copolymers particularly in an aqueous medium (Zhu and Liao 2015). Due to amphiphilic nature of polymeric micelles, di- or tri-square copolymers subsequently contain hydrophilic and hydrophobic units. These copolymers when exposed to an aqueous system, in a certain concentration, form structured micelles. Out of both units, hydrophobic unit of square copolymer establishes the centre of micelle, while hydrophilic unit shapes the shell of micelles. In this way, polymeric micelles have a centre/shell structure with a hydrophilic shell and a hydrophobic centre (Fig. 1.1d) (Biswas et al. 2016). It is reported that the hydrophobic centres of PMs allow the entanglement of hydrophobic medications and maintain release of drug from polymeric micelles. Be that as it may, the hydrophilic shell of polymeric micelles helps in stabilizing the centre, ensures the polymeric micelle solvency in water and maintains in vivo pharmacokinetics (Gothwal et al. 2016). The medication can be joined into polymeric micelles either by physical capture or by means of substance construction (Nakanishi et al. 2001). Various techniques utilized regularly for the development of polymeric micelles are co-solvent evaporation method, dialysis strategy, freeze-drying method, oil-in-water emulsion technique and solvent evaporation technique (Rapoport 2007).

Polymeric micelles have dual active and passive targeting of antitumour drugs. Aside from the underlying focus, the active focusing of malignancies with polymeric micelles is additionally conceivable by planning the more intelligent polymeric micelles (environment/stimulus-responsive polymeric micelles, i.e. receptive to changes in temperature, pH and so forth) or by altering the surface of polymeric micelles with tumour-focusing on ligand (Rapoport 2007). Polymer science advancement and adaptability of polymeric micelles shaping square copolymers make them an appealing transporter in the therapy of cancer. A couple of normally utilized copolymers for polymeric micelles are PEG-PLA (PEGylated polylactic acid), poloxamers, PEG-PAA (PEGylated polyaspartic acid), PEG-PCL (PEGylated polycaprolactone), PEGylated polyglutamic acid and PEG-PLGA (PEGylated poly (lactic-co-glycolic acid)) (Biswas et al. 2016). Numerous chemotherapeutic specialists including docetaxel, doxorubicin, paclitaxel, cisplatin and methotrexate have effectively been developed in polymeric micelles. A couple of studies representing PMs for drug-incorporated micelles are based on poloxamer (Ren et al. 2015), micelles of PEG-PLA (Li et al. 2009), PEGylated polyglutamic acid micelles (Hamaguchi et al. 2005) and micelles of PEG-PAA (Nakanishi et al. 2001).

1.2.1.5 Polymeric Nanoparticles (PNPs)

Over the most recent time, polymers have gained a lot of interest in the area of drug delivery system. Polymeric nanoparticles (PNPs) are 10–1000 nm colloidal particles of biodegradable polymers (Rao and Geckeler 2011; Bamrungsap et al. 2012). In light of its basic structure, polymeric nanoparticles (PNPs) can be arranged as either nanocapsules (frame type; Fig. 1.1e) or nanospheres (lattice type). PNPs may entrap or disperse the drug in a network of polymer. In PNPs, the medication is broken down/scattered in the fluid centre of oil or water exemplified by a strong polymeric film. In the categories of PNPs, the chemical conjugation or adsorption of drug on surface (capsule or matrix) is likewise conceivable (Prabhu et al. 2015).

There are two types of technique which are applied for formulation and development of PNPs: (1) preformed polymer dispersion and (2) monomer direct polymerization.

Type 1 technique strategies (preformed polymer dispersion) include dialysis, dispersion of preformed polymers as nanoprecipitation, salting out, supercritical fluid technology and solvent evaporation.

Type 2 technique strategies (monomer direct polymerization) include interfacial polymerization, monomer direct polymerization incorporating mini emulsion polymerization, microemulsion and emulsification-based polymerization, and controlled radical polymerization (Rao and Geckeler 2011).

Various biodegradable and biocompatible, both regular and engineered, polymers have been used for the development of PNPs. Due to biodegradable nature, these polymers degrade inside the body and are converted into single monomers and consequently eliminated through typical metabolic pathways. Mainly utilized engineered polymers incorporate PGA (polyglycolic acid), PLA (polylactic acid), PEG, PLGA, PCL (polycaprolactone), HPMA (N-(2-hydroxypropyl) methacrylamide), copolymer and polyaspartic acid. However, for the most part utilized common polymers are gelatin, heparin, collagen, chitosan, alginate, dextran and albumin (Wang et al. 2009).

Advantages of PNPs (Hu et al. 2010):

1. Better dependability on capacity and in vivo
2. Better medication payload
3. Molecule size appropriation is homogeneous
4. Better and controllable physicochemical properties
5. Sustained and progressively controlled drug delivery

It is additionally conceivable to make multifunctional PNPs for the treatment of cancer having ideal shapes, sizes and surface charges (Zhu and Liao 2015). Savvy polymers are proficient to change their physicochemical qualities in light of certain environmental signals, leading to precise and programmable drug conveyance in malignant growth treatment (Pérez-Herrero and Fernández-Medarde 2015). A couple of agent instances of studies, utilizing PNPs as a transporter for antineoplastic molecules, incorporate PLGA NPs (Khuroo et al. 2014; Wang et al. 2011), PLA NPs, 40 PCL NPs (Wang et al. 2015a, b), chitosan NPs (Anitha et al. 2014) and alginate NPs (Zhao et al. 2012).

1.2.1.6 Virus-Based Nanoparticles (VNPs)

As of late, virus-based nanoparticles (viruses as nanocontainers) were broadly investigated for the purposes including dosage form delivery, quality treatment, immunization, targeting and imaging (Pattenden et al. 2005). VNPs are nanosized (about <100 nm), self-accumulated core protein enclosures having well-characterized geometry and uniform structures (Fig. 1.1f) (Manchester and Singh 2006; Singh et al. 2007). VNPs confine from various sources like plant viruses tobacco mosaic infection (TMV), cowpea mosaic virus (CPMV), cowpea chlorotic mottle virus (CCMV) and red clover necrotic mosaic virus (RCNMV); group house virus (creepy crawly infections); bacteriophages or bacterial viruses (MS2, M13, Qβ); and individual viruses (polyomavirus, adenovirus) that have been explored for its application in nanotechnology (Ma et al. 2012).

VNPs offer several highlights such as:

1. Morphological consistency
2. Biocompatibility
3. Simple surface functionalization
4. Accessibility in different shapes and sizes

Acceptability of adaptable substance and genetic alterations in their surface enable virus-based nanoparticles to meet the requirements of prescription nanocarriers including hydrophilicity, biocompatibility and upgraded drug entanglement capacity that may improve drug-circulating time in the body (Jabir et al. 2012). There are two strategies accessible to capture drug into VNPs (Douglas and Young 2006).

1. Physical method: It is a very simple method; due to its simplicity it is economic. A basic and regular procedural step of supramolecular reassembly/self-assembling of viral protein capsid is utilized to load drug in VNPs. It is also shown that drug-loading capacity is lesser as compared to other methodologies available.
2. Chemical method: In this method, drug is loaded onto virus-based nanoparticles by means of covalent bonding of drug atoms to certain (normally existing or designed) responsive destinations on the cupid proteins. Due to covalent bonding, strength is higher as compared to physical method.

In general, VNPs loaded with chemotherapeutic specialists/antitumor agents have been explored for tumour-focusing in different and many investigations which are highly useful for human beings (Cao et al. 2014; Honarbakhsh et al. 2013).

1.2.2 Inorganic Nanocarriers

1.2.2.1 Carbon Nanotubes (CNTs)

The structure of carbon nanotubes (CNTs) involves a tube-like, nanosized, hollow assembly of particles of carbon which was founded by Iijima in 1991 (Iijima 1991). CNTs have been placed with a group of fullerenes (carbon of a third allotropic type). The shape is given by moving up of graphene sheet(s) into a cylinder-like structure (Bianco 2004). There are two types of CNTs. The first type is known as single-walled carbon nanotubes (SWCNTs) which are formulated by moving up a solitary graphene sheet. Second type is multi-walled carbon nanotubes (MWCNTs) which are shaped by folding up a limited number of concentric graphene sheets into a cylinder-like structure/gathering (Fig. 1.1g). CNTs have cross-sectional measurements (nm) in lengths and range that can stretch out thousand times to their widths. Ordinarily, it is reported that external widths of SWCNTs and MWCNTs are in the range of 0.4–2 nm and 2–100 nm, respectively (Madani et al. 2011). There are different techniques available to prepare CNTs such as arc discharge, chemical vapour deposition, and thermal and laser ablation (Yan et al. 2007).

CNTs have some important physicochemical and natural qualities which make them a suitable carrier for drug. A portion of these qualities include nano-needle shape, high thermal and electrical conductivities and their capacity for surface alterations, empty solid structure, high phase ratio (length:width >200:1), high mechanical quality, ultralight weight and ultrahigh surface region. There is a very simple technique of endocytosis utilized for transportation of drug molecule. The needle-like state of CNTs allows them to cross the membrane film by means of "endocytosis" or "needle-like infiltration"; this will help to go into the cell. There are certain serious issues faced with CNTs as a carrier. These are its low solubility in water and toxicity. However, CNTs can be surface functionalized. Its surface characteristics make them water dissolvable, biocompatible, non/less harmful and a serum-stable carrier (Vardharajula et al. 2012).

It is observed that apart from cell penetration capacities, there are certain special physicochemical properties to be kept in mind. CNTs are perfect nanocarriers for antitumour agents because of its suitable surface functionalization, intrinsic stability, structural flexibility and high drug-loading capacity. Since last two decades, CNTs have been broadly contemplated as an anticancer agent carrier (Iannazzo et al. 2013). In general, there are two simple ways utilized to encapsulate drug in CNTs: firstly anticancer drugs can be encapsulated in the inner cavity of CNTs (Ajima et al. 2008) and secondly it can bind either non-covalently or covalently to the outside of CNTs (Wu et al. 2009).

Besides, the attachment of various targeting agents to the surface-functionalized CNTs focused on the route of anticancer drug delivery (Fabbro et al. 2012). A couple of examples of utilization of CNTs in anticancer drug transport include incorporation of methotrexate (Das et al. 2013), mitomycin C (Levi-Polyachenko et al. 2009), paclitaxel (Lay et al. 2010), carboplatin (Arlt et al. 2010), cisplatin (Adeli et al. 2011; Bhirde et al. 2009) and doxorubicin (Adeli et al. 2012; Ji et al. 2012).

1.2.2.2 Mesoporous Silica Nanoparticles (MSNs)

The SiO_2 has seen expanded uses in the subject of biomedicine because of their simple blending methods and availability for large-scale manufacturing. Among silica materials, mesoporous silica is of specific importance in drug delivery as they can have a lot of opportunities due to honeycomb-like design with many pores available in it (Slowing et al. 2008). MSNs have the following excellent features (Li et al. 2017):

1. It has high biocompatibility.
2. It has high explicit surface area and pore volume.
3. It has high loading capacity.
4. It has controllable pore widths extending from 2 to 50 nm with desirable distribution chemical and thermal balance.
5. It is versatile in nature specifically in drug loading with lipophilic and hydrophilic attributes.

Above-listed features make them promising nanoscale drug carriers. What is more, simplicity of surface functionalization for controlled and focused drug delivery enables MSNs to reduce toxicity and increase therapeutic efficacy of drugs (Wang et al. 2015a, b).

Novel engineering and MSNs' attractive properties make these nanocarriers a perfect tool for anticancer drug delivery. It is also noted that drug targeting can be done by both modes of passive and active targeting. MSNs' mesoporous structure empowers to load high amount of anticancer drug as well as nanoscale molecule size range of MSNs helps them to gather in tumour tissues by means of passive targeting. MSN surface functionalization with various site-explicit directing empowers them to target tumour cells via active targeting. A wide range of anticancer medications, including methotrexate (Rosenholm et al. 2010), paclitaxel (Lu et al. 2007a, b), doxorubicin (Lebold et al. 2009) and camptothecin (Lu et al. 2007a, b), have been viably transported by means of MSNs.

1.2.3 Organic/Inorganic Hybrid Nanocarriers

These nanocarriers are created to join the upsides of inorganic and organic materials. Organic materials' specific functionalities at the outside of inorganic NPs are used to upgrade the selectivity and effectiveness of antitumour drugs. As an example, surface covering with polyethyleneimine (PEI) upgrades cell take-up of MSNs as well as produces cationic surface for effective nucleic acid delivery (Xia et al. 2009). It was accounted for that binding of hyperbranched PEI with MSNs brought about a high drug load and supported intracellular transport of short interfering RNA (siRNA) (Prabhakar et al. 2016). These MSNs/PEI hybrid nanocarriers effectively arrive at the tumour environment drifted via oozing from endosomes to the cytoplasm.

MSNs are used to retard early release of drug, prevent multidrug resistance and attain release of stimulus-responsive drug (Han et al. 2015). Rare models with anticancer drugs by this material are zoledronic acid for breast cancer (Desai et al. 2017) and lipid-capped doxorubicin-loaded MSNs with pH/redox-responsive release of drug (Han et al. 2015).

1.2.3.1 Cubosomes

The essential cubosome assembly incorporates honeycomb structure isolating the two inner watery channels alongside enormous interfacial territory. Cubosomes are discrete, submicron, nanostructured particles of the bi-continuous cubic fluid crystalline stage (Fig. 1.1h). The term cubosomes was coined by Larsson, which reflects the cubic atomic crystallography and closeness to liposomes. These are nanoparticles which are self-assembled fluid crystalline particles of specific surfactants with appropriate proportion of water with microstructure. The cubic stages have an exceptionally high strong-like thickness, which is a remarkable property due to their interesting bi-continuous structures which surround two distinct regions of water separated by a controlled bilayer of surfactant application.

Most likely cubosomes are made out of polymers, lipids and surfactants with polar and non-polar segments, thus said as amphiphilic. The amphiphilic particles are driven by the hydrophobic impact on polar dissolvability to incautiously recognize and collect into a fluid precious material of nanometre scale. In this way cubosomes are bi-continuous cubic fluid stage encasing two separate districts of water isolated by surfactant-controlled bilayers. Further these are like fluid crystalline substance with cubic crystallographic evenness and are optically isotropic and strong as well. The cubic stage can break and frame thermodynamic and colloidal stable particulate scatterings. In cubosomes dynamic substance constituent atoms are tied down through bonds to the polar leader of the phospholipids. The polymer and the individual medication compound structure a 1:1 or 2:1 complex relies upon the substance (Almeida et al. 1975).

Points of interest offered by cubosomes are the following:

1. Biocompatible and non-toxic
2. Simple methodology
3. Excellent bioadhesive properties
4. Thermodynamically stable
5. Ability of encapsulating hydrophilic, hydrophobic and amphiphilic drugs
6. Because of high internal surface territory and cubic crystalline structures, high drug-loading capacity

But along with all aforementioned advantages, the large-scale generation is troublesome because of phase behaviour and viscous properties.

1.2.3.2 Exosomes

In previous decades, vast research was done on exosomes. "Exosome" was first founded by Rose Johnstone in 2005. During the development of reticulocytes, he watched the arrangement of "an intracellular sac loaded up with little layer enclosed structure of almost uniform size" (Fig. 1.1i) (Johnstone 2005).

Exosomes began with little-size endosomes, going from 40 to 100 nm (Simons and Raposo 2009). An exosome is a "nanosphere" with a bilayer, containing different kinds of lipids and proteins. A portion of these proteins incorporate transport proteins, heat-shock proteins, proteins related with multivesicular body biogenesis (MVB) and tetraspanin. Notwithstanding proteins, exosomes are involved with various kinds of lipids, for example, sphingolipids, cholesterol, ceramides, phosphoglycerides and immersed unsaturated fat chains (Qin and Xu 2014). The creation of exosomes is basic since they fill in as a biomarker and give a sign of its capacity in biological procedures (Lai et al. 2013).

Major advantages:

Exosomes have certain advantages as follows (Petersen et al. 2014):

1. Exosomes as a copy of "nature's delivery systems" can be considered for the delivery of biological materials.
2. Not at all like common nanoparticulate systems, for example, polymeric nanoparticle liposomes, exosomes can conceivably maintain a strategic distance from the endosomal pathway and lysosomal destruction, and directly deliver material/drug into cytoplasm.
3. They are normally steady and have inherent focusing on properties relying upon the creation of the exosomes.
4. They have the capacity to cross BBB.

Major disadvantages:

1. Exosomes include heterogeneous parts and may indicate immunogenicity (immunosuppressive or immunostimulatory) impacts dependent on parental donor cell nature.
2. Exosomes that convey caspase-3 may likewise hinder cell passing by apoptosis or improve cancer cell survival by inhibiting anticancer drug accumulation.

1.2.3.3 Niosomes

Niosomes are considered as novel delivery systems; they can enhance the stability of pharmaceutical compounds. Numerous elements can affect noisome development, viz. the preparation method, type and amount of lipids, hydration temperature, drug entrapment and surfactant (Fig. 1.1j).

The main parts used for the preparation of niosomes are lipid mixture (cholesterol or L-α-soya phosphatidylcholine) and non-ionic surfactants. Mixture of lipids is used to give rigid nature and proper shape to niosomes. Niosomes are the competent medication transporters which require a bilayer structure made up of lipid mixture and non-ionic surfactant (L-α-soya phosphatidylcholine or cholesterol) in aqueous phase.

Niosomes are vesicles arranged by non-ionic surfactants. They have a hydrophilic head gathering and a hydrophobic tail which influence the entrapment proficiency of drug. Niosomes are composed from cholesterol and surfactant uncharged single chain, while liposomes are sorted out from twofold chain phospholipids.

1.2.3.4 Self-nanoemulsifying Drug Delivery System (SNEDDS)

SNEDDS are oil-in-water nanoemulsion, available as anhydrous isotropic blend of surfactants, drug and oil, which when brought into aqueous phase with gentle agitation get changed over into nanoemulsion. The stomach-related motility of gastrointestinal tract gives expected agitation to the development of nanoemulsions. The SNEDDS holds advantages related with nanoemulsions like increased bioavailability, expanded drug saturation, improved enzymatic and chemical stability and simplicity of scale-up and manufacturing. The SNEDDS of inadequate water-soluble drugs has demonstrated solubility enhancement. The result demonstrated desirable zeta potential, stability, globule size and sixfold increment in drug release when contrasted with marketed valsartan powder and tablet. From the outcomes, it might be presumed that SNEDDS of poorly soluble drugs shows a promising oral delivery for antihypertensive drugs. Likewise, another investigation has additionally demonstrated the drug release improvement of olmesartan and valsartan as SNEDDS (Gupta et al. 2011; Raval et al. 2012).

1.2.3.5 Proliposomes

Proliposomes are characterized as dry, free-streaming particles with a scattered framework that can quickly form a liposomal suspension when it comes in contact with water. Compared with liposomes, proliposomes display more points of interest in advancing drug absorption. Due to their strong properties, the physical soundness of these liposomes can be improved without affecting their intrinsic characteristics. Accordingly, proliposomes would be a potential vehicle to help improve the oral absorption of hydrophobic medications (Hildebrand et al. 2003). The pharmacokinetic parameters of isradipine proliposomes were assessed in male albino Wistar rats and they were found to give 2.4-fold enhancement in bioavailability when compared with control oral suspension (Bobbala and Veerareddy 2012). The improved oral bioavailability of valsartan-loaded proliposomes was also noted (Betageri and Nekkanti 2015).

1.3 Drug Targeting

Most of the drugs introduced to the clinical study exert their effects through concentration-dependent reversible interactions at specific receptor site(s). This may lead to obtaining of desirable therapeutic response by delivering optimum amount of drug to the site of action with subsequent control on drug release rate. The development over past decades indicates explicit progress in the area of controlled and targeted drug delivery. It comes to be much more relevant in the present context. It is apparent that most of the diseases treated by cytotoxic agent demand not only for controlled drug delivery but also for site-specific delivery at quantitative levels. The cell-related biological regular events occurring in high order of specificity and precision offer the basis for quantitative targeted drug delivery.

These involve a number of essential bioligands and biosignalling. These operate through bioports referred to as receptors. The ligand-receptor interactions are highly stereospecific. The ligand of receptors could be exploited for site/cell-specific drug delivery quantitatively in a well-defined manner. Let us discuss and define various facets of essentials of drug targeting. The practical realization of the concept shall be a great breakthrough in medical sciences. The precision programme and site specificity of the system are poised as a magic bullet. It will not only ensure site specificity, but also mitigate toxicity of drug(s) to non-target site(s) as a result of well-controlled and attenuated drug level.

Selective drug delivery or targeting seeks to improve the benefit/risk ratio associated with drugs. Ideally, a drug intended for clinical use should have high therapeutic index, which is a ratio of drug efficacy and drug toxicity. Many drugs, particularly chemotherapeutic agents, have a narrow therapeutic window and so their clinical use is limited and compromised by dose-limiting toxic side effects. Approaches are being adopted to control the distribution of drug either by incorporating it in a carrier system or by altering the structure of the drug at the molecular level or by controlling the absorption of the drug into the bioenvironment to ensure a programmed and desirable biodistribution. Rapid applications of the recent developments in molecular genetics are enabling both the diagnosis of disease and understanding of pathogenesis to the clockwork precision, while advents on designing of novel drugs and delivery systems offer opportunities for essential cure of diagnosed diseases.

The efforts to improve drug effectiveness in therapeutics have been assisted by parallel developments in molecular and cell biology. On the one hand, hybridoma and recombinant DNA technology have come strongly, and on the other, a number of cell membrane receptors and their interactions with respective ligands have been investigated and reported for many cell-related biological functions. Such expansions have supported the successful developments of target-oriented drug delivery systems.

The drug delivery technology has certainly infused new interests in seemingly traditional old drugs by providing new life specifically through their therapeutic targets. It is appreciated that target-oriented drug administration with improvements in therapeutic efficiency, reduction in side effects and optimized dosing regimen shall be the leading trend in the area of therapeutics. Various aspects of the target-oriented site-specific delivery at cellular, molecular or submolecular level are discussed in this chapter (Zhuxuan et al. 2019; Peng et al. 2018).

1.3.1 The Concepts of Targeting

The concept of designing specified delivery system to achieve selective drug targeting has been originated from the perception of "Paul Ehrlich", who proposed drug delivery as a "magic bullet". He described targeted drug delivery as an event where "a drug – carrier complex/conjugate, delivers drug exclusively to the preselected target cells in a specified manner".

Bangham's observation on phospholipid hexagonal liquid crystals, that they are selective to the ions in a manner similar to biomembrane, led to the discovery of artificial vesicular system based on phospholipid amphiphiles. Gregoriadis described drug targeting using novel drug delivery systems as "old drug in new clothes". In traditional drug delivery systems such as oral ingestion or intravascular injection, the drug is distributed throughout the body through the systemic blood circulation. For most therapeutic agents, only a small portion of the medication reaches the organ to be affected, such as in chemotherapy where nearly 99% of the drugs administered do not reach the tumour site. To deliver drug at the targeted site, we have to find a place to concentrate the medication in the tissues/organs in which we are interested, while decreasing the relative concentration of the medication in the remaining tissues/organs. This can be done by various ways like (a) avoiding the host's defence mechanisms and (b) inhibiting non-specific medication distribution in the liver and spleen, and finally a prepared system can reach the projected site of action in higher concentrations. It is believed that targeted medication delivery improves its efficacy as well as reduces side effects.

It was observed that when we want to introduce a targeted drug release system, the subsequent important points for such a system must be taken into consideration, such as (a) the route taken for the delivery of the drug, (b) the targeted site where the drug should be released, (c) drug properties, (d) side effects of the drugs, and finally (e) the disease.

1.3.2 Principle and Rationale of Drug Targeting

Controlled rate and mode of drug delivery to pharmacological receptor and specific binding with target cells, as well as bioenvironmental protection of the drug, are specific features of targeting. Every event stated contributes to higher drug concentration at the site of action and resultant lower concentration at non-target tissue where toxicity might crop up. The restricted distribution of the parent drug to the non-target site with effective accessibility to the target site could maximize the benefits of targeted drug delivery.

1.3.2.1 What Is Drug Targeting?

- To obtain a desired therapeutic response, the correct amount of drug should be transported and delivered to the site of action with subsequent control of drug absorption rate.
- To avoid distribution of drug to other tissues which seems to be unnecessary, wasteful and a potential cause of toxicity.

The therapeutic response of a drug depends upon the interaction of drug molecules with cell on cell membrane-related biological events at receptor sites in a concentration-dependent manner. Drug targeting means selective and effective localization of the pharmacologically active moiety at pre-identified target(s) in therapeutic concentration, as well as restricting its access to non-target(s) normal cellular linings. This will minimize toxic effects of drugs and maximize the therapeutic index. There are few desirable properties for the drug targeting such as drugs with less stability, low therapeutic index, less absorption, low specificity, large volume of distribution and finally short half-life.

1.3.2.2 Ideal Targeted Drug Delivery

The following are important properties of ideal targeted drug delivery:

1. Its method of preparation should be reasonably simple, reproductive and economically viable.
2. It must be non-toxic in nature to human body and environment also.
3. It must be biocompatible and physicochemically stable both in vivo and in vitro.
4. It must control drug distribution to target cells/tissues/organs.
5. It must have uniform capillary distribution.
6. It should obtain very controllable and predictable rate of drug release.
7. There must be minimum drug leakage during transit into body.
8. Carrier used for targeting drug delivery must be biodegradable in nature and it must readily eliminate from the body with no problem.

1.3.3 Carriers

Carrier is one of the most important entities essentially required for successful transportation of the loaded drug. They are drug vectors, which sequester, transport and retain the drug, which elute or deliver it within or in the vicinity of the target. Carriers can do so through either inherent or acquired characteristics (through structural modification), to interact selectively with biological targets, or otherwise they are engineered to release the drug in the proximity of the target cell lines demanding optimal pharmacological action (therapeutic index).

1.3.3.1 Requirement of Drug Carriers

An ideal drug carrier should have the following subsequent features:

- It must be competent to cross-anatomical barriers like tumour vasculature as far as tumour chemotherapy is concerned.
- It must be recognized specifically and selectively by the target cells and must maintain the need and specificity of the surface ligands.
- The linkage of the drug and therefore the directing unit (ligand) should be stable in plasma, interstitial and other biofluids.
- Carrier should be biodegradable particulate or macromolecule, non-immunogenic and non-toxic. Once recognition and internalization are completed, the carrier system should release the drug moiety inside the target organs/tissues or cells.
- The biomodules used for carrier navigation and site recognition must not be omnipresent; otherwise it should cross over the sites and should defeat the concept of targeting.

Based on the nature of their origin, carriers are categorized as endogenous (low-density lipoprotein, high-density lipoprotein, chylomicrons, serum albumin and erythrocytes) and exogenous (microparticulates, soluble polymeric and biodegradable polymeric drug carriers).

Various carrier systems used for targeted drug delivery are:

1. Colloidal carriers:
   (a) Vesicular systems: immunoliposomes, liposomes, virosomes, niosomes, pharmacosomes
   (b) Microparticulate systems: nanocapsules, microparticles, albumin microspheres, nanoparticles, magnetic microspheres
2. Cellular carriers: Resealed erythrocytes, serum albumin, antibodies, platelets, leukocytes

1.3.3.2 Various Types/Levels of Targeting (Zhuxuan et al. 2019; Vyas and Khar 2002)

1.3.3.2.1 Passive Targeting

Drug delivery systems (DDS) which are targeted to systemic circulation are characterized as passive delivery systems. In passive targeting, drug targeting is achieved because of the body's own natural response against physicochemical characteristics of the drug or drug carrier system.

Since the discovery of the EPR effect in 1980s by Maeda et al., a lot of efforts were done to understand the significance of this phenomenon in tumour targeting and appropriate DDS was developed. Some of these nanocarriers, such as the marketed Doxil© and Caelyx©, are now successfully used in clinics and the EPR effect became a golden standard in the design of passive tumour-targeted systems.

In passive targeting, the drug's success is directly related to circulation time. This is achieved by enveloping the nanoparticle with coating. It is possible to achieve this with several substances, for example polyethylene glycol (PEG). By adding PEG to the surface of the nanoparticle, it is rendered hydrophilic, thus allowing water molecules to bind to the oxygen molecules on PEG via hydrogen bonding. The result of this bond is a film of hydration around the nanoparticle which makes the substance antiphagocytic. The particles obtain this property due to the hydrophobic interactions that are natural to the reticular endothelial system (RES); thus the drug-loaded nanoparticle is able to stay in circulation for a longer period of time. To work in conjunction with this mechanism of passive targeting, nanoparticles of 10–100 nm in size have been found to circulate systemically for a longer period of time.

The EPR effect is highly dependent on the intrinsic tumour biology and in particular:

1. The degree of angiogenesis and lymph angiogenesis
2. The degree of perivascular tumour growth and the density of the stromal response
3. Intratumour pressure

All of these factors, together with the physicochemical characteristics of nanocarriers, will determine its drug delivery efficiency. Though the leakiness of newly formed tumour vessels influences the nanomedicine permeation, it contributes to the high interstitial pressure, which in contrast is able to inhibit the accumulation of drug carriers in tumour. Moreover, due to the disproportion of pro- and anti-angiogenic signalling inside different parts of the tumour, vessels become abnormal with dilated, tortuous and saccular channels, and disorganized patterns of interconnection and branching.

Due to passive targeting the following can be obtained:

- It is a natural type of biodistribution of the carrier.
- The colloids/particles which are engulfed/taken up by the RES are ideal vectors for passive targeting of medications to RES predominant compartments.
- Passive release of colloidal carriers by macrophages offers therapeutic opportunities for the delivery of anti-infective agents.
- Devices include drug-containing bilayer vesicular systems that also act as cellular carriers of micron or submicron size range.

It refers to the build-up of drug or drug carrier system at a selected site like anti-cancerous drug; we have kept certain important parameters to contribute to physicochemical or pharmacological factors of the disease. Hence, in case of cancer treatment, it is observed that the particle size and its surface properties of drug-delivering nanoparticles must be controlled. During treatment of cancer, specifically keep in mind to avoid the uptake by the RES in a view to maximize circulation times and targeting ability. The bottom line of passive targeting is named as misnomer which is a straightforward drug delivery system via blood circulation. Drug release or drug actions are limited to selective sites within the body like a tumour but it is not applicable for the liver. There are many other examples including targeting of antimalarial drugs for treatment of candidiasis, brucellosis and leishmaniasis.

1.3.3.2.2 Active Targeting

By utilizing both passive and active targeting, a drug-loaded nanocarrier has a heightened advantage over a conventional drug. It is able to circulate throughout the body for an extended period of time until it is successfully attracted to its target through the use of cell-specific ligands, magnetic positioning or pH-responsive materials. Because of these advantages, side effects from conventional drugs will be largely reduced as a result of the drug-loaded nanoparticles affecting only diseased tissue. However, an emerging field known as nanotoxicology has concerns that the nanoparticles themselves could pose a threat to both the environment and human health with side effects of their own.

Active targeting of drug-loaded nanocarrier enhances the effects of passive targeting to make the nanocarrier more specific to a target site. There are several ways that active targeting can be accomplished.

One way to actively target solely diseased tissue in the body is to know the nature of a receptor on the cell for which the drug will be targeted. Researchers can then utilize cell-specific ligands that will allow the nanocarrier to bind specifically to the cell that has the complementary receptor. This form of active targeting was found to be successful when utilizing transferrin as the cell-specific ligand. The transferrin was conjugated to the nanoparticle to target tumour cells that possess transferrin receptor-mediated endocytosis mechanisms on their membrane. This means targeting was found to increase uptake, as opposed to non-conjugated nanocarriers.

Active targeting can also be achieved by utilizing magnet liposomes, which usually serve as contrast agents in magnetic resonance imaging. Thus, by grafting these liposomes with a desired drug to deliver to a region of the body, magnetic positioning could aid with this process.

Furthermore, a nanocarrier can possess the capability to be activated by a trigger that is specific to the target site, such as utilizing materials that are pH responsive. Most of the body has a consistent, neutral pH. However, some areas of the body are naturally more acidic than others, and, thus, nanocarrier can take advantage of this ability by releasing the drug when it encounters a specific pH. Another specific triggering mechanism is based on the redox potential. One of the side effects of tumours is hypoxia, which alters the redox potential in the vicinity of the tumour. By modifying the redox potential that triggers the payload release the vesicles can be selective to different types of tumours.

In this approach (active targeting), with such novel carrier systems the drug reaches the target site. This is carried out on the basis of modification/alteration made on its surface apart from natural uptake by RES. Among various techniques available for surface modification are coating of surface with either a tissue antibody (i.e. monoclonal antibodies) or the use of bioadhesive, or non-ionic surfactant or specific cell, or carried out by albumin protein.

Active targeting means a particular ligand-receptor-type interaction for intracellular localization which occurs only after blood circulation and extravasations. This active targeting approach will be further classified into three different levels of targeting which are:

1. First-order targeting (organ compartmentalization) means very restricted/limited distribution of the drug carrier systems to the animal/human tissue of a predetermined target site, organ or tissue, e.g. compartmental targeting of medications into eyes and joints, lymphatics, cerebral ventricles, peritoneal cavity and plural cavity.
2. Second-order targeting (cellular targeting) means selective delivery of medications to specific cell types for example tumour cells but not normal cells, e.g. selective drug delivery to Kupffer cells within the liver.
3. Third-order targeting (intracellular targeting) means to deliver a drug specifically to the intracellular site of targeted cells, e.g. receptor-based ligand-mediated entry of a drug complex into a cell by endocytosis.

1.3.3.2.3 Ligand-Mediated Targeting (Peng et al. 2018; Zhang et al. 2017)

In biochemistry, a ligand is any molecule or atom which binds reversibly to a protein. A ligand can also be made synthetically in the laboratory. This is because the key properties of a ligand are found in its chemical structure. Covalent or non-covalent attachment of a targeting ligand on the surface of nanoparticles enables them to recognize the specific antigens or receptors on target cells which subsequently engulf the particles through endocytosis.

Ligands are carrier surface group(s), which can selectively direct the carrier to the pre-specified site(s) housing the appropriate receptor units to serve as "homing device" to the carrier/drug. Most of the carrier systems are colloidal in nature and can be specifically functionalized using various biologically relevant molecular ligands including antibodies, polypeptides, oligosaccharides, viral proteins and fusogenic residues. The ligands confer recognition and specificity upon drug carrier and endow them with an ability to approach the respective target selectivity and deliver the drug.

Types of Ligands

Unidentate ligands: Ligands with only one donor atom, e.g. NH_3, Cl^- and F^-

Bidentate ligands: Ligands with two donor atoms, e.g. ethylenediamine and C_2O_{42} (oxalate ion)

Tridentate ligands: Ligands which have three donor atoms per ligand, e.g. (dien) diethyl triamine

Examples of common ligands are the neutral molecules water (H_2O), ammonia (NH_3), carbon monoxide (CO), anion cyanide (CN^-), chloride (Cl^-) and hydroxide (OH^-). Occasionally, ligands can be cations (e.g. NO^+, N_2H^{5+}) and electron-pair acceptors also. Table 1.2 shows various ligand-mediated targeting.

1.3.3.2.4 Dual Targeting

Herein the transporter particle/molecule in the targeting method itself has its own individual therapeutic activity, thus proliferating the therapeutic result of drug. For instance, a carrier particle/molecule having its own antiviral activity are often loaded with antiviral agent and also the remaining synergistic effect of drug conjugate is observed.

Based on this method, drug conjugates are often organized through stimulated activity profile in contrast to viral replication. The virus replication method is commonly attacked at many places, eliminating the possibilities of resistant viral strain development.

The management of glioma has become a decent challenge because of the existence of brain barrier. So, to enhance a competent brain-targeting drug delivery

Table 1.2 Examples of ligand-mediated targeting

Ligands	Target	Tumour target
Folate	Folate receptor	Overexpression of folate receptor
Transferrin	Transferrin receptor	Overexpression of transferrin receptor
Galactosamine	Galactosamine receptors on hepatocytes	Hepatoma

system to importantly recover the brain penetrability of anticancer drugs, a totally unique brain-targeted glucose-vitamin C (Glu-Vc) derivative was designed and synthesized as a liposome ligand for preparing liposome to effectively deliver paclitaxel (PTX). This modified Glu-Vc liposome specified better targeting capability in the in vivo evaluation as compared with unprotected paclitaxel, non-coated, singly modified and co-modified by physical blending liposomes. The relative uptake efficiency of modified liposome was improved by 7.53-fold of uncovered/uncoated paclitaxel, while the concentration effectiveness was up to 7.89 times.

1.3.3.2.5 Double Targeting (Taghdisi et al. 2016)

Two methodologies are available for targeting: temporal and spatial. These two are combined to focus on a carrier system, and this type of targeting could also be called double targeting. Spatial placement typically relates to targeting drugs/molecules to particular organs/tissues/cells and perhaps to subcellular compartment, whereas temporal delivery of drug/molecule refers to controlling the proportion of delivery to target on-site.

It was noted that to achieve double-targeting effect various approaches are popular like site specificity of the drug, by virtue of targeting moiety, and a high-specificity module (mainly a photosensitizer) is linked to antibodies.

Clinical use of daunorubicin (Dau) in the management of leukaemia has been constrained as a result of its cardiotoxicity. Targeted delivery of anticancer drugs could decrease their off-target effects and enhance their efficacy. During this study a modified polyvalent aptamer (PA)-daunorubicin (Dau)-gold nanoparticle (AuNP) complex was fabricated and its efficacy was assessed in Molt-4 cells (human acute lymphoblastic leukaemia T-cell, target). The outcomes of flow cytometry analysis indicated that the PA-Dau-AuNP complex was efficiently internalized into target cells, but not into non-target cells. In inference, the fabricated drug delivery system had natural properties of efficient drug loading, tumour targeting, pH-dependent drug release and controllable delivery of Dau to tumour cells.

1.3.3.2.6 Combination Targeting

First time Petit and Gombtzb provided the term "combination targeting". The important aim is site-specific delivery of proteins and peptides. These type of targeting systems are promoted with carriers, polymers and homing devices of molecular specificity which may provide a right-away method to target on-site.

1.3.3.2.7 Inverse Targeting

It is believed that inverse targeting is one of the efficacious efforts to bypass and avoid passive uptake of colloidal carriers by reticuloendothelial system (RES). This procedure is mentioned as inverse targeting, because it effectively leads to the relapse of biodistribution trend of the carrier. Hence, this approach is useful to accomplish inverse targeting to decrease/reduce the function of RES by a pre-injection of an infinite quantity of blank colloidal carriers or macromolecules, for example dextran sulphate. By this approach it leads to blockage of RES and may lead to impairment of host defence system. There are alternative approaches available which include modification of the size, surface charge, composition, surface rigidity and hydrophilicity of carriers for required biofate.

Modification of the surface by imparting distinctive hydrophilicity to the carrier particles is as an efficient way of targeting of drug to non-RES organs. Phospholipid microspheres emulsified with poloxamer 338 indicated the lowest RES uptake in mouse peritoneal macrophages in vitro. Poloxamine 908 is another hydrophilic non-ionic surfactant, which focuses on normal RES uptake of coated emulsion and coated nanoparticles (polystyrene microsphere) to inflammatory locations in rabbits. The study recommends inverse targeting of medications to the sites aside from RES-rich organs by coating the lipid microemulsion (LM) with poloxamer 308. It has been suggested that surface hydrophilicity might decrease or even eradicate the adhesion of opsonin materials/HDL on the surface of lipid microemulsion, which is a vital step within the process of phagocytosis responsible for definitive uptake of LM by RES system.

1.3.3.2.8 Physical Targeting

In this type of drug targeting some important characteristics which are related to environment changes are used. Few examples are electric field, temperature, light intensity, pH and ionic strength. It was also reported that small and even specific stimuli like glucose concentration can be used to localize the drug carrier to specific site. This approach was found exceptional for tumour targeting and cytosolic delivery of entrapped/encapsulated drug or genetic material.

1.4 Critical Factors Affecting the PK/PD and Fate of Orally Administered Nanocarriers

After the administration of the drug, compound or new chemical entity (NCE) into the body, it undergoes ADME process. The study of the same is known as pharmacokinetics. NCE's pharmacokinetic properties' information is critical to its selection in a drug discovery programme. Following factors are critical for nanocarriers.

1.4.1 Particle Size

Particle size is a major factor not only for the residence time of nanoparticles in blood but also for further targeting performances. The smaller the nanoparticle the lesser chances of its recognition by the microphysiological systems (MPS) and elimination from the body (Liu et al. 1992). However, it has been demonstrated that nanosized particles (200 nm) have a major drawback, since it prevents nanocarriers from benefiting from the EPR effect. Table 1.3 shows some studies on how particle size can affect the distribution phase within the tumour tissue.

Table 1.3 Size affecting the distribution phases within the tumour tissues and it does matter indeed for tumour accumulation

Drug	Study design	Outcome	References
5-FU	Three different batches of stealth liposomes of 5-FU varying in size (i.e. 70–250 nm) in mice bearing resistant breast tumour	The smaller the liposomes the greater the tumour uptake	Fanciullino et al. (2014)
Doxorubicin	Liposome diameter on tumour distribution in mice bearing mammary carcinoma	Significantly reduced efficacy and accumulation were evidenced for bigger liposomes	Charrois and Allen (2003)

1.4.2 Composition

A proper cholesterol and lipid proportion is required to achieve the steadiest liposomes with ideal controlled release, as unstable nanoparticles will show expanded plasma clearance and decreased circulating time, when contrasted with stable nanoparticles (Briuglia et al. 2015). Cholesterol's consideration in the lipid bilayer of a liposome balances out its structure, and diminishes the leakage of drug and risk of opsonization followed by extending of the course time.

1.4.3 Electric Charge

Pharmacokinetics and stability of nanoparticles are highly influenced by zeta potential. Geng et al. (2014) contemplated the effect of cholesterol on doxorubicin PEGylated liposome strength by considering an electric charge as a major factor. Utilizing two cholesterol subsidiaries (emphatically charged and neutral), pharmacokinetic studies on rodents demonstrated that neutral cholesterol liposomes accomplished higher dependability than decidedly charged ones (Geng et al. 2014).

1.4.4 Particle Shape

Shape is another significant parameter for certain identified macrophages' take-up and ensuing nanodrug biodistribution (Toy et al. 2014). In macrophage-prompted phagocytosis, leeway relies upon the estimate as well as the geometry of the molecule.

Champion and Mitragotri (2006) portrayed the conduct of macrophages against six diverse moulded nanoparticles (spheres, prolate ellipsoids, oblate ellipsoids, rectangular plates, curved circles and unidentified flying object shapes). It was seen that the underlying edge contact between the macrophages and the nanoparticles impacts its phagocytosis and freedom (Champion and Mitragotri 2006).

1.4.5 Protein Corona

When in biological fluids, the nanoparticle surface attracts proteins and biomolecules which will form a dynamic layer around the liposome. This rapidly shaped layer is usually called protein corona. Its composition depends on the surface properties of the nanoparticle, the nature of the environment, the time of exposure and the tumour type (Colapicchioni et al. 2016). The protein corona can affect liposome specificity by modifying its surface properties (i.e. charge, size) or by hampering targeting agents (Salvati et al. 2013). However, the presence of specific proteins in the corona can also improve tumour uptake.

1.5 Main Causes for Nanoparticles' Pharmacokinetic Variability

1.5.1 Age

Because nanoparticles, especially the stealth ones, are expected to bypass liver uptake and to avoid renal elimination, age should theoretically not be a factor of variation. However, because age affects the MPS, it can in some respect change nanoparticle pharmacokinetics (Plowden et al. 2004).

1.5.2 Body Mass

Body mass is usually best described through body mass index (BMI) or body surface area (BSA). It can affect drug pharmacokinetics because V_d depends partly on hydrophilic or lipophilic profile (i.e. lipophilic drugs tend to accumulate in fat tissues). Considering the more specific distribution achieved with nanoparticles, limited impact is expected from changes in body mass. However, it has been suggested that body weight could modulate the clearance of nanoparticles via changes in the MPS (La-Beck et al. 2011).

1.5.3 Gender

It is well established that gender is a factor affecting drug pharmacokinetics, distribution and clearance due to variation in body mass distribution, enzyme activities, etc. in different gender (Harris et al. 1995).

1.5.4 Drug-drug Interactions

Because of reduced liver uptake and the fact that nanoparticles are not substrate of efflux transporters, they are less likely to be affected by inhibiting/inducting drugs.

However, liposomal doxorubicin (Doxil®) showed higher AUC and decreased clearance when co-administrated with paclitaxel or docetaxel (Briasoulis et al. 2004). Inversely, when given after cisplatin, the clearance of Doxil® was increased, although no clear underlying mechanisms have been identified yet (Petschauer et al. 2015).

1.5.5 Immunity

Nanoparticle clearance is being partly controlled by the immune system (Song et al. 2014).

1.5.6 Genetic Polymorphism

Germinal polymorphisms affecting genes coding for proteins implicated into ADME (liver enzymes, membrane transporters) can be major causes for pharmacokinetic variability.

1.6 Characterization of Nanocarriers (Ram and Agrawal 2015)

1.6.1 Mean Particle Size and Particle Size Distribution

These parameters are evaluated as they influence important properties of biphasic liquid nanosystems such as saturation solubility and dissolution velocity. Likewise, they play an important role in governing physical stability and biological performance of nanosuspensions, nanoemulsions and nanomicelles. The technologies used to measure mean particle size and particle size distribution are as follows:

1.6.1.1 Dynamic Light Scattering (DLS)

DLS is also known as photon correlation spectroscopy (PCS). In this method, dynamic fluctuations in the strength of scattered light are measured based on the velocity of particle movement in the medium. Many researchers have employed Malvern Zetasizer Nanoseries nano-ZS for determination of particle size in their nanosuspension. This technique has been employed to determine the average diameter of nanoemulsions and nanomicelles.

1.6.1.2 Laser Diffraction

Laser diffraction is one of the techniques used to determine particle size. It involves passing a laser ray through the given sample and measures the angular deviation in the intensity of scattered light. The larger the size of particles, the smaller the angular variation. This angular scattering intensity data is further analysed. The particle size of the sample is calculated by means of the Mie hypothesis of light scattering. This technique is employed particularly for characterization of solid lipid nanoparticles.

1.6.1.3 Coulter Counter

This instrument works on the Coulter principle. This involves immersing a cylinder by means of a tiny orifice on its wall into a beaker containing a particular sample. The sample is prepared by dispersing the particles (the size of which is to be analysed) in a low-concentration electrolyte. The tube is subjected to an electric field as the current is allowed to pass through the two electrodes, one of which is inside the aperture tube and the other outside the aperture tube, and the impedance connecting the electrodes can be calculated. In this process, as a particle/atom passes through the orifice, it displaces a volume of electrolyte corresponding to its volume resulting in short-term change in impedance across the orifice. This alteration can be calculated in a form of voltage pulse, where the pulse height is comparative to the volume of the sensed particle/atom.

1.6.2 Particle Morphology

1.6.2.1 X-ray Diffraction Analysis

X-ray diffraction provides accurate molecular geometry of crystals and is a key tool in the characterization of pharmaceutical solids and suspensions. The principle involves the use of productive interfering of monochromatic X-rays. Bragg's law is employed to study the lattice structure in the crystalline material.

1.6.2.2 Scanning Electron Microscopy (SEM)

Scanning electron microscopy (SEM) produces extremely high magnification images at high resolution up to nanorange and hence it is successfully used in characterizing nanocarriers. It works on the principle of scanning the surface of specimen by an electron beam that is generated by an electron cathode and the electromagnetic lens of the column. Because of the action of the scanning beam, electrons are emitted from the specimen, composed by an appropriately situated detector and reflected as an image on screen after suitable magnification.

1.6.3 Particle Charge (Zeta Potential)

The zeta potential (ζ) is a function of the surface charge which develops while some material is placed in any suitable medium. The zeta potential measures the degree of repulsion between adjacent, similarly charged particles in dispersion. Hence, it becomes a significant tool intended for understanding the status of the nanocarriers' surface and predicting as well as controlling the extended stability of the nanodispersions. The zeta potential values are calculated by determining the particle's electrophoretic mobility.

1.6.4 Determination of Drug Loading and Encapsulation Efficiency of Nanocarriers

The supernatant dispersion obtained after the centrifugation of the colloidal suspension was used to determine parameters like drug loading and encapsulation efficiency. In indirect method, concentration of entrapped drug is measured to determine drug loading and encapsulation efficiency (EE%) by any suitable analytical technique. The loading of drug and encapsulation efficiency were calculated using the following equation:

$$\text{Drug loading}(\%) = \frac{\text{Total amount of drug} - \text{Amount of entrapped drug}}{\text{Amount of nanocarriers recovered}} \times 100 \quad (1.1)$$

Drug entrapment efficiency was expressed as the ratio of the drug amount measured in the supernatant to the total drug:

$$\text{EE}(\%) = \frac{\text{Actual drug content in nanocarriers}}{\text{Total theoretical amount of drug}} \times 100 \quad (1.2)$$

1.6.5 Percentage Yield Value

The production yield of the nanocarriers was determined by the accurate calculation of the original mass of the raw material along with the weight of the nanocarriers obtained.

1.6.6 Evaluation of Polymer-Drug Interaction

Stability of nanocarriers is measured by studying the drug-polymer interaction using differential scanning calorimetry (DSC) and Fourier transform infrared spectroscopy studies. These techniques not only evaluate the drug-polymer interaction but also measure the physical state of polymer and drug in nanocarriers.

1.6.6.1 Differential Scanning Calorimetry

Differential scanning calorimeter is used to examine thermal actions throughout heating. Accurately weighed 5 mg samples were placed in open aluminium pan, heated from 25 to 300 °C at a rate of 10 °C/min and calibrated with alumina (Al_2O_3).

1.6.6.2 Fourier Transform Infrared Spectroscopy (FTIR)

A computerized Fourier, which transforms infrared spectroscopy, was used to obtain the spectra of various samples. The KBr discs are prepared by applying almost 2–3 tons of pressure for 2–5 min. The scanning range of 400–4000 cm^{-1} with the resolution of 1 cm^{-1} was utilized for the study.

1.6.7 In Vitro Release Studies

The in vitro drug release from the nanocarriers was calculated via the dialysis sack diffusion technique. It was noted that for all nanocarrier formulations, utmost care has to be taken to maintain sink conditions.

The dialysis sack reserves nanocarriers (donor compartment) and permits the diffusion of the drug instantaneously into the recipient compartment. A recipient compartment consists of dissolution medium which is pre-heated and maintained at temperature mentioned in pharmacopoeias/references used for the study. Then a fixed quantity of the medium was withdrawn at preset times by an automatic sampling system. An equal volume of fresh medium was added after each sampling. The amount of drug release in the medium was determined by any suitable analytical method.

1.6.8 In Vitro Drug Release Kinetics

The data received from release studies were fitted into different kinetic models like zero order, first order, Higuchi, Hixson-Crowell and Korsmeyer-Peppas, and the mechanism of drug release from nanocarriers is studied.

1.7 Regulatory Challenges to Nanomedicine Development

It is seen that numerous gravel may be present in the journey of nanocarriers before it reaches for existence like for clinic or at patient level. Separately commence the benchmark criteria designed for satisfactory safety and effectiveness as well as attractive pharmaceutical quality (e.g. stability, no difficulty in administration) which are applicable to most of the drugs. The ideal nanoparticle system or nanocarriers to be utilized for therapeutic function include the following reasons:

1. Well-intellectual capacity of basic segments and its collaborations
2. Recognizable proof of key attributes and its performance
3. Capacity to mimic significant attributes under manufacturing conditions
4. Simple to deliver in a sterile form
5. Power to target or combine at the place of movement by beating the biological boundaries
6. Great in stability, simple to store and to administer

The regulatory pathways for nanocarriers face a few obstacles because of the complexity and huge potential diversity of nanoparticle-based items. As of now, the

European Medicines Agency (EMA), Food and Drug Administration (FDA) and other regulatory agencies inspect every original nanoparticle-based drug on a product-by-product basis. There is usually an absence of principles in the assessment of nanocarriers as a distinctive kind of therapeutic agents.

It is noticed that we have entered late in the market of generic nanomedicines. Both nonexclusive medication makers and medication controllers will be considered with real difficulties in characterizing what type of study will be required to show that the conventional nanomedicine is bioequivalent near the innovator and to facilitate the products having the comparable physicochemical properties with protection and desirable feasibility.

- Variations in molecule size lead to poor reproducibility in the manufacturing procedure; particle size of nanocarriers is significant for successful dosage form formulation.
- Particle size distribution, stability and potency are significant qualities of any nanocarrier.
- A few liposomal type items contain the drug doxorubicin and amphotericin that have moved out of patent. Justifiably, the multifaceted nature of such products might require an alternate standard of "equivalence" testing than what is required for standard drugs.
- There is a critical requirement for the administrative organizations to build up a complete rundown of tests and a streamlined endorsement process that covers the entire scope of molecule interpretation, pharmacology and toxicology issues.

Because of the complex nature of nanomedicines, a progressively more contemporary degree of testing should be required to completely describe a nano-formulation.

1.8 Challenges for Scale-Up and Manufacturing

The fruitful scale-up and manufacturing of a nanomedicine show distinct difficulties in pharmaceutical advancement. Following are some important/critical points identified with scale-up and manufacturing:

1. Traditional pharmaceutical manufacturing does not commonly make three-dimensional multicomponent frameworks in the nanometre level and so every effect will be considered as restrictions for the scale-up of nanomedicines.
2. A large portion of nanoparticles are unpredictable, so full understanding of the components and its interactions are fundamental to characterize the important attributes of the formulation development. It is essential to optimize primary procedural steps and analytical parameters that will facilitate assurance reproducibility of the dosage form.
3. The nanoparticle process frequently includes varieties of unit operations such as utilization of organic solvents followed by evaporation, high-speed

homogenization, sonication, cross-linking process, centrifugation, filtration and lyophilization. During early advancement, at the lab or small scale, it is helpful to think about what approach may be precious if this operation turns for scale-up.

4. Characteristic evidence of significant procedure conditions is essential to accomplish important qualities and measurements. In this condition, it might include the proportion of polymers, drugs, targeting moieties, type of organic solvent, emulsifier, stabilizer, cross-linking agent, oil-to-water-stage proportion, blending time, temperature, weight and pH. Contingent upon conditions, the procedure may be developed for the scale-up.
5. A formulation method is required to be strong to guarantee elevated reproducibility, and be real rationalized to take into consideration the simplicity of scale-up design. The preparation of nanoparticles regularly requires a variety of procedure steps in multicomponent system. Although small-scale procedures may accomplish reproducibility with all-around described parts, the reproducibility and consistency build a steady challenge for the scale-up and manufacturing process.
6. When using multistep forms, evaluation of basic parameters during the process with a quick and dependable scientific technique is exceptionally useful regarding how well the procedure is controlled. Working up a database of statistics with reasonably targeted in-process tests might be fundamental to guarantee accomplishment at the manufacturing scale.
7. One more issue observed during the formulating nanoparticles is environmental safety. The management of dehydrated material in the nanometre size scale requests unique alert, as airborne nanoparticles disperse as aerosols. Lung deposition of such nanoparticles can prompt pneumonic toxicities. Some nanoparticles are equipped for penetrating the skin hindrance, making dermal introduction of a drug molecule, so sufficient security is essential.

1.9 Applications of Nanocarriers

1.9.1 Targeting Drug Delivery by Encapsulation

It observed that nano-encapsulation of drugs enhances its specificity, efficaciousness, tolerability as well as therapeutic performance. It safeguards from early degradation and interaction with the biological environment. Finally, it enhances absorption of drug into a certain tissue, increases bioavailability, improves retention time and ultimately improves intracellular penetration.

Nanomedicines of some diseases like cancer, AIDS, diabetes, malaria and tuberculosis are in various clinical trials for the testing, and it is good news that some of them are available in the market. Nanomedicine formulation depends on the selection of appropriate polymeric structure having superior encapsulation effectiveness which enhances bioavailability and drug retention time in the system, viz. insulin-loaded nanocarriers of PLGA, estradiol-encapsulated PLGA nanocarriers and progesterone-loaded PLA-PEG-PLA nanocarriers.

1.9.2 Nanocarriers for Oral Delivery of Peptides and Proteins

The GIT provides a diversity of physiological and morphological barriers against protein or peptide delivery like:

1. Proteolytic enzymes in the gut lumen, i.e. pepsin, trypsin and chymotrypsin
2. Proteolytic enzymes at the brush border membrane, i.e. endopeptidases
3. Bacterial gut flora
4. Mucus layer and epithelial cell lining itself

Even though many hurdles are present, polymeric nanocarriers allow encapsulation of these bioactive substances and prevent them against enzymatic and hydrolytic degradation, e.g. insulin-loaded nanocarriers for oral administration via encapsulation in folate (FA)-coupled polyethylene glycolylated (PEG) polylactide-coglycolide (PLGA) nanoparticles (FA-PEG-PLGA NPs).

1.9.3 Nanocarriers for Drug Delivery into the Brain

The blood-brain barrier (BBB) is the most significant preventive issue used for the improvement of novel drugs intended for the central nervous system owing to its comparatively impervious endothelial cells by means of tight junctions, enzymatic movement and dynamic efflux transportation systems. BBB is impervious for water-soluble drugs from blood circulation to CNS and it is selectively permeable for lipophilic as well as tiny-size molecules, for example PEG-coated NPs. PEGylation improves the pharmacokinetic profile of molecules by reducing opsonization, phagocytosis and clearance by the liver and reticuloendothelial system.

1.9.4 Nanocarriers for Ophthalmic Delivery

Nanosuspension of nanocarriers offers a good benefit of delayed residence time in cul-de-sac; this is the most essential design for the ocular diseases for successful management in addition to also sustaining tonicity particularly to the eye. It is observed that dissolution rate and intrinsic solubility lie on the lachrymal fluid. The intrinsic dissolution rate of the drug will vary because of the regular inflow and outflow of lachrymal fluids.

For example, improve stability and bioavailability of cloricromene in ophthalmic formulation and Eudragit RS100-loaded piroxicam nanocarriers as a safer controlled ocular delivery of anti-inflammation agents for inhibition of the uveitis symptoms.

1.9.5 Topical Formulations

Drug nanocarriers are able to be included into creams and water-free ointments. It was found that the nanocrystalline form leads in the direction of an improved saturation solubility of the drug inside the topical dosage form; therefore it enhances the diffusion of the drug inside the skin. Micellar nanoparticle is a talent appropriately designed for topical applications. These technologies allow more amount of drug to go through the skin and functionally generate a drug store in the stratum corneum and epidermis. It was observed that this way of delivering drug gives related advantages of patch expertise in avoiding both contact by means of the gastrointestinal tract and hepatic first-pass effects, and is cosmetically more acceptable to many patients. Solid lipid nanocarriers are used to treat several skin diseases like acne, atopic eczema, psoriasis due to its higher permeability throughout the skin, better surface area and strong spreading pattern. SLN formulations are applied on top of skin to decrease the total unfavourable effect of drugs. For example, vitamin A is incorporated in SLN and Estrasorb (17β-estradiol in Estrasorb), the world's first nano-engineered topical dosage form approved by FDA and developed by Novavax.

1.9.6 Nanocarriers for Diagnostic Applications

The nanocarriers can also be used in the direction of diagnosis of cancer. Magnetic nanocarriers, being a subfamily of nanomaterials, proved to be an extraordinarily innovative phenomenon such as superparamagnetic, high saturation field, extra anisotropy contributions or shifted loops after field cooling. These phenomena take place from restricted size and surface effects to facilitate the magnetic activities of nanocarriers. Small size gives useful surface area, low sedimentation velocity and tissue diffusion and reduces dipole-dipole moment. The magnetic property of nanocarriers offers a benefit to facilitate selective attachment to a functional molecule, grant magnetic properties in the direction of the target and permit exploitation and transportation to a preferred place through the control of a magnetic field created via an electromagnet or permanent magnet.

1.9.7 Cosmetic Applications

Solid lipid nanocarriers are one kind of nanoparticle formulated from physiological lipids. These lipids contain various natural properties like occlusive nature and ultraviolet ray protection which allows its application in cosmetics. Lipid-based nanocarriers followed by improved permeability allow skin hydration owing to its occlusive property.

1.9.8 Nanocarriers for Gene Delivery

Nanocarriers loaded with plasmid DNA and vaccines could release DNA at a constant rate via evasion from the endolysosomal section to cytoplasmic section. This gene delivery policy possibly will be useful to assist bone healing by means of PLGA nanocarriers containing therapeutic genes such as bone morphogenic protein.

1.9.9 Pulmonary Delivery

Unique size of nanocarriers may provide significant absorption of drug, controlled release as well as targeting of drugs to alveoli, bronchial tree or lungs. Anti-allergic, bronchodilators and steroidal anti-inflammatory agents are preferable drugs designed for this way of delivery.

1.10 Conclusions and Future Prospects

Nanotechnology is probably the most recent methodology that significantly resolves different acute and chronic conditions in the last decade. We discussed different nanocarriers which are used as developing measurement structure for the treatment of certain hazardous conditions. These nanocarriers have gotten a transformation disease medication conveyance by accomplishing objective to tumour cells unambiguously with advanced porousness and lesser poisonousness to neighbouring solid cells. These animating movements in malignant growth treatment and extraordinary improvement of various novel medication conveyance frameworks have exploded the desire for those engaging against tumours. There would be a minimum toxic impacts won't just upgrade the utilization of nanocarrier frameworks for antitumour medication delivery but in addition it would improve the patient consistence and comfort.

This advancement in innovation would guarantee the dynamic focusing to unhealthy state by getting together with the surface-connected ligand and the receptors on the favoured cells and tissues. In any case, it requires a couple of hindrances to crush; for instance, it requires a few obstacles to conquer, for example, an absence of potential skill, administrative obstacles, trouble in intersecting the cell film, cost adequacy and restricted helpful window of medications.

Unfortunately, the normal repeat of detailing-focused development has not accomplished the anticipated patient consistence and comfort, and still, at the end, nanocarriers can achieve the target set for malignant growth therapeutics, both in case of the customary and cutting-edge specialists. Unmistakable focus on nanocarriers has set up an improved helpful action in different animal models of tumours. Even more precisely, consistent clinical tasks with different invulnerable reactions containing nanocarrier subtleties are under assessment. So likewise, today the analysts can picture the class and territory of the tumour, which consequently provides the idea of proper remedial regimens. Also, if the tumour cells are of flowing sort as it occurred in lymphoma and leukaemia, a conveyance administration with long circulating half-life and high capacity of focusing on surface antigen is enjoyed. It is moreover trusted that soon the specialists would have the alternative to make concentrated subnuclear composites that may provoke improved medicinal results with less expenses.

In spite of the way that experts have investigated and developed different new medication conveyance frameworks to accomplish sedation effectiveness, specifically gathering of patients, only two or three of these promising preclinical prescription conveyance frameworks have shown up at the commercial centre. This may be attributed to the basic need of the difference in anticancer medication-stacked nanocarriers. Thus, it may be fundamental to change a bit of the customary models to sidestep such issues. Taking a gander at the issue, it is required to possess exceptional inconvenience shooting endeavours to decide a couple of issues on steady thought and to accomplish innocuous utilization of the as-of-late-made nanocarriers in the clinical programme. It consolidates the improvement of benchmark nano plans endorsed through in vitro and in vivo test for its efficaciousness, safety and potential toxicities.

References

Adeli M, Hakimpoor F, Ashiri M et al (2011) Anticancer drug delivery systems based on noncovalent interactions between carbon nanotubes and linear-dendritic copolymers. Soft Mater 7(8):4062–4070

Adeli M, Beyranvand S, Hamid M (2012) Noncovalent interactions between linear-dendritic copolymers and carbon nanotubes lead to liposome-like nanocapsules. J Mater Chem 22 (14):6947–6952

Ajima K, Murakami T, Mizoguchi Y (2008) Enhancement of in vivo anticancer effects of cisplatin by incorporation inside single-wall carbon nanohorns. ACS Nano 2(10):2057–2064

Almeida JD, Brand CM, Edwards DC et al (1975) Formation of virosomes from influenza subunits and liposomes. Lancet 2:899–901

Anitha A, Deepa N, Chennazhi KP et al (2014) Combinatorial anticancer effects of curcumin and 5-fluorouracil loaded thiolated chitosan nanoparticles towards colon cancer treatment. Biochim Biophys Acta 1840(9):2730–2743

Arlt M, Haase D, Hampel S (2010) Delivery of carboplatin by carbon-based nanocontainers mediates increased cancer cell death. Nanotechnology 21(33):335101

Bamrungsap S, Zhao Z, Chen T (2012) Nanotechnology in therapeutics: a focus on nanoparticles as a drug delivery system. Nanomedicine 7(8):1253–1271

Basu S, Sandanaraj BS, Thayumanavan S (2004) Molecular recognition in dendrimers. In: Mark HF (ed) Encyclopedia of polymer science and technology, 4th edn. Wiley, Hoboken, pp 385–424

Battaglia L, Gallarate M, Peira E (2014) Solid lipid nanoparticles for potential doxorubicin delivery in glioblastoma treatment: preliminary in vitro studies. J Pharm Sci 103(7):2157–2165

Betageri G, Nekkanti V (2015) Improved oral bioavailability of valsartan using proliposomes: design, characterization and in vivo pharmacokinetics. Drug Dev Indus Pharm 41 (12):2077–2088

Bhadra D, Bhadra S, Jain S et al (2003) A PEGylated dendritic nanoparticulate carrier of fluorouracil. Int J Pharm 257(1):111–124

Bhirde AA, Patel V, Gavard J (2009) Targeted killing of cancer cells in vivo and in vitro with EGF-directed carbon nanotube-based drug delivery. ACS Nano 3(2):307–316

Bianco A (2004) Carbon nanotubes for the delivery of therapeutic molecules. Expert Opin Drug Deliv 1(1):57–65

Biswas S, Kumari P, Lakhani PM et al (2016) Recent advances in polymeric micelles for anticancer drug delivery. Eur J Pharm Sci 83:184–202

Bobbala SKR, Veerareddy PR (2012) Formulation, evaluation, and pharmacokinetics of isradipine proliposomes for oral delivery. J Liposome Res 22(4):285–294

Briasoulis E, Karavasilis V, Tzamakou E et al (2004) Interaction pharmacokinetics of pegylated liposomal doxorubicin (Caelyx) on co-administration 35 with paclitaxel or docetaxel. Cancer Chemother Pharmacol 53:452–457

Briuglia ML, Rotella C, McFarlane A et al (2015) Influence of cholesterol on liposome stability and on in vitro drug release. Drug Deliv Transl Res 5(3):231–242

Cao J, Guenther RH, Sit TL et al (2014) Loading and release mechanism of red clover necrotic mosaic virus derived plant viral nanoparticles for drug delivery of doxorubicin. Small 10:5126–5136

Champion JA, Mitragotri S (2006) Role of target geometry in phagocytosis. PNAS 103 (13):4930–4934

Charrois GJR, Allen TM (2003) Rate of biodistribution of STEALTH® liposomes to tumour and skin: influence of liposome diameter and implications for toxicity and therapeutic activity. Biochim Biophys Acta 1609(1):102–108

Colapicchioni V, Tilio M, Digiacomo L et al (2016) Personalized liposome–protein corona in the blood of breast, gastric and pancreatic cancer patients. Int J Biochem Cell Biol 75:180–187

Das M, Datir SR, Singh RP et al (2013) Augmented anticancer activity of a targeted, intracellularly activatable, theranostic nanomedicine based on fluorescent and radiolabeled, methotrexate-folic acid-multiwalled carbon nanotube conjugate. Mol Pharm 10(7):2543–2557

Desai D, Zhang J, Sandholm J (2017) Lipid bilayer-gated mesoporous silica nanocarriers for tumour-targeted delivery of zoledronic acid in vivo. Mol Pharm 14(9):3218–3227

Douglas T, Young M (2006) Viruses: making friends with old foes. Science 312(5775):873–875

Fabbro C, Ali-Boucetta H, Ros TD et al (2012) Targeting carbon nanotubes against cancer. Chem Commun 48(33):3911–3926

Fanciullino R, Mollard S, Correard F et al (2014) Biodistribution, tumour uptake and efficacy of 5-FU-loaded liposomes: why size matters. Pharm Res 31:2677–2684

Geng S, Yang B, Wang G et al (2014) Two cholesterol derivative-based PEGylated liposomes as drug delivery system, study on pharmacokinetics and drug delivery to retina. Nanotechnology 25(27):5103

Gothwal A, Khan I, Gupta U (2016) Polymeric micelles: recent advancements in the delivery of anticancer drugs. Pharm Res 33(1):18–39

Gupta AK, Mishra DK, Mahajan SC (2011) Preparation and in-vitro evaluation of self-emulsifying drug delivery system of antihypertensive drug valsartan. Int J Pharm Life Sci 2(3):633–639

Hamaguchi T, Matsumura Y, Suzuki M (2005) A paclitaxel-incorporating micellar nanoparticle formulation, can extend in vivo anti-tumour activity and reduce the neurotoxicity of paclitaxel. Br J Cancer 92(7):1240–1246

Han N, Zhao Q, Wan L (2015) Hybrid lipid-capped mesoporous silica for stimuli-responsive drug release and overcoming multidrug resistance. ACS Appl Mater Interfaces 7(5):3342–3351

Harris RZ, Benet LZ, Schwartz JB (1995) Gender effects in pharmacokinetics and pharmacodynamics. Drugs 50:222–239

Hildebrand A, Beyer K, Neubert R et al (2003) Temperature dependence of the interaction of cholate and deoxycholate with fluid model membranes and their solubilization into mixed micelles. Colloids Surf B Biointerfaces 32(4):333–335

Honarbakhsh S, Guenther RH, Willoughby JA et al (2013) Polymeric systems incorporating plant viral nanoparticles for tailored release of therapeutics. Adv Healthc Mater 2(7):1001–1007

How CW, Rasedee A, Manickam S et al (2013) Tamoxifen-loaded nanostructured lipid carrier as a drug delivery system: characterization, stability assessment and cytotoxicity. Colloids Surf B Biointerfaces 12:393–399

Hu CM, Aryal S, Zhang L (2010) Nanoparticle-assisted combination therapies for effective cancer treatment. Ther Deliv 1(2):323–334
Iannazzo D, Piperno A, Pistone A et al (2013) Recent advances in carbon nanotubes as delivery systems for anticancer drugs. Curr Med Chem 20(11):1333–1354
Iijima S (1991) Helical microtubules of graphitic carbon. Nature 354(6348):56–58
Jabir NR, Tabrez S, Ashraf GM et al (2012) Nanotechnology-based approaches in anticancer research. Int J Nanomedicine 7:4391–4408
Ji Z, Lin G, Lu Q (2012) Targeted therapy of SMMC-7721 liver cancer in vitro and in vivo with carbon nanotubes based drug delivery system. J Colloid Interface Sci 365(1):143–149
Johnstone RM (2005) Revisiting the road to the discovery of exosomes. Blood Cells Mol Dis 34:214–219
Kakkar D, Dumoga S, Kumar R et al (2015) PEGylated solid lipid nanoparticles: design, methotrexate loading and biological evaluation in animal models. Med Chem Commun 6 (8):1452–1463
Khan DR (2010) The use of Nanocarriers for drug delivery in cancer therapy. J Cancer Sci Ther 2 (3):58–62
Khuroo T, Verma D, Talegaonkar S et al (2014) Topotecan–tamoxifen duple PLGA polymeric nanoparticles: investigation of in vitro, in vivo and cellular uptake potential. Int J Pharm 473 (1–2):384–394
La-Beck NM, Zamboni BA, Gabizon A et al (2011) Factors affecting the pharmacokinetics of pegylated liposomal doxorubicin in patients. Cancer Chemother Pharmacol 69:43–50
Lai PS, Lou PJ, Peng CL (2007) Doxorubicin delivery by polyamidoamine dendrimer conjugation and photochemical internalization for cancer therapy. J Control Release 122(1):39–46
Lai RC, Yeo RWY, Tan KH et al (2013) Exosomes for drug delivery—a novel application for the mesenchymal stem cell. Biotechnol Adv 31:543–551
Lay CL, Liu HQ, Tan HR et al (2010) Delivery of paclitaxel by physically loading onto poly (ethylene glycol) (PEG)-graft carbon nanotubes for potent cancer therapeutics. Nanotechnology 21(6):065101
Lebold T, Jung C, Michaelis J et al (2009) Nanostructured silica materials as drug-delivery systems for doxorubicin: single molecule and cellular studies. Nano Lett 9(8):2877–2883
Lee CC, Gillies ER, Fox ME (2006) A single dose of doxorubicin-functionalized bow-tie dendrimer cures mice bearing C-26 colon carcinomas. PNAS 103(45):16649–16654
Lee WH, Loo CY, Traini D et al (2015) Nano-and micro-based inhaled drug delivery systems for targeting alveolar macrophages. Expert Opin Drug Deliv 12(6):1009–1026
Levi-Polyachenko NH, Merkel EJ, Jones BT et al (2009) Rapid photothermal intracellular drug delivery using multiwalled carbon nanotubes. Mol Pharm 6(4):1092–1099
Li X, Yang Z, Yang K (2009) Self-assembled polymeric micellar nanoparticles as nanocarriers for poorly soluble anticancer drug ethaselen. Nanoscale Res Lett 4(12):1502–1511
Li Y, Li N, Pan W et al (2017) Hollow mesoporous silica nanoparticles with tunable structures for controlled drug delivery. ACS Appl Mater Interfaces 9(3):2123–2129
Liu D, Mori A, Huang L (1992) Role of liposome size and RES blockade in controlling biodistribution and tumour uptake of GM1-containing liposomes. Biochim Biophys Acta 1104:95–101
Lu J, Liong M, Sherman S (2007a) Mesoporous silica nanoparticles for cancer therapy: energy-dependent cellular uptake and delivery of paclitaxel to cancer cells. Nanobiotechnology 3 (2):89–95
Lu J, Liong M, Zink JI et al (2007b) Mesoporous silica nanoparticles as a delivery system for hydrophobic anticancer drugs. Small 3(8):1341–1346
Ma Y, Nolte RJ, Cornelissen JJ (2012) Virus-based nanocarriers for drug delivery. Adv Drug Deliv Rev 64(9):811–825
Madani SY, Naderi N, Dissanayake O et al (2011) A new era of cancer treatment: carbon nanotubes as drug delivery tools. Int J Nanomedicine 6:2963–2979

Malam Y, Loizidou M, Seifalian AM (2009) Liposomes and nanoparticles: nanosized vehicles for drug delivery in cancer. Trends Pharmacol Sci 30(11):592–599
Malik N, Evagorou EG, Duncan R (1999) Dendrimer-platinate: a novel approach to cancer chemotherapy. Anticancer Drugs 10(8):767–776
Manchester M, Singh P (2006) Virus-based nanoparticles (VNPs): platform technologies for diagnostic imaging. Adv Drug Deliv Rev 58(14):1505–1522
Mehnert W, Mader K (2001) Solid lipid nanoparticles: production, characterization and applications. Adv Drug Deliv Rev 47(2–3):165–196
Mishra B, Patel BB, Tiwari S (2010) Colloidal nanocarriers: a review on formulation technology, types and applications toward targeted drug delivery. Nanomedicine 6(1):9–24
Muller RH, Mader K, Gohla S (2000) Solid lipid nanoparticles (SLN) for controlled drug delivery—a review of the state of the art. Eur J Pharm Biopharm 50(1):161–177
Muller RH, Radtke M, Wissing SA (2002) Solid lipid nanoparticles (SLN) and nanostructured lipid carriers (NLC) in cosmetic and dermatological preparations. Adv Drug Deliv Rev 54:S131–S155
Nakanishi T, Fukushima S, Okamoto K (2001) Development of the polymer micelle carrier system for doxorubicin. J Control Release 74:295–302
Neubert RH (2011) Potentials of new nanocarriers for dermal and transdermal drug delivery. Eur J Pharm Biopharm 77(1):1–2
Patel MN, Lakkadwala S, Majrad MS (2014) Characterization and evaluation of 5-fluorouracil-loaded solid lipid nanoparticles prepared via a temperature-modulated solidification technique. AAPS PharmSciTech 15(6):1498–1508
Pattenden LK, Middelberg AP, Niebert M et al (2005) Towards the preparative and large-scale precision manufacture of virus-like particles. Trends Biotechnol 23(10):523–529
Peng Y, Zhao Y, Chen Y et al (2018) Dual-targeting for brain-specific liposomes drug delivery system: synthesis and preliminary evaluation. Bioorg Med Chem 26(16):4677–4686
Pérez-Herrero E, Fernández-Medarde A (2015) Advanced targeted therapies in cancer: drug nanocarriers, the future of chemotherapy. Eur J Pharm Biopharm 93:52–79
Petersen KE, Manangon E, Hood JL et al (2014) A review of exosome separation techniques and characterization of B16-F10 mouse melanoma exosomes with AF4-UV-MALS-DLS-TEM. Anal Bioanal Chem 406:7855–7866
Petschauer JS, Madden AJ, Kirschbrown WP et al (2015) The effects of nanoparticle drug loading on the pharmacokinetics of anticancer agents. Nanomed 10:447–463
Plowden J, Renshaw-Hoelscher M, Engleman C et al (2004) Innate immunity in aging: impact on macrophage function. Aging Cell 3:161–167
Prabhakar N, Zhang J, Desai D (2016) Stimuli-responsive hybrid nanocarriers developed by controllable integration of hyperbranched PEI with mesoporous silica nanoparticles for sustained intracellular siRNA delivery. Int J Nanomedicine 11:6591–6608
Prabhu RH, Patravale VB, Joshi MD (2015) Polymeric nanoparticles for targeted treatment in oncology: current insights. Int J Nanomedicine 10:1001–1018
Qin J, Xu Q (2014) Functions and applications of exosomes. Acta Pol Pharm 71:537–543
Qureshi OS, Kim HS, Zeb A (2017) Sustained release docetaxel-incorporated lipid nanoparticles with improved pharmacokinetics for oral and parenteral administration. J Microencapsul 34 (3):250–261
Ram ST, Agrawal R (2015) Application of Nanotechnology in Pharmaceutical Formulation Design and Development. Current Drug Therapy 10:20–34
Rao JP, Geckeler KE (2011) Polymer nanoparticles: preparation techniques and size-control parameters. Prog Polym Sci 36(7):887–913
Rapoport N (2007) Physical stimuli-responsive polymeric micelles for anti-cancer drug delivery. Prog Polym Sci 32:962–990
Raval C, Joshi N, Patel J et al (2012) Enhanced oral bioavailability of olmesartan by using novel solid self-emulsifying drug delivery system. Int J Adv Pharm 2(2):82–92

Ren J, Fang Z, Yao L (2015) A micelle-like structure of poloxamer–methotrexate conjugates as nanocarrier for methotrexate delivery. Int J Pharm 487:177–186
Rosenholm JM, Peuhu E, Bate-Eya LT et al (2010) Cancer-cell-specific induction of apoptosis using mesoporous silica nanoparticles as drug-delivery vectors. Small 6(11):1234–1241
Salvati A, Pitek AS, Monopoli MP et al (2013) Transferrin-functionalized nanoparticles lose their targeting capabilities when a biomolecule corona adsorbs on the surface. Nat Nanotechnol 8:137–143
Simons M, Raposo G (2009) Exosomes-vesicular carriers for intercellular communication. Curr Opin Cell Biol 21:575–581
Singh P, Prasuhn D, Yeh RM (2007) Bio-distribution, toxicity and pathology of cowpea mosaic virus nanoparticles in vivo. J Control Release 120(1–2):41–50
Slowing II, Vivero-Escoto JL, Wu CW et al (2008) Mesoporous silica nanoparticles as controlled release drug delivery and gene transfection carriers. Adv Drug Deliv Rev 60(11):1278–1288
Song G, Petschauer J, Madden A et al (2014) Nanoparticles and the mononuclear phagocyte system: pharmacokinetics and applications for inflammatory diseases. Curr Rheumatol Rev 10:22–34
Stiriba SE, Frey H, Haag R (2002) Dendritic polymers in biomedical applications: from potential to clinical use in diagnostics and therapy. Angew Chem Int Ed Engl 41(8):1329–1334
Sun T, Zhang YS, Pang B et al (2014) Engineered nanoparticles for drug delivery in cancer therapy. Angew Chem Int Ed Engl 53(46):12320–12364
Taghdisi SM, Danesh NM, Lavaee P et al (2016) Double targeting, controlled release and reversible delivery of daunorubicin to cancer cells by polyvalent aptamers-modified gold nanoparticles. Mater Sci Eng C Mater Biol Appl 61:753–761
Torchilin VP (2005) Recent advances with liposomes as pharmaceutical carriers. Nat Rev Drug Discov 4(2):145–160
Toy R, Peiris PM, Ghaghada KB et al (2014) Shaping cancer nanomedicine: the effect of particle shape on the in vivo journey of nanoparticles. Nanomedicine (Lond) 9:121–134
Uner M, Yener G (2007) Importance of solid lipid nanoparticles (SLN) in various administration routes and future perspectives. Int J Nanomedicine 2(3):289–300
Vardharajula S, Ali SZ, Tiwari PM (2012) Functionalized carbon nanotubes: biomedical applications. Int J Nanomedicine 7:5361–5374
Vyas SP, Khar RK (2002) Targeted and controlled drug delivery—novel carrier system, 1st edn. CBS Publisher and Distributers, New Delhi, pp 38–80
Wang X, Wang Y, Chen ZG et al (2009) Advances of cancer therapy by nanotechnology. Cancer Res Treat 41(1):1–11
Wang H, Zhao Y, Wu Y (2011) Enhanced anti-tumour efficacy by co-delivery of doxorubicin and paclitaxel with amphiphilic methoxy PEG-PLGA copolymer nanoparticles. Biomaterials 32 (32):8281–8290
Wang AZ, Langer R, Farokhzad OC (2012) Nanoparticle delivery of cancer drugs. Annu Rev Med 63:185–198
Wang W, Chen S, Zhang L (2015a) Poly (lactic acid)/chitosan hybrid nanoparticles for controlled release of anticancer drug. Mater Sci Eng C Mater Biol Appl 46:514–520
Wang Y, Zhao Q, Han N (2015b) Mesoporous silica nanoparticles in drug delivery and biomedical applications. Nanomedicine 11(2):313–327
Wong HL, Bendayan R, Rauth AM et al (2007) Chemotherapy with anticancer drugs encapsulated in solid lipid nanoparticles. Adv Drug Deliv Rev 59(6):491–504
Wu W, Li R, Bian X (2009) Covalently combining carbon nanotubes with anticancer agent: preparation and antitumour activity. ACS Nano 3(9):2740–2750
Xia T, Kovochich M, Liong M (2009) Polyethyleneimine coating enhances the cellular uptake of mesoporous silica nanoparticles and allows safe delivery of siRNA and DNA constructs. ACS Nano 3(10):3273–3286
Yan Y, Chan-Park MB, Zhang Q (2007) Advances in carbon-nanotube assembly. Small 3(1):24–42

Yuan H, Miao J, Du YZ et al (2008) Cellular uptake of solid lipid nanoparticles and cytotoxicity of encapsulated paclitaxel in A549 cancer cells. Int J Pharm 348(1–2):137–145

Zeb A, Qureshi OS, Kim HS (2017) High payload itraconazole-incorporated lipid nanoparticles with modulated release property for oral and parenteral administration. J Pharm Pharmacol 69 (8):955–966

Zhang C, Gu Z, Shen L et al (2017) A dual targeting drug delivery system for penetrating blood-brain barrier and selectively delivering siRNA to neurons for Alzheimer's disease treatment. Curr Pharm Biotechnol 8(14):1124–1131

Zhao D, Liu CJ, Zhuo RX et al (2012) Alginate/$CaCO_3$ hybrid nanoparticles for efficient codelivery of antitumour gene and drug. Mol Pharm 9(10):2887–2893

Zhu Y, Liao L (2015) Applications of nanoparticles for anticancer drug delivery: a review. J Nanosci Nanotechnol 15(7):4753–4773

Zhuo RX, Du B, Lu ZR (1999) In vitro release of 5-fluorouracil with cyclic core dendritic polymer. J Control Release 57(3):249–257

Zhuxuan J, Juan G, Jun Q et al (2019) Peptide ligand-mediated targeted drug delivery of nanomedicines. Biomater Sci 7:461–471

Role of Lipid Nanocarriers in Lymphatic Targeting: Promises and Safety Considerations

2

Jagruti L. Desai and Hetal P. Thakkar

Abstract

The lymphatic system is an added circulatory system present in the body and has a significant role in the identification of immune system and its reaction to a disease. The treatment of disease of the lymphatic system requires drug administration to desired delivery site so that adverse effects are minimized because of nonspecific spread throughout the body. Moreover, for the drugs undergoing extensive hepatic first-pass metabolism, promoting the absorption through the lymphatic route proves to be an effective way for oral bioavailability enhancement. Different types of nanocarriers have been investigated for their ability to target the drug to the lymphatics. In spite of promising applications of nanomedicine, there are certain issues such as the fate of nanoparticles after lymphatic absorption and the potential toxicity concerns, which need to be addressed. Since nanoparticles possess extremely small size and large surface area, there is a drastic change in the physicochemical properties as well as distribution pattern of drug in the body. Moreover, nonbiodegradable nanocarriers may accumulate in human tissues and organs leading to toxicity. This chapter focuses on the importance of lymphatic targeting, various nanocarrier-based approaches and their mechanisms, factors affecting lymphatic transport, and their safety and toxicity evaluation.

J. L. Desai
Ramanbhai Patel College of Pharmacy, Charotar University of Science and Technology, Changa, Gujarat, India
e-mail: jagrutidesai.ph@charusat.ac.in

H. P. Thakkar (✉)
Department of Pharmacy, Faculty of Pharmacy, The Maharaja Sayajirao University of Baroda, Vadodara, Gujarat, India
e-mail: hetal.thakkar-pharmacy@msubaroda.ac.in

N. Shah (ed.), *Nanocarriers: Drug Delivery System*,
https://doi.org/10.1007/978-981-33-4497-6_2

Keywords

Nanocarriers · Lymphatic delivery · Drug efficacy · Targeting delivery

2.1 Introduction

The lymphatic system is a complex network of blind-ended lymphatic capillaries found in almost all tissues connecting lymph nodes. The lymphatic capillaries merge to form collecting vessels that empty into large vascular trunks that finally drain to blood circulatory system via drainage point. This drainage point is the only connection between these two systems; otherwise the lymphatic system is entirely separate from the blood circulatory system. The lymphatic vessels are often seen in close juxtaposition to arteries and veins (Butler et al. 2009). Earlier, blood circulatory system was largely focused by scientists while lymphatic system was ignored. Since recent years, there has been a breakthrough in the investigations connecting lymphatic system with many diseases which unrevealed various aspects of the role of lymphatic system in diseases and treatment. Many discoveries on molecular, cellular, and genetic level as well as usage of modern imaging technologies have given an understanding of this system (Choi et al. 2012). The lymphatic system has been reported to have various significant functions. It is involved in maintaining tissue homeostasis and immune function (Maby-El Hajjami and Petrova 2008; Liao and Von Der Weid 2015), including removal of surplus extravascular fluid and conducting components like lipid-soluble nutrients and foreign bodies including pathogens from digestive tract to the immune system. Most drugs have great entry to the systemic circulation through the portal vein. But this results in first-pass metabolism of majority of drugs. Caliph et al. described that compounds with high lipophilicity have entry to the systemic circulation by uptake via lymph (Caliph et al. 2000). Hence, lymphatic delivery is useful for avoiding hepatic metabolism which eventually affects the bioavailability of such drugs (Singh et al. 2014). Lymphatic uptake would also prolong the time span of drug transport to the blood circulation, directly targeting lymphatic tissues and indirectly specific sites associated with low-density lipoprotein, which highlights the importance of lymphatic delivery of drug candidates (Porter et al. 2007). This variety of physiological functions of lymphatics has led to the understanding that they affect a broader array of diseases. The disease includes lymphedema, cancer and metastases, immune and inflammatory conditions like inflammatory bowel disease, rheumatoid arthritis and asthma, psoriasis, metabolic diseases such as obesity atherosclerosis and hypertension, liver disease and ascites, cardiovascular disease, infection concerning HIV, hepatitis, filariasis, and Ebola virus (Singh et al. 2014). The conventional systems have failed in providing effective drug delivery without any dose-limiting toxicities. Through latest developments in material technology, various formulations have been developed in order to deliver the drug to the lymphatic system. Some of them are illustrated in Table 2.1.

Table 2.1 Various drugs encapsulated in nanoformulation for lymphatic delivery

Drug	Formulation	Target	Model	Reference
Not applicable	LyP-1-conjugated PEG-PLGA nanoparticles	Lymphatic metastatic tumors	In vitro cellular uptake study	Luo et al. (2010)
Quetiapine fumarate	Solid lipid nanoparticles	Bioavailability enhancement by intraduodenal administration	In vivo pharmacokinetic study	Yasir et al. (2018)
Lopinavir	Solid lipid nanoparticles	Intestinal lymphatic targeting	Intestinal lymphatic transport study	Aji Alex et al. (2011)
Darunavir	Peptide-grafted lipid nanoparticles	CD4 receptor bearing T cells in lymphatic system	Uptake study by confocal microscopy, organ biodistribution study	Desai and Thakkar (2018)
Doxorubicin	LyP-1-conjugated PEGylated liposomes	Metastatic lymph nodes	In vitro cellular uptake and in vivo near-infrared fluorescence imaging, pharmacodynamic study, pathological examination	Yan et al. (2012)
Doxorubicin and vincristine	Nanostructured lipid carriers	Lymph cancer	Lymph cancer animal model	Dong et al. (2016)
Leflunomide	Nanostructured lipid carriers	Lymphatic uptake	Intestinal lymphatic uptake studies	Krishnan et al. (2018)
Melanoma antigen peptide Trp2	Cationic polyethylenimine (2k)-stearic acid (PSA) micelles	Lymph nodes	B16-F10 murine melanoma model	Zeng et al. (2015)
Doxorubicin and vinorelbine	PEG-PE polymeric micelles	Metastatic tumors	Microscopy imaging of live mice or tissue sections	Qin et al. (2013)
Tamoxifen	Alginate microparticles	Lymphatic system	Peyer's patch uptake	Coppi and Iannuccelli (2009)
Doxorubicin	Magnetic multi-walled carbon nanotube	Lymph node	Lymph node metastatic model	Ji et al. (2016)
Docetaxel	PEGylated polyglutamic acid and polyasparagine nanocapsules	Lymphatic system	In vivo biodistribution study	Abellan-Pose et al. (2016)
Doxorubicin	PEGylated polylysine dendrimer	Lymphatic system, lymph node metastases	Compartmental pharmacokinetic modeling	Ryan et al. (2013)

In this chapter, we first give a brief description of the lymphatic system, followed by description of various routes of drug administration to lymphatic system, fate of lipid carrier digestion in vivo, characterization of physicochemical properties of typical drug carriers and their influence on lymphatic uptake and transport, various nanocarrier-based approaches for lymphatic delivery, and then study of various models for in vitro and in vivo assessment of uptake of lipid nanocarrier by lymphatic system. At the end of this chapter, we review the safety and toxicity evaluation of lipid nanocarriers for lymphatic delivery.

2.2 Anatomy and Physiology of the Lymphatic System

The lymphatic system is a one-way, irreversible complex network of several organs, which includes the lymphatic vessels, lymph nodes, thymus, spleen, Peyer's patches, and tonsils (Singh et al. 2014).

2.2.1 Lymph

In normal conditions, the lymphatic vasculature carries cells as well as fluid from tissue interstitium, in the form of "lymph," to the blood circulation. During systemic transport, because of hydrostatic pressure difference, some amount of blood volume gets pushed into the surrounding tissue interstitium in the blood capillary walls. This surplus fluid forms the "interstitial fluid" which baths surrounding tissue cells and sooner or later diffuses into nearby lymphatic capillaries to form the lymph. This lymph is a clear interstitial fluid consisting of high amount of white blood cells (majorly lymphocytes) while protein content is about half that of blood protein content (Yoffey and Courtice 1970; Watson 2011).

2.2.2 Lymphatic Vessels

The lymphatic vessel system consists of lymphatic capillaries, lymphatic collecting vessels, lymphatic trunks, and lymphatic ducts. The lymphatic capillaries are the thinnest vessels among these four lymphatic vessels and comprise a single layer of thin-walled non-fenestrated lymphatic endothelial cells containing intercellular junctions of diameter between 10 and 60 μm. They are present throughout the body except for the nonvascular tissues (such as cartilage, epidermis, and cornea of eye) and central nervous system, and thereby facilitate absorption of protein molecules, cell debris, large foreign particles, and pathogens. These capillaries unite to form lymphatic vessels, which look like veins in structure but have thinner walls and more valves. The lymph fluids enter lymphatic system through lymphatic capillaries, followed by collecting vessels and after filtration through lymph nodes, the filtered lymph enters lymphatic trunks. The backward flow of lymph during any skeletal muscle relaxation is prevented through the semilunar valves that are present in the lymphatics. The lymphatic trunks carry lymph and drain through lymphatic ducts into right quadrant and remaining three-quarters of body, respectively, by connecting to subclavian veins (Mohrman and Heller 1997). Specialized lymphatic capillaries known as "lacteals" located in the central section of small intestine villi connect lymphatic capillaries from intestine, mucosa, and submucosa to form collecting lymphatic vessels.

2.2.3 Lymph Nodes

Another organ associated with lymphatic system is the lymph node—tiny, oval structures located in many connective tissues and extensively present in the body along the path of lymphatic vessels. They are majorly located in neck cervical, axillary, groin inguinal, vertebral column, and intestinal mesenteric regions. They are absent in central nervous system. They comprise lymphocytes and macrophages which promote to filter pathogens, foreign particulates, viruses, as well as bacteria present in the lymph fluid through immune response or phagocytosis mechanism (Watson 2011). Lymph nodes have been reported to play a crucial role in cancer metastasis as demonstrated in several types of human cancers, including breast (Cox et al. 2008), colon (Markl et al. 2012; Kojima et al. 2010), prostate (Datta et al. 2010), bladder (Guzzo et al. 2010), and head and neck cancers (Liao et al. 2012). These studies together demonstrate that regional lymph node metastasis is strongly associated with the diminished survival in cancer patients.

2.2.4 Lymph Organs

The spleen and the thymus are considered as lymphatic organs. The spleen is the biggest organ in the lymphatic system located just below the left rib cage. It contains tissue of two types: one is called as red pulp tissue that filters blood and other is known as the white pulp tissue which consists of immune cells—T cells and B cells. Spleen is involved in removal of foreign substances and damaged or old erythrocytes from the blood circulation. It also functions during hemorrhage by acting as an oxygen-rich blood reservoir to compensate for a loss of red blood cells (Godart and Hamilton 1963). Another lymph organ, thymus, is located in the anterior part of the chest, behind the sternum and in front of the heart. It produces cells known as progenitor cells that mature to form T cells. It is found to be most active during early neonatal and preadolescent periods though it remains active throughout life. The production of T-lymphocytes begins from bone marrow as precursors which migrate to thymus gland to mature and then transverse through the various organs of lymphatic system and blood circulatory system (Pearse 2006).

2.3 Lymphatic Drug Delivery Routes

In the management of chronic disease, it is essential to deliver the drug with high efficacy to the desirable site of absorption so that dose-related side effects are minimized. Hence, in disease conditions where lymphatic system plays a prominent role in the spread of disease (e.g., lymph organs acting as HIV reservoirs), it is important to deliver sufficient amount of dose to the lymphatic system for therapeutic action. Additionally, this system is not more reachable through IV administration. So, greater quantity of drug dose is generally required to reach the therapeutic concentration. To combat this difficulty, various alternative routes of drug administration to lymphatic system have been explored as discussed in this section.

2.3.1 Oral Route of Administration

Thus, in order to increase oral drug bioavailability, many novel lipid-based formulations have been developed which are said to deliver drug to systemic circulation via intestinal lymphatic uptake (Li et al. 2017; Desai and Thakkar 2016, 2019). Orally administered lipid formulation is first digested (digestive phase) and broken down into emulsion (of lipid droplets ~0.5 μm) of high surface area. The gastric lipase present in gastric juice acts on triglycerides (TGs) and cleaves into free fatty acids (FAs) and monoglycerides (MGs). The non-hydrolyzed TGs along with hydrolyzed entities enter the duodenum (Carey et al. 1983). In the intestine, TGs stimulate the pancreas to secrete enzyme lipase and co-lipase that form complex and act at hydrophilic/lipophilic interface of globules to hydrolyze TGs (Embleton and Pouton 1997). This ultimately leads to formation of mixed micelles (formed by interaction of FAs and MGs with bile salts) and vesicles (formed by TGs and FAs) (Ollivon et al. 1988). During this period of digestion, the drug released from the formulation gets resolubilized as micelles or mixed micelles (Fatouros et al. 2007). There are various factors that modify the solubilization of drug during this phase, like lipophilic nature of drug, chemical nature of drug, and nature of lipid (both endogenous and exogenous) involved in the formation of emulsion (Kossena et al. 2003). The digestive phase is followed by the absorptive phase wherein the colloidal species, so formed, are absorbed by enterocyte membrane cells through various mechanisms like passive diffusion, active transport, and facilitated diffusion. There exists another pathway for absorption which is known as chylomicron pathway. In this route, the free drug molecules absorbed are taken up by the intestinal lipoproteins (called as chylomicrons having large diameters of <1 μm) within the enterocyte. This eventually leads to uptake of lipophilic compounds by lymphatic capillaries from the intestine region (Chakraborty et al. 2009). Drug transport through lymphatic system therefore requires the presence of lipids to initiate the formation of lipoprotein (Charman et al. 1986). The mechanism of transport of drug to intestinal lymph is depicted in Fig. 2.1. As such, material absorbed from GI tract (small intestine) can be transported to both blood and lymphatic capillaries. The mode of transport through

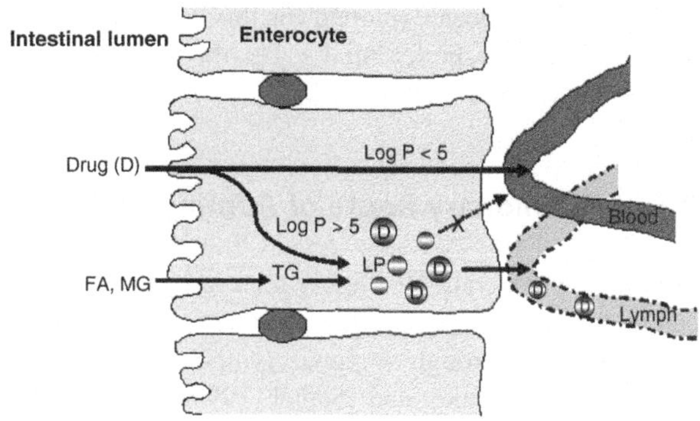

Fig. 2.1 Drug absorption via intestinal lymphatic system and portal vein (reproduced from Trevaskis et al. (2008) with permission)

any of these capillaries depends on lipophilicity of the compound. The molecules with log P greater than 5 have high possibility of uptake by lymphatic capillaries. Moreover, during this phase, the drug may also undergo metabolism by cytochrome P450 enzyme present in enterocytes. The studies reported have shown that the presence of lipids resulted in increased bioavailability of those drugs which are metabolized by these metabolizing enzymes (Trevaskis et al. 2006). Another pathway of absorption is through the M cells. After absorption, the TGs and phospholipids cannot be taken up directly by bloodstream because of their large molecular size. However, the lymphatic capillaries, being thin walled, allow easy entry of interstitial fluid inside. Additionally, the lymphatic vessels enable the uptake of high-molecular-weight substances (like chylomicrons). There are two main factors that govern the selection of either of the capillaries (blood vessels and lymphatic capillaries)—the size of lymph lipid precursor pool and FA chain length. It has been reported that intestinal lymph transport occurs when there is free FA of 12 carbons or higher chain length. Free FA of chain length $\leq$12 carbons is absorbed primarily by portal blood (Trevaskis et al. 2006). Also, high degree of unsaturation has shown to increase lymphatic uptake by producing large-size lipoproteins. Finally, the lymph fluid enters the subclavian vein via thoracic duct and thus bypasses first-pass metabolism.

2.3.2 Intestinal Route of Administration

Oral delivery of lipid excipient suffers from enzymatic degradation in gastric fluid so intestinal delivery is preferred over oral route. Intestinal absorption is the main pathway for absorption of many lipophilic compounds including fat-soluble vitamins and lipid from food into systemic circulation. Upon entry into the intestinal lumen, absorption and transport through intestinal membrane follow the same pathway as discussed in oral route of administration. Various drugs have been administered through intestinal route in order to increase oral bioavailability like quetiapine fumarate (Yasir et al. 2018), praziquantel (Mishra et al. 2014), and lopinavir (Aji Alex et al. 2011). Various lipid-based carriers like SMEDDS, SLNs, and NLCs have been exploited for this administration with a view to increase oral bioavailability by preferential entry through lymphatic system avoiding hepatic metabolism.

2.3.3 Pulmonary Route of Administration

Metastatic proliferation is highly seen in certain types of cancer like small cell lung carcinoma. In such cases, the malignant cells spread from hemithorax to the lymphatic system through regional lymph nodes and finally to blood circulation (Zangemeister-Wittke and Stahel 1999). Hence, drug administration to these regional lymph nodes via pulmonary route is of great importance. Besides this, there are other advantages of this route like reduction in systemic toxicity, avoidance of hepatic metabolism, noninvasive route of administration, and increased local drug concentration to various parts of lung. Further, particles having size less than 500 nm are generally taken up by alveolar macrophages, allowing them to reach lymphatics (Mcintire et al. 1998). Hence, nanosized particles can be effectively targeted to lymphatics via pulmonary route. In a study done by Videira et al., nebulized aerosol containing ^{99m}TC-HMPAO (hexamethyl propylene amine oxime) solid lipid nanoparticles (having mean particle diameter 200 nm) was administered by inhalation to adult male Wistar rats. The images showed that the nanoparticles were able to penetrate effectively to the interstitial alveolar space and high distribution was observed in periaortic, axillary, and inguinal lymph nodes (Videira et al. 2002).

2.3.4 Intraperitoneal Route of Administration

Cancer metastasis within abdominal and pelvic organs is a major culprit of failure in colorectal cancer. It is estimated that about 40% of colorectal cancer patients will develop peritoneal carcinomatosis (PC) as a result of the growth of tumor in exfoliation and shedding of malignant cells intraperitoneally. Moreover, during surgical procedure, tumor cells may also be released in the peritoneal cavity (Koppe et al. 2006). Thus, intraperitoneal administration is desirable that results in higher drug concentration in peritoneal cavity, thus improving the efficacy of the formulation. It has been reported that upon IP injection, molecules of molecular weight lesser than 20 KDa are taken up by peritoneal cavity into the systemic circulation. On the other hand, molecules of molecular weight greater than 20 KDa are taken up by the lymphatic vessel (Negrini et al. 1991). A commercialized liposomal formulation of paclitaxel (PTX) called as Lipusu® was injected through IP route in ovarian cancer xenografts (rat) and compared with paclitaxel solution (Ye et al. 2013). The result showed increased concentration of liposomal formation in lymph nodes compared to solution. The liposomal formulation significantly increased lymphatic drug targeting and tumor inhibition. In a study, paclitaxel-loaded PLGA nanoparticles were formulated and administered intraperitoneally to ovarian cancer xenografts (Lu et al. 2007). The result showed 20-fold higher uptake in pelvic lymph nodes 48 h after injection compared to paclitaxel solution. Moreover, the drug concentration in tumor enhanced threefold compared to solution formulation while tumor weight and ascites volume decreased significantly indicating better efficacy of nanoparticle formulation in ovarian cancer.

2.3.5 Subcutaneous Route of Administration

Nanocarriers, when administered through subcutaneous injection, get absorbed from the site of injection and reach interstitial area under dermis after which they are taken up by lymphatic capillaries followed by entry into lymphatic system along with peripheral lymph. However, variations in lymphatic drainage rate as well as subcutaneous blood flow largely affect the absorption rates which may vary regionally or patient-wise. Thus, corresponding variation might be seen in drug concentrations in vascular and lymphatic absorption pathways. In another study (Reddy et al. 2005), etoposide SLNs were studied for accumulation in tumor upon administration by IP, SC, and IV routes in Dalton's lymphoma-bearing mice. They found that the highest accumulation of drug carrier in tumor occurs through SC route postinjection and also after 24 h of injection.

2.4 Physicochemical Properties of Drugs/Carriers for Lymphatic Uptake

2.4.1 Molecular Weight of Molecules

Following subcutaneous administration, molecular weight is one of the factors playing a role in the selection of absorption pathways (blood capillaries and lymphatics) of molecules. Supersaxo et al. (1990) reported a direct connection between drug molecular weight and part of dose taken up by lymphatic capillaries upon SC administration. They found that molecules with molecular weight (MW) greater than 16,000 were absorbed primarily by the lymphatics while molecules with MW lesser than 1000 were hardly taken up by the lymphatics (Supersaxo et al. 1990). The difference between blood and lymphatic capillaries structurally (their endothelial cell lining) is a determining factor for the absorption of molecules. In case of blood capillaries, there is continuous and uninterrupted basement membrane below the endothelial lining, while the lymphatic capillaries do not possess this basement membrane. In addition, large macromolecules of up to 1 μm diameter may be transported via pores (of diameter 20–100 nm) present in endothelial cells of lymphatic capillaries (Tomlinson 1987).

2.4.2 Solubility of Drugs in Lipid and Partition Coefficient

The triglyceride (TG) solubility and log P value are considered as one of the key factors governing lymphatic transport of drug upon oral/intraduodenal administration. Charman and Stella have reported that the desirable property for high lymphatic uptake of drug was log P greater than 5 and TG solubility greater than 50 mg/mL (Charman and Stella 1986). They estimated these values based on the hypothesis that triglyceride-rich lipoprotein takes up the drug and thus transports the drug to lymphatic system. Studies on various compounds have supported these predictions. For example, in case of halofantrine which has log P of 8.5 and TG solubility greater than 50 mg/mL, 20% of dose was found in lymph upon co-administration of lipid carriers (Porter et al. 1996). In another study, dichlorodiphenyltrichloroethane (DDT) having log P of 6.19 and TG solubility of 80 mg/g showed a rise of 15% in lymphatic transport upon co-administration with lipid carriers. They also found that hexachlorobenzene which has enough log P of 6.53 but fails to fulfill the criteria of TG solubility (has a value of 7.5 mg/g) showed poor lymphatic uptake (Myers and Stella 1992). However, there are various reports that do not comply with these predictions of Charman and Stella. For example, penclomedine which has log P of 5.48 and TG solubility of 174 mg/mL showed only a small portion of the dose administered (3%) in lymphatic transport. Similarly, molecule Cl-176 that has log P of 5.83 and TG solubility greater than 100 mg/mL also showed less than 1% (of the dose administered) in lymphatic transport. These contradictions reflect that apart from lipid solubility and partition coefficient, other factors also determine the lymphatic uptake.

2.4.3 Properties of Lipids

Lipids are basically composed of TG which upon emulsification orients their polar head towards aqueous phase. The fatty acid chain length of TG has been shown to have an effect on lymphatic transport. Paliwal et al. formulated SLNs of methotrexate using various lipids (Compritol 888 ATO, stearic acid, monostearin, and tristearin). They found that among the four lipids, Compritol 888 ATO having the longest chain length showed the highest lymphatic uptake and hence better bioavailability compared to SLNs prepared using other lipids (Paliwal et al. 2009). These conclusions were further in line with the results obtained by Caliph et al. that compared the lymphatic transport and oral bioavailability of halofantrine upon co-administration with long-chain (C-18), medium-chain (C8-10), and short-chain (C4) fatty acid-based lipid vehicles. They found that the order of lymphatic uptake of halofantrine was formulation with lipid containing C18 (15.8% of dose) > C8-10 (5.5% of dose) > C4 (2.22% of dose) (Caliph et al. 2000). The comparison of digestive pathway for medium- (MCT) and long-chain triglycerides (LCT) is depicted in Fig. 2.2. It can be seen that medium-chain fatty acids are more hydrophilic compared to long-chain fatty acids that causes them to get absorbed even in the absence of bile salts. Digestion step is sometimes even not required for few MCT

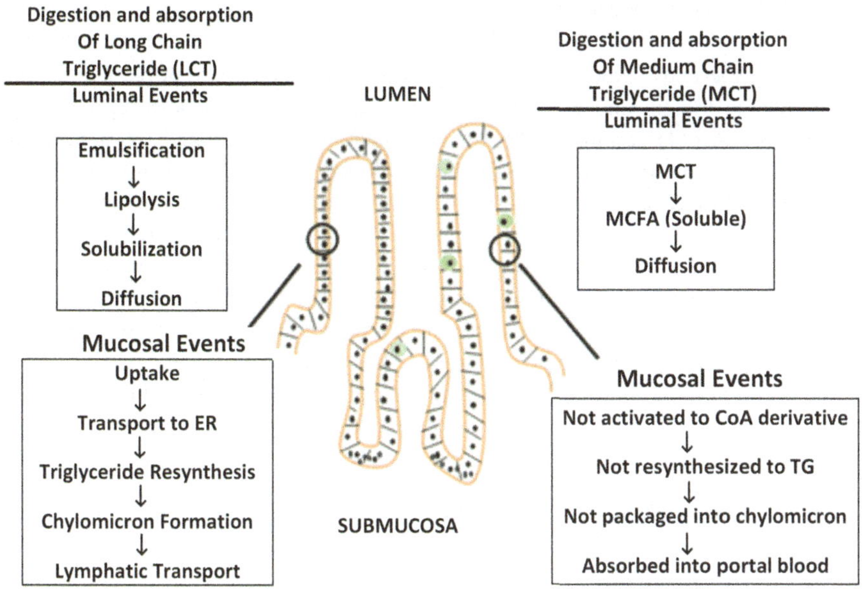

Fig. 2.2 Pathway of digestion for medium-chain triglycerides (MCT) and long-chain triglycerides (LCT) (reproduced from Sitrin 2014 with permission)

and they are directly released into blood circulation upon absorption by the intestine (Sitrin 2014).

2.4.4 Concentration of Emulsifiers Used in Preparation

Emulsifiers are important excipients used in the preparation of lipid formulation. Apart from its effect on particle/globule size of formulation, the drug partitioning is also affected. In the study done by Sanjula et al. (2009), carvedilol-loaded SLNs were prepared using the emulsifier poloxamer 188 (at varying concentrations of 5–15% w/v) (Sanjula et al. 2009) and the effect of variation in this concentration on entrapment efficiency of formulation was investigated. The authors found that the entrapment efficiency and lymphatic uptake decreased significantly when emulsifier concentration was increased. They hypothesized that the decrease in entrapment efficiency would be due to micelle formation that would increase the solubility of carvedilol in external aqueous phase. On the other hand, increase in emulsifier would reduce the hydrophobicity of SLNs that ultimately results in decreased lymphatic uptake of drug resulting in reduced oral bioavailability.

2.4.5 Size of Particles

The size of particles is most important while considering the fate of particle uptake by lymphatic system and its retention in lymph nodes. It has been reported that particles of size lesser than 100 nm are taken up by lymphatic system upon subcutaneous administration (Oussoren and Storm 2001). Above this range, the particles cannot easily traverse through the interstitial cells and thus stop at the injection site till they are removed by phagocytosis. On the other hand, below this range (that is less than 10 nm), the particles diffuse through blood capillaries and get entry into the systemic circulation. In another study done by Desai et al. (Desai and Thakkar 2016), solid lipid nanoparticles of darunavir were prepared with varying particle sizes and administered orally in rats. The results showed that both particles of sizes 100 and 200 nm (with no significant difference in bioavailability between them) showed higher bioavailability compared to particles of size 500 nm.

2.4.6 Surface Charge of Particles

Apart from size, surface charge also plays an important role in the uptake and localization of nanoparticles. Negatively charged drug carriers (such as liposomes, polymeric nanoparticles of PLGA, and dendrimers) have shown higher lymphatic uptake in comparison to positively charged neutral particles (Kaminskas and Porter 2011; Kaur et al. 2008; Rao et al. 2010). Moreover, negatively charged particles were reported to retain longer in the lymph nodes (Kaminskas and Porter 2011).

2.5 Various Nanocarrier-Based Approaches for Lymphatic Delivery

In an attempt to deliver drugs more effectively to the lymphatic system, various formulations have been described as follows.

2.5.1 Liposomes

Liposomes are small vesicles of spherical shape consisting of one or more phospholipid bilayers with size ranges from 25 to 500 nm. Liposomes possess high colloidal stability compared to biphasic emulsion formulation. In addition, liposomes can be coated using various polymers that further stabilize the liposomal bilayer for prolonged blood retention (Akbarzadeh et al. 2013). In the context of lymphatic delivery, liposomes are a promising carrier due to their proven biocompatibility, tunable size, and capacity of loading both hydrophilic and hydrophobic drugs (Akbarzadeh et al. 2013). In lymph nodes, particulate uptake is mainly by medullary sinuses and paracortex and liposomes have been reported to be taken up by littoral and reticular cells as well as by polymorphonuclear granulocytes. Moreover, liposomal surface modification using ethylene oxide-based copolymers has been shown to affect lymphatic uptake.

2.5.2 Self-Emulsifying Drug Delivery Systems (SEDDS)

SEDDS is a simple binary mixture of drug molecules, lipophilic phase (lipid), and surfactants. In vivo, this formulation spontaneously forms emulsion (oil in water) upon coming in contact with GI media and through agitation by gastrointestinal motility. SEDDS formulation is different from other lipid-based formulations in few aspects including concentration and type of excipients used, their ratio, and capacity of 100% drug-loading efficacy (Kohli et al. 2010). Many hydrophobic drugs like carvedilol (Han et al. 2020), ceftriaxone (Kanwal et al. 2019), candesartan (AboulFotouh et al. 2019), amphotericin B (Kontogiannidou et al. 2020), and myricitrin (Man et al. 2019) have been reported to show enhanced bioavailability upon administration through SMEDDS formulations. Supersaturated SMEDDS of hydrophobic drug saquinavir has been formulated and shown to have lymphatic absorption along with increased bioavailability (Jo et al. 2019). In another study, darunavir was delivered as SMEDDS formulation and delivered to male Wistar rats with and without lymphatic blocker (Pluronic F68). The result showed 1.74-fold increase in permeation of darunavir compared to that administered without lymphatic blocker indicating lymphatic absorption of the drug upon SMEDD formulation (Pokale and Bandivadekar 2016).

2.5.3 Solid Lipid Nanoparticles

Solid lipid nanoparticles (SLNs) are spherical nanosized particulate drug delivery system of size in the range of 50–1000 nm and which is solid at body temperature. The excipients used in the formulation of SLNs are biocompatible in nature, thus ensuring safety of the formulation (Lin et al. 2017). Variety of lipids such as waxes, fatty acids, steroids, and mono-, di-, or triglyceride mixtures and biocompatible surfactants (e.g., Tween 80, poloxamer 188, poloxamine 908, and sodium dodecyl sulfate) are used as emulsifiers. SLNs can be formulated using a variety of techniques including high-pressure homogenization, emulsification solvent evaporation, high shear homogenization, and supercritical fluid technology. The size and surface charge of SLNs are greatly influenced by the method adopted for the formulation (Mehnert and Mader 2001). Various drugs have been incorporated into SLN formulation and have showed enhanced drug delivery. Although SLNs have been widely used, it suffers from limitation of possibility of drug expulsion from SLN core resulting in inefficient drug release and subsequent undesired release characteristics of drug. To overcome this limitation, liquid lipid is incorporated along with solid lipid during formulation resulting in formation of nanostructured lipid carriers (NLCs). The stability of NLCs is more as compared to conventional SLN formulation (Beloqui et al. 2017). The delivery of SLNs to the lymphatic system is largely affected by parameters like mode of administration, size, and charge of nanoparticles (Cai et al. 2011). Curcumin-loaded solid lipid nanoparticle prepared using glyceryl monostearate as lipid and poloxamer 188 as surfactant showed 6.3-fold higher lymphatic uptake and 9.5-fold high oral bioavailability compared to curcumin solution (Baek and Cho 2017). Similarly methotrexate concentration in lymphatic system was enhanced by its SLN formulation. Compritol 888 ATO was used as lipid and soya lecithin as emulsifier. The authors reported that the formulated SLNs gained entry into the lymphatic system by probably two mechanisms: transport and uptake of intact SLNs (Paliwal et al. 2009).

2.6 Conclusion

Lymphatic circulation is a complex process of lymphatic absorption and transport offering potential for targeted drug delivery. Investigations in this area have made it clear that lymphatic circulation is no less than blood circulatory system while considering drug delivery for chronic diseases like cancer and HIV. Various nanocarriers exploited for lymphatic delivery were discussed in this chapter. By understanding lymphatic transport and their uptake one can design new therapeutic treatments for effective disease control.

References

Abellan-Pose R, Teijeiro-Valino C, Santander-Ortega MJ et al (2016) Polyaminoacid nanocapsules for drug delivery to the lymphatic system: effect of the particle size. Int J Pharm 509:107–117

Aboulfotouh K, Allam AA, El-badry M et al (2019) A self-nanoemulsifying drug delivery system for enhancing the oral bioavailability of candesartan Cilexetil: ex vivo and in vivo evaluation. J Pharm Sci 108:3599–3608

Aji Alex MR, Chacko AJ, Jose S et al (2011) Lopinavir loaded solid lipid nanoparticles (SLN) for intestinal lymphatic targeting. Eur J Pharm Sci 42:11–18

Akbarzadeh A, Rezaei-sadabady R, Davaran S et al (2013) Liposome: classification, preparation, and applications. Nanoscale Res Lett 8:102

Baek JS, Cho CW (2017) Surface modification of solid lipid nanoparticles for oral delivery of curcumin: improvement of bioavailability through enhanced cellular uptake, and lymphatic uptake. Eur J Pharm Biopharm 117:132–140

Beloqui A, Del Pozo-rodriguez A, Isla A et al (2017) Nanostructured lipid carriers as oral delivery systems for poorly soluble drugs. J Drug Deliv Sci Technol 42:144–154

Butler MG, Isogal S, Weinstein BM (2009) Lymphatic development. Birth Defects Res C Embryo Today 87:222–231

Cai S, Yang Q, Bagby TR et al (2011) Lymphatic drug delivery using engineered liposomes and solid lipid nanoparticles. Adv Drug Deliv Rev 63:901–908

Caliph SM, Charman WN, Porter CJ (2000) Effect of short-, medium-, and long-chain fatty acid-based vehicles on the absolute oral bioavailability and intestinal lymphatic transport of halofantrine and assessment of mass balance in lymph-cannulated and non-cannulated rats. J Pharm Sci 89:1073–1084

Carey MC, Small DM, Bliss CM (1983) Lipid digestion and absorption. Annu Rev Physiol 45:651–677

Chakraborty S, Shukla D, Mishra B et al (2009) Lipid—an emerging platform for oral delivery of drugs with poor bioavailability. Eur J Pharm Biopharm 73:1–15

Charman W, Stella V (1986) Estimating the maximal potential for intestinal lymphatic transport of lipophilic drug molecules. Int J Pharm 34:175–178

Choi I, Lee S, Hong YK (2012) The new era of the lymphatic system: no longer secondary to the blood vascular system. Cold Spring Harb Perspect Med 2:a006445

Coppi G, Iannuccelli V (2009) Alginate/chitosan microparticles for tamoxifen delivery to the lymphatic system. Int J Pharm 367:127–132

Cox CE, Kiluk JV, Riker AI et al (2008) Significance of sentinel lymph node micrometastases in human breast cancer. J Am Coll Surg 206:261–268

Datta K, Muders M, Zhang H et al (2010) Mechanism of lymph node metastasis in prostate cancer. Future Oncol 6:823–836

Desai JL, Thakkar HP (2016) Effect of particle size on oral bioavailability of darunavir-loaded solid lipid nanoparticles. J Microencapsul 33:669–678

Desai JL, Thakkar HP (2018) Darunavir-loaded lipid nanoparticles for targeting to HIV reservoirs. AAPS Pharm Sci Tech 19:648–660

Desai JL, Thakkar HP (2019) Enhanced oral bioavailability and brain uptake of Darunavir using lipid nanoemulsion formulation. Colloids Surf B Biointerfaces 175:143–149

Dong X, Wang W, Qu H, Han D et al (2016) Targeted delivery of doxorubicin and vincristine to lymph cancer: evaluation of novel nanostructured lipid carriers in vitro and in vivo. Drug Deliv 23:1374–1378

Embleton JK, Pouton CW (1997) Structure and function of gastro-intestinal lipases. Adv Drug Deliv Rev 25:15–32

Fatouros DG, Bergenstahl B, Mullertz A (2007) Morphological observations on a lipid-based drug delivery system during in vitro digestion. Eur J Pharm Sci 31:85–94

Godart SJ, Hamilton WF (1963) Lymphatic drainage of the spleen. Am J Physiol 204:1107–1114

Guzzo TJ, Resnick MJ, Canter DJ et al (2010) Impact of adjuvant chemotherapy on patients with lymph node metastasis at the time of radical cystectomy. Can J Urol 17:5465–5471

Han H, Li Y, Peng Z et al (2020) A Soluplus/Poloxamer 407-based self-nanoemulsifying drug delivery system for the weakly basic drug carvedilol to improve its bioavailability. Nanomedicine 27:102–199

Ji J, Liu M, Meng Y et al (2016) Experimental study of magnetic multi-walled carbon nanotube-doxorubicin conjugate in a lymph node metastatic model of breast cancer. Med Sci Monit 22:2363–2373

Jo K, Kim H, Khadka P et al (2019) Enhanced intestinal lymphatic absorption of saquinavir through supersaturated self-microemulsifying drug delivery systems. Asian J Pharm Sci 15:336–346

Kaminskas LM, Porter CJ (2011) Targeting the lymphatics using dendritic polymers (dendrimers). Adv Drug Deliv Rev 63:890–900

Kanwal T, Kawish M, Maharjan R et al (2019) Design and development of permeation enhancer containing self-nanoemulsifying drug delivery system (SNEDDS) for ceftriaxone sodium improved oral pharmacokinetics. J Mol Liq 289:111098

Kaur CD, Nahar M, Jain NK (2008) Lymphatic targeting of zidovudine using surface-engineered liposomes. J Drug Deliv 16:798–805

Kohli K, Chopra S, Dhar D et al (2010) Self-emulsifying drug delivery systems: an approach to enhance oral bioavailability. Drug Discov Today 15:958–965

Kojima T, Watanabe Y, Hashimoto Y et al (2010) In vivo biological purging for lymph node metastasis of human colorectal cancer by telomerase-specific oncolytic virotherapy. Ann Surg 251:1079–1086

Kontogiannidou E, Meikopoulos T, Virgiliou C et al (2020) Towards the development of self-Nano-emulsifying drug delivery systems (SNEDDS) containing trimethyl chitosan for the oral delivery of amphotericin B: in vitro assessment and cytocompatibility studies. J Drug Deliv Sci Technol 56:101524

Koppe MJ, Boerman OC, Oyen WJ et al (2006) Peritoneal carcinomatosis of colorectal origin: incidence and current treatment strategies. Ann Surg 243:212–222
Kossena GA, Boyd BJ, Porter CJ et al (2003) Separation and characterization of the colloidal phases produced on digestion of common formulation lipids and assessment of their impact on the apparent solubility of selected poorly water-soluble drugs. J Pharm Sci 92:634–648
Krishnan Y, Mukundan S, Akhil S et al (2018) Enhanced lymphatic uptake of Leflunomide loaded nanolipid carrier via chylomicron formation for the treatment of rheumatoid arthritis. Adv Pharm Bull 8:257–265
Li F, Hu R, Wang B et al (2017) Self-microemulsifying drug delivery system for improving the bioavailability of huperzine A by lymphatic uptake. Acta Pharm Sin B 7:353–360
Liao S, Von Der Weid PY (2015) Lymphatic system: an active pathway for immune protection. Semin Cell Dev Biol 38:83–89
Liao LJ, Lo WC, Hsu WL et al (2012) Detection of cervical lymph node metastasis in head and neck cancer patients with clinically N0 neck-a meta-analysis comparing different imaging modalities. BMC Cancer 12:236
Lin CH, Chen CH, Lin ZC et al (2017) Recent advances in oral delivery of drugs and bioactive natural products using solid lipid nanoparticles as the carriers. J Food Drug Anal 25:219–234
Lu H, Li B, Kang Y et al (2007) Paclitaxel nanoparticle inhibits growth of ovarian cancer xenografts and enhances lymphatic targeting. Cancer Chemother Pharmacol 59:175–181
Luo G, Yu X, Jin C et al (2010) LyP-1-conjugated nanoparticles for targeting drug delivery to lymphatic metastatic tumors. Int J Pharm 385:150–156
Maby-El Hajjami H, Petrova TV (2008) Developmental and pathological lymphangiogenesis: from models to human disease. Histochem Cell Biol 130:1063–1078
Man N, Wang Q, Li H et al (2019) Improved oral bioavailability of myricitrin by liquid self-microemulsifying drug delivery systems. J Drug Deliv Sci Technol 52:597–606
Markl B, Rossle J, Arnholdt HM et al (2012) The clinical significance of lymph node size in colon cancer. Mod Pathol 25:1413–1422
Mcintire GL, Bacon ER, Toner JL et al (1998) Pulmonary delivery of nanoparticles of insoluble, iodinated CT X-ray contrast agents to lung draining lymph nodes in dogs. J Pharm Sci 87:1466–1470
Mehnert W, Mader K (2001) Solid lipid nanoparticles: production, characterization and applications. Adv Drug Deliv Rev 47:165–196
Mishra A, Vuddanda PR, Singh S (2014) Intestinal lymphatic delivery of praziquantel by solid lipid nanoparticles: formulation design, in vitro and in vivo studies. J Nanotechnol 2014:351693
Mohrman DE, Heller LJ (1997) Cardiovascular physiology. McGraw-Hill, New York
Myers R, Stella V (1992) Factors affecting the lymphatic transport of penclomedine (NSC-338720), a lipophilic cytotoxic drug: comparison to DDT and hexachlorobenzene. Int J Pharm 80:51–62
Negrini D, Mukenge S, Del Fabbro M et al (1991) Distribution of diaphragmatic lymphatic stomata. J Appl Physiol 70:1544–1549
Ollivon M, Eidelman O, Blumenthal R et al (1988) Micelle-vesicle transition of egg phosphatidylcholine and octyl glucoside. Biochemistry 27:1695–1703
Oussoren C, Storm G (2001) Liposomes to target the lymphatics by subcutaneous administration. Adv Drug Deliv Rev 50:143–156
Paliwal R, Rai S, Vaidya B et al (2009) Effect of lipid core material on characteristics of solid lipid nanoparticles designed for oral lymphatic delivery. Nanomedicine 5:184–191
Pearse G (2006) Normal structure, function and histology of the thymus. Toxicol Pathol 34:504–514
Pokale R, Bandivadekar M (2016) Self micro-emulsifying drug delivery system for lymphatic uptake of darunavir. J Drug Discov Dev Deliv 3:1–7
Porter CJ, Charman SA, Charman WN (1996) Lymphatic transport of halofantrine in the triple-cannulated anesthetized rat model: effect of lipid vehicle dispersion. J Pharm Sci 85:351–356
Porter CJ, Trevaskis NL, Charman WN (2007) Lipids and lipid-based formulations: optimizing the oral delivery of lipophilic drugs. Nat Rev Drug Discov 6:231–248

Qin L, Zhang F, Lu X et al (2013) Polymeric micelles for enhanced lymphatic drug delivery to treat metastatic tumors. J Control Release 171:133–142

Rao DA, Forrest ML, Alani AW et al (2010) Biodegradable PLGA based nanoparticles for sustained regional lymphatic drug delivery. J Pharm Sci 99:2018–2031

Reddy HL, Sharma RK, Chuttani K et al (2005) Influence of administration route on tumor uptake and biodistribution of etoposide loaded solid lipid nanoparticles in Dalton's lymphoma tumor bearing mice. J Control Release 105:185–198

Ryan GM, Kaminskas LM, Bulitta JB et al (2013) PEGylated polylysine dendrimers increase lymphatic exposure to doxorubicin when compared to PEGylated liposomal and solution formulations of doxorubicin. J Control Release 172:128–136

Sanjula B, Shah FM, Javed A et al (2009) Effect of poloxamer 188 on lymphatic uptake of carvedilol-loaded solid lipid nanoparticles for bioavailability enhancement. J Drug Target 17:249–256

Singh I, Swami R, Khan W et al (2014) Lymphatic system: a prospective area for advanced targeting of particulate drug carriers. Expert Opin Drug Deliv 11:211–229

Sitrin MD (2014) Digestion and absorption of dietary triglycerides. In: Leung P (ed) The gastrointestinal system. Springer, Dordrecht, pp 159–178

Supersaxo A, Hein WR, Steffen H (1990) Effect of molecular weight on the lymphatic absorption of water-soluble compounds following subcutaneous administration. Pharm Res 7:167–169

Tomlinson E (1987) Theory and practice of site-specific drug delivery. Adv Drug Deliv Rev 1:87–198

Trevaskis NL, Porter CJ, Charman WN (2006) The lymph lipid precursor pool is a key determinant of intestinal lymphatic drug transport. J Pharmacol Exp Ther 316:881–891

Trevaskis NL, Charman WN, Porter CJ (2008) Lipid-based delivery systems and intestinal lymphatic drug transport: a mechanistic update. Adv Drug Deliv Rev 60:702–716

Videira MA, Botelho MF, Santos AC et al (2002) Lymphatic uptake of pulmonary delivered radiolabelled solid lipid nanoparticles. J Drug Target 10:607–613

Watson R (ed) (2011) Anatomy and physiology for nurses. Baillière Tindall Elsevier, Edinburgh

Yan Z, Wang F, Wen Z et al (2012) LyP-1-conjugated PEGylated liposomes: a carrier system for targeted therapy of lymphatic metastatic tumor. J Control Release 157:118–125

Yasir M, Gaur PK, Puri D et al (2018) Solid lipid nanoparticles approach for lymphatic targeting through intraduodenal delivery of quetiapine fumarate. Curr Drug Deliv 15:818–828

Ye L, He J, Hu Z et al (2013) Antitumor effect and toxicity of Lipusu in rat ovarian cancer xenografts. Food Chem Toxicol 52:200–206

Yoffey JM, Courtice FC (1970) Lymphatics, lymph and lymphomyeloid complex. Academic, London

Zangemeister-Wittke U, Stahel RA (1999) Novel approaches to the treatment of small-cell lung cancer. Cell Mol Life Sci 55:1585–1598

Zeng Q, Jiang H, Wang T et al (2015) Cationic micelle delivery of Trp2 peptide for efficient lymphatic draining and enhanced cytotoxic T-lymphocyte responses. J Control Release 200:1–12

Part II

Nanoparticulate Drug Delivery Carriers

Synthesis, Pharmacokinetics, and Toxicity of Nano-Drug Carriers

3

Noah Tedla, Jose Ruiz, Vicky Mody, Xinyu Wang, and Vishakha S. Bhave

Abstract

The use of nanoparticles in medicine has improved the options for diagnosis and therapy. Numerous diseases that were previously thought to be challenging are now addressed with the advent of nanoparticles. This is possible due to the ability of nanoparticles to selectively target desired tissue/cells and communicate with the cellular environment at nanoscale levels. Although nanomedicine has innumerous advantages over traditional medicine, the use of nanoparticles comes with its own baggage of special pharmacokinetic parameters and toxicity. Hence it becomes imperative to understand the development of nanoparticles from bench to clinic with emphasis on its pharmacokinetic properties and toxicity. With this objective in mind we have discussed the synthesis of clinically important metallic and nonmetallic nano-drug carriers and the process to alter their physicochemical properties for their application in therapy and/or diagnosis. This is followed by a discussion of the pharmacokinetic properties of selected nano-drug carriers that make them desirable for formulations as an efficient drug delivery system. Lastly, the chapter discusses potential toxicities that are associated with nanocarriers, mechanisms of toxicity, major target organs, and factors influencing these toxicities. We have concluded this chapter by discussion of two clinically used nano-drug carriers.

N. Tedla · J. Ruiz · V. Mody · X. Wang
Department of Pharmaceutical Sciences, Philadelphia College of Osteopathic Medicine School of Pharmacy, Suwanee, GA, USA
e-mail: nt7881@pcom.edu; jr259123@pcom.edu; vickymo@pcom.edu; xinyuwa@pcom.edu

V. S. Bhave (✉)
Exxat LLC, Warren, NJ, USA

Department of Pharmaceutical Sciences, Philadelphia College of Osteopathic Medicine School of Pharmacy, Suwanee, GA, USA
e-mail: vishakha.bhave@exxat.com

N. Shah (ed.), *Nanocarriers: Drug Delivery System*,
https://doi.org/10.1007/978-981-33-4497-6_3

Keywords

Nanotoxicity · Nanocarriers · Synthesis of nanocarriers · ADME · Pharmacokinetics · Toxicokinetic

3.1 Introduction

The International Organization of Standardization (ISO) has defined a nanoparticle (NP) as any discrete object with a three-dimensional size less than 100 nm (ISO 2017). In 2011, the European Standard of Commission stated that any object with any one dimension ranging in size from 1 to 100 nm is classified as NP even if the other dimension might be outside that range (ISO 2017). This size is also a characteristic parameter of a colloid which ranges in particle size (1–1000 nm). Thus, it is not uncommon to find literature that refers to NPs and colloidal particles in equal terms. The difference is essentially semantic for particles below 100 nm in size. In general, NPs can be broadly classified as metallic and nonmetallic NPs.

The metallic NPs mainly comprise the group of NPs such as iron oxide NPs, gold NPs, silver NPs, and nano-shells, whereas the nonmetallic NPs include dendrimers, liposomes, and polymeric NPs. Both metallic and nonmetallic NPs have attracted significant attention for their use as biosensors, microfluidics, drug delivery, and microarray tests to tissue engineering (Joseph and Venkatraman 2017; Patra et al. 2018; Krauel et al. 2004; Mirza and Siddiqui 2014). NPs have a high surface-to-volume ratio which makes them an excellent candidate for biomedical applications (Mody et al. 2010). Additionally, use of NPs for therapy requires that they are inert and can be directed to specific sites in the body after systemic administration. Hence selective targeting systems are designed by conjugating the NP with an appropriate ligand, which has a specific binding activity with target cells. NPs provide a platform to attach multiple moieties of therapeutic substance on it and increase the concentration of therapeutic and diagnostic substances at the pathological site. This increases both selectivity and specificity of the NPs. Thus NPs provide endless opportunities for molecular diagnosis and therapy (Praetorius and Mandal 2007). Having said that, NPs also bring with them unique challenges, particularly altered pharmacokinetics and potential toxicity. This presents a unique challenge for their use as theragnostic agents and hence has to be addressed in detail. Thus, in this chapter, we have discussed selected metallic and nonmetallic NPs, and their mode of synthesis along with modification techniques for their use in drug delivery. Additionally, we have also discussed the pharmacokinetics and toxicity of these different types of NPs.

3.2 Nanoparticle Types, Synthesis Methods, and Applications

3.2.1 Metallic Nanoparticles

3.2.1.1 Iron Oxide Nanoparticles

Iron (III) oxide (Fe_2O_3) NPs are superparamagnetic reddish brown, inorganic compounds synthesized by the reaction of Fe (II) and Fe (III) chloride under basic conditions. The more oxidized form of Fe_2O_3, Fe_3O_4, occurs naturally as the mineral

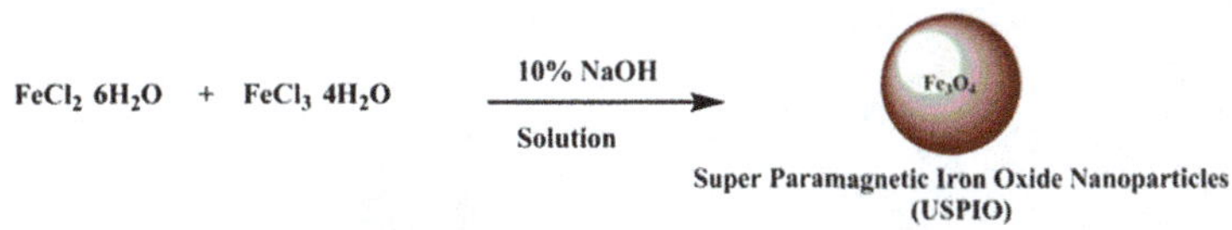

Fig. 3.1 Synthetic scheme for the development of USPIO

magnetite. They range in particle size from 10 to 200 nm. The particles with size in the range of 10–40 nm are classified as ultrasmall superparamagnetic iron oxide NPs (USPIO) and larger particles (60–150 nm) are classified as superparamagnetic iron oxide NPs (SPION). Both USPIO and SPION have found application as contrast agents for magnetic resonance imaging (MRI), targeted drug delivery and imaging, hyperthermia, gene therapy, stem cell tracking, molecular/cellular tracking, magnetic separation technologies (e.g., rapid DNA sequencing), and early detection of inflammation, cancer, diabetes, and atherosclerosis due to their ease of synthesis and biocompatibility (Morales et al. 2003; Babes et al. 1999; Yigit et al. 2008; Peng et al. 2008; Chertok et al. 2008; Gonzales-Weimuller et al. 2009; Wei et al. 2006; Elias and Tsourkas 2009; Moore et al. 2000; Kooi et al. 2003). In addition, they can also be conjugated with various drugs to target specific sites in the body to yield site-specific drug delivery agent.

Use of Fe_3O_4 NPs for targeted delivery and imaging requires the need of homogeneous NPs with particle size distribution in between 10 and 250 nm in diameter. Various methods have been proposed to develop homogeneously distributed magnetic NPs such as microemulsions, sol–gel synthesis, sonochemical reactions, hydrothermal reactions, hydrolysis and thermolysis of precursors, flow injection syntheses, and electrospray syntheses (Chin and Yaacob 2007; Albornoz and Jacobo 2006; Kim et al. 2005: Wan et al. 2005; Kimata et al. 2003; Alvarez et al. 2006; Basak et al. 2007). However, the most efficient method for the production of magnetite NPs is the chemical coprecipitation technique of Fe(II) and Fe(III) salts under basic conditions (Fig. 3.1) (Martinez-Mera et al. 2007; Morisson and Cahill 2005; Sun et al. 2004; Qiu et al. 2005; Lee et al. 2004).

This process yields homogeneously distributed NPs, in which the kinetic factors control the growth of the crystal. The coprecipitation method has been used to synthesize both USPIO and SPION with varying particle size. Figure 3.1 shows the synthetic scheme for the preparation of USPIO.

These Fe_3O_4 NPs can be characterized by transmission electron microscopy (TEM) for their particle size and have unique IR absorption spectrum as shown in Figs. 3.2 and 3.3, respectively. The TEM image shows that these NPs are spherical in shape with homogeneous size distribution, whereas the Fourier transform infrared spectroscopy (FTIR) spectrum of Fe_3O_4 shows a unique absorption at 571 cm^{-1}, due to stretching vibrations of Fe-O bond. The sharpness of these bands indicates lack of defects and crystal purity of Fe_3O_4.

3.2.1.2 Gold Nanoparticles

Gold NPs also known as colloidal gold are a nanosized suspension of metallic gold with a particle size of less than 100 nm. The use of colloidal solutions of gold NPs

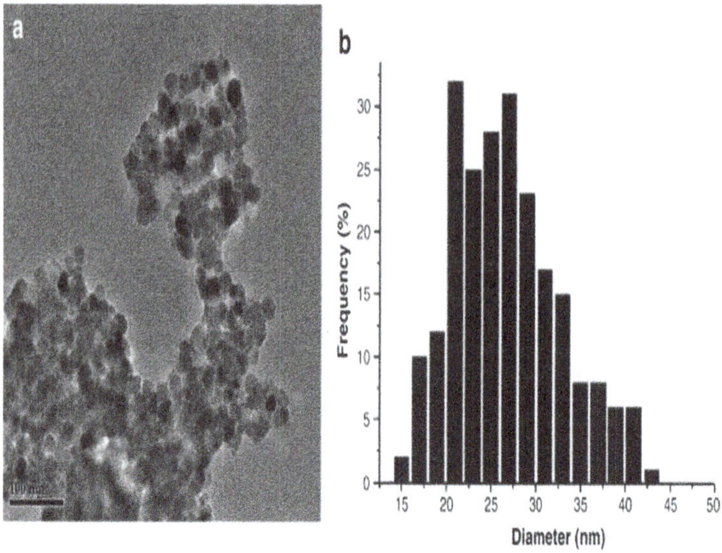

Fig. 3.2 TEM image of Fe_3O_4 NPs confirming the nanosize of these magnetic particles with a homogeneous size distribution. (**a**) Shows the TEM image at the scale of 100 nm and (**b**) shows the size distribution of NPs. It can be seen that the NPs are evenly distributed with average particle size of 24.9 nm [copyright from reference (Martinez-Mera et al. 2007)]

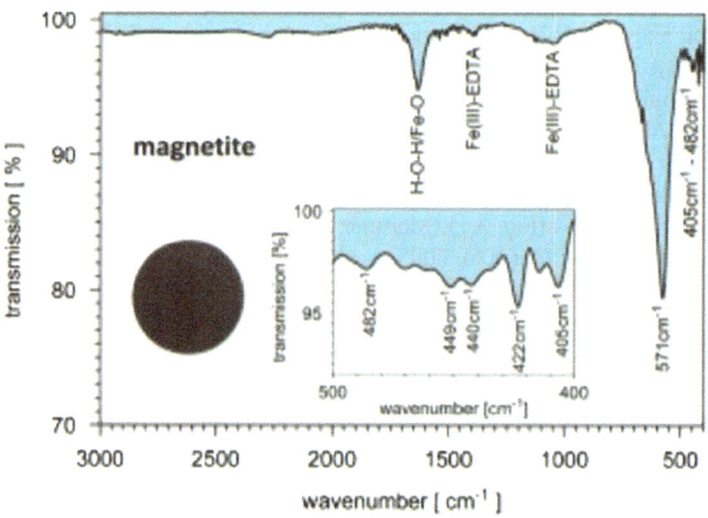

Fig. 3.3 FTIR spectrum of Fe3O4 NPs [copyright from reference (Ercuta and Chirita 2013)]

for decorative purposes has long been known; however, their use in medicinal application has recently gained tremendous attention (Giljohann et al. 2010). Michael Faraday pioneered the modern scientific evaluation of gold NPs when he observed that the colloidal gold solutions showed an intense red color (particle size <100 nm), whereas the larger particles showed dirty yellow color as shown in Fig. 3.4 (Murphy et al. 2008; Ling et al. 2009). The size-dependent optical properties of these gold NPs are due to their unique interaction of the surface plasmons with light (Jain et al. 2008). Gold nanospheres with particle size ~10 nm have a strong absorption maximum around 520 nm in aqueous solution due to their localized surface plasmon resonance (LSPR) (Fig. 3.5). LSPR is the surface phenomenon in which the free electrons on the surface of the metal NPs undergo resonance at a particular frequency in the presence of the oscillating electromagnetic field of light (Jain et al. 2006; Kelly et al. 2002; Kreibig and Vollmer 1995; Link and El-Sayed 2003). The effect of particle size on the LSPR is shown in Fig. 3.5. It can be seen that the increase in particle size changes the absorption of these NPs.

The properties and applications of colloidal gold NPs also depend upon its shape. Figure 3.4 shows that the difference in color of the particle solutions is more dramatic for rods than for spheres. For example, the rod-shaped NPs have two resonances due to plasmon oscillation along the short axis and long axis. Hence the plasmon oscillation of the nanorods strongly correlates to the change in length-to-width ratio also called as the nanorod aspect ratio (Murphy et al. 2005; Link et al. 1999). The absorption maxima of the gold nanorods show an absorption maximum close to infrared imaging (IR) region (>700 nm) and absorption wavelength increases as the length of the nanorod increases. Due to these unique optical properties of gold NPs, they have been significantly investigated for use as photothermal agents. Photothermal therapy (PTT) is a procedure in which a photo-sensitizer is excited with specific band light (mainly IR). This activation brings the surface plasmon to an excited state which returns to the ground state by releasing vibrational energy in the form of heat. The heat is the actual method of therapy that can be used to kill the targeted cells. One of the biggest recent successes in photothermal therapy is the use of gold NPs to kill cancer cells. The spherical gold NP absorptions have the peak absorptions at ~530 nm for 10 nm diameter, whereas skin, tissues, and hemoglobin have a transmission window from 650 to 900 nm. Hence spherical gold NPs have shown limited application in PTT. This was circumvented by the recent invention of gold nanorods by Murphy and coworkers, who were able to tune the absorption peak of these NPs from 550 nm up to 1 μm just by altering their aspect ratio (as shown in Fig. 3.3) (Ling et al. 2009; Busbee et al. 2003). These rod-shaped gold NPs absorb IR light to release heat which is enough to kill tumor cells. This potential application of gold nanorods separates them from other nanoprobes.

The synthesis of gold NPs is relatively straightforward and the chemical reduction of gold salts such as hydrogen tetrachloroaurate ($HAuCl_4$) using citrate or NaOH as the reducing agent is one of the most common methods for the synthesis of spherical gold NPs (Fig. 3.6) (Turkevich et al. 1951; Kimling et al. 2006; Frens 1972, 1973). This method produces monodispersed homogeneous spherical gold NPs in the range of 10–20 nm diameter. However, the synthesis of larger gold NPs with diameters between 30 and 100 nm was reported by Brown and Natan *via* seeding of Au^{3+} by hydroxylamine (Brown and Natan 1998). Subsequent research

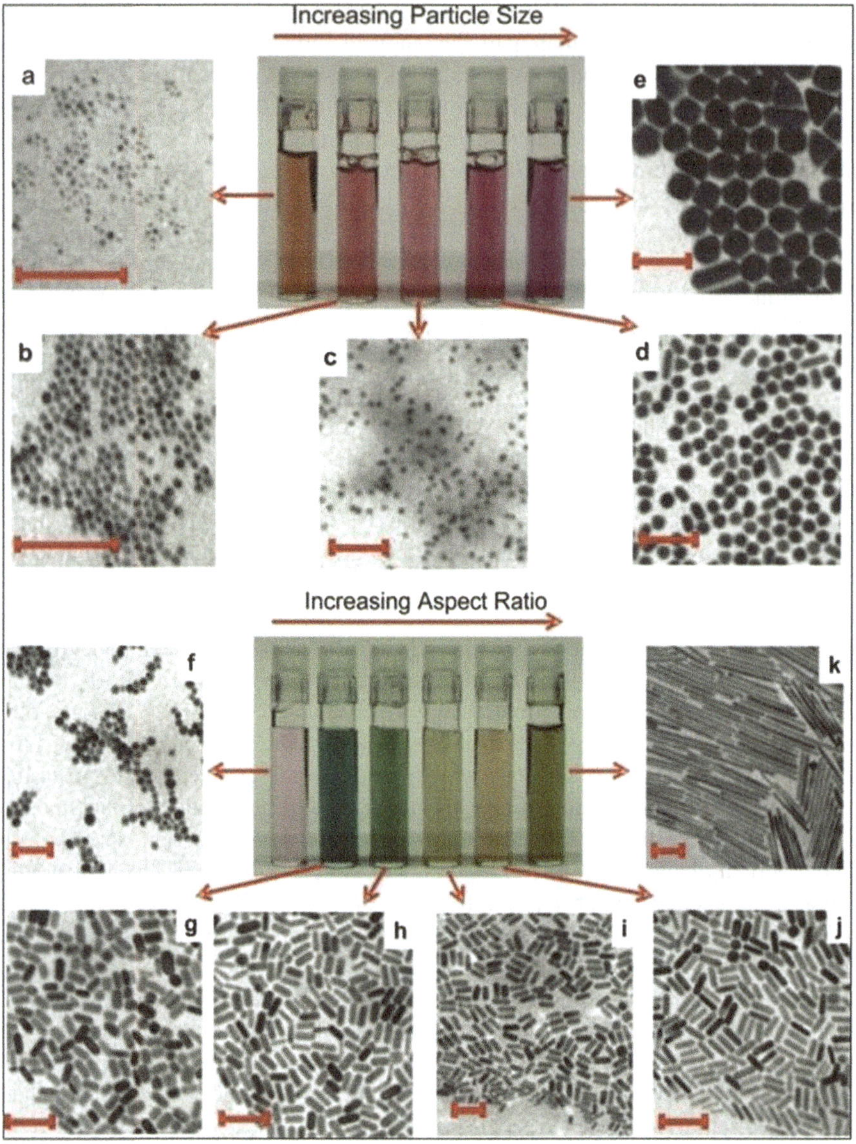

Fig. 3.4 Photographs of aqueous solutions of gold nanospheres (upper panels) and gold nanorods (lower panels) as a function of increasing dimensions. Corresponding transmission electron microscopy images of the NPs are shown; all scale bars 100 nm. The difference in color of the particle solutions is more dramatic for rods than for spheres. This is due to the nature of plasmon bands (one for spheres and two for rods) that are more sensitive to the size of rods compared with spheres. For spheres, the size varies from 4 to 40 nm (TEMs **a–e**), whereas for rods, the aspect ratio varies from 1.3 to 5 for short rods (TEMs **f–j**) and 20 (TEM **k**) for long rods [Copyright from reference (Murphy et al. 2008)]

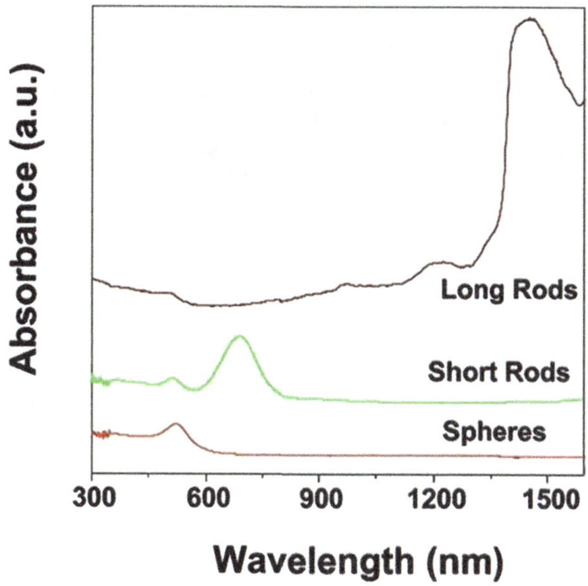

Fig. 3.5 Electronic absorption spectrum of spherical shape and rod-shape NPs. It can be seen that the spherical shape NPs have absorption maxima close to 525–530 nm whereas the rod-shape NPs show a red shift in absorption maxima. Additionally, the length of the rod-shape NPs has different absorption maxima and longer gold NPs show a bathochromic shift [Copyright from reference (Murphy et al. 2008)]

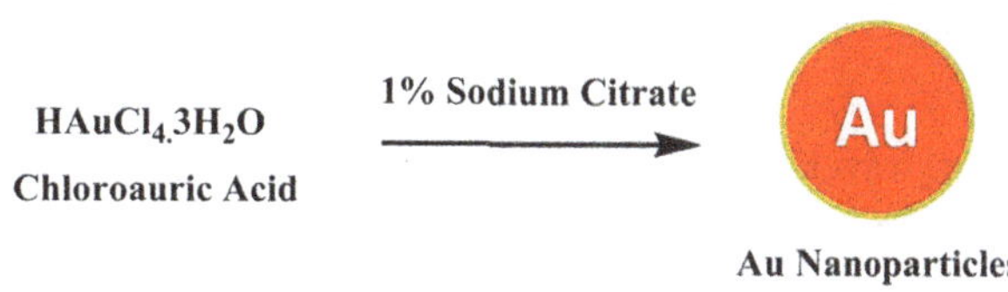

Fig. 3.6 Reaction scheme for the synthesis of Au NPs

led to the modification of the shape of these gold NPs resulting in rod, triangular, and polygonal rods and spherical particles under various conditions (Jana et al. 2001a, b; Kundu et al. 2007). Moreover, the gold surface offers a unique opportunity to conjugate ligands such as oligonucleotides, proteins, and antibodies containing functional groups such as thiols, mercaptans, phosphines, and amines, which demonstrate a strong affinity for gold surface (Alivisatos et al. 1996). Such gold nano-conjugates coupled with strongly enhanced LSPR have found applications in simpler but much powerful imaging techniques such as dark-field imaging, surface-enhanced Raman spectroscopy (SERS), and optical imaging for the diagnosis of various disease states (El-Sayed et al. 2005). However, their incompatibility with other high-resolution imaging techniques such as MRI and irreproducibility in shapes led to the invention of nanocages and nano-shells.

3.2.1.3 Nano-Shells and Nanocages

The need for multimodal imaging compatibility of the nanoparticles led to the invention of nano-shells. A typical nano-shell consists of two-layered NPs, a core which can be silica or Fe_2O_3 covered with an outer layer of a thin metal coating of gold (Fig. 3.7). It was theoretically calculated that a composite spherical particle that consists of a metallic shell and a dielectric core could give rise to LSPR modes with

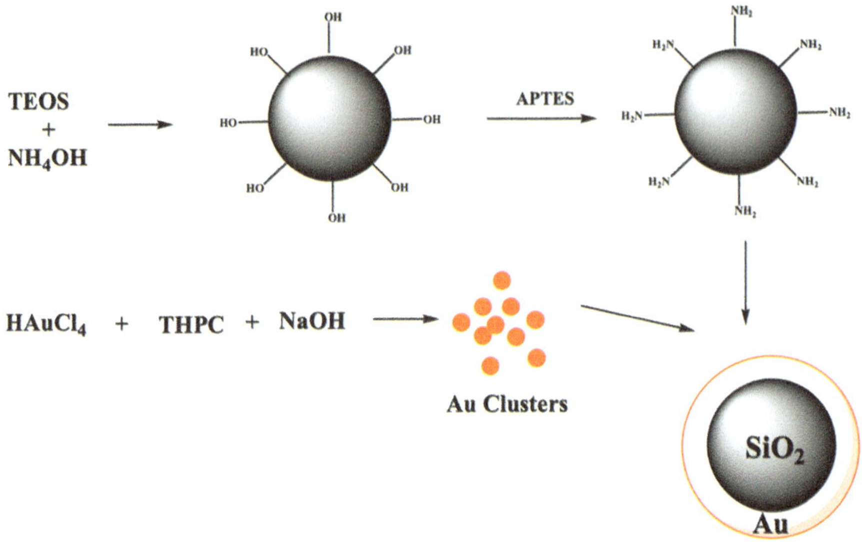

Fig. 3.7 Synthetic method for the development of gold nano-shells as developed by Oldenburg et al. (1998). *TEOS* tetraethyl orthosilicate, *THPC* tetra(hydroxymethyl) phosphonium chloride, *APTES* 3-aminopropyl-triethoxysilane

their wavelengths tunable over a broad range of the electromagnetic spectrum (Neeves and Birnboim 1989). This was later experimentally confirmed by Naomi Halas and Peter Nordlander when they demonstrated that absorption of silica–gold nano-shell particle takes place in the near-infrared (800–1300 nm) region (Oldenburg et al. 1998; Prodan et al. 2002; Hirsch et al. 2003). They developed silica–gold nanospheres by using freshly formed amine-terminated silica spheres. These amine-terminated silica spheres were then treated with a suspension of gold colloids (1–2 nm in size). Gold was then deposited *via* chemical reduction to cover the silica core. Figure 3.7 shows the synthetic steps involved in the development of silica-core gold nano-shells as reported by Oldenburh et al. (1998). Although this method is widely used, the intricacy involved in the control of thickness and smoothness of the metallic shells makes this method unsuitable for the routine synthesis of nano-shells. The use of nano-shells for whole-blood immunoassays was developed by Halas and West. They showed that the nano-shells when conjugated with antibodies act as recognition sites for a specific analyte. The analyte causes the formation of dimmers, which will modify the LSPR (Brongersma 2003). Subsequent work in this field led to the development of the multifunctional magnetic gold nano-shells (Mag-GNS) by Kim et al. based on Fe_3O_4 magnetic core. The presence of Fe_3O_4 allows the detection of NPs via MRI and the gold nano-shells enable photothermal therapy. The Mag-GNS can be targeted onto the cancer cells due to the presence of an antibody on the surface of nano-shells. Once localized, these particles enable the detection of cancer using MRI, whereas the photothermal therapy can be used to kill cancer cells (Kim et al. 2006).

Similar to gold nano-shells, gold nanocages represent a novel class of nanostructures that are hollow porous gold NPs that absorb light in the near-infrared range. They were first developed by Xia and coworkers *via* the reaction of silver NPs with chloroauric acid (HAuCl4) in boiling water (Chen et al. 2005). Since the reduction potential of AgCl/Ag is 0.22 V *vs.* standard hydrogen electrode (SHE) and the standard reduction potential of AuCl4−/Au is 0.99 V *vs.* SHE, the Ag nano-cubes serve as the template for the growth of the Au nanocages on the surface of the Ag nano-cubes. In general, this replacement reaction (Reaction 1) can be applied to any metal whose redox potential is more positive than the AgCl/Ag pair (Skrabalak et al. 2007, 2008).

Reaction 1 $3Ag_{(s)} + HAuCl_{4(aq)} \rightarrow Au_{(s)} + 3AgCl_{(s)} + HCl_{(aq)}$

Gold nanocages can also be used as an excellent contrast agent for a variety of optical imaging modalities such as OCT, two-photon, and multiphoton luminescence imaging. In fact, the gold nanocages with the edge length of ~35 nm offer a high absorption (approximately five orders of magnitude larger) than conventional dyes, thus demonstrating their capability as a contrast for both spectroscopic and conventional intensity-based OCT imaging. In addition, the strong absorption of gold nanocages in the NIR region makes them an excellent contrast agent for photoacoustic (PA) imaging. Furthermore, the hollow interiors can host small objects such as magnetic NPs to construct multifunctional hybrid nanostructures' diagnostic imaging and therapy (Mody et al. 2010). Due to these variations and structural properties nanocages have also found applications in drug delivery and/or controlled drug release.

3.2.1.4 Silver Nanoparticles

Silver NPs have been historically used to stain glasses but they are currently explored for a wide range of medical devices, including bone cement, surgical instruments, surgical masks, etc. Silver NPs have shown potential to treat infectious wound and are replacing silver sulfadiazine as an effective agent for the treatment of wounds (Yimin 2005; Atiyeh et al. 2007; AB 2006). Like gold NPs, silver NPs are synthesized by the reduction of a silver salt with a reducing agent like sodium borohydride and a colloidal stabilizer such as polyvinyl alcohol, poly (vinylpyrrolidone), bovine serum albumin (BSA), citrate, or cellulose. Other methods have employed the use of ß-D-glucose and starch as reducing agents for the synthesis of silver NPs (Stepanov et al. 2002). Elechiguerra et al. synthesized 1–10 nm sized silver NPs which undergo a size-dependent interaction with HIV-1 virus (Elechiguerra et al. 2005). Similarly, Furno and coworkers have developed novel antibacterial based on silicone-coated silver oxide NPs (Furno et al. 2004). The silver NPs have not emerged from the shadow of gold NPs or nanocages but they have shown great potential as an effective biomaterial due to which they are being continuously developed for newer applications.

3.2.1.4.1 Development of Metallic Nanoparticles for Drug Delivery

To use metallic NPs for drug delivery system they need to be surface functionalized by various ligands such as polyethylene glycol, dextran, amino silanes, or targeting ligands specific to target tissues such as cell-penetrating peptides and arginylglycylaspartic acid (RGD) (Mody et al. 2010). These surface ligands can be further conjugated with drug molecules as shown in Fig. 3.8. Once these drug-loaded NPs are injected into a patient, they are targeted to the desired sites due to the presence of tissue-targeting ligand to attain site-specific therapeutic effect. This increases the cellular uptake of these NPs for various biomedical applications (Chastellain et al. 2004; Mody et al. 2009). While traditional contrast agents distribute rather nonspecifically, targeted molecular imaging probes based on iron oxide NPs have been developed that specifically target body tissue or cells (Mody et al. 2009; Mazooz et al. 2005). Conroy and coworkers developed chlorotoxin (CTX), a biocompatible iron oxide NP for specifically targeting glioma tumors. This was confirmed via MRI which showed preferential accumulation of NPs onto gliomas (Conroy et al. 2008). Ferumoxytol is a carboxy-alkylated polysaccharide-coated iron oxide NP that was developed as an MRI contrast agent developed by AMAG Pharmaceuticals which has shown uptake by macrophage-rich plaques. These studies provide a new insight into the bioreactivity of engineered NPs, which can provide potential applications in medical imaging or drug delivery.

Farooq et al. recently investigated the use of gold NPs for two anticancer drugs, bleomycin and doxorubicin. They were able to demonstrate that the therapeutic efficacy of the novel formulation was strongly enhanced by the active targeting to HeLa cells with a significant decrease in effective drug concentration (Farooq et al. 2018). The use of gold nanorods to deliver small interfering RNA (siRNA) and target a drug-resistant ovarian cancer cell line is also well established (Wang et al. 2019). These applications of metallic NPs in biomedical imaging have opened a new avenue to target diseases which were previously thought untreatable. This could be accomplished due to the high surface-to-volume ratio of NPs, which increases the loading capacity of the drug or target ligands onto the NPs as compared to the larger particles (Mody et al. 2010). Hence metallic NPs have generated a lot of attention and various metallic nanocarriers will be developed in the near future which might show increased efficiency and specificity.

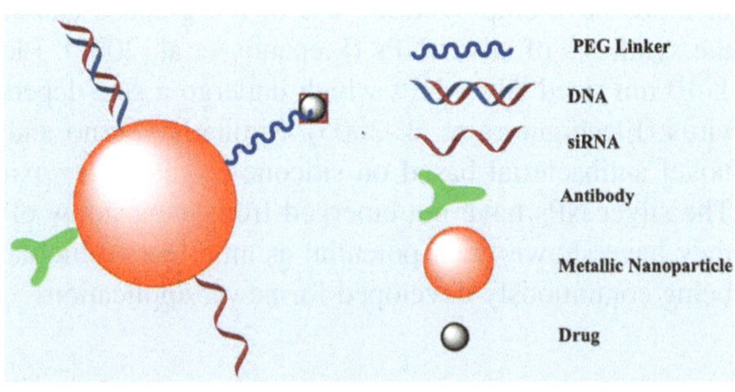

Fig. 3.8 General depiction of a drug- or antibody-loaded NPs

3.2.2 Nonmetallic Nanoparticles

Like metallic NPs, nonmetallic NPs are being studied for use in drug delivery as well. The use of nonmetallic NPs helps minimize system toxicity and increase bioavailability and therapeutic index of a drug, and can be used for targeted delivery. Three main types of nonmetallic NPs are frequently observed such as dendrimers, polymeric, and liposomes. Dendrimers are globular structures that consist of a core, a spacious interior with several generations of branches stemming from the core, and a dense membrane-like exterior. This structural feature can be used to synthesize various size of dendrimers with homogeneous molecular weight distribution in the range of 1–100 nm. The shape of the dendrimer also depends on the generation of branches and becomes spherical as the number of generations increases. Once synthesized the exterior surface can be functionalized with various groups or drugs that are responsible for interactions leading to adhesion and controlled release, at the targeted site (Madaan et al. 2014).

Dendrimers are mainly prepared via either divergent or convergent growth method. In divergent growth method a core (usually a nitrogen atom) is reacted with two or more moles of reagent containing at least two protecting branching sites. These branching sites then have their protecting groups removed to allow for more reagents to attach and create two or more new branches. The dendrimer will take on a globular structure at the fourth generation of branches, as the chains grow (Fig. 3.9), whereas the convergent growth method starts by forming the branches first and then attaching them to the core at the end. This method allows for more control over which generation dendrimer is produced.

Although dendrimers have the potential to be used for a wide range of possible applications, their cationic exteriors tend to interact with the anionic surfaces of biological membranes. This can result in membrane disruption and erosion which can potentially lead to vascular lesions (Markowicz-Piasecka et al. 2014). Additionally, positively charged dendrimers can affect hemostasis. However, a study shows that PEGylated dendrimers can avoid these challenges (Hu et al. 2019). Although not fully understood, it has been found that dendrimers of higher generations with a positive charge can contribute to aggregation of human platelets *in vitro*. Despite possible complications, some formulations of dendrimers have been clinically approved such as VivaGel (Starpharma). VivaGel is a microbicide which can be applied vaginally or rectally for the prevention of HIV and HSV. Although the drug is marketed in European countries, FDA is still awaiting additional confirmatory data for its approval in the US market.

Liposomes are small capsules made of a lipid bilayer that surrounds an aqueous center. This dual nature enables liposomes to carry both hydrophilic and hydrophobic components through the bloodstream. Even larger molecules like DNA, proteins, and imaging agents (Fig. 3.10) can also be easily transported via liposomes. Liposomes have shown promise as a drug delivery system because they were found to reduce toxicity and improve drug delivery to diseased tissue. Enhanced drug delivery to diseased tissue is due to the enhanced permeability and retention (EPR) effect of liposomes, which refers to liposomes slipping out of the bloodstream

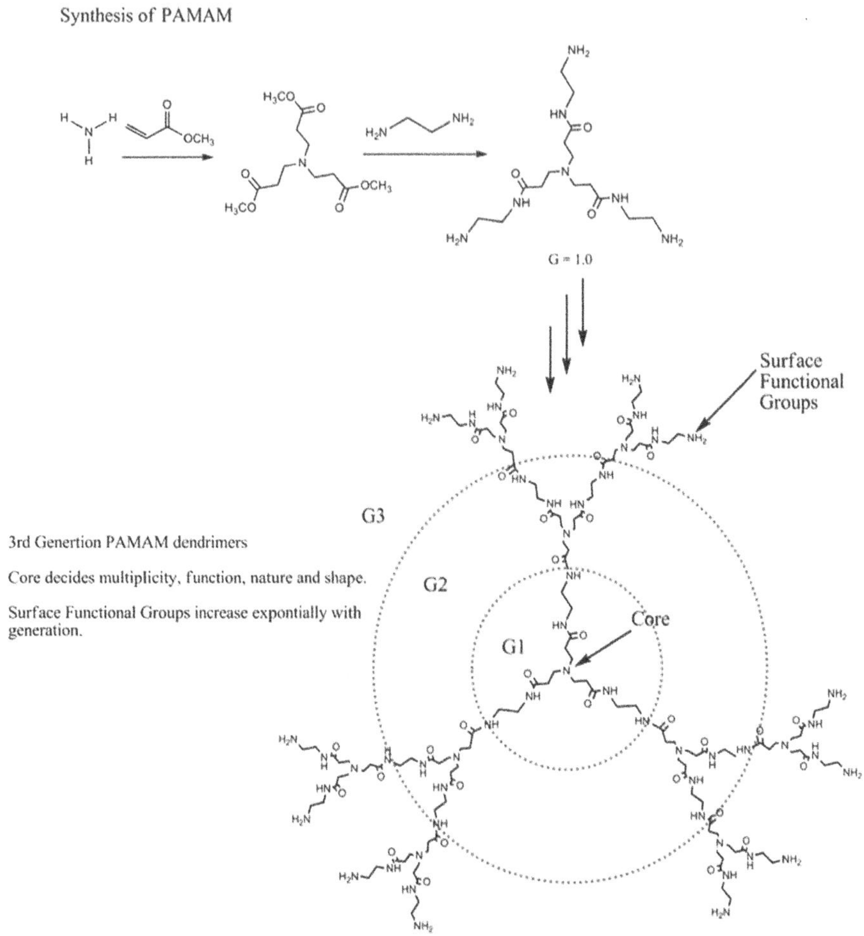

Fig. 3.9 Synthetic scheme for the preparation of G3 dendrimers [copyright from reference (Mody et al. 2009)]

through gaps in the vascular tissue commonly found in a tumor and inflammatory environment.

Structurally, liposomes can be divided into four types: conventional, PEGylated, ligand targeted, and theragnostic (Sercombe et al. 2015). The conventional liposome, the most studied and simplest liposome of the four, consists of a lipid bilayer composed of cationic, anionic, or neutral phospholipids and cholesterol (Stadnichenko et al. 2016). However, multiple studies show that these liposomes are prone to rapid elimination from the bloodstream. This is due to activation of the complement system which is a series of proteins found in the bloodstream. These proteins attach to the liposome and surround the liposome, a process known as opsonization. When the liposome is in this state, it can be attacked by immune cells

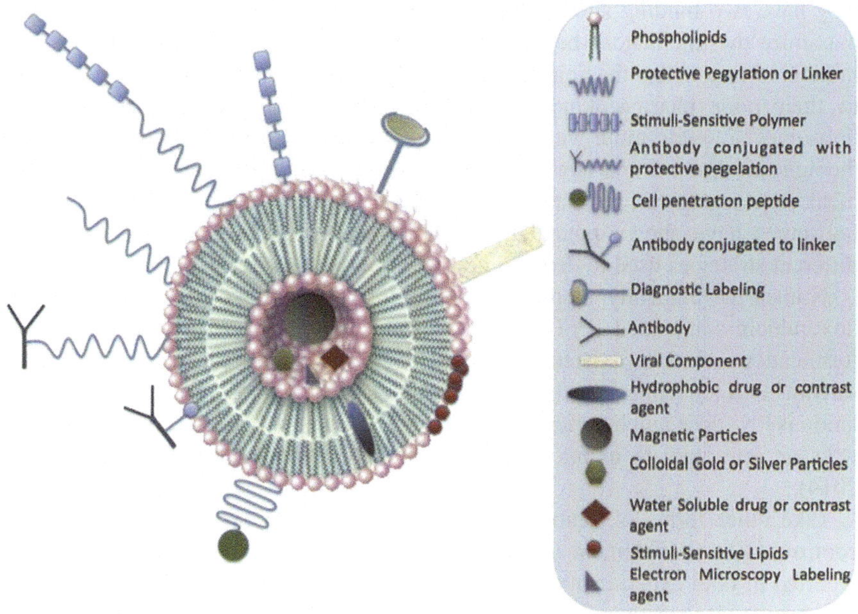

Fig. 3.10 Strategies that can be used to develop liposomal based drug delivery system [copyright from reference (Mody et al. 2009)]

or be taken up by the reticuloendothelial system (RES). Plasma proteins can initiate opsonization on liposomes at varying levels depending on the size, charge, and stability of the liposome. Smaller or positively charged liposomes tend to interact with the plasma proteins less than larger or negatively charged liposomes. Opsonization also allows for the RES to take in liposomes. The RES includes the liver, spleen, kidney, bone marrow, and lymph nodes. It is in these organs, mainly the liver and spleen, that liposomes are accumulated due to the vasculature of these organs where they are cleared by resident macrophages. To combat these issues, modifications were made to the exterior of the liposome to increase stability. These modifications lead to the development of the PEGylated liposome. The PEGylated liposome is a liposome surrounded by a polyethylene glycol (PEG) coat. This PEG coat forms a barrier for the liposome which translates to increased half-life in the bloodstream and decreased toxicity from the drug being transported. Increasing the liposome's ability to evade opsonization and RES also subjects those liposomes to increased EPR effect. However, this is slightly counterbalanced by the PEG barrier by decreasing interactions between the drug target and the drug.

The ligand-targeted liposomes are outfitted with a ligand that increases their targeting potential. They take advantage of the unique expression or overexpression of certain receptors of different cell types. These ligands can be peptides, carbohydrates, or antibodies. When antibodies are used as the ligand, it is referred to as an immunoliposome and is one of the most versatile of the liposome types as they have two binding sites. These liposomes can also conform in positioning to maximize the interactions between the ligand, drug, and receptor due to the high motional freedom that the lipid bilayer provides. However, they have limited use due to their poor pharmacokinetic profile. Additional modifications are allowing liposomes to earn more value in therapy. Current focus is primarily on the theragnostic liposome, a newer type of liposomes which contains a targeting element, an imaging component, and a therapeutic agent. All these different types of liposomes have shown potential in therapy and diagnosis and Fig. 3.9 shows different strategies used to develop liposomal based nanocarriers.

Numerous clinically approved liposome formulations are available. The doxorubicin-conjugated PEGylated liposome is one of the most common formulations which have been used to treat various types of cancer such as Kaposi's sarcoma and leukemia. Another example of a clinically approved liposome formulation is low-dose amphotericin B (AmBisome), used for CNS infections in patients who are refractory to or intolerant to amphotericin B deoxycholate (Stone et al. 2016).

Like other forms of nanocarriers liposomal formulations are also mired with controversies. Complement activation-related pseudo-allergy (CARPA) has been reported in both experimental and clinically approved liposomes. CARPA, a rapid hypersensitivity reaction, is thought to be mediated by the activation of the complement system. It can present as anaphylaxis, facial flushing and swelling, headache, and chills. General management of this involves lowering the infusion rate or outright stopping it along with standard allergy treatment. The characteristics thought to provoke a response from the complement system include having a surface charge, increase in size, presence of contaminants, and a high concentration of cholesterol in the lipid bilayer. Strategies to dampen this response have included methylation of anionic charges found on mPEG-phospholipid conjugates or switching to nonionic lipopolymers (Akbarzadeh et al. 2013). However, the immunogenic response can still cause a loss of efficacy and lead to more serious reactions in anaphylaxis. It is recommended that new formulations and treatments should be developed while considering ways to avoid CARPA.

Among the most diverse of the nonmetallic NPs, polymeric NPs have shown promise in the delivery of poorly soluble drugs. A polymeric NP (PNP) presents as either a nanosphere, or a solid particle of polymer in which a drug is evenly dispersed, or a nano-capsule where the polymer forms a membrane surrounding a drug-containing core. The synthesis of PNP provides an avenue to control their size, structure, and surface attributes. Particle size of PNPs has been observed to have a significant effect on drug release. A PNP's radius can range from 10 nm to 1 μm. Smaller PNPs will have a larger surface area-to-volume ratio than larger PNPs. This increase in surface area leads to faster drug release. Larger PNPs tend to aggregate less easily in storage, but some polymers degrade at a higher rate with larger sizes. For example, poly-lactic-co-glycolic acid was found to degrade faster with an increase in particle size. Therefore, a balance must be made between drug release and storage shelf life. Various forms of PNP such as nanosphere and nano-capsule can be developed based on the synthetic route. Numerous methods such as microemulsion, emulsion polymerization, solvent evaporation, and coprecipitation are developed for the synthesis of PNPs (Elias and Tsourkas 2009).

Microemulsion method has been used for the synthesis of poly-n-butyl cyanoacrylate (PBCA)-based NPs. Microemulsion is defined as a thermodynamically stable ternary mixture consisting of a polar solvent (usually water), a nonpolar solvent, and a surfactant in a certain ratio. The surfactant plays an important role in stabilizing the emulsion formed from water/oil mixture. The PBCA-based PNP thus synthesized can cross blood-brain barrier to deliver drug into the CNS (Moore et al. 2000). This method along with emulsion polymerization uses monomers to polymerize and prepare PNPs. Emulsion polymerization is one of the fastest methods for preparing PNPs and is easily scalable. The method makes use of an emulsion made from water, surfactants, and monomers. The emulsion can have either an aqueous continuous phase or an oil continuous phase, but an aqueous continuous phase is more common. An oil-in-water emulsion is formed when monomer is mixed into a solution of water and surfactant. From here, the polymerization process is initiated by an ionic form or free radical form of the monomer. The initiator can be introduced into the system, or it can be produced within the system by exposing the system to high-energy radiation such as ultraviolet (UV) light. The ion/radical then collides with other monomers to produce chains by anionic polymerization. Mini-emulsion is a different take on the emulsion polymerization method. It uses a low-molecular-mass compound as a co-stabilizer as well as a high-shear device, which makes use of high-frequency sound waves. This technique can make PNP with radii as little as 100 nm (Patra et al. 2018).

Solvent evaporation is a two-step method that makes use of a preformed polymer. First, the polymer is dissolved in a volatile solvent such as ethyl acetate. This solution is then mixed with an aqueous solution that contains surfactant, forming an oil-in-water emulsion. At this step, a high-shear device is used to produce a dispersion of nanodroplets. Step 2 involves evaporating the volatile solvent by reducing the air pressure of the system or by continuous magnetic stirring. Once the solvent has evaporated, the NPs form a suspension which can be collected by centrifuging. Once collected, the PNP can be washed and then lyophilized for long-term storage (Crucho and Barros 2017).

Nanoprecipitation is a single-step procedure where polymer is dissolved in a water-miscible solvent of intermediate polarity such as acetone. This organic solution is then added to an aqueous solution of water and surfactant. As the solvent diffuses through the aqueous phase, NPs form almost instantly to avoid water. These NPs would become nano-capsules with oil-based cores if nontoxic oil is mixed into the solvent mixture. Therefore, PNPs made using this method are ideal for transporting lipophilic drugs (Crucho and Barros 2017). All these methods have shown promise in the synthesis of PNPs but the use of the final method depends on the type of nanoparticle being prepared and ease of synthesis.

The development of both metallic and nonmetallic NPs has considerably advanced in the past three decades but for the formulation to be used by the body, the absorption, distribution, metabolism, and excretion of the nano-formulation play an important role and have to be systematically addressed to minimize toxicity.

Hence, in the next section we have discussed pharmacokinetics of various formulations which have been evaluated over the years.

3.3 Pharmacokinetics of Drug-Loaded Nanoparticles (NPs)

Absorption, distribution, metabolism, and excretion (ADME) are four essential processes related to what the body does to drugs after administration. Absorption is referred to as the process by which parent drug compound is transferred from the site of administration to systemic circulation. For example, orally administered drugs are absorbed through gastrointestinal tract and liver to reach bloodstream. Distribution is a reversible process of transferring a drug back and forth from systemic circulation to the surrounding tissues. For example, drugs are transported from the bloodstream into the lung. Metabolism is a biotransformation process by which parent drugs are converted to metabolic products (metabolites) by drug-metabolizing enzymes. The most common enzymes involved in drug metabolism are cytochrome P450 enzyme family. Excretion is the irreversible removal of drug molecules from the body. Major route of drug excretion occurs through kidneys.

Research on ADME of potential drug candidates is a prerequisite during drug discovery and development. A lot of lead compounds have shown great effects in cell culture models but failed in *in vivo* efficacy and safety testing due to limitation in ADME processes. For example, oral absorption of drugs is hindered due to extensive first-pass effects in the liver, where drug molecules are heavily metabolized before reaching the systemic circulation. In addition, many new drug candidates dissolve poorly in aqueous phase, therefore preventing the development of these compounds from moving forward and ultimately reducing the success rate of new therapeutic agents into the market. Research in recent years has been focusing on nanoparticles as a delivery tool to improve drug bioavailability given orally; enhance solubility of poorly soluble drugs in intravascular dosage form; minimize toxicity or side effects of drugs, especially anticancer agents; and achieve targeted delivery of therapeutic agents (Kalepu and Nekkanti 2016; Tran et al. 2017). In this section, we review recent advances on the application of NP-enhanced drug delivery and unique ADME aspects of NP-carried drugs compared to those of free drugs.

3.3.1 Absorption

Absorption of free drugs or NP-carried drugs from the site of administration to systemic circulation is the first step in the cascade of ADME processes. Routes of drug administration determine where drug molecules will be absorbed. For example, oral administration of drugs leads to drug absorption through the gastrointestinal (GI) tract, whereas drug molecules are absorbed from pulmonary system when being inhaled. Therefore, it is logical to discuss drug absorption based on corresponding administration routes. Drug-loaded NPs are commonly given through oral, intramuscular, pulmonary, dermal, and subcutaneous routes.

3.3.1.1 Oral Drug Administration and Absorption from GI Tract

Drug taken orally is the most common mode of drug delivery with greatest degree of patient compliance and acceptance because this route is simple, convenient, and self-administrable with flexible dosage regimen. In addition, tremendous surface area and relatively long transit time in the small intestine region have greatly enhanced drug absorption from the GI tract. Once drug molecules are pulled into systemic circulation after oral administration, they are distributed promptly from the site of absorption to the whole body by the bloodstream. This generates a sink condition that enables a large concentration gradient serving as the driving force for oral absorption of drugs. However, many anticancer or antibiotic drugs have very poor bioavailability when given orally. Major limiting factors are poor aqueous solubility of lipophilic drugs, membrane permeability of hydrophilic drugs, and metabolic stability of biologic drugs. In fact, close to 70% of new drug candidates in the drug development stage were ended due to insufficient oral bioavailability caused by these limiting factors (Gao et al. 2013). Drug-loaded NPs are orally absorbed through mechanisms of paracellular (through the tight junction between intestinal epithelial cells) transport, transcytosis (transport across the interior of a cell), and M cell (microfold cell) uptake (Des Rieux et al. 2006). One area of the GI tract, ileum portion of the small intestine, contains small masses of lymphatic tissue called Peyer's patches. These lymphoid follicles allow passage of particles in the size of nanometer and micrometer range (Florence 2005), therefore being responsible for the absorption of NP-carried drugs. One factor that affects the uptake of the NP carriers for drugs is the charged state. Positively charged NPs are reported to be absorbed to a greater extent than neutral and negatively charged NPs through the GI tract (Janes et al. 2001). In addition, small-size particles are absorbed more efficiently than large-size particles of the same type (Rizvi and Saleh 2018). Surface property of NPs is a determining factor that controls the stability of drug-loaded NPs in the GI tract. For example, poly-lactic acid (PLA)-based NPs are easily degraded in the GI fluid, whereas polymethyl-methacrylate NPs are relatively stable in the gastric juice and intestinal fluids (Lazzari et al. 2012).

Recent development of oral drug delivery focuses on NP targeting various regions of the GI tract. The term "targeting" here for oral drug delivery refers to different approaches to extend residence time of NP-loaded drugs and control the amount of drug release from NP carriers at the target sites such as stomach, small intestine, and colon. Stomach targeting by NPs provides several advantages including (1) treatment for illness in stomach such as *Helicobacter pylori* infections and consequential gastric disorders; (2) delivery of drugs mainly absorbed in the stomach such as metronidazole; and (3) delivery of drugs with poor solubility and stability in the intestinal region such as verapamil and captopril (Adebisi and Conway 2011, 2015). Small intestine is the major absorption site for many drugs. Small intestine targeting using NP carriers includes both passive and active approaches. Enteric polymers have been applied to NP formulations to resist acidic pH and harsh gastric milieu in the stomach. When reaching small intestine with an increase of pH, enteric polymers start to dissolve and enable release of drugs from NPs, which is considered targeting drug to the small intestine passively (Wang and Zhang 2012). NP surface modification to enhance interactions with enterocytes, goblet cells, and M cells is considered as targeting NP-loaded drug to the small intestine actively. These surface modifications include coating NPs with hydrophilic polyethylene glycol (PEG) and enzymes such as papain or thiols to improve the penetration of NPs through mucus and ligands to target surface receptors of the small intestine cells such as neonatal Fc receptors and lectins (Maisel et al. 2015; Muller et al. 2014; Kollner et al. 2015; Pridgen et al. 2013; Gabor et al. 2004). Colon belongs to the lower part of the GI tract. NP-loaded drugs designed to target this area have to remain intact and are not absorbed from the GI regions prior to colon. Colon-targeted drug delivery is designed for the treatment of local diseases such as inflammatory bowel disease, colon cancer, and irritable bowel syndrome (Patel et al. 2007; Hua et al. 2015). Another application of colon targeting is to deliver macromolecule drugs for systemic effects due to the advantages of low level of proteolytic activity, reduced amount of enzymes and transporters involved in macromolecule degradation, and relatively long transit time in colon (Sinha and Kumria 2003). One strategy to improve NP-targeted drug delivery to the colon is application of pH-sensitive polymers such as methacrylic acid copolymers (Eudragit) (Thakral et al. 2013).

3.3.1.2 Inhalation Drug Administration and Absorption from Respiratory System

Pulmonary drug delivery offers advantages over oral route including tremendous surface area and vascularization for fast drug absorption and bypass of hepatic first-pass effect (Sung et al. 2007). Inhalation medications have been available for many years for the treatment of lung diseases such as asthma, chronic obstructive pulmonary disorder, cystic fibrosis, lung infections, and lung cancers. They have also been studied as a possible route of administration for the treatment of systemic diseases including diabetes. The development of inhalation therapy that is efficacious and safe depends on not only a pharmacologically active molecule, but also well-designed delivery system and formulation. It is the optimization of the whole system (drug, drug formulation, and device) that is necessary for the successful development of inhalation therapies. The drug-device combination must aerosolize the drug in the appropriate particle size distribution and concentration to ensure optimal deposition and dose in the desired region of the lung. The lungs consist of conducting airway (trachea, bronchi, and bronchioles) and respiratory regions including alveoli which provide enormous surface area and a highly permeable membrane favoring systemic absorption of drugs. The efficacy of drug delivery through inhalation is determined by the particle size of drug molecules. Mechanism of drug particle deposition based on its size includes inertial impaction, gravitational sedimentation, and Brownian diffusion (Yang et al. 2008). The relationship between drug particle size and lung deposition is described below. Particles of size between 5 and 10 μm will get deposited in the primary bronchi area through impaction. Particles of size between 1 and 5 μm will get deposited in the secondary bronchi area through sedimentation. Particles of size between 1 and 3 μm will get deposited in the bronchioles area through sedimentation. Particles of size between 0.5 and 1 μm will get deposited in the alveoli area through Brownian diffusion.

When drug molecules are encapsulated into nanosized particles, they can penetrate the lungs more deeply and enter more into the alveolar region than traditional micron-sized particles. NP-based drug delivery systems applied for this route include solid lipid NPs and polymeric NPs. Solid lipid NPs are nanoscale colloidal carriers prepared from physiological lipids such as phospholipids and triglycerides, ranged from 50 to 1000 nm, and usually dispersed in aqueous solution. Advantages of solid lipid NPs include controlled release of drugs, application of biocompatible components, enhanced stability of drugs, and improved bioavailability (Mehnert and Mader 2001). Solid lipid NPs have been applied to deliver drugs through pulmonary route for therapeutic potential to treat lung infections using amikacin (Varshosaz et al. 2013), asthma using curcumin, and breast cancer with silibinin (Xu et al. 2013). Polymer-based NPs have gained attention for drug delivery through respiratory system due to several advantages such as modifiable surface properties, high loading efficiency of drugs, enhanced drug stability, and increased residence time in the body. Commonly used polymers for this purpose include chitosan, poly-lactic-*co*-glycolic acid (PLGA), and PLA (Makadia and Siegel 2011; Menon et al. 2014). Chitosan-based NPs were applied to deliver antitubercular drug, bedaquiline, for the treatment of tuberculosis (Rawal et al. 2018). PLGA NPs were recently used to carry ethionamide through respiratory route using dry-powder inhaler for the treatment of tuberculosis as well (Debnath et al. 2017).

3.3.1.3 Intramuscular Administration and Absorption

Intramuscular drug administration is considered as drug delivery into the body, indicating that the drug is given by routes other than going through GI tract. The application of this route for drug delivery is suitable for medication that needs relatively fast absorption but with prolonged action. Many drugs have been administered through this route including antibiotics, anticancer, anti-inflammation, and painkiller medications. NP-based drug delivery through intramuscular administration provides several advantages including controlled release and prolonged circulating time of drug. Moxifloxacin, an antibiotic drug used to fight bacterial infection in the body, was formulated into mesoporous silica nanoparticles (MSNs) and applied in a mouse model of pneumonic tularemia, an infection caused by an aerobic bacterium *Francisella tularensis*. Intramuscular delivery of MSN-loaded moxifloxacin was shown with better efficacy and improved pharmacokinetic profiles of this drug compared to administration of free drug and intravenous drug delivery (Clemens et al. 2019). Progesterone (PRG), a natural steroid hormone used to support women whose endogenous PRG level is low during pregnancy, is usually given through oral, vaginal, and intramuscular routes. Among all the routes, intramuscular injection of PRG formulated in lipophilic solution is a clinically preferred method. One downside of this administration route is that it requires repetitive dosing which may lead to severe pain and other side effects such as inflammation and tissue necrosis. Formulating PRG with hybrid NPs based on methoxy poly (ethylene glycol)-poly(ε-caprolactone) demonstrated greater plasma concentration of PRG at steady state compared to PRG formulated in traditional oil solution with reduced intramuscular dosing frequency in animal model. This sustained-release system enables less dosage and longer dosing interval of PRG than its traditional formulation (Xie et al. 2018). Ramizol, a new-class antibiotic designed to treat antibiotic-resistant bacteria, has been developed and loaded into PLGA NPs. PLGA NP carrier ramizol demonstrated higher absorption and longer elimination half-life, and therefore greater bioavailability than being given through oral and intravenous routes (Wright et al. 2018).

3.3.1.4 Transdermal Drug Delivery and Absorption

The potential of transdermal drug delivery has been well explored as an alternative route to oral and invasive administration of drug owing to its advantages to avoid hepatic first-pass effects, convenience to apply, improved patient compliance, extended duration of action, and controlled drug release (Brown et al. 2006). Transdermal drug delivery systems have been used initially to administer small-molecule drugs with lipophilic nature and given with low doses (Prausnitz and Langer 2008). The application of chemical enhancers and non-cavitation ultrasound enhanced drug delivery through skin, and iontophoresis provides additional benefit of controlled rate of drug release (Prausnitz and Langer 2008). Current advances in drug delivery through skin include microneedles and thermal ablation to administer macromolecule drugs including peptides, proteins, and vaccines (Prausnitz and Langer 2008). In addition, research efforts have been focused on NPs to improve transdermal delivery of large and hydrophilic drugs. NPs are usually made of polymer, metals, or lipids on the scale of nanometers. Drug molecules can be incorporated either on the surface or inside the core of these NPs. The challenge for transdermal drug delivery including NPs is to penetrate through the barrier of skin which is comprised of two layers, the epidermis and the dermis. The outermost epidermal layer, stratum corneum, and the tight junctions in the stratum granulosum are the major barriers preventing the permeability of most hydrophilic drugs and those with molecular size greater than 500 kDa (Bos and Meinardi 2000). Skin absorption of NP-loaded drugs is determined by many factors of NPs including size, shape, zeta potential, surface charge, and aggregation (Filon et al. 2015). In addition, when delivering NP carrier drugs to a damaged skin barrier, increased skin absorption of drug is expected through enhanced permeation of NPs. Transdermal delivery of NP-loaded drugs such as immunosuppressive calcineurin inhibitor tacrolimus is well studied for the treatment of inflammatory skin diseases psoriasis and atopic dermatitis. Greater penetration of tacrolimus using microemulsion system into human full-thickness skin was observed (Goebel et al. 2011). Improved skin targeting and alleviation of dermatitis were reported when using lipid NP carrier tacrolimus with the range of 20–100 nm on a mouse model of atopic dermatitis (Pople and Singh 2013).

3.3.1.5 Subcutaneous Administration and Absorption

Subcutaneous route of administration is commonly used for drugs that have low bioavailability when given orally and for the purpose of controlling drug release to achieve prolonged systemic exposure. Subcutaneous injection of NP-loaded drugs is suitable for delivery of long-acting drugs and vaccine. NP carrier drugs are absorbed from subcutaneous injected site into interstitial area underlying the dermis layer of the skin where they enter the lymphatic system and eventually reach systemic circulation. The rate and extent of absorption of NP-loaded drug from subcutaneous administration site depend on the size of drug-loaded NPs (Oussoren and Storm 2001). NPs with small size between 100 and 150 nm are usually absorbed into the bloodstream to a great extent, more than 60% of the administered dose through lymphatic capillaries and regional lymph nodes. Although large-sized NPs may have difficulty to carry drugs moving from subcutaneous site of injection into lymphatic capillaries, it is a better option than small-size NPs for drugs intended for sustained release. Solid lipid NPs are made from physiological lipids such as fatty acids, steroids, and waxes. These lipids remain in the solid state at room and body temperatures (Muller et al. 2000). Subcutaneous administration of chemotherapeutic drugs loaded into solid lipid NPs has shown greater tumor uptake of drugs in an animal model of Dalton's lymphoma compared to drug administration through the intraperitoneal and intravenous routes (Reddy et al. 2005). Vaccine delivery through subcutaneous administration is a very efficient and possibly the most appropriate route since antigens can be directly transported to lymphatic tissues where immunocompetent cells are located. To avoid the risk of certain antigens that may regain its pathogenicity, NPs have been used as a promising adjuvant for subcutaneous administration of vaccines. Advantages of NP carrier vaccines include improved immunization strategies, increased vaccine efficacy, and enhanced targeted delivery of vaccine (Pati et al. 2018). Recently, cancer vaccine using NPs to deliver tumor-associated antigens (TAA) has gained a lot attention (Kim et al. 2019). PLGA polymer NPs have been used to deliver imidazoquinoline-based toll-like receptor (TLR) 7/8 to enhance the retention of these potent cytokine inducers to stimulate TAA-specific CD8 T cells (Smith et al. 2018; Kim et al. 2019).

3.3.2 Distribution

The second stage of ADME process is distribution, which is where NPs are transferred via the assistance of blood flow after being absorbed in the body. Many factors determine the distribution pattern of NPs, including routes of administration, physicochemical properties of NPs (i.e., size, charge, shape, surface moiety), and dose (Griffin et al. 2016; Hamidi et al. 2013; Lin et al. 2015). A study has found that distribution of NPs varies based on routes of administration by tracking fluorescently tagged mesoporous silica nanoparticles (MSNs), with a size of 110 nm, given through different routes (IV, hypodermic, IM, and oral) to female rats at a dose of 50 mg/kg. Distribution of MSNs was monitored at 24 h and 7 days following treatment. The results showed that some MSNs were absorbed and accumulated in the liver after administration from hypodermal route. The proposed mechanism of MSNs' entry through skin included direct passage through biological skin barrier and phagocytosis by residential macrophages before transport to systemic circulation. On the other hand, the number of MSNs in the liver and spleen was higher after intravenous administration, compared to control group. After oral administration,

MSNs were absorbed from the GI tract and accumulated significantly in the liver through portal vein, where Kupffer cells phagocytized MSNs leading to their removal. The concentration of MSNs, however, was declined at the 7-day mark due to fecal elimination. Liver and spleen are the major organs of MSN accumulation. Lastly, the intramuscular administration of MSNs showed the least absorption and biodistribution compared to other routes. Among the physicochemical properties of NPs, the size of NPs plays a vital role in determining both absorption and distribution of NPs. A study, using 15, 50, 100, and 200 nm colloidal gold nanoparticles (AuNPs), showed size-dependent distribution, when given to male mice via intravenous route at a dose of 1 g/kg. The extent of distribution of AuNPs at a size of 15 nm follows the following order: liver > lung > kidney > brain > spleen. In general, it was observed that, when the size of AuNPs decreased, there was a tendency for these NPs to accumulate in the brain and lung. On the contrary, as size increased, these NPs tended to accumulate in the spleen and rarely in the brain. AuNPs at a size of 200 nm were found mostly in the liver, followed by the spleen, lung, kidneys, heart, and minimally brain; in addition to size-dependent distribution of AuNPs, surface modification of NPs alters their distribution pattern. Results showed that decreasing size of NPs and having more hydrophilic surface layer enable NPs to escape rapid clearance through reticuloendothelial system (RES), therefore increasing circulation and therapeutic effectiveness of targeted drug delivery. In addition, it showed that PEG-coated AuNPs of small size, especially 15 nm, result in large quantity of NP retention due to slow clearance (Sonavane et al. 2008). The effect of dose on the distribution of NPs was studied in mice using 12.5 nm AuNPs with doses of 40, 200, and 400 μg/kg through intraperitoneal route of administration. The results showed that the concentration of AuNPs in blood remained constant through 8 days following each dose. As the dose of AuNPs increased from 40 to 400 μg/kg, the amount of AuNPs in plasma decreased, indicating enhanced total body clearance. Interestingly, the results showed dose-dependent increase of AuNPs accumulated in different tissues following the order of liver > kidney > spleen. In addition, AuNPs were accumulated in the brain significantly higher than basal level, reflecting their ability to cross the blood-brain barrier in a non-saturable uptake manner (Lasagna-Reeves et al. 2010).

3.3.3 Metabolism

The function of drug metabolism enzymes is to convert lipophilic parent drugs to hydrophilic metabolites for the purpose of enhancing drug elimination from the body. These enzymes are categorized into phase I enzyme-metabolizing drugs through oxidation, reduction, or hydrolysis, and phase II enzymes through conjugation reaction with various hydrophilic moieties such as glucuronic acid and sulfate. These phase I and II reactions occur with xenobiotic substances when they enter the body especially through oral route. Orally administered medications in the GI tract travel through hepatic portal vein to the liver and undergo these reactions before entering the systemic circulation, which is named as first-pass effect. However, NPs are different than traditional xenobiotics mainly because there is no evidence to show that biological enzymes degrade pure NPs especially semimetal/metallic or metallic oxide-based NPs such as carbon/silicon, gold, silver, and iron oxide NPs. On the other hand, surface-modified/functionalized NPs, such as protein-coated, lipoprotein-coated, or PEG-coated NPs, have been shown to be degraded by biological enzymes. In addition, liposome-based NPs, owning to the nature of liposome structure, interact with biological membrane and disappear by merging into lipid membranes.

3.3.4 Excretion

Main routes of excretion for NPs are biliary and renal excretion. Fullerenes and single-walled carbon nanotubes have been reported to be cleared from the body through renal excretion, which is affected by not only the size but also the chemical properties of NPs as well (Hagens et al. 2007). It has been shown that polystyrene NPs were extracted by various cell components in the liver based on the size and eventually transported into the bile to be excreted (Hagens et al. 2007). However, more data and evidence such as animal studies are still required to better understand the excretion process of NPs. What was confirmed is that the excretion of NPs depends on their physicochemical properties such as size and chemical composition (Hamidi et al. 2013).

3.3.5 Elimination of NPs Through Opsonization

Elimination of NPs is often commonly associated with the process of opsonization by which foreign organisms or particles are encapsulated by endogenous plasma proteins or peptides. These plasma proteins include various components of the complement system, subclasses of immunoglobulins, fibronectin, and von Willebrand factors. Once opsonized, these NPs in the systemic circulation become easy target for macrophages. Once recognized and internalized by these phagocytic cells, NPs are subjected to rapid elimination from circulation and body through phagocytosis by RES, commonly found in liver (Kupffer cells), spleen, and bone marrow (Hamidi et al. 2013). RES activity on NPs can be strategically altered to enhance therapeutic effects. It has been shown that small-sized NPs tend to be cleared more slowly by RES than large-sized ones, and positively charged NPs are more likely to be taken up by cells and cleared by RES compared to neutral NPs. These findings about different clearance rates of NPs by RES allow for prolonged duration of NP carrier drugs in systemic circulation, leading to better clinical response (Hagens et al. 2007; Hamidi et al. 2013).

3.4 Toxicology of Nano-Drug Carriers

3.4.1 Factors Influencing Nanotoxicities

There are a number of factors that influence the type and extent of toxic effects generated by NPs. These have been summarized in Fig. 3.11.

3.4.1.1 Dose

The dose or concentration of NPs has been reported to produce various cytotoxicities; a threshold level of NPs sometimes is required for adverse effects to manifest (Prabhu et al. 2010; Nassimi et al. 2010). Solid lipid NPs (30 nm) showed toxic effects of elevated inflammation and pro-inflammation mediators in *in vivo* and *in vitro* studies on mice and lung tissue and after reaching a dose of 500 mcg/mL, but no toxic signs were seen below 200 mcg/mL (Nassimi et al. 2010). Silver NPs studied *in vivo* on Wistar rats showed dose-dependent toxicities, which included liver damage, ROS production, inflammation, imbalanced blood chemistry (WBC, platelet, and hemoglobin count), and DNA damage; biocompatible dose was determined to be less than 10 mg/kg, and any dose higher than 20 mg/kg showed toxic effects (Tiwari et al. 2011).

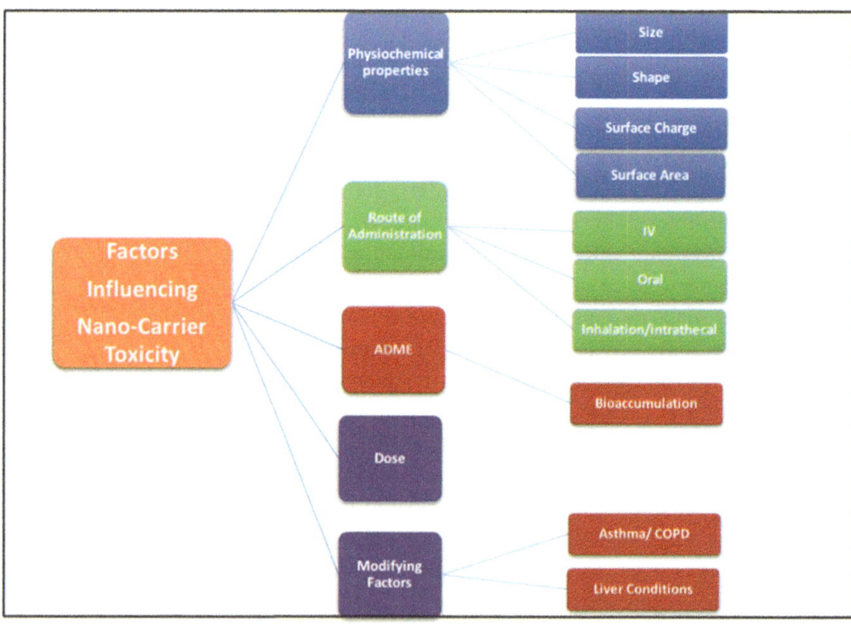

Fig. 3.11 Factors influencing NP drug carrier toxicities

3.4.1.2 Size/Surface Area

The size of NPs is a desired characteristic which makes them potential candidates as a therapeutic agent; however, this same quality plays a significant role in nanotoxicities; NP size is involved in absorption, distribution, metabolism, and elimination, and size-dependent nanotoxicity has been demonstrated (Griffin et al. 2016; Missaoui et al. 2018). *In vivo* and *in vitro* study of silica NPs showed size-dependent toxicities (Kim et al. 2015; Handa et al. 2017). Kim et al. studied silica NPs with diameter ranging from 20 to 200 nm and found size-related cell membrane disruption and oxidative stress; cytotoxicity increased as size decreased (Kim et al. 2015); similarly, Hinda et al. presented size-associated cytotoxicity of silica NPs (10–1000 nm) in mice, who were given intravenous doses, and toxicities such as thrombocytopenia, liver damage, and death increased as the size of silica NPs decreased (Handa et al. 2017). Likewise, various studies using cobalt-ferrite NPs, cobalt-chromium v, titanium-oxide NPs, cadmium telluride NPs, and silver NPs have established size-linked cytotoxicity which included elevated pro-inflammation mediators, production of more free radicals and superoxide leading to oxidative stress, genotoxicity and DNA damage, rapid cell proliferation, and presence of apoptotic makers and apoptosis (Colognato et al. 2007; Papageorgiou et al. 2007; Park et al. 2007; Zhang et al. 2007; Mishra et al. 2016). Biodegradable nonmetallic polymer NPs were studied *in vitro* using mammalian cells, by Kim et al., and size-dependent toxicity among five monodisperse polypyrrole (20, 40, 60, 80, 100 nm) was found, such as elevated oxidative stress, apoptosis or necrosis, and innate immune response (Kim et al. 2011).

3.4.1.3 Charge

The surface charge or zeta potential of NPs is an essential property which is known to play many critical roles. These roles include cellular uptake, biodistribution, aggregation/agglomeration, bio-interaction, drug targeting, and cytotoxicity/nanotoxicities (Honary and Zahir 2013; Schleh et al. 2012). Positively and negatively charged, sized polymer-based NPs (PHBHH) were studied *in vitro*, showing surface charge- and charge density-dependent cytotoxicity. Positively charged NPs were found to be more cytotoxic than negatively charged NPs, and higher charged NPs were more toxic than lower charged ones (Shao et al. 2015). Gold NPs coated with polymers showed surface charge-linked bio-internalization and cytotoxicity; positively charged gold NPs internalized into cell more and promoted more cytotoxicity when compared to negatively charged NPs (HuHn et al. 2013). Similarly, Chen et al. found that positively charged dendrimers were more toxic and produced a hemolytic effect than negatively charged dendrimers (Chen et al. 2004). Two different sized (45 and 90 nm) polymer-based NPs made from a combination of three polymers (PEG400-PHA-PEG400), each having both positive and negative surface charge, were studied *in vitro* on Caco-2-cells. The result showed size- and charge-dependent cytotoxicity. Positively charged polymer NPs with a size of 45 nm induced extensive cytotoxicity as compared to positively charged 90 nm NPs, such as mitochondrial membrane potential disruption, electron transfer chain uncoupling, reduction in ATP production, induction of ROS, and promotion of inflammation (elevated TNF-α); however, the negatively charged nano-polymer did not promote notable toxicity (Bhattacharjee et al. 2012).

3.4.1.4 Shape/Structure

There are various types of NPs and each type of NPs has multiple distinct shapes. For instance carbon-based NPs could have tubelike, sheetlike, spherical-like, cube-like, polygon-like shapes; likewise, polymer-based NPs could also have spherical-like, sheetlike, or triangular shaped-edged characteristics (Chen et al. 2004; Bhattacharjee et al. 2012). *In vitro* toxicity study on human epithelial lung cells (A549) was done using rod-shaped and sphere-shaped ZnO NPs. Controlling for size and surface area between rod-shaped and sphere-shaped NPs, the study found that rod-shaped ZnO NPs were more toxic than sphere-shaped ones, producing higher level of inflammation (marked elevated IL-8) and reduced cell viability (Hsiao and Huang 2011). Spherical and needle-shaped PLGA-PEG NPs, with similar volumes and surface charge and chemistry (zeta −23 mV), were studied on human liver cells (HepG2). Results showed that needle-shaped PLGA-PEG promoted significantly more cytotoxicity and genotoxicity, including expansion of lysosomes and distribution of lysosomal membrane, activation of apoptosis pathways due to lysosomal damage, DNA fragmentation, and cell death (Zhang et al. 2017). Different shapes of gold NPs, spherical (10 nm), flowerlike (370 nm), rodlike (41 nm), prism-like (160 nm), and starlike (240 nm), were studied in vitro on human cells, showing shape-related toxicity; the spherical and rod-shaped gold NPs were more toxic than other shapes, characterized by reduced cell viability (Wozniak et al. 2017).

3.4.1.5 Toxicokinetics

NPs have distinct absorption, distribution, metabolism, and excretion (ADME), and the resulting toxicities from this pharmacokinetics are still not fully characterized; however, some *in vivo* studies have examined bio-distribution of various NPs and their associated toxicities (Desai 2012; Lopez-Chaves et al. 2018). Lopez-Chaves et al. found that gold NPs follow specific pattern of bio-distribution, accumulation, and toxicity resulting from the same distributional pattern. *In vivo* study of three gold NPs (10, 30, and 60 nm) on Wistar rats after intraperitoneal injection resulted in size-dependent bio-distribution and accumulation of gold NPs in specific organs, such as intestine, spleen, and liver, initiating toxic response to these tissues: 10 and 30 nm gold NPs were found mostly in the intestine, followed by spleen, liver, and kidney, respectively, when compared with 60 nm gold NPs which were found almost exclusively in the spleen. The 10 nm gold NPs favored the intestine and 30 nm favored the liver and kidney. The result supported that smaller sized NPs were eliminated through kidney (10 and 30 nm) but larger sized NPs (60 nm) were accumulated in the spleen, where it was metabolized by the reticuloendothelial system. Interestingly, the 10 nm gold NPs were highly localized in the intestine due to the route of exposure; in addition, it was observed that fecal was the main elimination route compared to urine. Similarly, size-dependent toxicity was also observed. 10 nm sized gold NPs generated higher oxidative stress which led to damage to lipid, protein, and DNA. Silica NPs (20 and 80 nm), when studied *in vivo* on male ICR mice, after 30-day intravenous administration, showed bio-distribution-linked toxicity. Both sizes of silica NPs were evenly distributed and accumulated predominantly in the lung, liver, and spleen. Toxicity resulted from increased absorption due to extended stay in these tissues. This resulted in transient embolism in lung due to aggregation, higher deposition in liver and spleen due to phagocytosis by macrophages, and limited degradation after internalization leading to promotion of inflammation in portal triad leading to development of necrosis of hepatocytes (Xie et al. 2010). Cho et al. observed that PEG-coated 13 nm gold NPs, when administered intravenously to male BALB/c mice, accumulated mostly in liver and spleen, even 7 days after injection. Kupffer cells and macrophages in spleen endocytosed PEG-gold NPs and slowed down their clearance. The number of PEG-gold NPs found in other organs such as lung, kidney, brain, and testes decreased with time. Toxicity studies revealed that PEG-coated gold NPs caused acute inflammation and apoptosis of liver cells (Cho et al. 2009). Cho et al. observed that PEG-coated 13 nm gold NPs, when administered intravenously to male BALB/c mice, accumulated mostly in liver and spleen, even 7 days after injection. Kupffer cells and macrophages in spleen endocytosed PEG-gold NPs and slowed down their clearance. The number of PEG-gold NPs found in other organs such as lung, kidney, brain, and testes decreased with time. Toxicity studies revealed that PEG-coated gold NPs caused acute inflammation and apoptosis of liver cells (Li et al. 2015).

3.4.2 Mechanisms of Nanotoxicity

3.4.2.1 Reactive Oxygen Species (ROS)

The production of ROS occurs naturally within a biological system and has useful functionality. ROS, such as superoxide, hydroxyl radical, singlet oxygen, and hydrogen peroxide, are generated in various cellular organelles including mitochondria, lysosome, peroxisomes, and endoplasmic reticulum, as a subsequent by-product of normal function. The mitochondria yield ROS while making ATP, and lysosomes and peroxisomes generate ROS for the purpose of metabolism, and endoplasmic reticulum during biosynthesis of lipids (Manke et al. 2013; Abdal Dayem et al. 2017). The amount of ROS is kept in balance by ROS scavengers or antioxidant agents, which include superoxide dismutase (SOD), glutathione reductase (GR), glutathione peroxidase (GPX), catalase (CAT), NADPH oxidase (NOS), flavonoids, ascorbic acids, and vitamin E (Abdal Dayem et al. 2017; Fu et al. 2014). Imbalance between the amounts of ROS produced and neutralized via the antioxidant agents leads to overt amount of ROS in tissue which results in oxidative stress. Excess ROS activates many signaling pathways, which act to resolve the imbalance. NF-κB that is activated via ROS acts through IKK-dependent pathway and acts on genes which mount a response to oxidative stress, inflammation, and others. MAPK initiation due to ROS prompts extracellular signal-related kinases (ERK) and stress-activated MAPK, c-Jun NH2-terminal kinases (JNK), and p38 MAPK, which control gene expression for cell growth regulation, apoptosis, etc. PTP activation because of ROS sets off other signaling pathways responsible for growth, differentiation, and transformation. Src pathway is triggered by ROS and kick-starts non-tyrosine kinase pathways responsible for cell growth (Abdal Dayem et al. 2017; Fu et al. 2014; Park and Park 2009). There are two main ways (direct and indirect), by which NPs cause oxidative stress. NPs are able to generate ROS directly due to their physicochemical properties (size/surface area, charge, composition, etc.); secondly, NPs can generate ROS as a result of interacting with their environment (proteins, lipids, extracellular matrix) and ROS scavengers/antioxidants (inhibition or modification). The size of NPs is inversely related to their surface area; as size decreases the surface area of NP increases; similarly, as size decreases the uptake/internalization also increases, allowing more particles to enter cells or tissues that enable them to have ample space to interact. 15 nm Silver NPs when given at 50 μg/mL resulted in tenfold production of ROS as compared to 30 nm silver (Park and Park 2009; Pacurari et al. 2008) given at the same concentration. Study shows that large surface area enables NPs to perform various catalysis in biological environment or some NPs, like silicon oxide NPs, have been involved in releasing silicon oxide free radicals; likewise, in another study, silicon quartz in biological fluid has been known to produce ROS, such as oxygen singlet, hydrogen peroxide, or hydroxyl radical (Fu et al. 2014; Park and Park 2009). In some cases, metallic/metallic oxide NPs have been shown to produce ROS, undergoing Fenton and Haber-Weiss reactions making hydroxyl radicals (Fu et al. 2014; Park and Park 2009). Also, NPs can interact with cellular components which results in generation of ROS. NPs have been shown to enter mitochondria, and then block or modify oxygen electron chain pathway yielding more ROS. Some silver ion NPs have interacted with NADH dehydrogenase to generate free radicals (Fu et al. 2014). Other studies have highlighted the interaction of NPs with lipid membranes and polyunsaturated acids, which are key in generating ROS leading to toxicity (Fu et al. 2014; Park and Park 2009). Another noted interaction that NPs undergo is their involvement with immunological cells such as macrophages and neutrophils, which causes them to be more active in generating ROS (Fu et al. 2014; Park and Park 2009) (depicted in Figs. 3.12 and 3.13).

3.4.2.2 Inflammation

The relationship between NP-induced generation of ROS and development of inflammatory responses is well noted. Excess ROS in cells or tissues triggers various signaling pathways including pro-inflammatory cascades, activation of protein kinases, and gene transcription factors (Fu et al. 2014; Park and Park 2009; Pacurari et al. 2008). Study of silica NPs *in vitro* and *in vivo* showed dose- and time-dependent rise in the serum levels of IL-1β, TNF-α, and nitrogen oxide (NO) along with elevated expression of pro-inflammatory mediators like IL-1, IL-6, TNF-α, iNOS, and COX-2 (Park and Park 2009). Another study using single-walled carbon nanotubes (SWCNT) also showed the involvement of chemokine and cytokine activation as a result of ROS, mediating inflammation through MAPKs, AP-1, NF-κB, and AKT pathways (Pacurari et al. 2008). Alternatively, ROS-independent activation of pro-inflammatory mediators has been investigated by Deng et al. Poly (acrylic acid)-coated gold NPs were able to unfold fibrinogen,

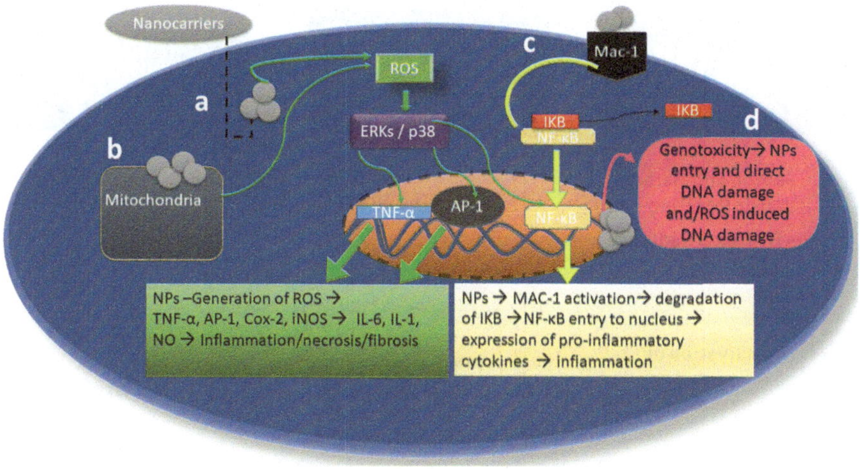

Fig. 3.12 Pictorial depiction of mechanisms of NP toxicity. (**a**) NPs can directly generate ROS via interaction with various intracellular macromolecules and through Fenton and Haber-Weiss reactions. (**b**) NPs can cause damage to mitochondria resulting in release of ROS from mitochondria. (**c**) ROS-independent toxicity mediated by NPs through Mac-1 pathway. (**d**) NPs can enter the nucleus and cause DNA damage directly and/or via generating ROS

which allowed Mac-1 receptor stimulation, leading to NF-κB signaling pathway activation and inflammatory response mediator expression. Chronic and extended inflammation can result in the development of fibrosis (Deng et al. 2011); Lam et al. showed that SWCNTs are able to induce fibrosis in mice (Lam et al. 2004) (as depicted in Figs. 3.12 and 3.13).

3.4.2.3 Genotoxicity

The toxic effect of NPs on genetic materials is mediated through various ways. Genotoxicity related to ROS production has been noted in studies; for instance, Shukla et al. found ROS-dependent toxicity involving DNA damage and dysfunction with micronucleus formation due to titanium NPs (Shukla et al. 2011). Using silver NPs on human Jurkat T cells, oxidative stress-linked DNA damage, cell cycle dysregulation, and apoptosis were observed via activation of p38 and MAP kinase pathway related to Nrf2 and NF-kappaB expression (Eom and Choi 2010). Multi-walled carbon nanotubes (MWCNTs) studied in vitro and in vivo showed DNA damage which was mediated via oxidative stress. ROS that was generated interacted with DNA, leading to the formation of 8-hydroxy-2′-deoxyguanosine (8-OHdG) (Kato et al. 2013) (as depicted in Figs. 3.12 and 3.13).

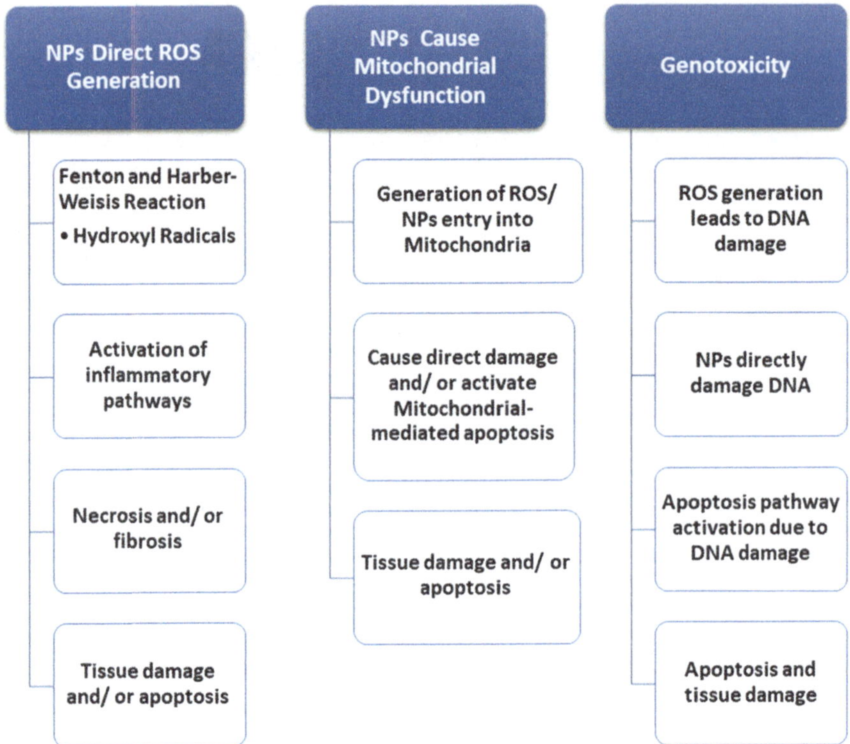

Fig. 3.13 Summary of proposed terminal mechanisms following NP toxicity

3.4.3 Major Target Organs Affected

3.4.3.1 Respiratory Tract/Pulmonary (Lung)

The induction of toxicity of NPs on the respiratory system has been examined *in vivo* and *in vitro* using copious model cell lines and living organisms (Card et al. 2008; Bonner 2010). Of note, carbon-based NPs', carbon nanotubes' (SWCNTs and MWCNTs), and carbon fullerenes' pulmonary toxicities have been examined (Lam et al. 2004; Chou et al. 2008; Qin et al. 2017; Muller et al. 2005; Mitchell et al. 2007). SWCNTs given intratracheally to mice showed dose-dependent pulmonary toxicities which included oxidative stress, inflammation, development of necrosis, markers of fibrosis responses, and bio-persistence (accumulation in the lung) even after 90 days of installation (Lam et al. 2004). Chou et al. demonstrated that SWCNTs induce oxidative stress, chronic inflammation, and activation of immune cells such as macrophages that are involved in the formation of serious granuloma (Chou et al. 2008). Modified carboxylated SWCNT administered intravenously to mice induced fibrosis formation which included activation of inflammation mediators (TNF-α and interleukins) as well as growth factors like transforming growth factor (TGF-1), leading to pathological tissue changes in the lung and surrounding capillaries due to accumulation (Qin et al. 2017). Likewise MWCNTs in *in vivo* and *in vitro* study showed various toxicities on the respiratory system; MWCNTs administered intratracheally to rats initiated inflammatory response, marked by elevation of TNF-α, persistent accumulation in the lung even with low dose, characteristic lesion formation in airway tracks, and fibrotic formation activities such as granulomas with excessive collagen in tissues around bronchial lumen (Muller et al. 2005). Mitchell et al. showed that inhaled MWCNTs can act as immunosuppressants via interfering with splenic function, such as reducing T-cell growth and development, lowering natural killer function, and elevating spleen-specific oxidative stress inflammation marker, like interleukin-10 (Mitchell et al. 2007). Another study compared the toxic effect of MWCNTs by using two different administration routes, inhalation and intratracheal. The result showed that MWCNTs given intratracheally demonstrated more severe toxicities due to aggregation of NPs near and around bronchial and alveolar lining, leading to a stronger inflammatory response and formation of lesions and fibrotic development. On the other hand, due to aerosolization and lower particle size of inhaled MWCNTs it showed reduced aggregation and lower pulmonary toxicity compared to intratracheal method (Li et al. 2007). Fullerenes, spherical shaped nanocarbons, upon intratracheal administration in male mice, induced pulmonary toxicity that included acute inflammation marked by pro-inflammatory mediators (cytokines and TNF-α). Moreover, possible toxicity of carbon-based NPs to other organs or tissues as a result of translocation after initial administration has also been examined. Inhaled MWCNTs translocated to subpleural membrane after single exposure of 30 mg in mice led to the development of fibrosis during the second and sixth weeks after exposure (Ryman-Rasmussen et al. 2009). Another study which also used mouse models showed that inhaled MWCNTs can translocate to other systemic organs such as heart, brain, liver, and kidney; the lymph nodes accumulated more translocated NPs than other organs (Mercer et al. 2013). Metal and metal oxide NPs also have demonstrated toxic effect on respiratory system. Ferric oxide NPs (22 and 280 nm), when administered intratracheally, caused inflammatory response, changes in cell proliferation, and dysregulation of clotting (delayed partial thromboplastin time (PTT) and activated partial thromboplastin time (aPTT)) (Zhu et al. 2008). Zinc oxide NPs instilled intratracheally to mice showed pH-induced inflammatory response upon internalization into macrophages and lysosomes, which was linked to the dissolution of zinc ions. Significant changes such as development of fibrosis and elevation of TGF-beta and eosinophilia were reported (Cho et al. 2011). The pulmonary toxicity of silicon oxide was studied on mice, following intratracheal instillation. Induction of oxidation marked by increased lipid oxidation and other mediators like chemokines (TNF-α, TGF-β1, IL-1β) led to pulmonary fibrosis (Hong et al. 2017). Polymer NPs have also been reported to have toxic effects on the lung. Polyamidoamine dendrimer (PAMAM) NPs caused lung injury in mice. Akt-TSC2-mTOR signaling pathway suppression leading to the death of autophagy has been proposed as a possible mechanism (Li et al. 2009).

3.4.3.2 Liver

Toxicological effects of NPs on liver have been reported in various studies, using different nanostructures and studied in *in vivo* and *in vitro* models. Silica NPs (14 nm; 25–200 μg/mL) on human liver hepatocellular carcinoma cell line (HepG2) induced toxicity which was dose dependent. This was mediated by excess ROS, partly due to reduction of antioxidant agents like glutathione and production of ROS through interaction with lipids. These mechanisms resulted in apoptosis that was mediated by p53, bax/bcl-2, and caspase pathways. Treatment with antioxidants like vitamin C alleviated toxicity (Ahmad et al. 2012). Another study using silicon NPs *in vivo* on rats and *in vitro* on buffalo rat liver cells showed that ROS-mediated inflammation, marked by elevation of TNF-α and NO in liver, was responsible for toxicity. There were changes in Kupffer cells including hyperplasia (Chen et al. 2013). Carbon-based NPs, such as graphene oxide and carbon nanotubes (MWCNTs), have also been reported to have toxic effects on the liver. Modified graphene oxide NPs, administered intravenously, consistently accumulated in liver, lung, and spleen 6 months after instillation. This was followed by marked ROS generation, inflammation, and liver injury (Wen et al. 2015). Another study showed that MWCNTs given intratracheally to rats translocated to other organs, such as liver and kidney, causing tissue injury, showing inflammatory and necrosis markers (Rama Narsimha Reddy et al. 2010). Polymer-coated NPs similarly show toxicity to liver; polymer iron oxide (polyacrylic acid coating), intravenously given (20 or 50 mg/kg), was found to accumulate in liver (hepatic macrophages) and spleen, inducing inflammatory and oxidative stress activities, such as elevating neutrophil and lymphocyte counts and increased lipid peroxidation (Couto et al. 2016). Titanium oxide NPs (10 nm; given IP 50 mg/kg) on mice tended to accumulate in liver and lung, and caused oxidative DNA damage, accompanied with tissue damages initiated through oxidative stress, inflammation, and apoptosis (Li et al. 2016). Furthermore, zinc oxide NPs, when given orally to mice (300 mg/kg; 14 days), caused oxidative stress-induced liver injuries which included DNA damage, apoptosis, lesions on liver tissue, and changes in serum ALT and ALP (Sharma et al. 2012). PEG-coated gold NPs (13 nm) showed heavy accumulation in the liver and spleen that were associated with tissue injuries; NPs were found in macrophages, vesicles, and lysosomes, leading to initiation of inflammation and apoptosis (Cho et al. 2009). Another study using gold NPs (10, 20, and 50 nm) produced size-dependent injuries to liver, causing Kupffer cells' pathological growth and dysfunction to hepatocytes, as well as inducing oxidative related toxicities, such as inflammation and necrosis (Abdelhalim and Jarrar 2011). Silver NPs *in vitro*, on human liver cells, showed various toxicities, such as increasing ROS generation, DNA damage, and activation of apoptotic path that was mediated by injury to mitochondria (Bax, Bcl-2, and caspase 9/3) (Piao et al. 2011) (as shown in Figs. 3.12 and 3.13).

3.4.3.3 Spleen

NPs tend to accumulate in the spleen and liver. Gold NPs of sizes 10, 50, 100, and 250 nm, when studied on rats, showed size-dependent accumulation. Smaller gold NPs were observed in the spleen, liver, lung, and kidney, whereas larger sized gold

NPs selectively aggregated in the spleen and liver (De Jong et al. 2008). Likewise, another study demonstrated that gold NPs accumulated in spleen and liver, after repeated intraperitoneal administration with variable doses (40, 200, 400 μg/kg; 8 days) on mice (Lasagna-Reeves et al. 2010). Furthermore, PEG-coated gold NPs (13 nm) heavily accumulated in the spleen, leading to acute inflammation (Cho et al. 2009). Titanium oxide NPs in mice showed splenic injury (doses 2.5, 5, and 10 mg/kg; 90 days). Toxic effects on the spleen included expression of inflammation mediators (NF-kB, TNF-α, interleukins, and suppression of Bcl-2 and heat-shock protein 70), pathological tissue changes (lesions), apoptosis, and necrosis. Significant depression of immunological cells such as CD3, CD4, CD8, B cell, and natural killer cell was also observed (Sang et al. 2012). Silver NPs (20 nm), administered intravenously to rats, repeatedly for 28 days (6 mg/kg), demonstrated capacity to suppress immunological function, causing increased spleen weight and lowering thymus size, while also decreasing KLH-specific IgG levels (Vandebriel et al. 2014). Zinc oxide NPs given orally for 4 weeks to mice caused toxicity to spleen and thymus. This was mediated through induction of pro-inflammatory mediators such as cytokines, TNF-α, INF-γ, CD3, CD11b, heme oxygenase (HO-1), and pathological tissue changes in spleen and thymus (Abass et al. 2017). Polyacrylic acid-coated iron oxide was shown to preferably accumulate in spleen and liver, causing serious injuries, resulting in oxidative stress and initiation of inflammation as indicated by marked elevation in neutrophils and lymphocytes (Couto et al. 2016). Modified graphene oxide NPs were reported to accumulate and remain in spleen, liver, and lung even after 6 months of intravenous administration in mice, leading to tissue injuries, characterized by marked moderate-to-chronic inflammation (Wen et al. 2015). Silica NPs (70 nm) also accumulate in megakaryocytes, and increase inflammation (Nishimori et al. 2009) (as shown in Figs. 3.12 and 3.13).

3.4.3.4 Kidney

Toxic effect of NPs on renal system is still under investigation (Iavicoli et al. 2016). Silver NPs, orally administered to male and female mice, resulted in noticeable adverse toxicities at middle-dose (125 mg/kg) range. Gender-dependent accumulation of silver NPs in the kidney was observed. Female mice had two times more silver NP accumulation in the kidney as compared to male mice (Kim et al. 2010). Another study, using oral silver NPs on female mice, with doses below toxic levels (50 and 200 ppm; 60 days), showed acute dysfunction of renal system, evidenced by elevated serum creatinine, KIM-1, and clustering, which are indicators of acute kidney damage. Moreover, activation of inflammatory mediators and chemokines (TNF-α, interleukins) was demonstrated to cause interference with apoptotic pathways (that remove damaged microstructures) leading to the development of necrosis (Tiwari et al. 2011). Cadmium-coated silica NPs revealed renal toxicity in mice, after intratracheal administration (1 mg dose/rat). This was mediated by significant apoptosis in cortex and medulla and pathological tissue changes in the glomerulus and tubules of kidney following 7- and 30-day administration (Coccini et al. 2013). Another study using mesoporous silica NPs on mice found that high dose (600 mg/kg) leads to severe renal impairment, mainly induced through the generation of ROS, and development of fibrosis. Suppression of the NF-κB pathway showed reduction in fibrosis *in vitro* (Chen et al. 2015). Gold NPs (10 nm), administered intraperitoneally in mice (50 μL for 7 days), showed signs of renal dysfunction, indicated by increased BUN, urea, and creatinine. Co-administration of quercetin and arginine, known for their antioxidative and anti-inflammation activities, resulted in significant reduction in all renal dysfunction biomarkers (Abdelhalim et al. 2018). Copper NPs (100 and 200 mg/kg; for 5 days) showed dose-dependent toxic effects on kidney and liver. The 200 mg dose caused more toxicity, which included notable development of necrosis in renal tubules, generation of ROS, and abnormality with the mitochondria function, such as oxidative respiration, urea cycle, and lipid production, among others (Lei et al. 2015) (as shown in Fig. 3.12).

3.4.3.5 CNS/Brian

Toxic effects of NPs on the brain were reported when gold NPs (20 nm) administered intraperitoneally to male rats for 3 days (20 μg/kg) showed toxicities induced as a result of oxidative stress (generation of ROS and reduction/interference with antioxidants such as glutathione). This in turn led to oxidative DNA damage, apoptosis, elevation of IFN-gamma level, and changes in dopamine and serotonin levels (Siddiqi et al. 2012). Another study elucidated that titanium oxide NPs (21 nm) when instilled intravenously for 4 weeks (5, 25, 50 mg/kg) on mice led to activation of pro-inflammatory mediators NF-κB (p65), HSP 60, p38, nitric oxide, IFN-γ, and TNF-α. This was accompanied by reduction in antioxidants, glutamic acid level, acetylcholinesterase function, as well as oxidative stress-induced DNA damage and apoptosis (Meena et al. 2015).

3.4.4 Toxicities from Selected FDA-Approved Nanoparticle-Loaded Drugs

While most of the toxicity-related data comes from studies performed in *in vivo* and *in vitro* animal models, there are a few studies that elucidate the toxic effects of nanocarriers in clinical trials and case studies.

3.4.4.1 Liposome-Based Nano-Drug: PEGylated Liposomal Doxorubicin (Doxil® or Caelyx®)

Liposomal based drug delivery systems such as a PEGylated doxorubicin have numerous advantages which make them ideal candidates; the fact that liposomes are made from lipids (cholesterols and various forms of phospholipids) and also come from natural sources, plants and animals (egg yolk/soybean oil), gives liposomes a more enhanced safe characteristic (Slingerland et al. 2012). In addition, attachment of PEG polymers to conventional liposomes has made it possible to escape opsonization and reticuloendothelial system leading to increased time in the body, which prevents frequent dosing regimens and allows drugs loaded into the liposomes to induce more effect on targeted sites in the body (Slingerland et al. 2012;

Sercombe et al. 2015). In the case of PEGylated doxorubicin, clinical studies have shown that compared to free doxorubicin, PEGylated doxorubicin exhibits better pharmacokinetic properties which include increased AUC at least by 60-fold, more tendency to be located in tumor tissues, reduced clearance, and improved toxicities (significantly reduced cardiac toxicities). Although PEGylated doxorubicin has many advantages, specific toxicities unique to liposome formulation only have been observed in clinical studies: more pronounced hypersensitivity reaction and palmar-plantar erythrodysesthesia (PPE) (more commonly known as hand-foot syndrome or hand-to-foot syndrome) (Lorusso et al. 2007; Slingerland et al. 2012). Studies have showed that as many as 30.8% of people experience hypersensitivity reactions with PEGylated liposomal formulation showing accompanying changes in blood pressure, respiration, flashing, and rash, and a choking feeling; similarly, PPE or hand-foot syndrome is linked with the use of PEGylated formulation of doxorubicin, where skin-related reaction such as skin eruptions, diffused follicular rash, and red painful inflamed patches are observed, and this unique toxicity is associated with the ability of liposome-based formulation's extended circulation time and its ability to get deposited in tissue due to its lipophilic nature (Lorusso et al. 2007; Lotem et al. 2000).

3.5 Conclusion

Through this chapter, we discussed key principles governing the synthesis of commonly used nanocarriers in research as well as drug development. We also summarized different pharmacokinetic aspects of nano-drug carriers, from absorption of drug-loaded NPs through various routes of administration, their distribution to other organs, metabolism, and excretion from the body. Furthermore, we discussed the physicochemical factors affecting ADME of NP carrier drugs with evidence through various cases reported in the literature. With regard to the toxicity of nanocarriers, most of the available literature is based on *in vitro* and *in vivo* toxicity studies in rodents with few cases of toxicities observed after clinical use. As nanocarriers find more use in therapy, new toxicities may emerge. These toxicities need to be monitored closely in order to improve and enhance the utilization of nanocarriers in therapy.

References

Ab L (2006) Silver in health care: antimicrobial effects and safety in use. Curr Probl Dermatol 33:17–34

Abass MA, Selim SA, Selim AO et al (2017) Effect of orally administered zinc oxide nanoparticles on albino rat thymus and spleen. IUBMB Life 69:528–539

Abdal Dayem A, Hossain M, Lee S et al (2017) The role of reactive oxygen species (ros) in the biological activities of metallic nanoparticles. Int J Mol Sci 18:120

Abdelhalim MAK, Jarrar BM (2011) Gold nanoparticles administration induced prominent inflammatory, central vein intima disruption, fatty change and Kupffer cells hyperplasia. Lipids Health Dis 10:133

Abdelhalim MAK, Qaid HA, Al-Mohy Y et al (2018) Effects of quercetin and arginine on the nephrotoxicity and lipid peroxidation induced by gold nanoparticles in vivo. Int J Nanomedicine 13:7765

Adebisi A, Conway BR (2011) Gastroretentive microparticles for drug delivery applications. J Microencapsul 28:689–708

Adebisi AO, Conway BR (2015) Modification of drug delivery to improve antibiotic targeting to the stomach. Ther Deliv 6:741–762

Ahmad J, Ahamed M, Akhtar MJ et al (2012) Apoptosis induction by silica nanoparticles mediated through reactive oxygen species in human liver cell line hepg2. Toxicol Appl Pharmacol 259:160–168

Akbarzadeh A, Rezaei-Sadabady R, Davaran S et al (2013) Liposome: classification, preparation, and applications. Nanoscale Res Lett 8:102

Albornoz C, Jacobo SE (2006) Preparation of a biocompatible magnetic film from an aqueous ferrofluid. J Magn Magn Mater 305:12–15

Alivisatos AP, Johnsson KP, Peng X et al (1996) Organization of 'nanocrystal molecules' using DNA. Nature 382:609–611

Alvarez GS, Muhammed M, Zagorodni AA (2006) Novel flow injection synthesis of iron oxide nanoparticles with narrow size distribution. Chem Eng Sci 61:4625–4633

Atiyeh BS, Costagliola M, Hayek SN et al (2007) Effect of silver on burn wound infection control and healing: review of the literature. Burns 33:139–148

Babes L, Denizot B, Tanguy G et al (1999) Synthesis of iron oxide nanoparticles used as mri contrast agents: a parametric study. J Colloid Interface Sci 212:474–482

Basak S, Chen DR, Biswas P (2007) Electrospray of ionic precursor solutions to synthesize iron oxide nanoparticles: modified scaling law. Chem Eng Sci 62:1263–1268

Bhattacharjee S, Ershov D, Fytianos K et al (2012) Cytotoxicity and cellular uptake of tri-block copolymer nanoparticles with different size and surface characteristics. Part Fibre Toxicol 9:1–19

Bonner JC (2010) Nanoparticles as a potential cause of pleural and interstitial lung disease. Proc Am Thorac Soc 7:138–141

Bos JD, Meinardi MM (2000) The 500 dalton rule for the skin penetration of chemical compounds and drugs. Exp Dermatol 9:165–169

Brongersma ML (2003) Nanoscale photonics: nanoshells: gifts in a gold wrapper. Nat Mater 2:296–297

Brown KR, Natan MJ (1998) Hydroxylamine seeding of colloidal Au nanoparticles in solution and on surfaces. Langmuir 14:726–728

Brown MB, Martin GP, Jones SA et al (2006) Dermal and transdermal drug delivery systems: current and future prospects. Drug Deliv 13:175–187

Busbee BD, Obare SO, Murphy CJ (2003) An improved synthesis of high-aspect-ratio gold nanorods. Adv Mater 15:414–416

Card JW, Zeldin DC, Bonner JC et al (2008) Pulmonary applications and toxicity of engineered nanoparticles. Am J Physiol Lung Cell Mol Physiol 295:L400–L411

Chastellain M, Petri A, Gupta A et al (2004) Superparamagnetic silica-iron oxide nanocomposites for application in hyperthermia. Adv Eng Mater 6:235–241

Chen HT, Neerman MF, Parrish AR et al (2004) Cytotoxicity, hemolysis, and acute in vivo toxicity of dendrimers based on melamine, candidate vehicles for drug delivery. J Am Chem Soc 126:10044–10048

Chen J, Saeki F, Wiley BJ et al (2005) Gold nanocages: bioconjugation and their potential use as optical imaging contrast agents. Nano Letters 5:473–477

Chen Q, Xue Y, Sun J (2013) Kupffer cell-mediated hepatic injury induced by silica nanoparticles in vitro and in vivo. Int J Nanomedicine 8:1129

Chen X, Zhouhua W, Jie Z et al (2015) Renal interstitial fibrosis induced by high-dose mesoporous silica nanoparticles via the nf-κb signaling pathway. Int J Nanomedicine 10:1

Chertok B, Moffat BA, David AE et al (2008) Iron oxide nanoparticles as a drug delivery vehicle for mri monitored magnetic targeting of brain tumors. Biomaterials 29:487–496

Chin AB, Yaacob II (2007) Synthesis and characterization of magnetic iron oxide nanoparticles via w/o microemulsion and Massart's procedure. J Mater Process Technol 191:235–237

Cho WS, Cho M, Jeong J et al (2009) Acute toxicity and pharmacokinetics of 13 nm-sized peg-coated gold nanoparticles. Toxicol Appl Pharmacol 236:16–24

Cho WS, Duffin R, Howie SE et al (2011) Progressive severe lung injury by zinc oxide nanoparticles; the role of zn 2+ dissolution inside lysosomes. Part Fibre Toxicol 8:1–16

Chou CC, Hsiao HY, Hong QS et al (2008) Single-walled carbon nanotubes can induce pulmonary injury in mouse model. Nano Letters 8:437–445

Clemens DL, Lee BY, Plamthottam S et al (2019) Nanoparticle formulation of moxifloxacin and intramuscular route of delivery improve antibiotic pharmacokinetics and treatment of pneumonic tularemia in a mouse model. ACS Infect Dis 5:281–291

Coccini T, Barni S, Manzo L et al (2013) Apoptosis induction and histological changes in rat kidney following cd-doped silica nanoparticle exposure: evidence of persisting effects. Toxicol Mech Methods 23:566–575

Colognato R, Bonelli A, Bonacchi D et al (2007) Analysis of cobalt ferrite nanoparticles induced genotoxicity on human peripheral lymphocytes: comparison of size and organic grafting-dependent effects. Nanotoxicology 1:301–308

Conroy S, Omid V, Jonathan G et al (2008) In vivo MRI detection of gliomas by chlorotoxin-conjugated superparamagnetic nanoprobes. Small 4:372–379

Couto D, Freitas M, Costa VM et al (2016) Biodistribution of polyacrylic acid-coated iron oxide nanoparticles is associated with proinflammatory activation and liver toxicity. J Appl Toxicol 36:1321–1331

Crucho CIC, Barros MT (2017) Polymeric nanoparticles: a study on the preparation variables and characterization methods. Mater Sci Eng C 80:771–784

De Jong WH, Hagens WI, Krystek P et al (2008) Particle size-dependent organ distribution of gold nanoparticles after intravenous administration. Biomaterials 29:1912–1919

Debnath SK, Saisivam S, Omri A (2017) PLGA ethionamide nanoparticles for pulmonary delivery: development and in vivo evaluation of dry powder inhaler. J Pharm Biomed Anal 145:854–859

Deng ZJ, Liang M, Monteiro M et al (2011) Nanoparticle-induced unfolding of fibrinogen promotes Mac-1 receptor activation and inflammation. Nat Nanotechnol 6:39–44

Des Rieux A, Fievez V, Garinot M et al (2006) Nanoparticles as potential oral delivery systems of proteins and vaccines: a mechanistic approach. J Control Release 116:1–27

Desai N (2012) Challenges in development of nanoparticle-based therapeutics. AAPS J 14:282–295

Elechiguerra J, Burt J, Morones J et al (2005) Interaction of silver nanoparticles with HIV-1. J Nanobiotechnol 3:1–10

Elias A, Tsourkas A (2009) Imaging circulating cells and lymphoid tissues with iron oxide nanoparticles. Hematology Am Soc Hematol Educ Program 2009:720–726

El-Sayed IH, Huang X, El-Sayed MA (2005) Surface plasmon resonance scattering and absorption of anti-EGFR antibody conjugated gold nanoparticles in cancer diagnostics: applications in oral cancer. Nano Lett 5:829–834

Eom HJ, Choi J (2010) P38 MAPK activation, DNA damage, cell cycle arrest and apoptosis as mechanisms of toxicity of silver nanoparticles in Jurkat T cells. Environ Sci Technol 44:8337–8342

Ercuta A, Chirita M (2013) Highly crystalline porous magnetite and vacancy-ordered maghemite microcrystals of rhombohedral habit. J Cryst Growth 380:182–186

Farooq MU, Novosad V, Rozhkova EA et al (2018) Gold nanoparticles-enabled efficient dual delivery of anticancer therapeutics to HeLa cells. Sci Rep 8:1–12

Filon LF, Mauro M, Adami G et al (2015) Nanoparticles skin absorption: new aspects for a safety profile evaluation. Regul Toxicol Pharmacol 72:310–322

Florence AT (2005) Nanoparticle uptake by the oral route: fulfilling its potential? Drug Discov Today Technol 2:75–81
Frens G (1972) Particle size and sol stability in metal colloids. Colloid Polym Sci 250:736–741
Frens G (1973) Controlled nucleation for regulation of particle size in monodisperse gold suspensions. Nat Phys Sci 241:20–22
Fu PP, Xia Q, Hwang HM et al (2014) Mechanisms of nanotoxicity: generation of reactive oxygen species. J Food Drug Anal 22:64–75
Furno F, Morley KS, Wong B et al (2004) Silver nanoparticles and polymeric medical devices: a new approach to prevention of infection? J Antimicrob Chemother 54:1019–1024
Gabor F, Bogner E, Weissenboeck A et al (2004) The lectin-cell interaction and its implications to intestinal lectin-mediated drug delivery. Adv Drug Deliv Rev 56:459–480
Gao L, Liu G, Ma J et al (2013) Application of drug nanocrystal technologies on oral drug delivery of poorly soluble drugs. Pharm Res 30:307–324
Giljohann DA, Seferos DS, Daniel WL et al (2010) Gold nanoparticles for biology and medicine. Angew Chem Int Ed Engl 49:3280–3294
Goebel AS, Neubert RH, Wohlrab J (2011) Dermal targeting of tacrolimus using colloidal carrier systems. Int J Pharm 404:159–168
Gonzales-Weimuller M, Zeisberger M, Krishnan KM (2009) Size-dependant heating rates of iron oxide nanoparticles for magnetic fluid hyperthermia. J Magn Magn Mater 321:1947–1950
Griffin BT, Guo J, Presas E et al (2016) Pharmacokinetic, pharmacodynamic and biodistribution following oral administration of nanocarriers containing peptide and protein drugs. Adv Drug Deliv Rev 106:367–380
Hagens WI, Oomen AG, De Jong WH et al (2007) What do we (need to) know about the kinetic properties of nanoparticles in the body? Regul Toxicol Pharmacol 49:217–229
Hamidi M, Azadi A, Rafiei P et al (2013) A pharmacokinetic overview of nanotechnology-based drug delivery systems: an ADME-oriented approach. Crit Rev Ther Drug Carrier Syst 30 (5):435–467
Handa T, Hirai T, Izumi N et al (2017) Identifying a size-specific hazard of silica nanoparticles after intravenous administration and its relationship to the other hazards that have negative correlations with the particle size in mice. Nanotechnology 28:135101
Hirsch LR, Stafford RJ, Bankson JA et al (2003) Nanoshell-mediated near-infrared thermal therapy of tumors under magnetic resonance guidance. Proc Natl Acad Sci U S A 100:13549–13554
Honary S, Zahir F (2013) Effect of zeta potential on the properties of nano-drug delivery systems-a review (part 2). Trop J Pharm Res 12:265–273
Hong Y, Wu QY, Li MY et al (2017) Pulmonary toxicity in rats caused by exposure to intratracheal instillation of SiO2 nanoparticles. Biomed Environ Sci 30:264–279
Hsiao IL, Huang YJ (2011) Effects of various physicochemical characteristics on the toxicities of ZnO and TiO2 nanoparticles toward human lung epithelial cells. Sci Total Environ 409:1219–1228
Hu X, Chai Z, Lu L et al (2019) Bortezomib dendrimer prodrug-based nanoparticle system. Adv Funct Mater 29:1807941
Hua S, Marks E, Schneider JJ et al (2015) Advances in oral nano-delivery systems for colon targeted drug delivery in inflammatory bowel disease: selective targeting to diseased versus healthy tissue. Nanomedicine 11:1117–1132
HuHn D, Kantner K, Geidel C et al (2013) Polymer-coated nanoparticles interacting with proteins and cells: focusing on the sign of the net charge. ACS Nano 7:3253–3263
Iavicoli I, Fontana L, Nordberg G (2016) The effects of nanoparticles on the renal system. Crit Rev Toxicol 46:490–560
ISO 2017. Nanotechnologies. https://www.iso.org/obp/ui/#iso:std:iso:tr:18401:ed-1:v1:en
Jain PK, Lee KS, El-Sayed IH et al (2006) Calculated absorption and scattering properties of gold nanoparticles of different size, shape, and composition: applications in biological imaging and biomedicine. J Phys Chem B 110:7238–7248

Jain PK, Huang X, El-Sayed IH et al (2008) Noble metals on the nanoscale: optical and photothermal properties and some applications in imaging, sensing, biology, and medicine. Acc Chem Res 41:1578–1586

Jana NR, Gearheart L, Murphy CJ (2001a) Seed-mediated growth approach for shape-controlled synthesis of spheroidal and rod-like gold nanoparticles using a surfactant template. Adv Mater 13:1389–1393

Jana NR, Gearheart L, Murphy CJ (2001b) Wet chemical synthesis of high aspect ratio cylindrical gold nanorods. J Phys Chem B 105:4065–4067

Janes KA, Fresneau MP, Marazuela A et al (2001) Chitosan nanoparticles as delivery systems for doxorubicin. J Control Release 73:255–267

Joseph RR, Venkatraman SS (2017) Drug delivery to the eye: what benefits do nanocarriers offer? Nanomedicine 12:683–702

Kalepu S, Nekkanti V (2016) Improved delivery of poorly soluble compounds using nanoparticle technology: a review. Drug Deliv Transl Res 6:319–332

Kato T, Totsuka Y, Ishino K et al (2013) Genotoxicity of multi-walled carbon nanotubes in both in vitro and in vivo assay systems. Nanotoxicology 7:452–461

Kelly KL, Coronado E, Zhao LL et al (2002) The optical properties of metal nanoparticles: the influence of size, shape, and dielectric environment. J Phys Chem B 107:668–677

Kim EH, Lee HS, Kwak BK et al (2005) Synthesis of ferrofluid with magnetic nanoparticles by sonochemical method for MRI contrast agent. J Magn Magn Mater 289:328–330

Kim J, Park S, Lee JE et al (2006) Designed fabrication of multifunctional magnetic gold nanoshells and their application to magnetic resonance imaging and photothermal therapy. Angew Chem Int Ed Engl 45:7754–7758

Kim YS, Song MY, Park JD et al (2010) Subchronic oral toxicity of silver nanoparticles. Part Fibre Toxicol 7:20

Kim S, Oh WK, Jeong YS et al (2011) Cytotoxicity of, and innate immune response to, size-controlled polypyrrole nanoparticles in mammalian cells. Biomaterials 32:2342–2350

Kim IY, Joachim E, Choi H et al (2015) Toxicity of silica nanoparticles depends on size, dose, and cell type. Nanomedicine 11:1407–1416

Kim H, Griffith TS, Panyam J (2019) Poly(d,l-lactide-co-glycolide) nanoparticles as delivery platforms for TLR7/8 agonist-based cancer vaccine. J Pharmacol Exp Ther 370:715–724

Kimata M, Nakagawa D, Hasegawa M (2003) Preparation of monodisperse magnetic particles by hydrolysis of iron alkoxide. Powder Technol 132:112–118

Kimling J, Maier M, Okenve B et al (2006) Turkevich method for gold nanoparticle synthesis revisited. J Phys Chem B 110:15700–15707

Kollner S, Dunnhaupt S, Waldner C et al (2015) Mucus permeating thiomer nanoparticles. Eur J Pharm Biopharm 97:265–272

Kooi ME, Cappendijk VC, Cleutjens KB et al (2003) Accumulation of ultrasmall superparamagnetic particles of iron oxide in human atherosclerotic plaques can be detected by in vivo magnetic resonance imaging. Circulation 107:2453–2458

Krauel K, Pitaksuteepong T, Davies NM et al (2004) Entrapment of bioactive molecules in poly (alkyl cyanoacrylate) nanoparticles. Am J Drug Deliv 2:251–259

Kreibig U, Vollmer M (1995) Optical properties of metal clusters, vol 25. Springer Science & Business Media, Berlin

Kundu S, Panigrahi S, Praharaj S et al (2007) Anisotropic growth of gold clusters to gold nanocubes under UV irradiation. Nanotechnology 18:75712

Lam CW, James JT, Mccluskey R et al (2004) Pulmonary toxicity of single-wall carbon nanotubes in mice 7 and 90 days after intratracheal instillation. Toxicol Sci 77:126–134

Lasagna-Reeves C, Gonzalez-Romero D, Barria M et al (2010) Bioaccumulation and toxicity of gold nanoparticles after repeated administration in mice. Biochem Biophys Res Commun 393:649–655

Lazzari S, Moscatelli D, Codari F et al (2012) Colloidal stability of polymeric nanoparticles in biological fluids. J Nanopart Res 14:920

Lee SJ, Jeong JR, Shin SC et al (2004) Synthesis and characterization of superparamagnetic maghemite nanoparticles prepared by coprecipitation technique. J Magn Magn Mater 282:147–150

Lei R, Yang B, Wu C et al (2015) Mitochondrial dysfunction and oxidative damage in the liver and kidney of rats following exposure to copper nanoparticles for five consecutive days. Toxicol Res 4:351–364

Li JG, Li WX, Xu JY et al (2007) Comparative study of pathological lesions induced by multiwalled carbon nanotubes in lungs of mice by intratracheal instillation and inhalation. Environ Toxicol 22:415–421

Li C, Liu H, Sun Y et al (2009) Pamam nanoparticles promote acute lung injury by inducing autophagic cell death through the AKT-TSC2-MTOR signaling pathway. J Mol Cell Biol 1:37–45

Li L, Liu T, Fu C et al (2015) Biodistribution, excretion, and toxicity of mesoporous silica nanoparticles after oral administration depend on their shape. Nanomedicine 11:1915–1924

Li Y, Yan J, Ding W et al (2016) Genotoxicity and gene expression analyses of liver and lung tissues of mice treated with titanium dioxide nanoparticles. Mutagenesis 32:33–46

Lin Z, Monteiro-Riviere NA, Riviere JE (2015) Pharmacokinetics of metallic nanoparticles. Wiley Interdiscip Rev Nanomed Nanobiotechnol 7:189–217

Ling T, Qingshan W, Alexander W et al (2009) Gold nanorods as contrast agents for biological imaging: optical properties, surface conjugation and photothermal effects. Photochem Photobiol 85:21–32

Link S, El-Sayed MA (2003) Optical properties and ultrafast dynamics of metallic nanocrystals. Annu Rev Phys Chem 54:331–366

Link S, Mohamed MB, El-Sayed MA (1999) Simulation of the optical absorption spectra of a gold nanorods as a function of their aspect ratio and the effect of the medium dielectric constant. J Phys Chem B 103:3073–3077

Lopez-Chaves C, Soto-Alvaredo J, Montes-Bayon M et al (2018) Gold nanoparticles: distribution, bioaccumulation and toxicity. In vitro and in vivo studies. Nanomedicine 14:1–12

Lorusso D, Di Stefano A, Carone V et al (2007) Pegylated liposomal doxorubicin-related palmar-plantar erythrodysesthesia ('hand-foot' syndrome). Ann Oncol 18:1159–1164

Lotem M, Hubert A, Lyass O et al (2000) Skin toxic effects of polyethylene glycol–coated liposomal doxorubicin. Arch Dermatol 136:1475–1480

Madaan K, Kumar S, Poonia N et al (2014) Dendrimers in drug delivery and targeting: drug-dendrimer interactions and toxicity issues. J Pharm Bioallied Sci 6:139–150

Maisel K, Ensign L, Reddy M et al (2015) Effect of surface chemistry on nanoparticle interaction with gastrointestinal mucus and distribution in the gastrointestinal tract following oral and rectal administration in the mouse. J Control Release 197:48–57

Makadia HK, Siegel SJ (2011) Poly lactic-co-glycolic acid (PLGA) as biodegradable controlled drug delivery carrier. Polymers (Basel) 3:1377–1397

Manke A, Wang L, Rojanasakul Y (2013) Mechanisms of nanoparticle-induced oxidative stress and toxicity. Biomed Res Int 2013:942916

Markowicz-Piasecka M, Luczak E, Chalubinski M, Broncel M et al (2014) Studies towards biocompatibility of PAMAM dendrimers—overall hemostasis potential and integrity of the human aortic endothelial barrier. Int J Pharm 473:158–169

Martinez-Mera I, Espinosa ME, Perez-Hernandez R et al (2007) Synthesis of magnetite (Fe3O4) nanoparticles without surfactants at room temperature. Mater Lett 61:4447–4451

Mazooz G, Mehlman T, Lai TS et al (2005) Development of magnetic resonance imaging contrast material for in vivo mapping of tissue transglutaminase activity. Cancer Res 65:1369–1375

Meena R, Kumar S, Paulraj R (2015) Titanium oxide (TiO 2) nanoparticles in induction of apoptosis and inflammatory response in brain. J Nanopart Res 17:49

Mehnert W, Mader K (2001) Solid lipid nanoparticles: production, characterization and applications. Adv Drug Deliv Rev 47:165–196

Menon JU, Ravikumar P, Pise A et al (2014) Polymeric nanoparticles for pulmonary protein and DNA delivery. Acta Biomater 10:2643–2652
Mercer RR, Scabilloni JF, Hubbs AF et al (2013) Extrapulmonary transport of MWCNT following inhalation exposure. Part Fibre Toxicol 10:1–14
Mirza AZ, Siddiqui FA (2014) Nanomedicine and drug delivery: a mini review. Int Nano Lett 4:94
Mishra AR, Zheng J, Tang X et al (2016) Silver nanoparticle-induced autophagic-lysosomal disruption and NLRP3-inflammasome activation in HepG2 cells is size-dependent. Toxicol Sci 150:473–487
Missaoui WN, Arnold RD, Cummings BS (2018) Toxicological status of nanoparticles: what we know and what we don't know. Chem Biol Interact 295:1–12
Mitchell LA, Gao J, Wal RV et al (2007) Pulmonary and systemic immune response to inhaled multiwalled carbon nanotubes. Toxicol Sci 100:203–214
Mody VV, Nounou MI, Bikram M (2009) Novel nanomedicine-based MRI contrast agents for gynecological malignancies. Adv Drug Deliv Rev 61:795–807
Mody V, Siwale R, Singh A et al (2010) Introduction to metallic nanoparticles. J Pharm Bioallied Sci 2:282–289
Moore A, Marecos E, Bogdanov A et al (2000) Tumoral distribution of long-circulating dextran-coated iron oxide nanoparticles in a rodent model. Radiology 214:568–574
Morales MP, Bomati-Miguel O, De Alejo RP et al (2003) Contrast agents for MRI based on iron oxide nanoparticles prepared by laser pyrolysis. J Magn Magn Mater 266:102–109
Morisson SA, Cahill CL, Carpenter E et al (2005) Atomic engineering of mixed ferrite and core–shell nanoparticles. J Nanosci Nanotechnol 5:1323–1344
Muller RH, Mader K, Gohla S (2000) Solid lipid nanoparticles (SLN) for controlled drug delivery—a review of the state of the art. Eur J Pharm Biopharm 50:161–177
Muller J, Huaux F, Moreau N et al (2005) Respiratory toxicity of multi-wall carbon nanotubes. Toxicol Appl Pharmacol 207:221–231
Muller C, Perera G, Konig V et al (2014) Development and in vivo evaluation of papain-functionalized nanoparticles. Eur J Pharm Biopharm 87:125–131
Murphy CJ, Sau TK, Gole AM et al (2005) Anisotropic metal nanoparticles: synthesis, assembly, and optical applications. J Phys Chem B 109:13857–13870
Murphy CJ, Gole AM, Stone JW et al (2008) Gold nanoparticles in biology: beyond toxicity to cellular imaging. Acc Chem Res 41:1721–1730
Nassimi M, Schleh C, Lauenstein H et al (2010) A toxicological evaluation of inhaled solid lipid nanoparticles used as a potential drug delivery system for the lung. Eur J Pharm Biopharm 75:107–116
Neeves AE, Birnboim MH (1989) Composite structures for the enhancement of nonlinear-optical susceptibility. J Opt Soc Am B 6:787–796
Nishimori H, Kondoh M, Isoda K et al (2009) Histological analysis of 70-nm silica particles-induced chronic toxicity in mice. Eur J Pharm Biopharm 72:626–629
Oldenburg SJ, Averitt RD, Westcott SL et al (1998) Nanoengineering of optical resonances. Chem Phys Lett 288:243–247
Oussoren C, Storm G (2001) Liposomes to target the lymphatics by subcutaneous administration. Adv Drug Deliv Rev 50:143–156
Pacurari M, Yin XJ, Zhao J et al (2008) Raw single-wall carbon nanotubes induce oxidative stress and activate MAPKS, AP-1, NF-κb, and AKT in normal and malignant human mesothelial cells. Environ Health Perspect 116:1211–1217
Papageorgiou I, Brown C, Schins R et al (2007) The effect of nano-and micron-sized particles of cobalt–chromium alloy on human fibroblasts in vitro. Biomaterials 28:2946–2958
Park EJ, Park K (2009) Oxidative stress and pro-inflammatory responses induced by silica nanoparticles in vivo and in vitro. Toxicol Lett 184:18–25
Park J, Bauer S, Von Der Mark K et al (2007) Nanosize and vitality: TiO2 nanotube diameter directs cell fate. Nano Lett 7:1686–1691

Patel M, Shah T, Amin A (2007) Therapeutic opportunities in colon-specific drug-delivery systems. Crit Rev Ther Drug Carrier Syst 24:147–202

Pati R, Shevtsov M, Sonawane A et al (2018) Nanoparticle vaccines against infectious diseases. Front Immunol 9:2224

Patra JK, Das G, Fraceto LF et al (2018) Nano based drug delivery systems: recent developments and future prospects. J Nanobiotechnol 16:71

Peng XH, Qian X, Mao H et al (2008) Targeted magnetic iron oxide nanoparticles for tumor imaging and therapy. Int J Nanomedicine 3:311–321

Piao MJ, Kang KA, Lee IK et al (2011) Silver nanoparticles induce oxidative cell damage in human liver cells through inhibition of reduced glutathione and induction of mitochondria-involved apoptosis. Toxicol Lett 201:92–100

Pople PV, Singh KK (2013) Development and evaluation of colloidal modified nanolipid carrier: application to topical delivery of tacrolimus, part II—in vivo assessment, drug targeting, efficacy, and safety in treatment for atopic dermatitis. Eur J Pharm Biopharm 84:72–83

Prabhu BM, Ali SF, Murdock RC et al (2010) Copper nanoparticles exert size and concentration dependent toxicity on somatosensory neurons of rat. Nanotoxicology 4:150–160

Praetorius NP, Mandal TK (2007) Engineered nanoparticles in cancer therapy. Recent Pat Drug Deliv Formul 1:37–51

Prausnitz MR, Langer R (2008) Transdermal drug delivery. Nat Biotechnol 26:1261–1268

Pridgen EM, Alexis F, Kuo TT et al (2013) Transepithelial transport of Fc-targeted nanoparticles by the neonatal Fc receptor for oral delivery. Sci Transl Med 5:213ra167

Prodan E, Lee A, Nordlander P (2002) The effect of a dielectric core and embedding medium on the polarizability of metallic nanoshells. Chem Phys Lett 360:325–332

Qin Y, Li S, Zhao G et al (2017) Long-term intravenous administration of carboxylated single-walled carbon nanotubes induces persistent accumulation in the lungs and pulmonary fibrosis via the nuclear factor-kappa b pathway. Int J Nanomedicine 12:263

Qiu J, Yang R, Li M et al (2005) Preparation and characterization of porous ultrafine Fe2O3 particles. Mater Res Bull 40:1968–1975

Rama Narsimha Reddy A, Krishna DR, Narsimha Reddy Y et al (2010) Translocation and extra pulmonary toxicities of multi-wall carbon nanotubes in rats. Toxicol Mech Methods 20:267–272

Rawal T, Patel S, Butani S (2018) Chitosan nanoparticles as a promising approach for pulmonary delivery of bedaquiline. Eur J Pharm Sci 124:273–287

Reddy HL, Sharma RK, Chuttani K et al (2005) Influence of administration route on tumor uptake and biodistribution of etoposide loaded solid lipid nanoparticles in Dalton's lymphoma tumor bearing mice. J Control Release 105:185–198

Rizvi SA, Saleh AM (2018) Applications of nanoparticle systems in drug delivery technology. Saudi Pharm J 26:64–70

Ryman-Rasmussen JP, Cesta MF, Brody AR et al (2009) Inhaled carbon nanotubes reach the subpleural tissue in mice. Nat Nanotechnol 4:747–751

Sang X, Zheng L, Sun Q et al (2012) The chronic spleen injury of mice following long-term exposure to titanium dioxide nanoparticles. J Biomed Mater Res A 100:894–902

Schleh C, Semmler-Behnke M, Lipka J et al (2012) Size and surface charge of gold nanoparticles determine absorption across intestinal barriers and accumulation in secondary target organs after oral administration. Nanotoxicology 6:36–46

Sercombe L, Veerati T, Moheimani F et al (2015) Advances and challenges of liposome assisted drug delivery. Front Pharmacol 6:286

Shao XR, Wei XQ, Song X et al (2015) Independent effect of polymeric nanoparticle zeta potential/surface charge, on their cytotoxicity and affinity to cells. Cell Prolif 48:465–474

Sharma V, Singh P, Pandey AK et al (2012) Induction of oxidative stress, DNA damage and apoptosis in mouse liver after sub-acute oral exposure to zinc oxide nanoparticles. Mutat Res Genet Toxicol Environ Mutagen 745:84–91

Shukla RK, Sharma V, Pandey AK et al (2011) Ros-mediated genotoxicity induced by titanium dioxide nanoparticles in human epidermal cells. Toxicol In Vitro 25:231–241
Siddiqi NJ, Abdelhalim MK, El-Ansary AK et al (2012) Identification of potential biomarkers of gold nanoparticle toxicity in rat brains. J Neuroinflammation 9:123
Sinha VR, Kumria R (2003) Microbially triggered drug delivery to the colon. Eur J Pharm Sci 18:3–18
Skrabalak SE, Au L, Li X et al (2007) Facile synthesis of Ag nanocubes and Au nanocages. Nat Protoc 2:2182–2190
Skrabalak SE, Chen J, Sun Y et al (2008) Gold nanocages: synthesis, properties, and applications. Acc Chem Res 41:1587–1595
Slingerland M, Guchelaar HJ, Gelderblom H (2012) Liposomal drug formulations in cancer therapy: 15 years along the road. Drug Discov Today 17:160–166
Smith M, Garcia-Martinez E, Pitter MR et al (2018) Trial watch: Toll-like receptor agonists in cancer immunotherapy. Oncoimmunology 7:e1526250
Sonavane G, Tomoda K, Makino K (2008) Biodistribution of colloidal gold nanoparticles after intravenous administration: effect of particle size. Colloids Surf B Biointerfaces 66:274–280
Stadnichenko OV, Krasnopolsky YM, Yarnykh TG (2016) The study of the lipid membrane charge effect when creating liposomes with oxaliplatin. Bull Pharm 4:34–37
Stepanov AL, Popok VN, Hole DE (2002) Formation of metallic nanoparticles in silicate glass through ion implantation. Glass Phys Chem 28:90–95
Stone NR, Bicanic T, Salim R et al (2016) Liposomal amphotericin B (Ambisome®): a review of the pharmacokinetics, pharmacodynamics, clinical experience and future directions. Drugs 76:485–500
Sun S, Zeng H, Robinson DB et al (2004) Monodisperse MFe_2O_4 (M= Fe, Co, Mn) nanoparticles. J Am Chem Soc 126:273–279
Sung JC, Pulliam BL, Edwards DA (2007) Nanoparticles for drug delivery to the lungs. Trends Biotechnol 25:563–570
Thakral S, Thakral NK, Majumdar DK (2013) Eudragit: a technology evaluation. Expert Opin Drug Deliv 10:131–149
Tiwari DK, Jin T, Behari J (2011) Dose-dependent in-vivo toxicity assessment of silver nanoparticle in Wistar rats. Toxicol Mech Methods 21:13–24
Tran S, Degiovanni PJ, Piel B et al (2017) Cancer nanomedicine: a review of recent success in drug delivery. Clin Transl Med 6:44
Turkevich J, Stevenson PC, Hillier J (1951) A study of the nucleation and growth processes in the synthesis of colloidal gold. Discuss Faraday Soc 11:55–75
Vandebriel RJ, Tonk EC, De La Fonteyne-Blankestijn LJ et al (2014) Immunotoxicity of silver nanoparticles in an intravenous 28-day repeated-dose toxicity study in rats. Part Fibre Toxicol 11:21
Varshosaz J, Ghaffari S, Mirshojaei SF et al (2013) Biodistribution of amikacin solid lipid nanoparticles after pulmonary delivery. Biomed Res Int 2013:136859
Wan J, Chen X, Wang Z et al (2005) A soft-template-assisted hydrothermal approach to single-crystal Fe3O4 nanorods. J Cryst Growth 276:571–576
Wang XQ, Zhang Q (2012) Ph-sensitive polymeric nanoparticles to improve oral bioavailability of peptide/protein drugs and poorly water-soluble drugs. Eur J Pharm Biopharm 82:219–229
Wang J, Thomas M, Lin P et al (2019) siRNA delivery using dithiocarbamate-anchored oligonucleotides on gold nanorods. Bioconjug Chem 30:443–453
Wei W, Xu C, Wu H (2006) Magnetic iron oxide nanoparticles mediated gene therapy for breast cancer—an in vitro study. J Huazhong Univ Sci Technol Med Sci 26:728–730
Wen KP, Chen YC, Chuang CH et al (2015) Accumulation and toxicity of intravenously-injected functionalized graphene oxide in mice. J Appl Toxicol 35:1211–1218
Wozniak A, Malankowska A, Nowaczyk G et al (2017) Size and shape-dependent cytotoxicity profile of gold nanoparticles for biomedical applications. J Mater Sci Mater Med 28:92

Wright L, Rao S, Thomas N et al (2018) Ramizol® encapsulation into extended release PLGA micro- and nanoparticle systems for subcutaneous and intramuscular administration: in vitro and in vivo evaluation. Drug Dev Ind Pharm 44:1451–1457

Xie G, Sun J, Zhong G et al (2010) Biodistribution and toxicity of intravenously administered silica nanoparticles in mice. Arch Toxicol 84:183–190

Xie B, Liu Y, Guo Y et al (2018) Progesterone PLGA/mPEG-PLGA hybrid nanoparticle sustained-release system by intramuscular injection. Pharm Res 35:62

Xu P, Yin Q, Shen J et al (2013) Synergistic inhibition of breast cancer metastasis by silibinin-loaded lipid nanoparticles containing TPGS. Int J Pharm 454:21–30

Yang W, Peters JI, Williams RO III (2008) Inhaled nanoparticles—a current review. Int J Pharm 356:239–247

Yigit MV, Mazumdar D, Lu Y (2008) MRI detection of thrombin with aptamer functionalized superparamagnetic iron oxide nanoparticles. Bioconjug Chem 19:412–417

Yimin Q (2005) Silver-containing alginate fibres and dressings. Int Wound J 2:172–176

Zhang Y, Chen W, Zhang J et al (2007) In vitro and in vivo toxicity of CdTe nanoparticles. J Nanosci Nanotechnol 7:497–503

Zhang B, Lung PS, Zhao S et al (2017) Shape dependent cytotoxicity of PLGA-PEG nanoparticles on human cells. Sci Rep 7:1–8

Zhu MT, Feng WY, Wang B et al (2008) Comparative study of pulmonary responses to nano-and submicron-sized ferric oxide in rats. Toxicology 247:102–111

Nanostructured Lipid Carriers (NLCs): A Modern Versatile Drug Delivery Vehicle

4

Nirmal Shah, Dipti Gohil, and Snehal Patel

Abstract

A suitable drug delivery is an essential element in achieving efficient therapeutic response of drug molecule. Conventional dosage forms have been utilized since several decades but are experiencing certain debilitating issues. Developments of some novel pharmaceutical innovations are required to acquaint into the drug delivery for providing safety and efficacy of existing medications. Amongst all, nanostructured lipid carriers (NLCs), composed of solid lipid and liquid lipid, have emerged as promising drug delivery vehicles. These lipid carriers, inferable from their high surface area-to-volume proportion, can possibly modify fundamental properties and bioactivity of drug. Essential features, such as sheltering of encapsulated drug with expanded stability, site-specific delivery by means of different targeting approaches with lowest side effects, better entrapment, improved pharmacokinetics and biodistribution, compatibleness with biological system, controlled delivery, low toxicity, and some more traits, make them a choice of ultimate delivery system for the improvement in therapeutic benefit of drug molecules. Such uniqueness of NLCs offers a fortunate intent to achieve maximum bioavailability with safe and successful drug delivery. This chapter boons an extensive overview of NLCs with an emphasis on their types, manufacturing technologies, advanced research studies, marketed products, and patents.

N. Shah (✉) · D. Gohil · S. Patel
Department of Pharmacy, Sumandeep Vidyapeeth Deemed to be University, Piparia, Vadodara, Gujarat, India
e-mail: nirmalshah.dop@sumandeepvidyapeethdu.edu.in; diptigohil.dop@sumandeepvidyapeethdu.edu.in

N. Shah (ed.), *Nanocarriers: Drug Delivery System*,
https://doi.org/10.1007/978-981-33-4497-6_4

Keywords

Nano-lipid carriers · Enhanced stability · Potential drug delivery · Site-specific transport · Bioavailability

4.1 Introduction

Lipid nanocarriers are fabricated from natural and biological ancestries' constituents that make them distinguishable to other polymeric nanoparticles. SLNs are the dominating lipid-based nanocarriers which are manufactured with pure lipids that may lead to crystallization in a perfect lattice.

In general, active pharmaceutical ingredient is assimilated amongst the chains of fatty acid or lipid layers in crystal inadequacies inside the SLN environment. Drug may have very less space for the entrapment due to the perfect crystalline matrix created by crystallization of pure lipid utilized for the preparation of SLNs. Instantaneously after the fabrication of SLNs, such pure lipids may undergo crystallization process in elevated energetic lipid modifications, α and β′. Conversely, formation of less energetic modifications, βi and β, may take place in ageing period as a part of time-dependent rearrangement process of the lipid molecules. Such perfect crystalline arrangement may lead to discharge of medicament (Westesen et al. 1997). This may result in the availability of lesser space for the accommodation of drug. Therefore, even though lipid is an effective nanodrug carrier, low drug loading and drug expulsion in storage make scientists to go for other efficient drug carriers. Hence, nanostructured lipid carriers (NLCs) have been evolved that significantly modifies the therapeutic constraints (Carmona-Ribeiro 2010).

In the incident of NLCs, identical dissimilar lipid molecules are mixed to form a lipid unit matrix as imperfect as conceivable. Usually, solid lipid (long chain) and liquid (oil) lipid (short chain) are mixed in a ratio of 70:30 up to a ratio of 99.9:0.1 to produce NLCs. Resultant matrix of lipid particles shows a melting depression compared to original solid lipid, even though they are still solid at room and body temperatures (Chen et al. 2010). Owing to numerous imperfections in NLCs, the enrichment in drug entrapment and reduction in drug expulsion in ageing period may be observed.

Henceforth, NLCs shall be defined as an upcoming generation of SLNs, with solid and liquid lipid blend that forms a less ordered or imperfect arrangement (Radtke et al. 2005; Muller et al. 2002a, b; Das and Chaudhury 2011). This promotes significant improvement in loading of drug and minimizing drug expulsion from the matrix during storage (Shah et al. 2016, 2017; Radtke et al. 2005; Muller et al. 2002a, b). Recently, along with therapeutic applications, NLCs have also been explored for dermatological preparations in cosmetics.

4.1.1 Significance of NLCs (Radtke et al. 2005; Muller et al. 2002a, b)

1. Significantly better drug entrapment compared to other nanocarriers
2. Higher stability
3. Feasibility of encapsulating both types of drugs, lipophilic and hydrophilic
4. Excellent biocompatibility due to the biodegradable property of lipids
5. Easiness in large-scale production and sterilization
6. More affordable than polymeric or surfactant-based nanocarriers
7. Feasible validation and regulatory approval

4.1.2 Restraints of NLCs

1. Sometimes possibility of drug degradation due to auto-oxidation of lipid
2. Probability of particle growth during ageing
3. Lipid interaction with drug, during and after NLC preparation
4. Unexpected dynamic of polymeric transition

4.2 Categories of NLCs

The categories of NLCs have been influenced by various experimental and processing parameters such as physicochemical properties of drug, solid/liquid lipids and emulsifiers, solubility of drug in lipids, concentration of emulsifier, and temperature of the process. Basically NLCs are categorized into three types as shown in Fig. 4.1 (Qianwen et al. 2017; Sharma and Baldi 2018; Ganesan and Narayanasamy 2017):

1. Imperfect/disordered type NLCs
2. Amorphous type NLCs
3. Multiple type NLCs

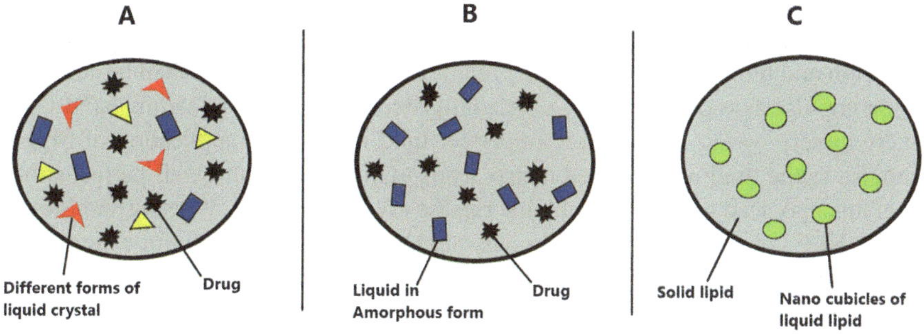

Fig. 4.1 Categories of NLCs: (**a**) Imperfect/disordered, (**b**) amorphous, and (**c**) multiple (Qianwen et al. 2017)

The nature of matrix formed in type 1 NLCs is imperfectly organized solid matrix. This system is made up of an environment with numerous voids that can accommodate the larger amount of drug. This can be formed by blending spatially altered lipids that may turn to disorder or imperfections in the crystal matrix. The environment of NLCs is unable to produce a significantly ordered arrangement due to different fatty acid chain length and the blend of mono-, di-, and triacylglycerols. This may result in the formation of imperfect arrangement inside the matrix which can accommodate larger amount of drug (Sharma and Baldi 2018; Ganesan and Narayanasamy 2017).

The nature of matrix formed in type 2 NLCs is unorganized solid amorphous matrix which is also known as noncrystalline phase. Such type of NLCs are prepared by blending of well-ordered mixture of special types of solid and liquid lipids (IPM, dibutyl adipate, hydroxy stearate) that do not recrystallize (Gaba et al. 2015) during cooldown process. Such materials generate an amorphous environment, hence diminishing drug expulsion during shelf life.

Development of multiple type 3 NLCs is based on the fact that solubility of drug is higher in liquid lipids than solid lipids. Thus, a larger amount of liquid lipid is mixed with solid lipid. Oil globules in fewer amounts are effortlessly dispersed into the lipid medium. More amount of oil may lead to partitioning of phases producing small oil nano-compartments which are coated with solid lipid. Such type of system may allow controlled release with hindering in drug discharge (Ganesan and Narayanasamy 2017).

4.3 Modulation of Drug Release from NLCs

Lipid nanoparticles follow a basic principle for releasing drug molecules. Drug's partition coefficient has a significant impact on the release of drug. Usually, drug release is inversely proportional to its partition coefficient (Nandvikar et al. 2019). The size reduction of drug to nano-range has a major influence on drug release owing to its larger surface area. Once the drug is uniformly dispersed in the lipid medium at that time deliberate amount of drug release shall be accomplished. Drug release from lipid nano-units ensues by diffusion and concurrently by lipid particle degradation in the human biological system (Nandvikar et al. 2019; Wissing et al. 2001). In few instances, there might be a need for a controlled faster drug release going further than diffusion and biological degradation. Preferably, such drug release shall be activated by an impulse as soon as the NLCs are administered. Larger drug loading in NLCs is likely merely owing to its highly imperfect lipid matrix. An application of trigger impulse to the lipid unstructured matrix leads to conversion into well-perfect structure; this may lead to initiation of burst release (Wei et al. 2017; Wissing and Muller 2003). NLCs of definite type may be activated by such an approach: for example, an application of nanoparticle-loaded cream to the skin (Dubey et al. 2012; Muller et al. 2002a, b). Two factors, namely rise in temperature and water evaporation, may tend to improve drug release rate as shown in Fig. 4.2.

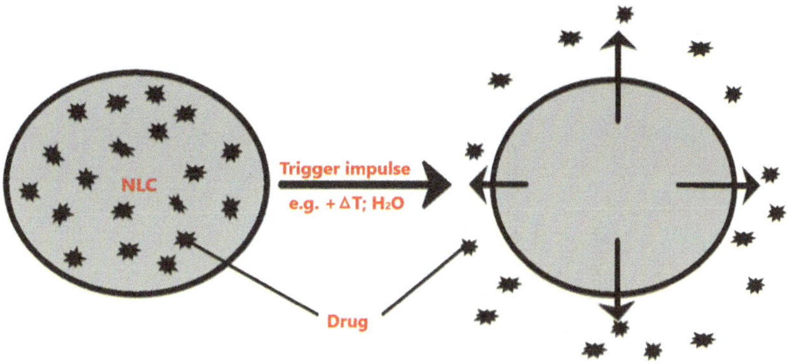

Fig. 4.2 Modulation of drug release from NLCs (Selvamuthukumar and Velmurugan 2012)

4.4 Composition of NLCs

The therapeutic effectiveness and stability of NLCs are reliable on the lipids used for fabrication. The lipid has a significant role in drug entrapment and sustained/ controlled release of nanocarriers (Shah et al. 2015). Almost all lipids are permitted as generally recognized as safe (GRAS) with some exception and biodegradable in physiological environment. Various lipids and emulgents used in preparation of NLCs are shown in Table 4.1 (Sharma and Baldi 2018; Natarajan et al. 2017).

4.5 Methods of Preparation for NLCs

Numerous manufacturing methodologies are available for NLCs. These methodologies have been executed from polymeric nanoparticle manufacturing procedures. The different techniques utilized for the small-scale and large production of NLCs are as given below (Asif Iqbal et al. 2012; Qianwen et al. 2017; Sharma and Baldi 2018):

4.5.1 High-Energy Approaches

4.5.1.1 High-Pressure Homogenization Method

The most frequently utilized large-scale manufacturing technique is the high-pressure homogenization method (HPH) that employs high-temperature as well as high-pressure process parameters for making nanosize emulsion of entrapped drug. Commercially, jet stream and piston-gap homogenizers are accessible for the production of NLCs. The two different approaches mainly used to formulate NLCs by HPH are described underneath (Kadam et al. 2014; Purohit et al. 2016; Radtke et al. 2005):

Table 4.1 List of lipids and emulgents used for NLCs

Type of lipid		Type of emulgent		
Solid	Liquid	Hydrophilic	Lipophilic	Amphiphilic
• *Fatty acids* Stearic acid Palmitic acid • Monoglycerides Glyceryl monostearate (Geleol™ Mono and Diglycerides NF) Glyceryl behenate (Compritol® 888 ATO) • *Diglycerides* Glyceryl palmitostearate (Precirol® ATO 5) • Triglycerides *Glycerol tricaprate (Tricaprin)* *Glyceryl trimyristate (Trimyristin)* *Glyceryl tristearate (Tristearin)* • *Waxes* Cetyl palmitate	• Natural and synthetic oils Mustard oil Castor oil Cod liver oil Oleic acid Soya bean oil Isopropyl myristate (IPM) Linoleic acid Decanoic acid Paraffin oil Propylene glycol dicaprylocaprate (Labrafac) Medium-chain triglycerides (MCT)/caprylic and capric triglycerides	Tween 20 Tween 40 Tween 80 Polyvinyl alcohol (PVA) Poloxamer 188 (Pluronic® F68) Poloxamer 407 (Pluronic® F127) Sodium deoxycholate Sodium glycocholate Sodium oleate	Span 20 Span 40 Span 60 Monoglyceride (Myverol® 18-04K)	Egg lecithin Soya lecithin Phosphatidylcholines Phosphatidylethanolamines Stearoyl polyoxylglycerides NF (Gelucire® 50/13)

(a) Hot homogenization
The hot homogenization procedure is performed constantly at a temperature above the melting point of lipids used in the formulation. In this method, lipid and aqueous phases are heated separately. The melted lipid phase is dispersed uniformly into the hot aqueous phase containing emulgent by continuous stirring to form an emulsion followed by homogenization in HPH at similar temperature to achieve hot nanoemulsion consisting of particles in nanosizes. Probe or bath sonication can also be performed to achieve particles in nano-range. Solidification can be performed by cooling down the dispersion at room temperature.

(b) Cold homogenization
The cold homogenization method involves mixing of drug into molten lipid mixture at 5–10 °C above the temperature of lipid melting point. This melt is solidified by liquid nitrogen or dry ice and grounded to achieve lipid micron-sized particles followed by addition of this micropowder into hydrophilic emulgent dispersion medium to obtain pre-suspension. This dispersion is subjected to high-pressure homogenization at same or below room temperature to obtain nanosized particles.

4.5.2 Low-Energy Approaches

4.5.2.1 Microemulsion Technique

In this method, the lipid phase (comprises drug, lipid, and lipophilic emulgent) and the aqueous phase (comprises hydrophilic emulgent) are heated separately. The aqueous phase is subjected for addition into the lipid phase at same temperature under constant stirring. Subsequently, a transparent microemulsion is formed when the mixing of all components is done in precise ratio. This clear microemulsion is dispersed into the cold aqueous medium with stirring to precipitate fine nanosized particles (Natarajan et al. 2017).

4.5.2.2 Double-Emulsion Approach

The hydrophilic drug molecules and peptides can be successfully encapsulated into the nanocarriers by using double-emulsion approach. The first step is formation of stable primary W/O emulsion by blending aqueous solution of medicament and molten lipid, using an oleaginous stabilizer. This primary mixture is emulsified with a solution of aqueous emulgent by continuous stirring to make double-W/O/W emulsion. Nanosized particles are precipitated by cooling the multiple emulsion. The drawback of this technique is production of relatively larger sized particles (Ganesan and Narayanasamy 2017).

4.5.2.3 Membrane Contactor Technique

The membrane contactor technique works by using a cylindrical membrane unit and it is scalable as per the requirement. This membrane is constructed with internal channels and pores. The size of particles can be manipulated by using membranes with variable size pores. An aqueous solution comprises a surfactant that is disseminated in the internal channel where lipid melt is forced through membrane pores that results in the formation of minute droplets. These droplets are collected at the other end of the membrane. This dispersion is allowed to cool at room temperature for the precipitation of NLCs (Charcosset et al. 2005; Ganesan and Narayanasamy 2017).

4.5.2.4 Film Ultrasonication Method

The mixture consisting of lipid melt, drug, and polymer is dispersed uniformly into the surfactant aqueous solution followed by ultrasonication using probe sonicator. The nanosized particles are collected after solidification and filtration (Salvi and Pawar 2019).

4.5.3 Approaches with Organic Solvents

4.5.3.1 Solvent Diffusion Method

This method for the preparation of NLCs is quite simple and easy to operate and needs basic requirements of instruments. The solid and liquid lipids, surfactant, and drug were dissolved at higher temperature in organic phase. The hot organic phase is quickly injected by needle into the aqueous phase consisting of dissolved surfactant/stabilizer with continuous stirring through mechanical stirrer for a specific period of time to evaporate organic solvent. The resultant NLC dispersion shall be placed in vacuum desiccators for a period of 24 h for the complete removal of residual organic solvent (Natarajan et al. 2017; Salvi and Pawar 2019). This method of preparation of NLCs can improve entrapment efficiency of medicament when the saturated drug solution is employed in dispersed phase. This technique is not feasible for large-scale production and also suffers from the drawback of inability of complete removal of residual organic solvent (Qianwen et al. 2017).

4.5.3.2 Emulsification Solvent Evaporation Technique

This technique has significant benefits over other approaches as it produces high drug-loaded nano-lipid particles with minimal polydispersity index. The large-scale production of lipid particles is possible by assembling this process fully automated (Jaiswal et al. 2004).

The organic phase comprises soluble lipid substances uniformly dispersed in a definite volume of an aqueous solution by means of a high-speed homogenizer to produce coarse-sized pre-emulsion. This mixture is instantaneously passed through a high-pressure homogenizer to make nanosized emulsion followed by stirring overnight for complete removal of organic solvent. Subsequently, precipitation takes place and solidified drug-loaded nano-lipid particles are collected by filtration (Pedersen et al. 2006; Ganesan and Narayanasamy 2017).

4.6 Approaches to Resolve Stability Concerns

NLCs are one of the colloidal carriers which may suffer from stability issue. For the duration of longer period of stability, clumping of particles may arise owing to particle collision that turns to cohesion followed by flocculation (Khosa et al. 2018). The dispersion with more amounts of NLCs may show pearl-like impact due to network foundation that controls the collision, cohesion, and flocculation between particles. Followed by oral intake, the dispersion may get diluted with gastric fluid and therefore the system is distressed and releases distinct nanosize particles (Selvamuthukumar and Velmurugan 2012). The NLC formulations shall conserve their physicochemical characteristics during stability period, viz. particle size and stability against bacterial attack. Certain strategies discussed underneath may be selected to control the physical instability issue.

4.6.1 Lyophilization Technique

Lyophilization has been utilized for long-duration stability of a preparation having hydrolyzable medicaments (Nandvikar et al. 2019). Transformation into the solid state would prevent Ostwald ripening and prevent hydrolytic reactions. Conversely, if lyophilization is carried out without the cryoprotectant, then the resultant product may become aggregated. Several cryoprotectants are used in pharmaceutical industries such as trehalose, Microcelac®, sucrose, sorbitol, Avicel® RC591, glucose, mannose, and maltose (Zhuang et al. 2010; Ranpise et al. 2014; Shete and Patravale 2013; Varshosaz et al. 2012).

4.6.2 Addition of a Preservative Agent

For various types of fluid and semisolid preparations comprising water as the main component, preservatives play a vital role in maintaining the physical strength of the product. On the other hand sometimes, preservatives may also make dispersion unstable; therefore it is of a prime necessity to understand the impact of the preserving agent before the addition into dispersion. Many agents are used for preserving NLCs such as propylene glycol, Rokonsal™ PB5, Euxyl® PE1090, Euxyl® K700, caprylyl glycol, ethanol, Phenonip™, pentylene glycol, and MultiEx Naturotics™. Certain factors like water repellant property of particles, association of preservative with particle surface, decline in zeta potential, and type of stabilizer govern the stability of NLCs. Model preservatives shall have good hydrophilicity along with nonionic nature so that zeta charge cannot be affected (Obeidat et al. 2010; Khosa et al. 2018).

4.6.3 Addition of a Stabilizing Agent

Stabilizers make significant contribution to maintaining the dispersion stability of NLCs. Pluronic® F-68 (Poloxamer 188) may expressively enhance the mechanical stability and keep the desirable impact on rheological properties. Pluronic® F-127 (Poloxamer 407) with certain organic solvents makes the system thermodynamically stable by converting into two liquid crystal structures (Rajinikanth and Chellian 2016; Swidan et al. 2018).

Usually, stabilizers such as polyethylene glycol (PEG) may be utilized for coating of nanoparticles to modify its surface. This may provide good physical strength and dispersibility and prolong availability of nanoparticles into the systemic circulation (Karmakar et al. 2018).

4.7 Reviews of Earlier Research in the Area

Yung-Chih Kuo and Jui Fang Chung have fabricated nevirapine-encapsulated SLNs and NLCs and these were coated by human serum albumin. The oleic acid was chosen as liquid lipid for the preparation of NLCs. This lipid diminished the thermal resistance of NLCs and thereby enhanced nevirapine release from the optimized NLCs. Human serum albumin-coated nanoconstructs shall be an efficient dosage form for the administration of nevirapine in viral therapy (Kuo and Chung 2011).

In earlier research, it was found that NLCs with particle size lesser than 230 nm were beneficial for the dermal permeation of coenzyme Q10. Schwarza and coworkers have prepared NLCs even with lesser particle size of 80 nm. The purpose of this study was to check the skin permeation and penetration along with physical and chemical stability of NLCs. The optimized formulation of NLCs has shown impactful enhanced dermal delivery of coenzyme Q10. This showed the usefulness of NLC formulation in topical delivery (Schwarz et al. 2013).

Shah and coworkers have fabricated raloxifene drug-incorporated NLCs by solvent diffusion technique employing glyceryl monostearate as solid lipid and Capmul MCM C8 as liquid lipid. A statistical approach has been utilized to validate the impact of two independent variables, namely solid-to-liquid lipid ratio and amount of stabilizer, on drug-loading capacity in formulated NLCs. Solid-state studies showed conversion of drug from crystal nature to amorphous form. Optimized NLCs showed 3.75 times bioavailability improvement compared to plain drug suspension. Such findings prove the efficacy of solid and liquid lipid-composed nanocarriers in the potential delivery of poorly soluble molecules (Shah et al. 2016).

The bioavailability of testosterone undecanoate is found to be only 7% in men. Muchow and coworkers have prepared NLCs encapsulated with this drug to develop nanocrystals of drug. An in vivo experiment was performed for the evaluation of bioavailability of optimized NLC formulation and commercial market. The NLCs showed twofold enhancements in bioavailability than commercial product. This statement keeps NLCs as a better delivery system for the said drug (Muchow et al. 2013).

Date and coworkers have studied the capability of Gelucire 50/13, a stabilizer, for enhancement in oral delivery of poorly soluble drug repaglinide by constructing lipid-based nanosize NLCs using Precirol ATO 5 as a solid lipid and some liquid lipids. The study was conducted between Gelucire 50/13-based NLCs and commercially available repaglinide tablets. Gelucire-based optimized NLCs showed considerable good antidiabetic property compared to commercial products (Date et al. 2011).

Chen and coworkers have carried out a study to evaluate the efficiency of lovastatin-loaded NLCs in improving oral bioavailability using the blend of Precirol and squalene. Over 70% of drug was loaded in nanocavities of NLCs; this was more in comparison with SLNs. Drug-loaded nanosize particles have size in the range of 180–290 nm that was important in defining the release pattern of drug molecules. Myverol-containing NLCs exhibited significant enhancement in drug's pharmacokinetics as far as bioavailability is concerned. It drastically improves bioavailability from 4 to 24% with good storage stability which shows the potential of NLCs in enhancing the efficacy of drugs (Chen et al. 2010).

Das and coworkers have carried out a comparison study between SLNs and NLCs by taking clotrimazole as the model drug. Various experimental parameters including solid-state study, surface morphology, and release profile were compared for SLNs and NLCs. The drug release was found to be more rapid in NLCs compared to that in SLNs. The optimized formulation of NLCs demonstrated better stability than that of SLNs. During storage period, SLN formulation exhibited significant deviation in drug release when compared to initial release pattern, while NLCs did not show any changes. As a result, it can be concluded that NLCs have better superiority than SLNs (Das et al. 2012).

Tran and coworkers have investigated fenofibrate-loaded NLCs for improving bioavailability. They have fabricated nanocarriers using hot homogenization technique and subsequently an ultrasonication approach. NLCs showed drug entrapment up to 99%, which indicated the ability of NLCs to encapsulate more amount of drug than other nanocarriers. The in vivo investigation showed that drug-loaded NLCs revealed four times enhancement in plasma concentration compared to plain drug suspension. These results depicted that NLCs have greater potential in extending drug response duration in the body and therapeutic effect of drug (Tran et al. 2014).

Araujo and coworkers have investigated triamcinolone acetonide-loaded NLCs for intravitreal targeting. NLCs were prepared by high-pressure homogenization approach using Precirol® ATO5 (solid lipid), squalene (liquid lipid), and Lutrol® F68 (surfactant). The optimized formulation showed less than 200 nm particle size, up to 95% of drug entrapment, and discrete spherical particles. Thermal study depicted amorphous matrix of NLCs. No ocular toxicity was found under in vivo Draize test (Araujo et al. 2010).

The intravenous delivery of poorly water-soluble artemether for the treatment of malaria was studied by Joshi and coworkers by constructing NLCs (Nanoject) using microemulsion template approach. The in vivo evaluation for antimalarial activity was carried out in mice carrying *Plasmodium berghei* by doing a comparison between Nanoject and commercial artemether injection. The antimalarial activity was found considerably high in Nanoject formulation than the commercial product, which remained for a longer period of time even after 20 days. This gives the indication that Nanoject may remain for a longer circulating period in vivo. The optimized formulation exhibited expressively good survival rate at 31 days than commercial product (Joshi et al. 2008).

NLCs encapsulated with ascorbyl palmitate have been investigated for the evaluation of its chemical stability by Teeranachaideekul and coworkers. None of the colloidal dispersion has revealed long-period chemical stability of this drug. The study was conducted by considering all experimental parameters that may affect the stability of drug. The study demonstrated great improvements in the chemical stability of the drug after incorporating NLCs (Teeranachaideekul et al. 2007).

Zhang and coworkers have studied etoposide NLC formulation for the evaluation of its oral potential. NLCs were formulated through the emulsification approach followed by solidification at low temperature. The drug loading capacity and particle size of prepared NLCs were noted as about 57.9–89.7% and 125.9–91.2 nm, respectively. The diffusion chamber was employed for the evaluation of absorption to intestine. It was investigated that drug-loaded tiny-size NLCs produced easy passage for drug from mucosal to serosal portion. The optimized formulation given orally to rats demonstrated up to 3.5 times enhancement in bioavailability compared to plain drug suspension. Such remarkable improvement in pharmacokinetic study was observed with NLC formulation making it a suitable and potential carrier for oral delivery (Zhang et al. 2011).

Oxybenzone-loaded NLCs have been prepared by Sanad and coworkers to study sun screening efficacy and safety using solvent diffusion approach. Full 23 factorial design statistical tool was utilized to investigate the significance of experimental variables such as type and concentration of liquid lipid and drug concentration on the particle size, drug entrapment, and drug release. The study showed significant enhancement of in vitro sun protection factor by six times and erythemal UVA protection factor by eight times with lesser potential for irritation (Sanad et al. 2010).

NLCs containing isoliquiritigenin model drug were fabricated by Zhang and team to find out antitumor efficiency and immunomodulation properties in mice model having sarcoma 180 and murine hepatoma by intraperitoneal route. The biodistribution study was also carried out for optimized formulation. The study found sphere-shaped particles with average nanosize of 160.73 nm and entrapment capacity of 96%. Isoliquiritigenin-loaded NLCs considerably hinder the tumor growth and biodistribution study revealed 2.5 times more concentration of drug compared to plain drug suspension. The pharmacokinetic parameters also indicated higher value than the drug suspension. It is concluded from the study that drug-encapsulated NLCs are innovative and promising nanocarriers in showing anticancer effect and immune properties of drug (Zhang et al. 2013).

Chen and coworkers have evaluated tumor-targeting capability of stearyl-2-amino-2-deoxyglucose, a broad tumor-targeting ligand, by fabricating NLCs with paclitaxel anticancer drug using melted ultrasonic technique. The study was conducted in MCF-7 tumor cells as well as in tumor-containing mice. The result depicted that selected tumor-targeting ligand was successfully and dynamically accumulated by the side of tumor site. Anticancer drug-loaded NLCs exhibited moral effectiveness against tumor with less toxicity in MCF-7 tumor-containing mice. The research showed the extensive prospective of NLCs for tumor diagnosis along with possible targeted chemotherapy (Chen et al. 2012).

4.8 Commercially Available Products of NLCs

Many commercial products of NLCs are available exclusively for the cosmetic market as shown in Table 4.2 (Muller et al. 2007).

Table 4.2 List of marketed formulations of NLCs

Brand name	Name of manufacturer
Intensive Serum Nano Repair Q10	Dr. Rimpler GmbH, Germany
Cutanova Cream Nanorepair Q10	
Cutanova Cream Nanovital Q10	
NLC Deep Effect Eye Serum	Beate Johnen, Germany
NLC Deep Effect Repair Cream	
NLC Deep Effect Reconstruction Cream	
NLC Deep Effect Reconstruction Serum	
NanoLipid Q10 CLR	Chemisches Laboratorium Dr. Kurt Richter GmbH, Germany
NanoLipid Restore CLR	
Super Vital Extra Moist Softener	Amorepacific Corporation, South Korea
Super Vital Extra Moist *Emulsion*	
Super Vital *Eye Cream*	
SURMER Creme Legere Nano protection	Isabelle Lancray, Paris
SURMER Creme Riche Nano Restructurante	
SURMER Elixir de Beauté Nano Vitalisant	
SURMER Masque Crème Nano Hydrant	

4.9 Patent Status of NLCs

Various patents for NLCs are disclosed as shown in Table 4.3.

Table 4.3 Patent status of NLCs

Patent/ publication number	Title of patent	Name of inventor/s	Year of publication	Reference
EP 2 229 936 B1	Nanosized testosterone formulations for improved bioavailability	Keck et al.	2010	Keck and Muchow (2009)
WO2008/000448 A3	Nanostructured lipid carriers containing riluzole and pharmaceutical formulations containing said particles	Bondi et al.	2008	Bondi et al. (2008)
US 2013/ 0017239 A1	Lipid nanoparticle capsules	Viladot Petit et al.	2013	Viladot Petit et al. (2013)
CN101366697A	Novel nano-lipid carrier for injection embodying paclitaxel series substances and preparation method thereof	Liu et al.	2009	Liu et al. (2009)
CN101129335A	Use of nanostructured lipid carrier drug feeding system	Jian et al.	2010	Jian et al. (2010)
CN102688152A	Composite anti-screening agent nanostructured lipid carrier and preparation method thereof	Qiang et al.	2012	Qiang and Xueyang (2012)
CN102935077A	Bionic lovastatin nanostructured lipid carrier and preparation method thereof	Jianping et al.	2013	Jianping and Ji (2013)
CN101658468A	Coenzyme Q nanostructured lipid carrier and preparation method thereof	Summer et al.	2013	Summer and Hongxia (2013)
JP2016523884A	Lipid nanoparticles for wound healing	Lafuente et al.	2016	Lafuente et al. (2016)
CN102283809B	Preparation of nanostructured lipid carrier (NLC) method and products made	Ismail et al.	2016	Ismail et al. (2016)
KR101777616B1	Nanostructured lipid carrier comprising α-tocopherol and preparing method thereof	Geun et al.	2017	Geun et al. (2017)
US 10245322 B2	Nanostructured carriers for guided and targeted on-demand substance delivery	Lal et al.	2019	Lal et al. (2019)

4.10 Conclusion

In current circumstances, NLCs have been into consideration for all investigators in lieu of extensive range of applications with improved bioavailability and extraordinary blood level. The matrix of solid lipid and liquid lipid of NLCs is not static in structural arrangement. Therefore, the structure is flexible with significant good entrapment efficiency and certainly with no drug leakage. Thereby, NLCs are reflected as one of the most projecting tiny carriers as they justify necessary requirements of an ideal dosage form carrier. Amongst all nanocarriers, NLCs have revealed distinct potential in the competent delivery of therapeutics via several routes of drug administration, for example oral, parenteral, pulmonary, ocular, topical, and intranasal. NLCs have also made known impact on protein-, peptide-, and gene-based delivery. In addition, the components utilized for manufacturing of NLCs are biologically compatible and degradable, nonirritating, and approved with GRAS eminence. NLCs are informal for large-scale production with possibility of making desired modifications such as prolonged and controlled drug release, particle size, and drug entrapment. Various researches have shown remarkable dispersion stability of NLCs for an extended duration of time with all fundamental physicochemical characteristics in steadied form. Henceforward, it can be said that NLCs, as a novel generation of lipid nanoparticles, shall be utilized as a proficient carrier in improving therapeutic effectiveness of various drugs with wide prospects for advance development.

References

Araujo J, Gonzalez-Mira E, Egea MA et al (2010) Optimization and physicochemical characterization of a triamcinolone acetonide-loaded NLC for ocular antiangiogenic applications. Int J Pharm 393(1–2):168–176

Asif Iqbal MD, Shadab MD, Sahni JS (2012) Nanostructured lipid carriers system: recent advances in drug delivery. J Drug Deliv 20(10):813–830

Bondi ML, Giammona G, Craparo EF et al (2008) Nanostructured lipid carriers containing riluzole and pharmaceutical formulations containing said particles. WO2008000448 A3

Carmona-Ribeiro AM (2010) Biomimetic nanoparticles: preparation, characterization and biomedical applications. Int J Nanomed 5:249–259

Charcosset C, El-Harati A, Fessi H (2005) Preparation of solid lipid nanoparticles using a membrane contactor. J Control Release 108:112–120

Chen CC, Tsai TH, Huang ZR et al (2010) Effects of lipophilic emulsifiers on the oral administration of lovastatin from nanostructured lipid carriers: physicochemical characterization and pharmacokinetics. Eur J Pharm Biopharm 74(3):474–482

Chen J, Chen H, Cui S et al (2012) Glucosamine derivative modified nanostructured lipid carriers for targeted tumor delivery. J Mater Chem 22(12):5770–5783

Das S, Chaudhury A (2011) Recent advances in lipid nanoparticle formulations with solid matrix for oral drug delivery. AAPS Pharm Sci Technol 12(1):62–76

Das S, Ng WK, Tan RB (2012) Are nanostructured lipid carriers (NLCs) better than solid lipid nanoparticles (SLNs): development, characterizations and comparative evaluations of clotrimazole-loaded SLNs and NLCs? Eur J Pharm Sci 47(1):139–151

Date AA, Vador N, Jagtap A et al (2011) Lipid nanocarriers (GeluPearl) containing amphiphilic lipid Gelucire 50/13 as a novel stabilizer: fabrication, characterization and evaluation for oral drug delivery. Nanotechnology 22(27):275102

Dubey A, Prabhu P, Kamath JV (2012) Nano structured lipid carriers: a novel topical drug delivery system. Int J Pharm Tech Res 4(2):705–714

Gaba B, Fazil M, Ali A et al (2015) Nanostructured lipid (NLCs) carriers as a bioavailability enhancement tool for oral administration. Drug Deliv 22:691–700

Ganesan P, Narayanasamy D (2017) Lipid nanoparticles: different preparation techniques, characterization, hurdles, and strategies for the production of solid lipid nanoparticles and nanostructured lipid carriers for oral drug delivery. Sustain Chem Pharm 6:37–56

Geun KB, Jun PD, Heaven JJ (2017) Nano-structured lipid carrier comprising α-tocopherol and preparing method thereof. Korean Patent 101777616B

Ismail R, Qiuhong L, Linan HL et al (2016) Preparation of nanostructured lipid carriers (NLC) method and products made. Chinese Patent 102283809B

Jaiswal J, Gupta SK, Kreuter J (2004) Preparation of biodegradable cyclosporine nanoparticles by high-pressure emulsification-solvent evaporation process. J Control Release 96:169–178

Jian DY, Fuqiang YH, Yuan H (2010) Use of nano structured lipid carrier drug feeding system. Chinese Patent CN101129335A

Jianping L, Ji W (2013) Bionic lovastatin nano-structured lipid carrier and preparation method thereof. Chinese Patent 102935077A

Joshi M, Pathak S, Sharma S (2008) Design and in vivo pharmacodynamic evaluation of nanostructured lipid carriers for parenteral delivery of artemether: Nanoject. Int J Pharm 364:119–126

Kadam VB, Dhanawade KB, Salunkhe VA et al (2014) Nanoparticle—novel drug delivery system. J Curr Pharm Res 4:1318–1335

Karmakar G, Nahak P, Guha P et al (2018) Role of PEG 2000 in the surface modification and physicochemical characteristics of pyrazinamide loaded nanostructured lipid carriers. J Chem Sci 130(4):42–52

Keck C, Muchow M (2009) Nanosized testosterone formulations for improved bioavailability. European Patent 2229936 B1

Khosa A, Reddi S, Saha RN (2018) Nanostructured lipid carriers for site-specific drug delivery. Biomed Pharmacother 103:598–613

Kuo YC, Chung JF (2011) Physicochemical properties of nevirapine-loaded solid lipid nanoparticles and nanostructured lipid carriers. Colloids Surf B Biointerfaces 83(2):299–306

Lafuente EG, Lucea GG, Rincon SV et al (2016) Lipid nanoparticles for wound healing. Japanese Patent 523884A

Lal R, Landon PB, Mo A (2019) Nanostructured carriers for guided and targeted on-demand substance delivery. US Patent 10245322 B2

Liu D, Ning R, Song J et al (2009) Novel nano-lipid carrier for injection embodying paclitaxel series substances and preparation method thereof. Chinese Patent 101366697A

Muchow M, Maincent P, Muller RH et al (2013) Testosterone undecanoate-increase of oral bioavailability by nano structured lipid carriers (NLC). J Pharm Technol Drug Res 2(1):1–10

Muller RH, Radtke M, Wissing SA (2002a) Nanostructured lipid matrices for improved microencapsulation of drugs. Int J Pharm 242(1–2):121–128

Muller RH, Radtke M, Wissing SA (2002b) Solid lipid nanoparticles (SLN) and nanostructured lipid carriers (NLC) in cosmetic and dermatological preparations. Adv Drug Deliv Rev 54 (1):131–155

Muller RH, Petersen RD, Hommoss A et al (2007) Nanostructured lipid carriers (NLC) in cosmetic dermal products. Adv Drug Deliv Rev 59:522–530

Nandvikar NY, Lala RR, Shinde AS (2019) Nanostructured lipid carrier: the advanced lipid carriers. Int J Pharm Sci Res 10(12):5252–5265

Natarajan J, Karri VVSR, De A (2017) Nanostructured lipid carrier (NLC): a promising drug delivery system. Glob J Nano 1(5):120–125

Obeidat WM, Schwabe K, Muller RH et al (2010) Preservation of nanostructured lipid carriers (NLC). Eur J Pharm Biopharm 76(1):56–67
Pedersen N, Hansen S, Heydenreich AV et al (2006) Solid lipid nanoparticles can effectively bind DNA, streptavidin and biotinylated ligands. Eur J Pharm Biopharm 62:155–162
Purohit DK, Nandgude TD, Poddar SS (2016) Nano-lipid carriers for topical application: current scenario. Asian J Pharm 10:1–9
Qiang X, Xueyang D (2012) Composite anti-screening agent nanostructured lipid carrier and preparation method thereof. Chinese Patent 102688152A
Qianwen L, Tiange C, Yinghong H (2017) A review of the structure, preparation, and application of NLCs, PNPs, and PLNs. Nanomaterials 7:122
Radtke M, Souto EB, Muller RH (2005) Nanostructured lipid carriers: a novel generation of solid lipid drug carriers. Pharm Technol Eur 17(4):45–50
Rajinikanth PS, Chellian J (2016) Development and evaluation of nanostructured lipid carrier-based hydrogel for topical delivery of 5-fluorouracil. Int J Nanomedicine 11:5067–5077
Ranpise NS, Korabu SS, Ghodake VN (2014) Second generation lipid nanoparticles (NLC) as an oral drug carrier for delivery of lercanidipine hydrochloride. Colloids Surf B Biointerfaces 116:81–87
Salvi VR, Pawar P (2019) Nanostructured lipid carriers (NLC) system: a novel drug targeting carrier. J Drug Deliv Sci Technol 51:255–267
Sanad RA, Abdel Malak NS, Bayoomy TS et al (2010) Formulation of a novel oxybenzone-loaded nanostructured lipid carriers (NLCs). AAPS Pharm Sci Tech 11:1684–1694
Schwarz JC, Baisaeng N, Hoppel M et al (2013) Ultra-small NLC for improved dermal delivery of coenzyme Q10. Int J Pharm 447(1–2):213–217
Selvamuthukumar S, Velmurugan R (2012) Nanostructured lipid carriers: a potential drug carrier for cancer chemotherapy. Lipids Health Dis 11:159
Shah R, Eldridge D, Palombo E et al (2015) Lipid nanoparticles: production, characterization and stability. Briefs Pharm Sci Drug Dev 1:11–23
Shah NV, Seth AK, Balaraman R et al (2016) Nanostructured lipid carriers for oral bioavailability enhancement of raloxifene: design and in vivo study. J Adv Res 7(3):423–434
Shah N, Gohil D, Seth AK (2017) Nanostructured lipid carriers: as an efficient drug delivery carrier. Glob J Nano 3(3):82–84
Sharma A, Baldi A (2018) Nanostructured lipid carriers: a review. J Dev Drugs 7(1):191
Shete H, Patravale V (2013) Long chain lipid based tamoxifen NLC part I: preformulation studies, formulation development and physicochemical characterization. Int J Pharm 454(1):573–583
Summer S, Hongxia W (2013) Coenzyme Q nanostructured lipid carrier and preparation method thereof. Chinese Patent 101658468A
Swidan SA, Mansour ZN, Mourad ZA et al (2018) DOE, formulation, and optimization of repaglinide nanostructured lipid carriers. J Appl Pharm Sci 8(10):8–16
Teeranachaideekul V, Muller RH, Junyaprasert VB (2007) Encapsulation of ascorbyl palmitate in nanostructured lipid carriers (NLC) - effects of formulation parameters on physicochemical stability. Int J Pharm 340:198–206
Tran TH, Ramasamy T, Truong DH et al (2014) Preparation and characterization of fenofibrate-loaded nanostructured lipid carriers for oral bioavailability enhancement. AAPS Pharm Sci Tech 15(6):1509–1515
Varshosaz J, Eskandari S, Tabbakhian M (2012) Freeze-drying of nanostructure lipid carriers by different carbohydrate polymers used as cryoprotectants. Carbohydr Polym 88(4):1157–1163
Viladot Petit JL, Delgado Gonzalez R, Fernandez Botello A (2013) Lipid nanoparticle capsules. US Patent 0017239 A1
Wei H, Huating D, Houjiu W et al (2017) Preparation and characterization of nobiletin-loaded nanostructured lipid carriers. J Nanomater 2898342:1–10
Westesen K, Bunjes H, Koch MHJ (1997) Physicochemical characterization of lipid nanoparticles and evaluation of their drug loading capacity and sustained release potential. J Control Release 48(2–3):223–236

Wissing SA, Muller RH (2003) Cosmetic applications for solid lipid nanoparticles (SLN). Int J Pharm 254(1):65–68

Wissing S, Lippacher A, Muller R (2001) Investigations on the occlusive properties of solid lipid nanoparticles (SLN). J Cosmet Sci 52(5):313–324

Zhang T, Chen J, Zhang Y et al (2011) Characterization and evaluation of nanostructured lipid carrier as a vehicle for oral delivery of etoposide. Eur J Pharm Sci 43:174–179

Zhang XU, Qiao H, Ni JM et al (2013) Preparation of isoliquiritigenin-loaded nanostructured lipid carrier and the in vivo evaluation in tumor-bearing mice. Eur J Pharm Sci 49(3):411–422

Zhuang CY, Li N, Wang M et al (2010) Preparation and characterization of vinpocetine loaded nanostructured lipid carriers (NLC) for improved oral bioavailability. Int J Pharm 394 (1):179–185

Nanocapsules

5

Chintan J. Aundhia

Abstract

Nanoparticles are structures of sizes ranging from 1 to 100 nm in at least one dimension. Nanocapsules along with nanospheres are the most common forms of nanoparticles. Polymeric nanoparticles are named nanocapsules when they contain a polymeric wall composed of non-ionic surfactants, macromolecules, phospholipids and an oil core. Nanocapsules are known to offer various advantages in the domain of drug delivery. They are known to be container systems in which hydrophilic and hydrophobic drugs can be incorporated. This accommodating ability enables to carry, store and release many drugs and trackers or reporter molecules for sophisticated applications. This chapter highlights various formulation strategies of nanocapsules and formulation ingredients along with their various applications.

Keywords

Nanocapsules · Improved bioavailability · Novel drug carrier · Controlled release

5.1 Introduction

Nanotechnology systems have come a long way since the term was first used by Japanese scientist Norio Taniguchi. It has paved the way for nanomedicine which is a key science of the twenty-first century. There is an increasing optimism that nanotechnology will go a long way in making substantial advances in the treatment and diagnosis of diseases (De Jong and Borm 2008). The way in which

C. J. Aundhia (✉)
Department of Pharmacy, Sumandeep Vidyapeeth Deemed to be University, Vadodara, Gujarat, India
e-mail: chintanaundhia.dop@sumandeepvidyapeethdu.edu.in

N. Shah (ed.), *Nanocarriers: Drug Delivery System*,
https://doi.org/10.1007/978-981-33-4497-6_5

nanotechnology can contribute to the benefit of the mankind has been researched in detail only from 1990s (Krukemeyer et al. 2015). Thus, in a nutshell, nanomedicine and nano-drug delivery systems are still in its infantile stage but rapidly developing science.

As far as drug delivery is concerned, the most important nanoparticle platforms are liposomes, polymer conjugates, metallic nanoparticles (for example AuNPs), polymeric micelles, dendrimers, nanoshells and protein- and nucleic acid-based nanoparticles (Silva et al. 2014). All the delivery systems hold tremendous potential as an effective delivery system as established in many research works.

It has been well established that nanoscale-sized particles exhibit unique structural, chemical, mechanical, magnetic, electrical and biological properties. According to the definition from the National Nanotechnology Initiative (NNI), nanoparticles are structures of sizes ranging from 1 to 100 nm in at least one dimension. However, the prefix 'nano' is commonly used for particles that are up to several hundred nanometres in size (Wilczewska et al. 2012). Nanoparticles have been extensively investigated as drug carriers because of their unique ability to facilitate various drug delivery objectives.

A capsule refers to a dosage form comprising a hollow shell into which the medicament or drug is incorporated for effective delivery of drug (Jones 2016; Lachman et al. 1986). Polymeric nanoparticles are named nanocapsules when they contain a polymeric wall composed of non-ionic surfactants, macromolecules, phospholipids and an oil core (Jäger et al. 2007; Béduneau et al. 2006). The earlier nanocapsules to be investigated extensively were the fullerenes which led to the formulation of carbon nanocapsules which have the potential to be a wonder technology in various fields. Metals like cobalt and iron have also been encapsulated in carbon nanocapsules (Masuda et al. 2000; Saito and Masuda 1995).

Nanocapsules are known to offer various advantages in the domain of drug delivery. They are known to be container systems in which hydrophilic and hydrophobic drugs can be incorporated. This accommodating ability enables to carry, store and release many drugs and tracker or reporter molecules for sophisticated applications of nanotechnology. The polymeric shell augments the lifetime of encapsulated drugs by shielding from extreme factors such as pH, enzyme activity and tissue irritation whereas it also circumvents the side effects of loaded molecules to the internal environment by decreasing hasty release of the drug. These outstanding benefits make nanocapsules the unanimous choice for a drug carrier. Some other important technological advantages are high stability, high drug loading and facilitating of administration of drugs from various routes right from oral to nasal. Manipulation of these parameters aids the formulation scientist in enhancing the bioavailability, reducing the dosing frequency and increasing the patient compliance. Concisely, nanocapsules fulfil various conditions of an ideal carrier for drug delivery like biodegradability, sufficient circulation in blood flow, controlled and sustained release of drugs and complete clearance from the body after the payload is released (Iyisan and Landfester 2019).

5.2 Structure of Nanocapsules

Nanocapsules, characteristic class of nanoparticles, are made up of one or more active materials (core) and a protective matrix (shell) in which the therapeutic substance may be confined (Kothamasu et al. 2012). The core usually consists of liquid which can be an aqueous or oily substance whereas the shell essentially consists of a thin polymer matrix.

Nanospheres are the homogenously solid counterparts of nanocapsules. In other words, nanocapsules are vesicular systems in which a drug is confined to a cavity surrounded by a polymer membrane, whereas nanospheres are matrix systems in which the drug is physically and uniformly dispersed (Vinothini and Rajan 2019). In fact, solid lipid nanoparticles and lipid nanocarriers bear a large similarity to nanospheres and nanocapsules in this regard.

However, nanocapsules are pharmaceutically attractive from the point of view that they allow greater encapsulation of the drug and are more efficient in delivering lipophilic materials. The basic structure of nanocapsules is shown in Fig. 5.1.

5.3 Formulation Components of Nanocapsules: Requirements and Candidates

5.3.1 Shell

Shell materials play a vital role in the development of nanocapsules and nanospheres. As discussed earlier many of the advantages which the nanocapsules offer are heavily dependent on the polymer selected to form the matrix. The shell material directly impacts the stability, entrapment efficiency, release profile and absorption of nanocapsules.

Biodegradable polymers (e.g. PLA, PLGA) are the materials of choice for formulating nanocapsules. However non-biodegradable but biocompatible materials (e.g. PEG, PVA) have also been extensively accepted for the fabrication of nanocapsules. Biocompatibility is defined as the ability to produce an appropriate reaction in the host without any adverse effects. Depending upon their varied

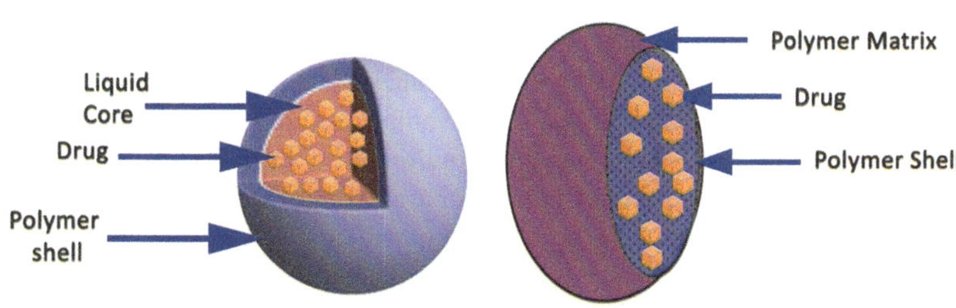

Fig. 5.1 Structure of nanocapsule and nanospheres

applications, polymers are classified as natural and synthetic based on their source (Deng et al. 2020).

5.3.2 Natural Polymers

Polysaccharides are macromolecules composed of sugar units linked by glycosidic bonds. (Torres et al. 2019). They offer a host of advantages over synthetic polymers such as natural abundance, biocompatibility, mucoadhesion and ease of manipulation (Wurm and Weiss 2014). As polysaccharides are typically hydrophilic and water soluble, formulating these polymers as nanoparticles in aqueous dispersion requires either making it hydrophobic prior to formulation or cross-linking the polymer to form nanoscaled hydrogels. When the cross-linking reaction is conducted at the interface of a droplet, nanocapsules with liquid core can be generated. When hydrophobic agents are used, it can be used to lead to impermeable shells (Wurm and Weiss 2014).

Starch, a low-cost biodegradable polymer, can be used to formulate nanocapsule shells. In a study, aqueous-core nanocapsules were prepared via the inverse mini-emulsion technique, by interfacial polymerization between gelatinized potato starch and 2,4-toluene diisocyanate. The nanocapsules demonstrated higher stability, sustained drug release, minimum leakage and high encapsulation efficiency (Steinmacher et al. 2017).

Chitosan, a deacetylated derivative of chitin, is widely used due to its non-toxic, biocompatible, biodegradable and mucoadhesive properties (Lertsutthiwong and Rojsitthisak 2011; Torres et al. 2019). Additionally chitosan provides a cationic surface which can facilitate the formation of cross links in the polymer. In a study, it was found that chitosan played a significant role in the recovery of the active pharmaceutical ingredient by forming cross links (Lertsutthiwong and Rojsitthisak 2011). Thus, chitosan provides the added benefit of a cationic surface which can facilitate the formation of cross links in the polymer. In yet another study, the mucoadhesive property of chitosan was emphasized (Frank et al. 2014).

Alginic acid and sodium and potassium alginates (ALG) have emerged as one of the most extensively explored mucoadhesive biomaterials owing to very good cytocompatibility and biocompatibility, biodegradation, sol-gel transition properties and chemical versatility (Sosnik 2014). As alginate is a pH-sensitive polymer, it can be used for effective drug targeting in the intestine (Mukhopadhyay et al. 2015).

Likewise, gelatin also offers various advantages like biocompatibility, biodegradation and low immunogenicity. Another advantage offered by the amino acid side chains of the gelatin matrix molecule is the option of multiple further modifications. This could be used for coupling of ligands to improve targeting of diseased tissues, enhance specific cellular uptake or affect intracellular distribution (Schwick and Heide 1969).

Lignin is one of the most abundant renewable biopolymers. There are very few examples which use lignin in nanoparticle formulations. In a study, hollow degradable lignin nanocapsules with an aqueous core were synthesized. These were found to be an attractive long-term release system (Yiamsawas et al. 2014).

Another polysaccharide used in the delivery of drugs is hyaluronic acid. Being a simple polysaccharide, hyaluronic acid exhibits an astonishing array of biological functions. Hyaluronic acid interacts with various proteins or proteoglycans to organize the extracellular matrix and to maintain tissue homeostasis. The unique physical and mechanical properties of hyaluronic acid contribute to the maintenance of tissue hydration, the mediation of solute diffusion through the extracellular space and the lubrication of certain tissues (Dicker et al. 2014).

5.3.3 Proteins

Proteins are natural macromolecules which have garnered great attention in the formulation of nanocapsules. As they are biological in origin, they are biocompatible and their properties can be modified.

Human serum albumin has been frequently used as an encapsulating material in nanocapsules. Besides controlling drug permeation rate, albumin reduces the immunogenicity of nanoparticles and thus assists them to escape from the reorganization of reticuloendothelial system (Deng et al. 2020). Due to its biogenic properties, HSA-based nanocapsules can be effectively used for targeted drug deliveries. Folic acid-functionalized human serum albumin nanocapsules were used for targeted drug delivery to chronically activated macrophages (Rollett et al. 2012).

Zein is a prolamine isolated from corn. As with glutamine, proline and asparagine, it contains large amount of hydrophobic amino acids. Due to its solubility in binary solvents, hydrophobicity and biodegradability, it is used in the formulation of many nanoparticles. Its hydrophobicity also provides a moisture barrier due to which it is also studied in the formulation of colloidal systems. Drug-loaded nanocapsules have demonstrated various advantages like mucoadhesive properties, organ targeting and improved drug stability (Elzoghby et al. 2015).

5.3.4 Synthetic Materials

Synthetic materials have some distinct advantages over natural materials as they give reproducible results, the reason being that they are of the same quality. Furthermore, adjustment of the mechanical property, degradation and solubility is also possible according to the application.

Synthetic homopolymers used in the formulation of nanocapsules include polylactide, poly (lactide-co-glycolide), poly (epsilon-caprolactone), poly (isobutyl cyanoacrylate), poly (isohexyl cyanoacrylate) and poly (n-butyl cyanoacrylate).

Polylactic acid (PLA) is classified as an aliphatic polyester because of the ester bonds that connect the monomer units. PLA naturally degrades in situ through hydrolysis mechanism: water molecules break the ester bonds that constitute polymer backbone. Other advantages include eco-friendliness, biocompatibility and better processability (Casalini et al. 2019). Aqueous-core PEG-coated PLA nanocapsules were formulated which helped in efficient entrapment of water-soluble anticancer drug (Cosco et al. 2015).

Poly (lactide)-poly (ethylene glycol) (PLA-PEG), poly (lactide-co-glycolide)-poly (ethylene glycol) (PLGA-PEG), poly (epsilon-caprolactone)-poly (ethylene glycol) (PCL-PEG), etc. are the various copolymers which are used in the synthesis of the polymeric shell.

Polymethacrylate polymers, commonly available under the trade name Eudragit, are synthetic polymers having variable ratios of acrylate and methacrylate polymers. These polymers exhibit pH-dependent water solubility, making them suitable coating agents for drug delivery systems (da Silva Júnior et al. 2017).

5.3.5 Core Materials

The core of the nanocapsules can be liquid, hollow or even solid, thereby facilitating it to carry different drugs. The ideal properties of solvent include solubilizing of the active ingredients, releasing of the drug according to the desired profile, non-toxicity and inertness. However no solvent is an ideal candidate.

The most common solvents used are toluene, benzene, chloroform, dichloromethane, etc. These solvents are quite toxic and the search for 'green alternatives' has garnered a lot of attention.

Among a wide variety of potential solvents, natural oils, extracted from various plant parts such as fruits, seeds, roots and leaves, have emerged as viable alternatives. These solvents provide an added benefit of antioxidant and antimicrobial properties. Thymol, eugenol, linalool, menthone, carvone, basil oil, benzaldehyde, etc. are some of the essential oils and essential oil extracts which can be used as alternative solvents (Marturano et al. 2018).

5.4 Methods of Preparation

Generally, there are six classical methods for the preparation of nanocapsules: nanoprecipitation, emulsion–diffusion, double emulsification, emulsion-coacervation, polymer coating and layer by layer. However other methods such as emulsion evaporation have also been used in the preparation of nanocapsules (Mora-Huertas et al. 2010).

5.4.1 Solvent Displacement Method

Solvent displacement method is also referred to as nanoprecipitation method or interfacial deposition method. The advantages of this method are that it is easy, less complex and widely acceptable with the use of any complex additive (Hornig et al. 2009). Essentially, nanoprecipitation is a rapid desolvation of the polymer which leads to the precipitation of nanoparticles. To aid precipitation, a non-solvent is added to the solvent.

Thus, essentially, this method requires two phases: a solvent and a non-solvent phase. The solvent phase can be a single solvent or a mixture of solvents into which the drug and the polymer are incorporated.

Alternatively, the non-solvent consists of a single solvent or a mixture of solvents for aiding the film formation process which is enhanced with surface active agents. Generally, the solvent is referred to as organic solvent and the non-solvent is referred to as aqueous solvent. However, both the organic and aqueous solvents can be used interchangeably as a solvent and non-solvent phase.

Subsequently, the solvent phase is added gradually to the non-solvent phase with constant stirring. This leads to rapid desolvation of the polymer and leads to precipitation of nanocapsules.

The proposed mechanism involves four major steps, viz. supersaturation, nucleation, condensation and particle growth. The crucial phenomenon is nucleation, which involves supersaturation aided by interfacial tension between the solid particles and the solvent. The core concept is that diffusion of solvent leads to the deposition of molecules into the other phase and forms miniature supersaturation areas which results in formation of the particles. Finally, a suspension of the nanocapsules is obtained. The nanocapsules can be recovered by various methods, the most common of which are ultracentrifugation followed by lyophilization. Factors which influence the supersaturation process are the concentration of polymer, ratio of solvents and miscibility of the polymer whereas the stirring rate influences the deposition of polymer (Lince et al. 2008).

Nanocapsules produced by this method are shown to release the drug through two phases: an initial burst-release phase followed by a steady sustained-release phase (Wang and Tan 2016).

It may be noteworthy to mention that the presence of stabilizers is imperative to ensure non-aggregation and avoid clumps of nanocapsules from aggregating after formulation and also on long-time storage. The underlying mechanism involves decreasing the particle-particle interactions which arise due to a balance in favour of van der Walls and hydrophobic interactions which in turn influences the collision frequency. Polymers such as PVA and poloxamers have been commonly used to stabilize nanocapsule formulations (Abdelwahed et al. 2006; Salatin et al. 2017).

The scaling up of nanoprecipitation method remains a challenge since it has been demonstrated on numerous occasions that the size of the nanocapsules prepared in the laboratory is significantly less as compared to the nanocapsules when scaled up (Rivas et al. 2017). The crucial variables involved in this are the mixing time and mixing rate which are easy to achieve at the laboratory scale but difficult in the larger scale. However, some methods such as flash nanoprecipitation are used to alter the method which helps to decrease the particle size on a large scale (Zhang et al. 2013).

However, this method is not ideal because of several reasons: (1) likelihood of presence of residual, potentially toxic monomers or oligomers; (2) chances of cross-reaction between the drug and polymer leading to destruction of the drug activity; and (3) polymer aggregation which is frequently observed when working with high polymer concentrations or a low organic solvent/water ratio (Quintanar-Guerrero et al. 1998).

5.4.2 Emulsion–Diffusion Method

Three phases, i.e. organic or water-miscible solvent, aqueous and dilution phases, are used to encapsulate drugs via emulsion–diffusion method. The whole process of nanocapsule synthesis using this method can be divided into two steps which are the emulsification step and the diffusion step. The sequence of events for preparing nanocapsules is as follows: (a) The drug, polymer and oil are dissolved in the organic or the water-miscible layer like benzyl alcohol or propylene carbonate and saturated with water. (b) The aqueous layer is used which also contains the stabilizer like polyvinyl alcohol or poloxamer. (c) An o/w emulsion is fabricated using the two layers (emulsification step). (d) The dilution phase is added to the emulsion which leads to the formation of nanocapsules (diffusion step).

During the diffusion step, the separation of the polymer and the oil takes place which is brought about by the diffusion of the organic solvent into the aqueous phase leading to its removal. This results in the formation of small droplets and each droplet forms a particle of small size.

In this method, the stabilizer performs the dual role of preventing the formation of polymer aggregates and stabilizing the nuclei which subsequently leads to the formation of nanocapsules.

5.4.3 Double-Emulsification Method

Double emulsions are rightfully termed as 'emulsions of emulsions'. These are complex systems in which the droplets of the dispersed phase themselves consist of smaller dispersed phase. Double emulsion is a unique process encompassing the advantage of encapsulating both lipophilic and hydrophilic drug molecules (Iqbal et al. 2015).

This method takes place in two steps. In the first step, the hydrophilic drugs are added to the aqueous layer and lipophilic drugs or polymers are added to the oil phase. Both the phases are homogenized to form the primary emulsion which is further emulsified by adding it to an external aqueous phase containing the stabilizer, thus forming a w/o/w double emulsion. Subsequently, the organic solvent is removed by the means of evaporation, thus resulting in particle hardening and precipitation. Water can be added at specific intervals to ensure that the entire quantity of the organic solvent has been eliminated (Bilati et al. 2005). In this method, sonication can be used for forming the primary emulsion and multiple emulsion.

This method has numerous advantages in the fact that it allows the encapsulation of thermosensitive drugs, the particle size distribution of the obtained nanocapsules is in a narrow range, it has excellent batch-to-batch reproducibility, the solvents used are non-toxic and the scale-up process is easy.

However, the downside of this method is the possibility of organic solvent remains, poor entrapment of hydrophilic drugs, requirement of larger volume for formulation and lengthier agitation time of the solvent to facilitate complete removal of organic solvent (Campos et al. 2013; Hao et al. 2013).

In this method, the common organic solvents used are ethyl acetate, methylene chloride, dichloromethane, etc. whereas the surface active agents like Span and Tween are preferred (Mora-Huertas et al. 2010).

5.4.4 Emulsion Coacervation Method

The word coacervation comes from the Latin word ‘acervus’ meaning aggregation. Coacervation can be defined as the separation of a macromolecular solution into two immiscible liquid phases: a dense coacervate phase and a dilute equilibrium one. Hence, the coacervation-phase separation method uses a coacervation-inducing agent to attain the coacervation-phase separation after solvent evaporation to form nanocapsules (Kas and Oner 2000).

This method principally consists of three key steps: (1) emulsification of the organic layer containing the active substance and oil with the addition of the aqueous phase containing polymer and stabilizing agent aided by constant stirring; (2) precipitation of the emulsion thus formed onto the dispersed droplets by lowering the solubility of the hydrocolloid by means such as addition of non-solvent, pH change, temperature change or electrolyte; and (3) stabilization and hardening of the microcapsules by adding a cross-linking agent such as formaldehyde, glutaraldehyde or transglutaminase (Salaün 2016).

Several factors such as stirring speed, cooling rate, concentration of phase separation-inducing agent and viscosity/molecular weight of polymer have a direct influence on the encapsulation efficiency of the drug. This fact was proved in a study involving water-soluble drugs like nitenpyram and chlorpromazine hydrochloride (Wieland-Berghausen et al. 2002).

This method has numerous advantages as it is an inexpensive method and allows incorporation of thermosensitive drugs but the disadvantage is that it can lead to the degradation of the polymer under acidic conditions.

5.4.5 Polymer Coating Method

Different approaches can be used to deposit a thin layer of polymer on the exterior surface of the nanocapsules. This can be accomplished by adsorbing the polymer onto the initially formed uncoated nanocapsules. The coating of the polymer is dependent on the stirring speed and the stirring time at which the system is maintained (Calvo et al. 1997).

An alternative method involves two steps—the first, formation of nanoemulsion and second, incubation of nanoemulsion with a polymer solution leading to coating by surface deposition (Prego et al. 2006).

However, as the research pertaining to this method is restricted, it becomes difficult to establish a general overall step-by-step procedure for this method.

5.4.6 Layer-by-Layer Method

This method involves including a colloidal preformed prototype onto which the polymer layer is adsorbed by adding a polymer solution. The mechanism of depositing the polymer layer includes lowering the polymer solubility by adding a cosolvent or by incubation (Radtchenko et al. 2002). The thickness of the shell wall can be tuned in the nanometre range by varying the adsorption conditions and the number of layers (Antipov et al. 2002).

The principle behind this method of coating is the polarity gradient across the capsule wall. Capsule creation makes use of electrostatic interaction and can involve many substances as layer constituents, such as synthetic polyelectrolytes, proteins, nucleic acids, lipids and multivalent dyes (Radtchenko et al. 2002; Sukhorukov et al. 1998).

The driving forces of the adsorption are electrostatic interaction between the adjacent layers and the entropy increase upon the capsule formation (Antipov et al. 2002).

Further, it is commonly observed that salt has the strongest impact on the quantity of the polymer deposited per cycle. Polymer concentration, molecular weight and deposition time are known to be less significant variables (Dubas and Shlenoff 1999).

As reported in different research works, the layer-by-layer method makes use of polycations such as polylysine, chitosan, gelatin B, poly(allylamine) (PAA), poly (ethyleneimine) (PEI), aminodextran and protamine sulphate. The polyanions used are poly(styrene sulphonate) (PSS), sodium alginate, poly(acrylic acid), dextran sulphate, carboxymethyl cellulose, hyaluronic acid, gelatin A, chondroitin and heparin (Mora-Huertas et al. 2010).

5.5 Merits and Applications of Nanocapsules

Nanocapsules are extensively used as delivery systems for the encapsulation of a wide variety of drugs and other drugs for various biomedical applications such as anti-inflammation therapy, anticancer therapy and immunotherapy (Deng et al. 2020). The main applications of nanocapsules in the domain of drug delivery are as follows.

Nanocapsules have the ability to shield various encapsulated substances like peptides, proteins, enzymes, hormones, metabolites, genes or reporter molecules from hostile environment to prevent biological and chemical degradation. This in turn leads to increase in the bioavailability of the drug.

Various agents in stimulus-responsive nanocapsules like pH change, reducing agents and heat are used to control the intracellular release of drugs (Kim et al. 2010). The shell of the nanocapsules, composed of polymer, has been shown to selectively target cancer cells or other life-threatening cells on contact. Self-assembled hyaluronan nanocapsules have demonstrated effective intracellular distribution of antineoplastic drug docetaxel (Cadete et al. 2019). The targeting of drugs to tumour cells has been attributed to the leaky tumour vasculature which can enhance the permeability to macromolecules (Alasvand et al. 2017). According to recent studies, ligand-targeted nanoparticles have been able to achieve path-breaking development towards effective oral delivery of nanomedicines that can overcome the intestinal epithelial cellular barrier (Yameen et al. 2014). In yet another study, prednisolone-loaded colon-targeted nanocapsules were formulated using Eudragit S100 as a polymer (Kshirsagar et al. 2012). In yet another study a polymeric self-emulsifying nanocapsule formulation of curcumin was successfully developed to target the colon (Wadhwa et al. 2014). In other words, nanocapsules have improved the biological behaviours of encapsulated drugs by enhancing drug efficiency even as toxicity is avoided. This property makes nanocapsules a convenient option for oral drug delivery.

Sustained release of drugs is the key property required when the drugs are hastily eliminated from the body and have a short half-life. Sustaining the release of drug from the dosage form allows the retention of the drug in the bloodstream or the organ for a prolonged period of time (Alasvand et al. 2017). The choice of materials used in the formulation of nanocapsules has been found to have a direct correlation with the kinetics of drug release from nanocapsules. It is found that the characteristics of the nanocapsule oil core can also influence the drug release kinetics when the drug is dispersed within the lipophilic core (Jäger et al. 2009). A stable injectable formulation of the antimalarial drug halofantrine based on nanocapsules was prepared from biodegradable polymers. It was found that the formulation was long lasting and could be effectively used for treating severe malaria (Mosqueira et al. 2006). Tretinoin-loaded lipid-core nanocapsules have also shown promise as a parenteral nanomedicine for the treatment of acute promyelocytic leukaemia (Ourique et al. 2010). Chitosan hydrogel-containing capsaicinoid-loaded nanocapsules were also found to be highly effective in controlling the release of capsaicin and dihydrocapsaicin (Contri et al. 2010).

Solubility and permeability are the two major attributes required for the proper absorption of drugs from the respective absorption windows. There are numerous instances in which encapsulation of drugs in nanocapsules has brought about a marked increase in the solubility of drugs. Formulation of curcumin-loaded PLGA nanocapsules made it possible to overcome the poor solubility of curcumin and exhibit enhanced values of minimum inhibitory concentrations (Gao et al. 2020). Similar studies have shown increase in the solubility of BCS class II drugs like fenofibrate, catechin and platinum-based anticancer drugs (Hamelers and De Kroon 2007; Shelake et al. 2018; Monika et al. 2014).

Thus, incorporation of drugs into nanocapsules offers a bouquet of advantages like improving poor aqueous solubility, stabilizing drugs by protecting the molecule from the environment, providing the desired pharmacokinetic profile, allowing controlled release, as well as facilitating oral, parenteral, topical and transdermal administration.

References

Abdelwahed W, Degobert G, Fessi H (2006) A pilot study of freeze drying of poly (epsilon-caprolactone) nanocapsules stabilized by poly (vinyl alcohol): formulation and process optimization. Int J Pharm 309:178–188

Alasvand N, Urbanska AM, Rahmati M et al (2017) Therapeutic nanoparticles for targeted delivery of anticancer drugs. Multifunctional systems for combined delivery, biosensing and diagnostics. Elsevier

Antipov AA, Sukhorukov GB, Leporatti S et al (2002) Polyelectrolyte multilayer capsule permeability control. Colloids Surf A Physicochem Eng Asp 198:535–541

Béduneau A, Saulnier P, Anton N et al (2006) Pegylated nanocapsules produced by an organic solvent-free method: evaluation of their stealth properties. Pharm Res 23:2190–2199

Bilati U, Allémann E, Doelker E (2005) Strategic approaches for overcoming peptide and protein instability within biodegradable nano- and microparticles. Eur J Pharm Biopharm 59:375–388

Cadete A, Olivera A, Besev M et al (2019) Self-assembled hyaluronan nanocapsules for the intracellular delivery of anticancer drugs. Sci Rep 9:1–11

Calvo P, Vila-Jato JL, Alonso MJ (1997) Evaluation of cationic polymer-coated nanocapsules as ocular drug carriers. Int J Pharm 153:41–50

Campos EVR, De Melo NFS, De Paula E et al (2013) Screening of conditions for the preparation of poly (-caprolactone) nanocapsules containing the local anesthetic articaine. J Colloid Sci Biotechnol 2:106–111

Casalini T, Rossi F, Castrovinci A et al (2019) A perspective on polylactic acid-based polymers use for nanoparticles synthesis and applications. Front Bioeng Biotechnol 7:259

Contri RV, Katzer T, Pohlmann AR et al (2010) Chitosan hydrogel containing capsaicinoids-loaded nanocapsules: an innovative formulation for topical delivery. Soft Mater 8:370–385

Cosco D, Paolino D, De Angelis F et al (2015) Aqueous-core PEG-coated PLA nanocapsules for an efficient entrapment of water soluble anticancer drugs and a smart therapeutic response. Eur J Pharm Biopharm 89:30–39

Da Silva Júnior WF, De Oliveira Pinheiro JG, Moreira CD et al (2017) Alternative technologies to improve solubility and stability of poorly water-soluble drugs. Multifunctional systems for combined delivery, biosensing and diagnostics. Elsevier

De Jong WH, Borm PJ (2008) Drug delivery and nanoparticles: applications and hazards. Int J Nanomedicine 3:133

Deng S, Gigliobianco MR, Censi R et al (2020) Polymeric nanocapsules as nanotechnological alternative for drug delivery system: current status, challenges and opportunities. Nanomaterials 10:847

Dicker KT, Gurski LA, Pradhan-Bhatt S et al (2014) Hyaluronan: a simple polysaccharide with diverse biological functions. Acta Biomater 10:1558–1570

Dubas ST, Shlenoff JB (1999) Factors controlling the growth of polyelectrolyte multilayers. Macromolecules 32:8153–8160

Elzoghby AO, Elgohary MM, Kamel NM (2015) Implications of protein-and peptide-based nanoparticles as potential vehicles for anticancer drugs. Elsevier, Advances in protein chemistry and structural biology

Frank LA, Sandri G, D'autilia F et al (2014) Chitosan gel containing polymeric nanocapsules: a new formulation for vaginal drug delivery. Int J Nanomedicine 9:3151

Gao M, Long X, Du J et al (2020) Enhanced curcumin solubility and antibacterial activity by encapsulation in PLGA oily core nanocapsules. Food Funct 11:448–455
Hamelers IH, De Kroon AI (2007) Nanocapsules: a novel formulation technology for platinum-based anticancer drugs. Future Lipidol 2:445–453
Hao S, Wang B, Wang Y et al (2013) Preparation of Eudragit L 100-55 enteric nanoparticles by a novel emulsion diffusion method. Colloids Surf B Biointerfaces 108:127–133
Hornig S, Heinze T, Becer CR et al (2009) Synthetic polymeric nanoparticles by nanoprecipitation. J Mater Chem 19:3838–3840
Iqbal M, Zafar N, Fessi H et al (2015) Double emulsion solvent evaporation techniques used for drug encapsulation. Int J Pharm 496:173–190
Iyisan B, Landfester K (2019) Modular approach for the design of smart polymeric nanocapsules. Macromol Rapid Commun 40:1800577
Jäger A, Stefani V, Guterres SS et al (2007) Physico-chemical characterization of nanocapsule polymeric wall using fluorescent benzazole probes. Int J Pharm 338:297–305
Jäger E, Venturini CG, Poletto FS et al (2009) Sustained release from lipid-core nanocapsules by varying the core viscosity and the particle surface area. J Biomed Nanotechnol 5:130–140
Jones DS (2016) FASTtrack pharmaceutics dosage form and design. Pharmaceutical Press, London
Kas HS, Oner L (2000) Microencapsulation using coacervation/phase separation: an overview of the technique and applications. Marcel Dekker, Inc., New York
Kim E, Kim D, Jung H et al (2010) Facile, template-free synthesis of stimuli-responsive polymer nanocapsules for targeted drug delivery. Angew Chem Int Ed 49:4405–4408
Kothamasu P, Kanumur H, Ravur N et al (2012) Nanocapsules: the weapons for novel drug delivery systems. Bioimpacts 2:71
Krukemeyer M, Krenn V, Huebner F et al (2015) History and possible uses of nanomedicine based on nanoparticles and nanotechnological progress. J Nanomed Nanotechnol 6
Kshirsagar SJ, Bhalekar MR, Patel JN et al (2012) Preparation and characterization of nanocapsules for colon-targeted drug delivery system. Pharm Dev Technol 17:607–613
Lachman L, Lieberman HA, Kanig JL (1986) The theory and practice of industrial pharmacy. Lea & Febiger, Philadelphia
Lertsutthiwong P, Rojsitthisak P (2011) Chitosan-alginate nanocapsules for encapsulation of turmeric oil. Die Pharmazie 66:911–915
Lince F, Marchisio DL, Barresi AA (2008) Strategies to control the particle size distribution of poly-ε-caprolactone nanoparticles for pharmaceutical applications. J Colloid Interface Sci 322:505–515
Marturano V, Bizzarro V, De Luise A et al (2018) Essential oils as solvents and core materials for the preparation of photo-responsive polymer nanocapsules. Nano Res 11:2783–2795
Masuda M, Maeda K, Kobayashi T et al (2000) Synthesis, crystal structure and magnetic properties of iron particles encaged in carbon nanocapsules. Jpn J Appl Phys 39:L733
Monika P, Basavaraj B, Murthy K et al (2014) Development of sustained release nanocapsules of catechin rich extract for enhanced bioavailability. Int J Pharm Pharm Sci 6:331–337
Mora-Huertas CE, Fessi H, Elaissari A (2010) Polymer-based nanocapsules for drug delivery. Int J Pharm 385:113–142
Mosqueira VCF, Legrand P, Barratt G (2006) Surface-modified and conventional nanocapsules as novel formulations for parenteral delivery of halofantrine. J Nanosci Nanotechnol 6:3193–3202
Mukhopadhyay P, Chakraborty S, Bhattacharya S et al (2015) pH-sensitive chitosan/alginate core-shell nanoparticles for efficient and safe oral insulin delivery. Int J Biol Macromol 72:640–648
Ourique A, Azoubel S, Ferreira C et al (2010) Lipid-core nanocapsules as a nanomedicine for parenteral administration of tretinoin: development and in vitro antitumor activity on human myeloid leukaemia cells. J Biomed Nanotechnol 6:214–223
Prego C, Fabre M, Torres D et al (2006) Efficacy and mechanism of action of chitosan nanocapsules for oral peptide delivery. Pharm Res 23:549–556

Quintanar-Guerrero D, Allémann E, Doelker E et al (1998) Preparation and characterization of nanocapsules from preformed polymers by a new process based on emulsification-diffusion technique. Pharm Res 15:1056–1062

Radtchenko IL, Sukhorukov GB, Möhwald H (2002) A novel method for encapsulation of poorly water-soluble drugs: precipitation in polyelectrolyte multilayer shells. Int J Pharm 242:219–223

Rivas CJM, Tarhini M, Badri W et al (2017) Nanoprecipitation process: from encapsulation to drug delivery. Int J Pharm 532:66–81

Rollett A, Reiter T, Nogueira P et al (2012) Folic acid-functionalized human serum albumin nanocapsules for targeted drug delivery to chronically activated macrophages. Int J Pharm 427:460–466

Saito Y, Masuda M (1995) Crystallographic structure and magnetic properties of Co fine particles encaged in carbon nanocapsules. Jpn J Appl Phys 34:5594

Salatin S, Barar J, Barzegar-Jalali M et al (2017) Development of a nanoprecipitation method for the entrapment of a very water soluble drug into Eudragit RL nanoparticles. Res Pharm Sci 12:1

Salaün F (2016) Microencapsulation technology for smart textile coatings. In: Active coatings for smart textiles, Woodhead Publishing, Elsevier, pp 179–220

Schwick H, Heide K (1969) Immunochemistry and immunology of collagen and gelatin. In: Hässig A et al (eds) Modified gelatins as plasma substitutes. Karger Publishers, Basel, pp 111–125

Shelake S, Patil S, Patil S (2018) Formulation and evaluation of fenofibrate-loaded nanoparticles by precipitation method. Indian J Pharm Sci 80:420–427

Silva J, Fernandes AR, Baptista PV (2014) Application of nanotechnology in drug delivery. In: Sezer AD (ed) Application of nanotechnology in drug delivery, pp 128–154

Sosnik A (2014) Alginate particles as platform for drug delivery by the oral route: state-of-the-art. Int Sch Res Notices 2014:926157

Steinmacher FR, Baier G, Musyanovych A et al (2017) Design of cross-linked starch nanocapsules for enzyme-triggered release of hydrophilic compounds. Processes 5:25

Sukhorukov GB, Donath E, Lichtenfeld H et al (1998) Layer-by-layer self-assembly of polyelectrolytes on colloidal particles. Colloids Surf A Physicochem Eng Asp 137:253–266

Torres FG, Troncoso OP, Pisani A et al (2019) Natural polysaccharide nanomaterials: an overview of their immunological properties. Int J Mol Sci 20:5092

Vinothini K, Rajan M (2019) Mechanism for the nano-based drug delivery system. In Mohapatra SS, et al (ed) Characterization and biology of nanomaterials for drug delivery. Elsevier, pp 219–263

Wadhwa J, Asthana A, Gupta S et al (2014) Development and optimization of polymeric self-emulsifying nanocapsules for localized drug delivery: design of experiment approach. Scientific World Journal 2014:516069

Wang Y, Tan Y (2016) Enhanced drug loading capacity of 10-hydroxycamptothecin-loaded nanoparticles prepared by two-step nanoprecipitation method. J Drug Deliv Sci Technol 36:183–191

Wieland-Berghausen S, Schote U, Frey M et al (2002) Comparison of microencapsulation techniques for the water-soluble drugs nitenpyram and clomipramine HCl. J Control Release 85:35–43

Wilczewska AZ, Niemirowicz K, Markiewicz KH et al (2012) Nanoparticles as drug delivery systems. Pharmacol Rep 64:1020–1037

Wurm FR, Weiss CK (2014) Nanoparticles from renewable polymers. Front Chem 2:49

Yameen B, Choi WI, Vilos C et al (2014) Insight into nanoparticle cellular uptake and intracellular targeting. J Control Release 190:485–499

Yiamsawas D, Baier G, Thines E et al (2014) Biodegradable lignin nanocontainers. RSC Adv 4:11661–11663

Zhang X, Zhang X, Wang S et al (2013) Surfactant modification of aggregation-induced emission material as biocompatible nanoparticles: facile preparation and cell imaging. Nanoscale 5:147–150

6 Nanodiamonds for Theragnostic: Manufacturing and Biomedical Applications

Dhrumi Patel and Sarika Wairkar

Abstract

Nanodiamonds (NDs) are nanosized form of carbon allotropes having distinct optical and mechanical properties. The production of NDs is a three-step process which commences with synthesis followed by processing and modification. Several methods are reported for synthesis of NDs from basic detonation to newly evolved methods like eradiation of carbon onions by electrons. Further processing of NDs involves three stages, viz. purification, disaggregation and fractionation. The modification step involves their doping and surface modification. Doping focuses on controlled introduction of atoms like silicon, nitrogen or boron. Surface modification techniques are focused towards incorporation of chemical moieties to impart essential characteristics like improved stability, lower degradation at physiological conditions and target-specific delivery. NDs have a broad array of diagnostic as well as therapeutic applications. Diagnostic applications involve the use of NDs in super-resolution imaging, photoacoustic imaging and many more. Therapeutic applications of NDs are extended to globally prevalent diseases like cancer, bone disorders and inflammatory diseases, attributed to its flexibility in surface modification. NDs have also found application in the delivery of amino acids and proteins as well as cosmetic purposes like anti-ageing and sunscreen formulations. This chapter gives an insight into production, modification, application, toxicity and advancements in NDs.

D. Patel · S. Wairkar (✉)
Shobhaben Pratapbhai Patel School of Pharmacy and Technology Management SVKMs NMIMS, Mumbai, Maharashtra, India
e-mail: sarika.wairkar@nmims.edu

N. Shah (ed.), *Nanocarriers: Drug Delivery System*,
https://doi.org/10.1007/978-981-33-4497-6_6

Keywords

Nanodiamonds · Surface modification · Diagnostic applications · Therapeutic applications · Biocompatibility

6.1 Introduction

Nanotechnology has seen an array of revolutions in last decades. It has been considered as an alternative solution to problems of traditional drug delivery systems (Ho et al. 2015). The construction of carbon nanomaterials from "element of life" has been a catalyst for research and technical findings. The interest was further fuelled up by the Nobel Prizes awarded in years 1996 and 2010 for discovering fullerene and graphene which are two different forms of carbon. Later in the year 2012 the Noble Prize was awarded for describing the route of synthesis of nanodiamonds (NDs) (Prabhakar and Rosenholm 2019). Diamonds are the most valued gems attributing to their unique optical properties. Hiding in the sparkle are unique characteristics of diamonds including being inert and hardest material, having distinct optical properties and thermal conductivity, and being chemically resistant. These attributes help diamonds never to lose its charm and make it a record bearer in material science. The difference in crystal lattice of diamonds is a primary reason for their distinguishing properties. The nanosized form of diamond ranging from 2 to 10 nm is also known as NDs that manage to be an eye-catcher in various biomedical applications (Laan et al. 2018).

The history of NDs dates back to 1960s where NDs were produced using detonation technique in the Union of Soviet Socialist Republics (USSR); however, the discovery remained unknown till 1980s (Danilenko 2004). The late 1990s gave various innovations which highlighted the use of NDs to a wider scientific community. The first breakthrough was of NDs' colloidal suspension with a size diameter of 4–5 nm (Ozawa et al. 2007). Secondly, NDs, attributing to their non-toxic properties, found application as a tool for biomedical imaging (Chang et al. 2008; Mochalin and Gogotsi 2009). The third research involved developing magnetic sensors using NDs (Maze et al. 2008). Later, the surface modifications possible on NDs were explored using various strategies of wet and gas chemistry to generate NDs having distinct characteristics suitable to incorporate a broad array of drugs and biomolecules (Krueger 2008; Shimkunas et al. 2009). The current research in NDs focuses on the development of novel environmental friendly techniques capable of generating tailor-made NDs in bulk at lower costs (Shenderova et al. 2011a, b).

When compared to traditional carbon nanocarriers like fullerene and carbon nanotubes, NDs are advantageous for the primary reasons like their inert nature and their ability to be converted to nanometre sizes with narrow size distribution. Due to protected crystal lattice structure, they do not bleach and also offer chances for surface functionalization. The effect of drugs can be sustained with suitable surface modification. NDs can be easily fabricated using detonation, a high-temperature and -pressure method, making its manufacturing flexible. The NDs offer a wide range of applications starting from diagnostics to biomedical applications. NDs are helpful in the improvement of therapeutic efficacy as well as targeted drug delivery can be made successful by using NDs. Although there are no commercial drug formulations of NDs, substantial research is being carried out in this emerging area for getting formulations in the market.

This chapter gives an insight into the fundamentals of NDs, synthesis, applications, characterization, associated toxicity, recent patents on NDs and future aspects.

6.2 Types of NDs

NDs are classified into three broad classes based on their average size and structural characteristics, namely monocrystalline diamonds, ultrananocrystalline diamonds and diamondoids.

6.2.1 Monocrystalline Diamonds

They are NDs in a size range of 10–100 nm and are synthesized from monocrystalline diamond particles or polycrystals (Arnault 2014). Micron-sized monocrystalline particles are processed to obtain nanosized monocrystalline particles. The processing steps include grinding for size reduction, purification to remove impurities and fractionation for segregating particles of desired size range. The edges of monocrystalline diamond are sharper in comparison to other diamond particles. These type of NDs are spherical in shape, have extremely low impurities and are in purest form and are therefore used for bio-labelling purposes. Polycrystalline NDs are platelet shaped, have higher contamination and thereby are not recommended for use in biomedical purposes (Bhosale et al. 2013).

6.2.2 Ultrananocrystalline Diamonds

Ultrananocrystalline diamonds are commonly produced using detonation technique. They are spherical in shape and in a size range of 3.5–6 nm. This type of NDs are commonly used as carriers for proteins, genes and nucleic acids (Bhosale et al. 2013). Eidelman and co-workers synthesized ultrananocrystalline diamonds using detonation technique in conjunction with media milling using microbeads (Eidelman et al. 2005). The NDs had average size of 4 nm and unexpected higher viscosity and resulted in black-coloured suspension.

6.2.3 Diamondoids

They are rigid and easily modifiable structures termed as higher diamondoids that are comparable to adamantine moiety comprising ten carbons and ultrananocrystalline diamonds in sizes greater than 2 nm (Bhosale et al. 2013). The extraction of higher diamondoids is carried out from petroleum and they have shapes of helices, pyramids, etc. The extraction of lower diamondoids like adamantane is easier when compared to higher diamondoids (Marchand 2012).

6.3 Method of Preparation

Depending on the method of synthesis, the appropriate starting material would be required for preparation of NDs. Graphite, TNT (2-methyl-1,3,5-trinitrobenzene) and RDX (1,3,5-trinitroperhydro-1,3,5-triazine), carbon black and gun powder are few of the starting materials used to produce NDs. Various other agents containing cyanide, carboxylic acid, nitro, amino, lactone, etc. can be utilized to produce NDs depending upon the type of surface functionalization. The linkers like siloxanes and alkyl chains may be incorporated to form stable NDs (Tinwala and Wairkar 2019). Surface modification of NDs can also be achieved using polymers like polyethylene glycol (PEG) and poly(vinylidene fluoride) (Zhang et al. 2012a; Silva et al. 2017).

Despite the fact that the existence of NDs was reported several years ago in meteorites, crude oil or sediment layers of earth, the first NDs having a size range of 4–10 nm were synthesized using detonation technique where carbon explosives were used as starting materials (Van Thiel et al. 1987). The three basic steps in manufacturing NDs are synthesis, processing and modification as demonstrated in Fig. 6.1.

6.3.1 Synthesis of NDs

The synthesis of NDs is by either static or dynamic process. Static process is carried out at extreme high-temperature and high-pressure conditions whereas dynamic process is a non-equilibrium process where these extreme conditions exist only for short duration usually in microseconds. There are various methods to synthesize NDs starting from the very basic detonation technique to newly evolved techniques like eradiation of carbon onions by electrons (Tinwala and Wairkar 2019). Each technique offers unique merits which form the basis of selecting appropriate method to synthesize NDs. The size of NDs considerably depends on the technique used for their manufacturing.

6.3.1.1 NDs Formed by Detonation Shock Wave-Assisted Synthesis

Detonation techniques can be classified into two categories based on different starting materials used in the synthesis. The former technique uses high-energy

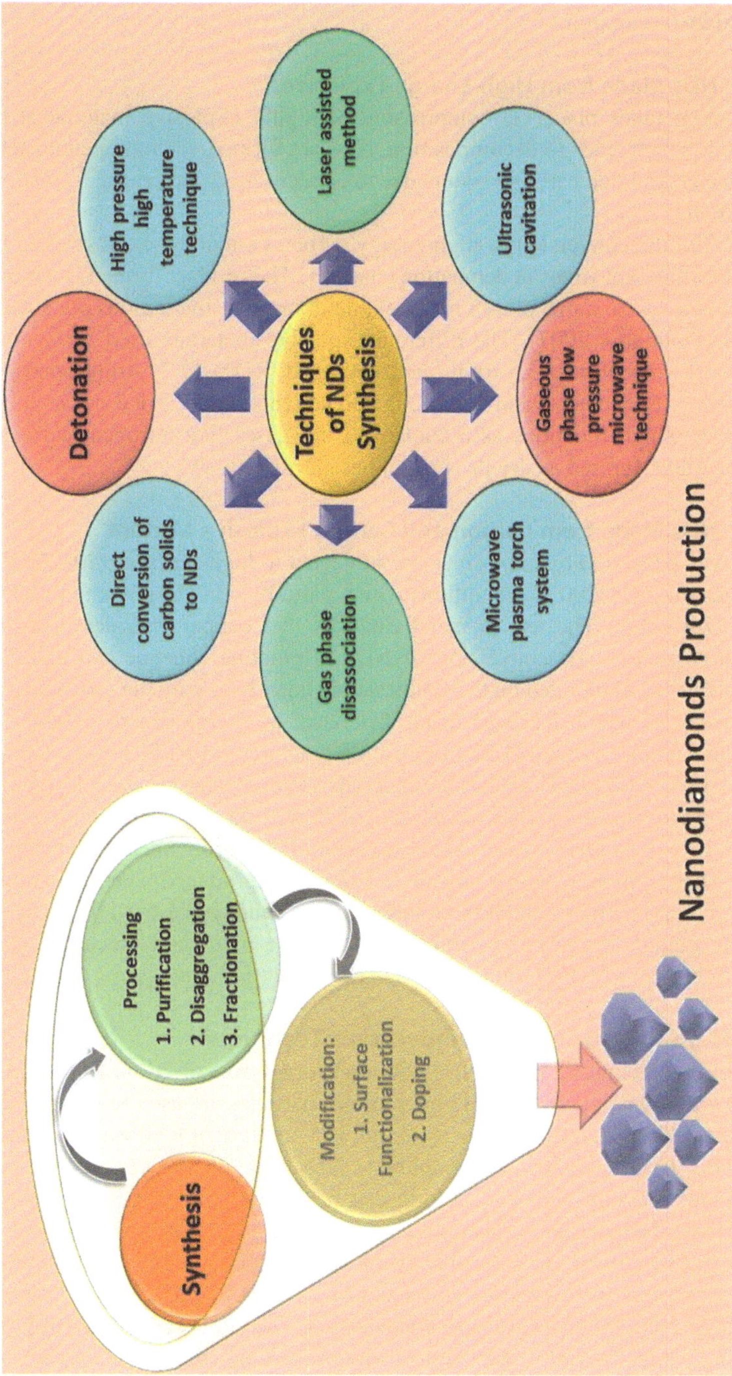

Fig. 6.1 Manufacturing steps of NDs and different synthesis techniques for production

explosives as starting materials whereas the latter involves the use of carbon and carbon explosive mixtures.

6.3.1.1.1 NDs Made from High-Energy Explosives

ND formation takes place via compression of highly explosive material at high temperature which leads to decomposition. NDs are formed via homogenous nucleation by supersaturated carbon vapours due to condensation and crystal formation of liquid carbon.

Practically, the conversion of high-energy carbon-containing explosives to NDs takes by heating explosives in detonation chamber. The product formed is a mixture of NDs having a size range of 4–5 nm, metallic impurities and various other carbon allotropes (Dolmatov 2007). The different explosive materials used for synthesis include TNT (2-methyl-1,3,5-trinitrobenzene) and RDX (1,3,5-trinitroperhydro-1,3,5-triazine) (Shenderova and Nunn 2017). The reaction is carried out in an inert environment which functions as a coolant, using gases like nitrogen and carbon dioxide or water, termed as dry or wet synthesis, respectively.

6.3.1.1.2 NDs Made from Carbon and Carbon Explosive Mixture

Shock waves generated by explosives are subjected to high-pressure (20–200 GPa) and -temperature (>2000 K) conditions which partially convert graphite to NDs of ≈20 nm size range. The use of metal catalysts like copper provides faster heat dissipation which avoids conversion of NDs to graphite (Shenderova and Nunn 2017). Tatsii et al. elaborated on the synthesis of Dalan NDs with the use of different starting materials which had significant difference in charge density (Tatsii et al. 2009). The findings suggested that the use of graphite and RDX having a charge density of 1.61–1.67 g/cm^3 as starting materials yielded diamond particles in a size range of 1–3 μm. However when the mixture of carbon black and RDX with a charge density of 1.34–1.39 g/cm^3 was used as precursor, the NDs obtained were in the size range of 20–80 nm. NDs with a size of 1–3 nm were produced when a mixture of gunpowder, carbon black and RDX having a charge density of 1.67–1.75 g/cm^3 was used as precursor. It can be concluded that the use of carbon precursor greatly influences the size of NDs produced.

6.3.1.2 High-Pressure High-Temperature Technique

The use of this technique yields NDs with an average size of 20 nm. This method describes top-down approach where micron-size particles are used to generate NDs having sizes below 10 nm. The process requires a pressure of 6 GPa and a temperature of 1500 °C for conversion of graphite to NDs. The altered process conditions, that is, lower pressure and temperature, can be used when fullerene and carbon nanotubes are designated as precursors to generate NDs (Cao et al. 2001). The technique has simplified mechanism where at high temperature and pressure tubular structure of precursors breaks and converts to form spheroidal networks leading to formation of NDs (Shenderova and Nunn 2017). Davydov et al. described high-temperature high-pressure technique when starting materials naphthalene, octafluoronaphthalene and their mixtures were used to generate NDs (Davydov et al. 2015). The mechanism suggested for formation of NDs was carbonization of starting materials under pressure. High-pressure high-temperature technique is expensive but is used when controlled doping is to be achieved.

6.3.1.3 Laser-Assisted Method

The use of laser-assisted methods has gained attention during the last decade. It offers numerous advantages like elimination of chemical precursors, reduction in the number of steps of reaction and generation of relatively pure NDs due to elimination of contamination generated from intermediates and reactants. The commonly used methods for ND formation are pulsed laser ablation and light hydrodynamic pulse. Pulsed laser ablation involves focusing of pulsed laser beam to solid target which is immersed in liquid. This results in explosive decomposition of heated surface and formation of bubbles in the surrounding liquid due to boiling of liquid. The bubbles collapse and lead to formation of NDs because of nucleus formation, growth and solidification. The NDs are formed due to occurrence of high pressure and temperature during collapse of bubbles (Yang et al. 2007). This approach gives very less yield of NDs and hence it is not widely used for the preparation of NDs. Zousman and Levinson described light hydrodynamic pulse technique for production of NDs (Zousman and Levinson 2014). A solid target made of carbon source is focused by laser beam in transparent liquid. In this technique, laser beam is not directly focused on target surface. A shock wave created by carbon source generates pressure and temperature conditions sufficient for production of NDs. The selection of optimum laser radiation criteria is essential to generate NDs in desired size ranges. Using light hydrodynamic pulse technique, yield was increased and average particle size of NDs obtained ranged from 4 to 5 nm or 250 to 300 nm.

6.3.1.4 Other Techniques

Ultrasonic cavitation is one of the primitive techniques for the synthesis of NDs. The collapse of bubbles results in extreme high temperature up to 1000 K and pressure conditions up to ~105–106 bar which assists in the formation of NDs (Khachatryan et al. 2008). Galimov used the concept of ultrasonic cavitation for the synthesis of NDs by benzene cavitation destruction. The NDs obtained were in the size range of 10–30 nm. This approach proved the feasibility of ultrasonic cavitation at a small scale (Galimov 1973).

Frenklach et al. (1989) elaborated on the technique of ND synthesis using gaseous-phase low-pressure microwave reactors. The synthesis did not require any substrates. Carbon sources used were mixtures of methane, chloroform, acetylene and dichloromethane. The NDs formed had an average size of 50 nm. The effect of adding heteroatom to carbon sources was studied via diborane addition and the size range of NDs formed was within 5–450 nm (Frenklach et al. 1989). Similar study was performed by Buerki et al. where laser-induced decomposition of ethylene was performed under reduced pressure and temperature yielding diamond particle with a size of 6–18 μm (Buerki et al. 1990).

Ting et al. (2007) developed a technique for ND synthesis using microwave plasma torch system which helped in the elimination of use of high pressure. The synthesis could be efficiently carried out under atmospheric pressure conditions using inert gases like argon or nitrogen. The average size of NDs produced using this technique was 25–50 nm (Ting et al. 2007).

Kumar et al. developed NDs based on gas-phase disassociation of ethanol vapours leading to formation of NDs in a significantly smaller size of 3 nm (Kumar et al. 2013). The addition of hydrogen gas assists in stabilization of NDs formed. The method provides a unique basis for synthesis of NDs of reduced size.

Another technique focused on direct conversion of carbon solids to NDs. The experimental studies reveal that the use of heavy ions or electron irradiation helps in induction of diamond crystallites positioned concentrically in carbon fullerenes. A study reported successful conversion of concentric graphite onion corer to NDs using electron irradiation (energy: 1.2 MeV and temperature: 900 K) (Banhart and Ajayan 1996).

6.3.2 Processing of NDs

The attributes of NDs are dependent on the method of synthesis; therefore additional post-synthesis processes like purification, disaggregation, crushing and fractionation are required (Tinwala and Wairkar 2019).

6.3.2.1 Purification

NDs offer many biomedical applications and thus more emphasis is given on its purity. The purification step is very critical when detonation method is used for production of NDs (Baidakova and Vul 2007). The purity of crude NDs made using denotation is within 30–75%. The NDs contain several metal and metal oxide impurities (1–8%) and non-diamond carbon (25–85%). The primary objectives of purification are elimination of non-diamond carbon and elimination of inorganic contaminants formed by metal and metal oxides (Galli 2010). The metal impurities arise due to ignition chamber, reaction chamber, coolants and reagents. Inductively coupled mass spectroscopy helps in efficient monitoring of these metal impurities (Mitev et al. 2007). The removal of non-diamond carbon contaminants is carried out using selective oxidation. Non-diamond carbons are very reactive and get easily oxidized. The mixture of sulphuric acid, hydrofluoric acid, hydrochloric acid, nitric acid and potassium dichromate in sulphuric acid is commonly used as a liquid-oxidizing agent (Chiganov 2004). Schrand and co-workers compared the efficiency of oxidizing agents used for purification of NDs (Schrand et al. 2009). The best amongst all was the mixture of hydrofluoric and hydrochloric acid which resulted in non-diamond carbon content of 0.2%. The underlying reason for significant reduction in non-diamond carbon content is high solubility of metal oxides in hydrofluoric acid. Also average size of NDs decreased twofold when this mixture was used as an oxidizing agent. An additional step of ion exchange and membrane filtration is a requirement in few cases for complete elimination of non-diamond carbon contamination.

Microwave-based technique for elimination of non-diamond carbon contamination was suggested by Mitev and co-workers (Mitev et al. 2014). The findings suggested the use of microwave-based technique in combination with chelating agents like ethylenediaminetetraacetic acid (EDTA) for elimination of non-diamond contamination. This technique is more advantageous when compared to liquid oxidization as it reduces the cost, is less hazardous and does not require any waste pretreatment. An effective environment-friendly technique of purification of NDs is reported by using ozone gas at elevated temperature (Shenderova et al. 2011a). This technique substantially eliminates the use of hazardous liquid-oxidizing agents. Oxidation is done in the presence of air subjected to elevated temperature of 400–430 °C which is also being looked upon as a purification technique for NDs (Osswald et al. 2006). Selective thermal oxidation of non-diamond carbon contamination in the presence of air using boric anhydride was suggested by Chiganov (2004). The purification techniques assist in removal of impurities like metal oxides and non-diamond carbon impurities and yield NDs with high purity. However the need to break aggregates of NDs remains unaddressed and thereby subsequent step after purification is disaggregation of NDs which helps to de-aggregate NDs.

6.3.2.2 Disaggregation or De-aggregation

In the prior steps of synthesis and purification, NDs have a tendency to form aggregates. There is a need to de-aggregate purified NDs to attain the desired size of NDs. The commonly used techniques include ball milling, jet milling, and thermal and chemical treatments. Table 6.1 enlists key features of various disaggregation techniques. Fractionation is carried out upon successful disaggregation.

6.3.2.3 Fractionation

Disaggregation is less efficient to yield NDs in desired size range; hence an additional step of fractionation is required. To achieve fractionation, NDs formed are dispersed in suitable solvents like sucrose solution followed by centrifugation of dispersion (Yoichi et al. 2008). Higher colloidal stability of NDs is a preliminary criterion for centrifugation. Density gradient centrifugation is proved to be successful in achieving NDs with desired size range (Peng et al. 2013).

6.3.3 Modification

Surface characteristics of NDs influence their therapeutic applications as surface is the immediate point of contact for body's physiological environment. Surface modification thus enables effective targeting. Surface functionalization and doping are commonly used modification techniques for NDs.

6.3.3.1 Surface Functionalization

The adaptable surface chemistry of NDs opens doors for a variety of functionalization schemes. A broad array of functional groups can be incorporated

Table 6.1 Key features of various disaggregation techniques

Disaggregation technique	Key features	References
Ball milling	• Commonly used due to low cost • Can be carried out in wet as well as dry state • Increases contamination and heat arises from technique	Kru et al. (2005), Pentecost et al. (2010)
Jet milling	• Not efficient for reduction in sizes below 1 μm • Used in combination with ball milling • Minimum contamination	Boudou et al. (2009)
Thermal oxidation in air and reduction of hydrogen	• Removal of non-diamond carbon contamination causes size reduction • Introduction of impurities	Etzold et al. (2014)
Chemical treatment	• Uses sulphuric acid, hydrofluoric acid, hydrochloric acid, nitric acid and potassium dichromate in sulphuric acid for eliminating impurities due to metal oxides, thereby causing size reduction up to twofold • Not used widely due to hazardous nature	Schrand et al. (2009)
Ultrasonic disaggregation	• Works on the principle of ultrasonic cavitation • Powerful dispersion technique • Lower yields	Uchida et al. (2007)
Bead-assisted sonic disaggregation	• Uses a combination of ultrasonic and milling media • Introduces contamination, thereby requiring strong chemical treatments	Ozawa et al. (2007)
Matrix-assisted milling	• Eliminates use of costly beads • Reduced chances of contamination • Significant amount of heat is generated	Pentecost et al. (2010)

on the surface of NDs to impart essential characteristics like improved stability, improved solubility, lower degradation at physiological conditions and target-specific delivery (Xing and Dai 2009). Attributing to oxidative processes involved in the production of NDs majority of carbons on the surface of NDs are modified with carbonyl or hydroxyl groups (Huang et al. 2008). A variety of functional groups can be attached on the surface of NDs (Tinwala and Wairkar 2019).

Various moieties containing terminal amino group can be attached on the surface of NDs. To impart stability to these conjugates linker groups like amino methyl, alkyls and siloxanes are used. The Kaiser test type of colorimetric analysis has been developed for quantitation of aminated NDs (Jarre et al. 2014). The surface of NDs can be modified with borane and lithium to increase hydroxyl and carbonyl groups which further enable immobilization of larger biomolecules (Jarre et al. 2014). Air oxidation can be done to oxidized surface carbons of NDs. The reduction of carbonyl groups by microwave-assisted techniques helps in the production of hydrogenated

NDs. Esterification of hydrogenated NDs using acyl chlorides yields NDs with increased alkyl groups (Spitsyn et al. 2006).

Surface functionalization via biomolecules like biotin, peptides, amino acids, biocompatible polymers and proteins can also be carried out for use of biomedical purposes. Hens and co-workers synthesized aminated NDs with the use of aluminium hydride. Aminated NDs were used for conjugation with derivates of N-hydroxysuccinimide and biotin (Hens et al. 2008). Strong interactions between carboxyl groups and amino groups of bovine serum albumin, lysozyme, myoglobin and cytochrome C resulted in the absorption of proteins on the surface of NDs (Huang and Chang 2004). Poly(amino acid)-functionalized NDs were synthesized by Xu et al. by polymerization of alpha-amino acid due to ethylene glycol and glutamic acid (Xu et al. 2018). The functionalized NDs exhibited improved dispensability and lower toxicity. The use of polymers in surface functionalization is for improving hydrophilicity, stability and target-specific delivery and imparting properties to withstand physiological conditions of body. Polyethylene glycol (PEG) is the most widely used polymer for surface modification of NDs (Huang et al. 2017). PEG has several limitations like its tendency to oxidize in the presence of enzymes and immune reactions. To solve these limitations, polymers like thiol-containing PEG and dextran polysaccharides are being used as alternatives (Huang et al. 2018; Wu et al. 2015). The attachment of polymers to the surface of NDs is via non-covalent adsorption where interactions are spontaneous or due to multiple steps of process. Another mechanism for surface functionalization with polymers is termed as covalent grafting where sophisticated techniques of polymerization like atom transfer radical polymerization (ATRP) and reversible addition-fragmentation chain transfer (RAFT) polymerization have been used (Liu et al. 2004; Chen et al. 2018). Surface functionalization of NDs with different groups is shown in Fig. 6.2.

6.3.3.2 Doping

Defects in the internal structure of jewels cause significant reduction in their costs. However for NDs creation of internal defects helps to broaden the scope of applications of NDs. Doping involves controlled introduction of foreign atoms like silicon, nitrogen, transition metals or boron on the surface of NDs. These foreign atoms create optically active centres which are responsible for different applications of NDs (Gaillou et al. 2014). In a study comparing nitrogen content of NDs during doping, it was observed that nitrogen is not completely utilized in the synthesis process and it gets trapped into the core of NDs, thereby creating optically active centres termed as nitrogen vacancy (NV) (Shenderova et al. 2011a, b). Silicon vacancy is seeking attention and can be created using silicon as doping agent. The silicon vacancy centres have a unique feature of photoluminescence and may be used as the potential candidate in bioimaging (Vlasov et al. 2009). This silicon vacancy can be generated by chemical vapour deposition or ion implantation (Neu et al. 2013; Wang et al. 2006). The intensities of silicon vacancy centres are enhanced upon adding nitrogen gas during the growth phase of NDs (Zhang et al. 2014). Boron-doped NDs are widely used in the electronic industry; however the low boron content (80 ppm) of NDs causes lower conductivity, thus limiting the applicability

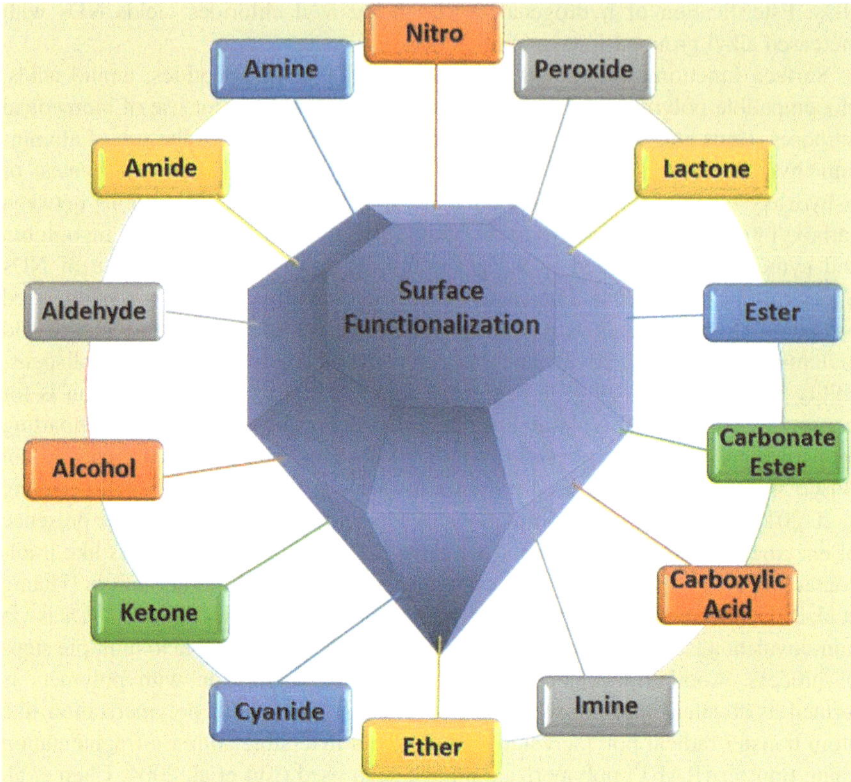

Fig. 6.2 Surface functionalization of NDs with various functional groups

of boron-doped NDs. To overcome this limitation direct high-pressure high-temperature synthesis with organo-boron compound as starting material or an alternative technique like solid-state diffusion is suggested (Ekimov et al. 2015; Kondo et al. 2015). For production of doped NDs, diffusion techniques based on thermal and electric field have also been developed (Krutko et al. 2000) (Prelas et al. 2014). To attain highly doped NDs in greater quantities still remains an unaddressed challenge at a commercial scale.

6.4 Characterization of NDs

Characterization of NDs is of prime importance due to their different applications including in the field of biomedical sciences. The characterization of NDs focuses on the determination of particle size and size distribution, identification of impurities and ratio of sp^3 and sp^2 hybridization of carbon.

6.4.1 Structural Characterization

The preliminary structural characterization of NDs is done using Fourier transform infrared spectroscopy (FTIR) for identification of functional groups. FTIR gives basic insight into surface chemistry of NDs and is advantageous when surface-modified NDs are to be characterized (Petit and Puskar 2018). NV centres can be identified by two broad bands seen in the region, 1100–2500 cm^{-1}. Functional groups like hydroxyl and amine exhibit peaks in the range of 3200–3600 cm^{-1} and 3000–3300 cm^{-1}, respectively (Ji et al. 1998). Amine bands are less intense and sharper when compared to hydroxyl bands. Nuclear magnetic resonance (NMR) helps in the determination of molecular structure of NDs by measuring the interaction of nuclear spins when placed in an external magnetic field (Fang et al. 2009). Alternatively electron spin resonance (ESR) measures the behaviour of electrons in NDs and helps in the identification of unpaired electrons which helps in detecting NV centres undergoing strong spin-sensitive optical transitions and transition metal impurities like iron can be well identified using this technique (Baranov et al. 2009). Raman spectroscopy is the commonly used technique for characterization of carbon systems. It aids in the determination of structure and composition of NDs. It works on the principle of inelastic scattering of monochromatic light. The data obtained can be correlated with sp^3 content, elastic constants and chemical constitution of NDs, thereby illustrating purity of NDs (Ferrari et al. 2004).

6.4.2 Particle Characterization

Scanning electron microscopy (SEM) studies give an insight into composition, morphology and topography of NDs. High-resolution transmission electron microscopy (HRTEM) is an imaging technique which helps in assessing the atomic structure of NDs, crystallinity and presence of contaminants (Alexenskii et al. 2002). X-ray diffraction can be used to demonstrate the particle size, atomic arrangement and crystallinity of NDs (Hawelek et al. 2008).

6.5 Hurdles for Transit of NDs to Market

The introduction of NDs in drug delivery and bioimaging has created a significant impact on nanoscience domain due to their distinctive properties which offer versatility in structural modification, enhancement in biocompatibility and reduction in toxicity. Despite advancements in manufacturing, and improvement in safety and efficacy of drugs, there still remains a gap between clinical translations of NDs. After successfully addressing these hurdles, the purpose of improvement in treatment outcomes for globally dominant diseases like cancer and tissue repair could be easily bridged with the use of this multifunctional carrier, NDs. Unlike other carriers which have only therapeutic applications, NDs extend its application to effective diagnostic agents as well. The biggest hurdle in the transit of NDs is the cost involved in developing new delivery systems or diagnostic aids. Therefore, the development of NDs should focus on identification of compounds suitable for production of scalable NDs which have characteristic features of being safe, efficacious and cost effective. The methods used for synthesis of NDs should be giving an edge in manner which can be easily correlated to production at commercial scale. The development of scalable techniques for doping which yield highly doped and purified NDs still remains a challenge, thereby limiting the application of NDs. The lack of clarity pertaining to pharmacokinetic parameters of NDs acts as a major constraint for its clinical translation. The clear regulatory requirements for nanomedicine may act as a catalyst for advance developments in the field of NDs. Thus, overcoming these obstacles using robust validation procedures would surely take NDs a step closer to clinical translation (Vaijayanthimala et al. 2015).

6.6 Applications of NDs

The advancement in the field of nanoscience and nanotechnology has led to the synthesis and discovery of a wide range of nanoparticles. NDs are the revolutionary nanoparticles with unique mechanical, biological, chemical and optical properties, which make them important for several diagnostic, imaging and therapeutic applications, as summarized in Fig. 6.3 (El-say 2014; Gogotsi et al. 2015).

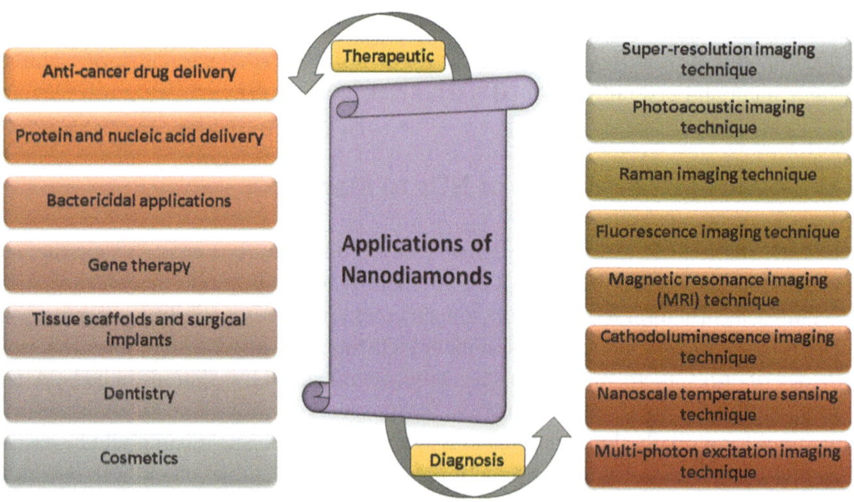

Fig. 6.3 Therapeutic and diagnostics applications of NDs

6.6.1 NDs as Diagnostic Agents

NDs are effective in diagnosis due to their sharp Raman signal peaks, high refractive index and other optical properties. These properties make them superior to the fluorescent nanoparticles which possessed demerits, such as light scattering from tissue, cell autofluorescence and colour fading (Barnard et al. 2009). NDs are non-fluorescent and optically inactive in the near-infrared region in their pure crystalline form. However, before using them as diagnostic agents, they are made optically active by chemical conjugation or creating colour centres.

The colour centres in the NDs can be used to make them fluorescent. This is made by creating vacancies utilizing atoms like silicon, nitrogen or europium (Su et al. 2013; Vlasov et al. 2009). NV centres are commonly studied as nitrogen is the most commonly found impurity in NDs (Merson et al. 2013; Miguel and Jacobs David 2012). However, silicon vacancy centres are more beneficial as compared to NV centres because of weak and narrow vibronic sideband, narrow photon emission maximum at 739 nm and shorter excited-state lifetime (Andrea et al. 2016). The immovable and structurally unstable vacancy centres are produced by irradiating the NDs with high-energy helium, protons, gamma radiations and electrons (Lin and Su 2014; Davies et al. 1992).

Although NDs can be effectively made fluorescent using widely used colour centre method, chemical conjugation can also be used to make NDs fluorescent. Various chemical groups, such as amine, ketone, esters, hydroxyl and octadecyl, have a significant effect in imparting fluorescence to NDs (Wahab et al. 2017; Pentecost et al. 2017).

6.6.2 Application and Utilization of FNDs in Bioimaging

Stable bright fluorescence with the absence of photon bleaching makes FNDs useful as a diagnostic agent along with several other imaging techniques. The recent development in microscopy, i.e. stimulated emission depletion microscopy (STED), is useful in studying FNDs delivered through various methods (Yu et al. 2005). A resolution of up to 40 nm was obtained which enabled differentiation of each FND particle which was precoated with bovine serum albumin up to the size of 30 nm and which also differentiated them from the endosomal trapped FNDs (Tzeng et al. 2011). STED could help in the resolution of NV centres in the FND size of about 40–250 nm. The resolution by STED microscopy was about −10 nm without photobleaching (Arroyo-Camejo et al. 2013).

Photoacoustic imaging is the latest technique in biomedical applications which provides high-resolution images without any invasive procedures. The working of photoacoustic imaging technique involves the illumination of the sample with short and focused laser pulses (Ntziachristos 2010). A study by Wang et al. indicated that NDs administered through subcutaneous route were detected using photoacoustic imaging in mouse (Wang and Hu 2013). Diamond gives a sharp peak at 1332 cm^{-1}

which indicates that FNDs can be used effectively in Raman imaging technique as a Raman label (Cheng et al. 2014).

FNDs are fluorescent in nature which aids in studying the interaction of FNDs and their conjugates with various constituents of cells and are useful in drug tracking (Weng et al. 2009). Recent studies regarding the management of FNDs at the cellular level reported that FNDs localize inside cells. It also reported that the endocytosis and exocytosis of FNDs occur simultaneously and are useful for cell tracking for a prolonged period of time (Prabhakar et al. 2017).

The presence of a unique electronic spin and a ground state which is magnetically sensitive in the NV centres in NDs makes rapid and ultra-sensitive detection. The modifications in the fluorescence signal were observed during the transition from ground state by Gruber et al. This suggested the usefulness of NV centres in nanoscale magnetic resonance imaging (MRI) (Gruber et al. 2014). A novel MRI contrast substance has a negligible amount of cell toxicity and advanced imaging capabilities. These novel NDs are produced using Fe ion (Lin et al. 2018).

Cathodoluminescence imaging technique involves the detection of the image when the electron strikes the luminescent material causing the emission of photons. FNDs are stable with a high fluorescent intensity which opens new possibilities for cathodoluminescent imaging. A study by Glenn et al. indicated that 40–80 nm FNDs of red and blue colour are useful in a cathodoluminescence imaging process with a resolution of about 5 nm (Glenn et al. 2012).

The temperature changes occurring in and around various cells can be detected using nanoscale temperature-sensing technique and its correlation with the biological process is helpful in the diagnosis of several diseases (Jaque and Vetrone 2012). FNDs are proved to be effective nanothermometers by Kucsko et al. who in their studies detected temperature changes in human embryonic fibroblasts (Kucsko et al. 2013).

Optical imaging of the thick tissues and lower damage to the cells are significant merits of this multiphoton excitation imaging technique. Hui et al. proved that the FND particles of the size of about 40 nm can be detected using multiphoton excitation imaging technique (Hui et al. 2010).

6.6.3 NDs in Therapeutic Applications

Numerous studies have been carried out to evaluate the therapeutic effectiveness of NDs as a drug delivery system by loading various therapeutic agents. These drug-loaded NDs are then administered and introduced into the biological system of a living organism and bioavailability and biocompatibility of the ND drug delivery system were evaluated. Varying class of drugs, such as anticancer drugs, antibiotics and anti-inflammatory drugs, have been loaded on NDs (Ezzeddine and Khashab 2014). The primary reason for use of NDs as drug vehicles is their non-toxic nature and the absence of immune response. They are also found to be easily binding to several drug molecules and delivering them at the target site. The characteristics of the carrier and its interaction with different components of cells are important aspects to be considered in the drug delivery system. NDs deliver the drug by diffusing through the plasma membrane and reaching the target site present intracellularly (Boudou et al. 2008). In another study, clathrin-mediated endocytosis process was adopted by NDs of the size of 46 nm approximately after an incubation period of 120 min to enter the HeLa cells. But micropinocytosis was adopted by the NDs (100 nm) to enter HeLa cells after an incubation period of 240 min (Boudou et al. 2008). Study by Perevedentseva et al. noted the interaction of NDs with various cell lines including the cancerous cell lines, i.e. A549 human lung carcinoma cells and non-cancerous cell lines, i.e. Beas-2b human bronchial epithelial cells and HFL1 human lung fibroblast cells (Perevedentseva 2013). Confocal microscopy was used to detect the effects of ND drug delivery system and it was found that cancerous cells are attacked by the NDs loaded with the drug at a very large extent as compared to non-cancerous cells (Liu et al. 2007; Faklaris et al. 2009).

6.6.3.1 Anticancer Drug Delivery

Several studies suggested that NDs are useful in delivering the anticancer drug, such as paclitaxel, doxorubicin, cisplatin and gemcitabine (Lim et al. 2016; Li et al. 2016). ND-paclitaxel is found to have an enhanced therapeutic effect with better cellular uptake of the drug. ND-doxorubicin conjugate is found to have enhanced drug loading, controlled release of drug, better distribution, higher tumour cell apoptosis and better inhibition of tumour (Master and Sen Gupta 2012).

The effectiveness of NDs in the modulation of drug release was also evaluated in several studies. A study indicated that steady and continuous-release pattern of doxorubicin for a period of 1 month was observed from a topical patch containing ND-doxorubicin conjugate (Lam et al. 2008). The development of resistance of tumour cells against the existing mode of treatment and chemotherapeutic agents is the major obstacle in the path of treatment. However, studies have reported the therapeutic effects of ND-doxorubicin on resistant cancer cells and indicated that the conjugate enhanced apoptosis and inhibited the growth of cancer cells (Chow et al. 2011).

The active constituents of plants have gained significant importance as anticancer agents, and thereby their conjugation with NDs and effects on targeting the tumour are of importance. Studies were performed to evaluate the effectiveness of the ND conjugate with a secondary plant metabolite called citropen compared with the effects of free citropen. It was found that the conjugate was better in inhibiting the tumour cell cycle and reducing the tumour growth (Gismondi et al. 2016).

A monoclonal antibody, cetuximab, is beneficial in the targeted delivery of drug for the management of colon cancer. NDs were used in delivering cetuximab and cisplatin combination which enhances the specific recognition and selective cell cytotoxicity (Li et al. 2017). NDs are suitable in delivering several small-molecule drugs and can be used as a carrier for many drugs used in the treatment of cancer (Tinwala and Wairkar 2019).

Applications and advantages of NDs for cancer treatment are depicted in Fig. 6.4.

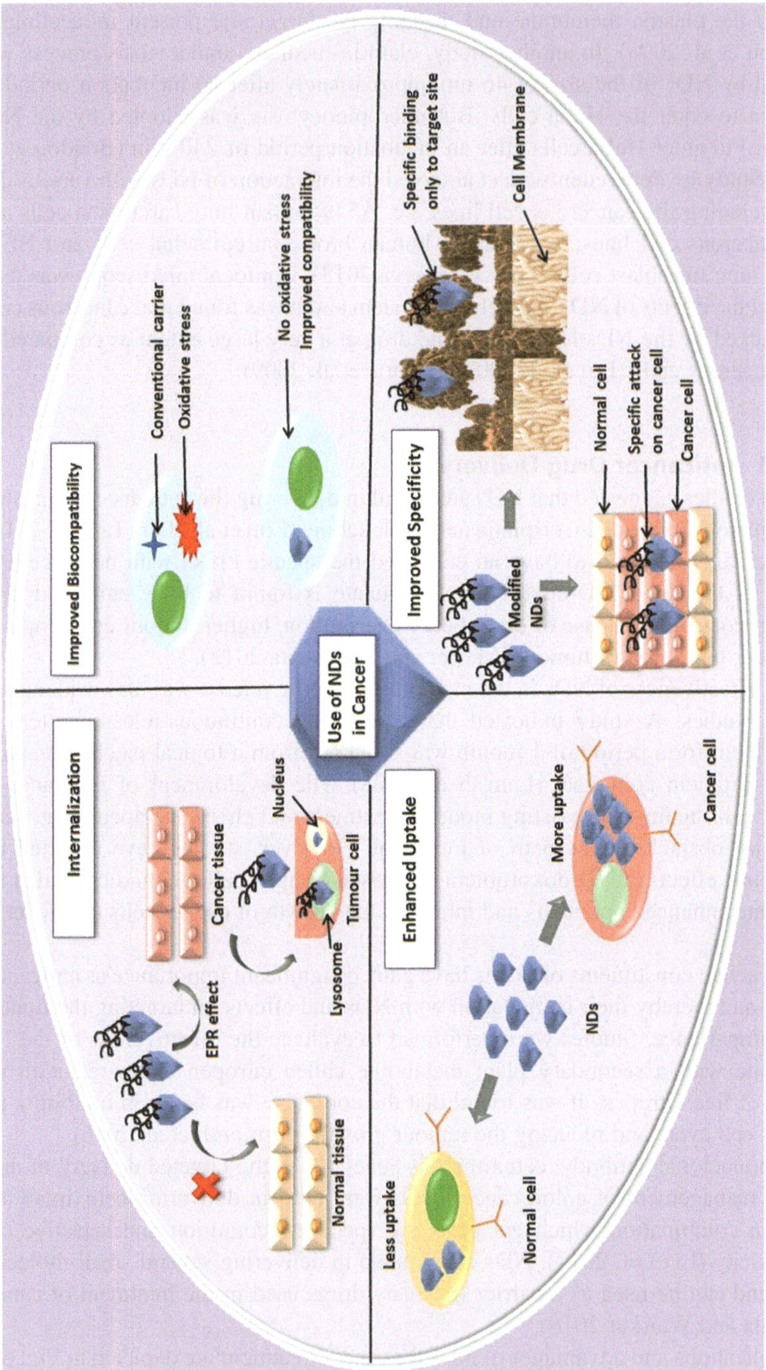

Fig. 6.4 Applications and advantages of NDs for cancer treatment

6.6.3.2 Protein and Nucleic Acid Delivery

Peptides and proteins have both anionic and cationic groups on their surfaces which supports their loading on NDs through charge-charge interaction and hydrogen bonding also facilitates physical adsorption (Mochalin et al. 2011). A recent study also suggested that unconjugated NDs are also helpful in delivering protein to the site of action. In this case, NDs do not need any surface modification for the effective targeting of fibroblast growth factors (Balek et al. 2018).

The nucleic acid is found to have great potential as a therapeutic agent in the treatment of several chronic diseases. However, certain limitations like inefficient intracellular delivery of the drug due to the high molecular weight and excessive surface negative charges hinder the application of this agent (Wiethoff and Middaugh 2003). This created a need to develop a vector which could carry the genetic material like nucleic acids inside the cells. NDs are found to fit into this requirement of the mobile vector and are also free of the drawbacks of conventional vectors such as low efficiency and high toxicity (Whitehead et al. 2009).

6.6.3.3 Bactericidal Applications

Plain NDs can act as antibacterial agents, effective against gram-negative organism called *Escherichia coli.* The ability of *Escherichia coli* to form colony is significantly reduced by plain NDs and a concentration of 50 μg/mL causes 100% inhibition of the bacteria (Konopa and Beranova 2012). Later, it was also found that glycosylated NDs are useful in the detection and elimination of various bacteria from water (Hartmann et al. 2012). Surface modification of NDs with oxygen-containing functional groups is shown to be effective in the inhibition of gram-positive and gram-negative bacteria. The bactericidal activity is linked to partially oxidized and negative surface charge present in functional groups (Wehling et al. 2014).

Lysozyme immobilization by adsorption or non-covalent bonding on NDs was investigated. It was observed that lysozyme immobilized on NDs exhibited catalytic action and caused lyses of gram-positive and gram-negative bacteria (Mogilnaya and Bondar 2012). The effect of menthol-modified NDs was studied on gram-positive (*Staphylococcus aureus*) and gram-negative (*Escherichia coli*) bacteria. These NDs were found to be non-toxic, and showed improved anti-biofilm action. Menthol-modified NDs were found to be more efficient in the reduction of biofilm formation when compared to ampicillin, free menthol and oxidized NDs (Turcheniuk et al. 2014).

6.6.3.4 Gene Therapy

Gene therapy is promising in the treatment of inherited disorders since it is effective and safe. NDs are immobilized with 800 Da polyethyleneimine (PEI800) and covalent attachment of amine groups. The modified NDs were noncytotoxic and exhibited high transfection efficiency. They also offer a platform for incorporating functional groups essential for cell-specific targeting and these carriers could easily form complexes with deoxyribonucleic acid (DNA) (Zhang et al. 2009).

The entry of NDs in cells is by endocytosis followed by intracellular translocation, leading to quicker endosomal escape and stable residence in cytoplasm.

Minimal cytotoxicity is observed to be associated with cytosolic release of NDs. The unique features of NDs like higher cellular uptake and exhibiting of cytosolic release facilitate its application to gene delivery. Green fluorescent gene was taken as an example to demonstrate cytosolic delivery of genes using NDs as carrier (Chu et al. 2015).

6.6.3.5 Tissue Scaffolds and Surgical Implants

Enhanced mechanical characteristics, potential surface modification and biocompatible nature of NDs make them attractive carriers for bone tissue engineering and use in surgical implants. A fluorescent composite bone scaffold was produced using biodegradable polymer, poly(L-lactic acid) and octadecylamine-functionalized NDs. An increase in the hardness of scaffold up to 200% and Young's modulus up to 800% was observed on addition of 10% w/w octadecylamine-functionalized NDs. There were no negative effects in cell proliferation observed in murine osteoblast for a period of 1 week. This investigation thus opens up avenues for use of NDs as tissue scaffolds, regenerative medicine and fixation of musculoskeletal tissue (Zhang et al. 2011). A tissue scaffold suitable for bone tissue engineering was produced using poly(lactic-co-glycolic acid) (PLGA) loaded with NDs based on phospholipid compound. In comparison to free PLGA matrix, an enhancement in hardness (550%) and Young's modulus (100%) and significant decrease in water contact angle from 80 to 55° were noted. In vivo biocompatibility studies in mice revealed that these carriers were highly biocompatible (Zhang et al. 2015). The enhancement of mechanical properties as well as biocompatible nature makes NDs a potential carrier in bone tissue engineering.

ND-based magnesium implants helped in reduction of corrosion in physiological environment. The corrosion rate was reduced 4.5 times in these implants. The underlying reasons for enhancement in corrosion resistance may be physical adsorption and chemical interaction of NDs. The in vitro biocompatibility tests indicated that ND-based magnesium implants possess high biocompatibility. These could thus prove promising candidates for developing biodegradable surgical implants (Gong et al. 2015).

6.6.3.6 Dentistry

Root canal therapy is used to address infected pulp tissue and aids in protection against subsequent infection. This therapy involves removal of dental pulp, decontaminating infected tissue and root canal sealing using fillers wherein gutta-percha is the commonly used filler material. Lee and co-workers developed NDs embedded with gutta-percha functionalization with amoxicillin (Lee et al. 2015). Material characterization studies indicated improvement in mechanical properties in comparison to conventional gutta-percha. The bacterial growth inhibition assays confirmed functionalization of NDs with amoxicillin. The study had promising treatment outcomes and could aid in further advancements in dentistry. A reinforced dental filler was produced using silver-loaded polycation-modified NDs. The NDs synthesized had excellent mechanical properties and also aided in combining bactericidal effect of silver (Cao et al. 2018).

6.6.3.7 Cosmetics

NDs offer excellent skin absorption and can be used in a variety of skin care products like deodorants, cleansers, dermal strips, creams and soaps. NDs bind easily to skin surface and help to keep skin hydrated and prevent ageing of skin. NDs also assist in protection of skin from harmful ultraviolet rays of sun and are recommended for use in sunscreens (Namdar and Nafisi 2018). ND-based gel for use in microdermabrasion was prepared which showed no allergic response and erythema (Janaszczyk et al. 2017). A ND-based sunscreen for providing blockage of ultraviolet radiation was prepared with an average size of NDs >60 nm. Further, NDs were doped with nitrogen and incorporated in sunscreen. These nitrogen-doped NDs exhibited enhancement in absorption of sunscreen, attributed to changes created in internal structure (Shenderova and Grichko 2016).

6.7 Biocompatibility and Toxicity of NDs

Lampblack, diamond and charcoal are some of the finely divided carbon particles which have several useful applications since ancient times. With the advancement in nanoscience and nanotechnology, carbon-based nanomaterials like NDs have been found to have effective therapeutic properties with low chemical reactivity, making them a useful material in numerous diagnostic and therapeutic applications. In the emerging era of ND application, it becomes necessary to evaluate its adverse effects, biocompatibility and toxicity (Schrand et al. 2006). Puzyr et al. conducted an in vitro study showing that NDs alter the kinetics of the formation of reactive oxygen species destroying the leucocytes and haemolysis of red blood cells (Puzyr et al. 2004). Subsequently ND-based sterile hydrosols were prepared exhibiting higher colloidal stability and biocompatibility.

Another study by Yuan et al. evaluated the bio-distribution and fate of NDs in vivo (Yuan et al. 2009). An accumulation of NDs (50 nm) was observed in the liver after injection through the intravenous route in the mouse. Lungs and spleen were also found to be other major targets of ND particles. Raman spectroscopy and electron micrograph imaging of the solutions prepared from the digested organs proved the entrapment of ND particles for a prolonged period in the liver and lungs of the mouse. The concentration of NDs in the urine and faeces sample was negligible when measured through Raman spectroscopy. Since NDs were entrapped in vital organs of the body, the adverse events associated with these particles should be evaluated before using them for any biological purposes in humans.

A study by Schrand et al. published in 2006 assessed the cytotoxicity of NDs using cell viability assays such as luminescent ATP production and mitochondrial function (Schrand et al. 2006). The results indicated that NDs were non-toxic to several types of cells and were safe as they produced an insignificant amount of reactive oxygen species. Thus, NDs are ideal materials for many biological purposes and in different cell types. Schrand et al. continued the study which reported the differential biocompatibility of carbon nanotubes and NDs (Schrand et al. 2007). In this study, they evaluated the lung and neuronal cell lines for biocompatibility in

NDs, carbon black and single- and multi-walled carbon nanotubes. The results suggested that NDs had unique biocompatibility which was greater than that of other materials. In addition, the amount of reactive oxygen species generated in the case of NDs was insignificant as compared to other materials, proving its safety and nontoxicity. Carboxylated nanometre-sized diamonds (cNDs) are used for conjugation of molecules like DNA and protein after surface modification. A study by Liu et al. reported that atomic force microscopy, confocal microscopy and flow cytometry are effective tools in the identification and detection of cND interacting with HDL-1 normal fibroblasts and lung A549 epithelial cells (Liu et al. 2007). NDs can be used in biomedical applications based on various characteristics, such as detectability, level of toxicity to the human cells and uptake ability. In another study, it was indicated that the cells take NDs by a process like clathrin-mediated endocytosis pathways and macropinocytosis (Liu et al. 2009). The endocytic ND particles were not found to affect the cell growth ability of both 3T3-L1 embryonic fibroblasts and A549 lung cancer cells. NDs were found to be safe in cell division and differentiation which makes them useful in labelling as well as tracking of cancerous cells. NDs even did not affect the gene and protein expression of the regular cell cycle.

Zhang et al. (2012a, b) published a comparative study to evaluate and compare the cellular uptake, safety and toxicity of graphene oxide, multi-walled carbon nanotubes and ND (Zhang et al. 2012a, b). The result of this study suggested that NDs have the greatest cellular uptake and least toxicity as compared to other materials. In another study, ND-based hydrogels were evaluated for safety and biocompatibility and it was concluded that NDs were safe and biocompatible; however emphasises was laid on conducting genotoxicity studies for NDs (Huang et al. 2007).

Although the above-stated studies proved that NDs are safe and biocompatible and the cytotoxicity data was also evaluated, the genotoxicity study was still necessary. In the genotoxicity of NDs, the embryonic stem cells were incubated and evaluated with NDs (Xing et al. 2011). Surface chemistry-specific genotoxicity was noted as the raw NDs produced lesser DNA damage as compared to the oxidized NDs. However, both raw NDs and oxidized NDs produced lesser DNA damage as compared to the multi-walled carbon nanotubes. The effects of carboxylated NDs on different cell lines including liver, kidney, lung and intestine indicated that NDs are neither cytotoxic nor genotoxic at a concentration of up to 250 μg/mL. This study was performed on six cell lines which makes NDs a potential agent for the diagnostic and therapeutic applications in humans along with an effective negative control agent in numerous nanotoxicological studies (Paget et al. 2014).

In vivo bio-distribution of labelled NDs, such as those labelled with 18F, was performed using positron emission tomography (Rojas et al. 2011). It was found that the radiolabelled NDs accumulated primarily in the liver, lungs and spleen. These NDs were found to be primarily eliminated through the excretory system, mainly through the urinary tract. When surfactants were added to this formulation, it did not lead to any change in the effects except for a slight reduction in the urinary excretion of the NDs.

The aforesaid studies provide useful information regarding the safety, biocompatibility, bio-distribution, genotoxicity and cytotoxicity, which can help in the development of novel strategies and design of newer methods for the efficient delivery of drugs to the target site with least amount of toxicity to the human cells.

6.8 Recent Advances in NDs

The current advances in NDs are focused on different techniques for production of NDs, role of surface-functionalized NDs, and applications of NDs as a carrier for drug delivery for treating a broad spectrum of diseases and as a potential candidate in bioimaging techniques.

Table 6.2 summarizes recent advancements in the emerging sector of NDs.

In diagnostics, the current fluorescence-based imaging agents utilized in in vivo imaging show poor photostability and biocompatibility. To overcome these issues of present diagnostic agent, FNDs were prepared to target vascular endothelial growth factor receptors in tumour cells. Recently, click chemistry technique was used in the preparation of FNDs where vascular endothelial growth factor sites were conjugated with cyclooctene. The FNDs showed high affinity for vascular endothelial growth factor receptors and selectivity for tumour cells (Torelli et al. 2019). NDs were used in ^{13}C MRI because maximum nuclear polarizations were attained without rapid spin relaxation. ^{13}C polarization in NDs was achieved using acid cleaning and air oxidation and thus these NDs can be explored as a potential bioimaging agent (Boele et al. 2019). In another case, NDs doped with carbon aerogel opened a platform for creation of polyatomic defects in ND structure using target-specific molecular precursors, thereby enhancing the specificity of NDs in diagnostic applications (Borja 2019).

In therapeutic applications, functional supraparticles were produced using alkyl amine-modified NDs for treatment of cancer. Camptothecin, widely studied anticancer drug, was loaded on NDs which demonstrated improvement in efficacy and biocompatibility. The NDs loaded with camptothecin exhibited excellent passive targeting potential due to enhanced permeability and retention effect as well as offered flexibility in modifying the size of NDs by change in alkyl chains (Yu et al. 2019). NDs dispersed in polycaprolactone-based scaffold were prepared and results indicated improvement in wound healing with ND-based scaffold due to increase in proliferation of epithelial cells and restriction of microbial activity (Houshyar et al. 2019). Chitosan is a widely used biopolymer for drug delivery and the presence of amino group in chitosan structure aids in the formation of composites with NDs. These ND-based composites can be explored as matrix for development of bionanocomposites due to their superior mechanical and optical properties (Sukhova et al. 2019).

Table 6.2 Recent advancements in NDs

Title	Conclusion	Reference
Targeting fluorescent nanodiamonds to vascular endothelial growth factor receptors in tumour	• Selective tumour targeting • Enhanced biocompatibility	Torelli et al. (2019)
Nanodiamonds for optical imaging and beyond	• Gives an insight into applications of NDs in biomedical imaging • Elaborates on techniques like STED and MRI	Prabhakar and Rosenholm (2019)
Tailored nanodiamonds for hyperpolarized 13C MRI	• Illustrates use of NDs in hyperpolarized 13C MRI • 13C MRI is of significance in metabolic imaging of heart, brain and prostate	Boele et al. (2019)
Anticancer drug delivery to cancer cells using alkyl amine-functionalized nanodiamond supraparticles	• Camptothecin-loaded NDs were prepared for use in osteosarcoma • NDs caused greater tumour suppression	Yu et al. (2019)
Nanodiamonds formed with defect centres using doped carbon aerogel	• Defect was created using nitrogen • Can be used in bioimaging	Borja (2019)
Surface engineering and applications of nanodiamonds in cancer treatment and imaging	• Describe the role of surface-functionalized NDs in cancer treatment and diagnosis • NDs coated with bioactive compounds, synthetic polymers and NDs/inorganic composites are elaborated	Lai et al. (2019)
Nanodiamond/poly-ε-caprolactone nanofibrous scaffold for wound management	• Detonation NDs were synthesized using caprolactone nanofibrous scaffold which exhibited wound-healing potential • The preparation had improved wicking and moisture absorption properties	Houshyar et al. (2019)
Synthesis and characterization of porphyrin-functionalized nanodiamonds	• Illustrates synthetic pathway for functionalization of NDs with porphyrin using thermal treatment • Functionalized NDs showed higher stability when compared to non-functionalized NDs	Picollo et al. (2019)
Preparation and properties of chitosan–nanodiamond dispersions and composite films	• Chitosan ND films were prepared which showed improved mechanical properties	Sukhova et al. (2019)
Polyelectrolyte-stabilized nanodiamond dispersions	• Stable dispersions of negatively charged NDs were synthesized using amphiphilic, cationic block copolymers • Polyelectrolyte complexing of NDs prevents aggregation and clustering of dispersions	Tiainen et al. (2019)

Table 6.3 Recent patents on nanodiamonds

Patent number	Applicant	Title	Publication date	Summary
8,865,916	Beaujuge; Pierre M.El Tall; OmarRaja; Inam U.	Functionalized diamond nanoparticles	Jan 9, 2014	Illustrates the functionalization of NDs with dienophile
9,636,650	Zousman; Boris	Method and system for controlled synthesis of nanodiamonds	Jun 12, 2014	Gives an insight into synthesis of NDs via irradiation energy beam
9,359,282	University of Saskatchewan	Functionalized nanodiamonds as delivery platforms for nucleic acids	Oct 23, 2014	Illustrates the method for functionalization of NDs using nucleic acids
10,391,172	Commissariat A L'Energie Atomique Et Aux Energies Alternative	Use of nanodiamonds for generating free radicals for therapeutic purposes under radiation	Jul 2, 2015	Describes the use of NDs in treating tumours
9,486,163	Verily Life Sciences LLC	Silicon vacancy-doped nanodiamonds for molecular and cellular imaging	Aug 27, 2015	Describes doped NDs as useful imaging agents
9,435,791	Verily Life Sciences LLC	Method for using nanodiamonds to detect nearby magnetic nanoparticles	Dec 31, 2015	Gives an insight into the use of NDs for diagnostic purpose in cancer
9,889,076	International Technology Center Inc.Adamas Nanotechnologies Inc.	Light-attenuating formulations	Jun 16, 2016	Elaboration on NDs providing light or radiation attenuation between about 190 and 800 nm
9,573,815	National Tsing Hua University	Thiolation method for modifying nanodiamonds	Mar 17, 2016	Elaborates on steps to generate thiolated NDs
10,294,162	Kobe Steel, LTD.	Detonation-mediated carbon particle production method	Nov 3, 2016	Gives insight into the use of detonation technique for producing NDs
10,081,546	Hyundai Motor Company	Nanodiamond, method of	Mar 2, 2017	Elaborates on the method for

(continued)

Table 6.3 (continued)

Patent number	Applicant	Title	Publication date	Summary
		manufacturing the same, and nano-fluid using the same		production of NDs and subsequent oxidation of surface of NDs
10,287,824	Baker Hughes IncorporatedDiamond Innovations, Inc	Methods of forming polycrystalline diamond	Sep 7, 2017	Discusses the method to form polycrystalline NDs using group VIII metals, aluminium and stabilizer
10,406,500	Arnold; Christopher J.	Composition comprising nucleated nanodiamond particles	Dec 28, 2017	Elaborates on the method for non-detonation synthesis of NDs
10,240,251	North Carolina State University	Synthesis and processing of pure and NV nanodiamonds and other nanostructures for quantum computing and magnetic sensing applications	Dec 28, 2017	Describes synthesis and processing involved in nitrogen vacancy NDs

6.9 Patents Reported

Variety of methods for synthesis of NDs and applications of ND carriers in biomedical and diagnostic purposes are reported in patents and Table 6.3 enlists an overall summary of patents of NDs.

6.10 Conclusion

NDs have found a broad array of applications in therapeutic as well as diagnostic field. With a research focusing on advancements and simplification in synthesis, purification and modification of NDs, the transit to market would fasten for this novel carbon-based carrier. NDs have already proven to be a suitable carrier for various anticancer, antibacterial and protein-based drugs as well as are used for a variety of applications in bioimaging. The key reasons for these vivid applications can be attributed to flexibility in surface modification, biocompatibility and safety of NDs. However pharmacokinetics and pharmacodynamics of NDs remain unclear which limits clinical translation of NDs. The future scope of NDs should be focused on rigorous evaluation on the fate of NDs in physiological environment. Also, research should be encouraged to develop commercially scalable manufacturing processes. The current milestones achieved by NDs are commendable and its application as a sparkling drug carrier and diagnostic agent is anticipated with a promising potential for clinical translation.

References

Alexenskii AE, Baidakova MV, Shames AI et al (2002) Defects and impurities in nanodiamonds: EPR, NMR and TEM study. J Phys Chem Solids 63:1993–2001

Andrea B, Daniel S, Kay D et al (2016) Nanodiamonds carrying silicon-vacancy quantum emitters with almost lifetime-limited linewidths nanodiamonds carrying silicon-vacancy quantum emitters with almost lifetime-limited linewidths. New J Phys 18:073036–0730314

Arnault JC (2014) Surface modifications of nanodiamonds and current issues for their biomedical applications. In: Novel aspects of diamonds, pp 85–122

Arroyo-Camejo S, Adam MP, Besbes M et al (2013) Stimulated emission depletion microscopy resolves individual nitrogen vacancy centers in diamond nanocrystals. ACS Nano 7 (12):10912–10919

Baidakova M, Vul A (2007) New prospects and frontiers of nanodiamond clusters. J Phys D Appl Phys 40:6300–6311

Balek L, Buchtova M, Kunova M et al (2018) Nanodiamonds as "artificial proteins": regulation of a cell signalling system using low nanomolar solutions of inorganic nanocrystals. Biomaterials 176:106–121

Banhart F, Ajayan PM (1996) Carbon onions as nanoscopic pressure cells for diamond formation. Nature 382:433–435

Baranov PG, Soltamova AA, Kidalov SV et al (2009) Electron spin resonance detection and identification of nitrogen centers in nanodiamonds. JETP Lett 89:473–477

Barnard AS, Snow AW, Barnard AS (2009) Diamond standard in diagnostics: nanodiamond biolabels make their mark. Analyst 134:1729–1940

Bhosale RR, Osmani RA, Ghodake PP et al (2013) Nanodiamonds: a new-fangled drug delivery system. Indo Am J Pharm Res 3:1395–1403

Boele T, Waddington DEJ, Gaebel T et al (2019) Tailored nanodiamonds for hyperpolarized 13 C MRI. Phys Rev B 101:155416

Borja L (2019) News & analysis materials news. News Anal Mater News 44:527

Boudou J, Faklaris O, Garrot D et al (2008) Detection of single photoluminescent diamond nanoparticles in cells and study of the internalization pathway. Small 4:2236–2239

Boudou J, Curmi PA, Jelezko F et al (2009) High yield fabrication of fluorescent nanodiamonds. Nanotechnology 20:235602

Buerki PR, Leutwyler S, Buerki PR et al (1990) Homogeneous nucleation of diamond powder by CO2 laser-driven gas-phase reactions homogeneous nucleation gas-phase reactions of diamond powder by CO, -laser-driven. J Appl Phys 69:3739–3744

Cao L, Gao C, Sun H et al (2001) Synthesis of diamond from carbon nanotubes under high pressure and high temperature. Carbon N Y 39:311–314

Cao W, Wang X, Li Q et al (2018) Mechanical property and antibacterial activity of silver-loaded polycation functionalized nanodiamonds for use in resin-based dental material formulations. Mater Lett 220:104–107

Chang Y, Lee H, Chen K et al (2008) Mass production and dynamic imaging of fluorescent nanodiamonds. Nat Nanotechnol 3:284–288
Chen J, Liu M, Huang Q et al (2018) Facile preparation of fluorescent nanodiamond-based polymer composites through a metal-free photo-initiated RAFT process and their cellular imaging. Chem Eng J 337:82–90
Cheng C, Perevedentseva E, Tu J et al (2014) Direct and in vitro observation of growth hormone receptor molecules in A549 human lung epithelial cells by nanodiamond labeling direct and in vitro observation of growth hormone receptor molecules in A549 human lung epithelial cells by nanodiamond label. Appl Phys Lett 90:98–101
Chiganov AS (2004) Selective inhibition of the oxidation of nanodiamonds for their cleaning. Phys Solid State 46:620–621
Chow EK, Chow EK, Zhang X et al (2011) Nanodiamond therapeutic delivery agents mediate enhanced chemoresistant tumor treatment. Sci Transl Med 3:21–31
Chu Z, Miu K, Lung P et al (2015) Rapid endosomal escape of prickly nanodiamonds: implications for gene delivery. Sci Rep 5:1–8
Danilenko VV (2004) On the history of the discovery of nanodiamond synthesis. Phys Solid State 46:595–599
Davies G, Lawson SC, Collins T et al (1992) Vacancy-related centers in diamond. Phys Rev B 46:157–170
Davydov VA, Rakhmanina AV, Agafonov V et al (2015) On the nature of simultaneous formation of nano- and micron-size diamond fractions under pressure-temperature-induced. Carbon N Y 90:231–233
Dolmatov VY (2007) Detonation-synthesis nanodiamonds: synthesis, structure, properties and applications. Russ Chem Rev 76:339–360
Eidelman ED, Siklitsky VI, Sharonova LV et al (2005) A stable suspension of single ultrananocrystalline diamond particles. Diam Relat Mater Relat Mater 14:1765–1769
Ekimov EA, Kudryavtsev OS, Khomich AA et al (2015) High-pressure synthesis of boron-doped ultrasmall diamonds from an organic compound. Adv Mater 27:5518–5522
El-say K (2014) Nanodiamond as a drug delivery system: applications and prospective nanodiamond as a drug delivery system. J Appl Pharm Sci 1:29–39
Etzold BJM, Neitzel I, Kett M et al (2014) Layer-by-layer oxidation for decreasing the size of detonation nanodiamond. Chem Mater 26:3479–3484
Ezzeddine A, Khashab NM (2014) Applications of nanodiamonds in drug delivery and catalysis. J Nanosci Nanotechnol 14:332–343
Faklaris O, Joshi V, Irinopoulou T et al (2009) Photoluminescent diamond nanoparticles for cell labeling: study of the uptake mechanism in mammalian cells. ACS Nano 3:3955–3962
Fang X, Mao J, Levin EM et al (2009) Nonaromatic core—shell structure of nanodiamond from solid-state NMR spectroscopy. J Am Chem Soc 131:1426–1435
Ferrari AC, Robertson J, Trans P et al (2004) Raman spectroscopy of amorphous, nanostructured, diamond—like carbon, and nanodiamond. Philos Trans R Soc A 362:2477–2512
Frenklach M, Kematick R, Huang D et al (1989) Homogeneous nucleation of diamond powder in the gas phase homogeneous nucleation of diamond powder in the gas phase. J Appl Phys 66:395–399
Gaillou E, Post JE, Byrne KS et al (2014) Study of the blue moon diamond. Gems Gemol 50:280–287
Galimov EM (1973) Possibility of natural diamond synthesis under conditions of cavitation, occurring in a fast-moving magmatic melt. Nature 243:389–391
Galli G (2010) Structure, stability and electronic properties of nanodiamonds. In: Colombo L, Fasolino A (eds) Computer-based modeling of novel carbon systems and their properties. Carbon materials: chemistry and physics, vol 3. Springer, Dordrecht, pp 37–56
Gismondi A, Nanni V, Reina G et al (2016) Nanodiamonds coupled with 5, 7- dimethoxycoumarin, a plant bioactive metabolite , interfere with the mitotic process in B16F10 cells altering the actin organization. Int J Nanomedicine 11:557–574

Glenn DR, Zhang H, Kasthuri N et al (2012) Correlative light and electron microscopy using cathodoluminescence from nanoparticles with distinguishable colours. Sci Rep 865:1–6
Gogotsi O, Gogotsi Y, Mochalin V et al (2015) Nanodiamonds for drug delivery and other biomedical applications. In: 4th international conference nanobiophysics: fundamental and applied aspects, 1–4 October 2015, Kyiv, Ukraine, pp 1–2
Gong H, Anasori B, Dennison CR et al (2015) Fabrication, biodegradation behavior and cytotoxicity of Mg-nanodiamond composites for implant application. J Mater Sci Mater Med 26:110–119
Gruber A, Drabenstedt A, Tietz C et al (2014) Scanning confocal optical microscopy and magnetic resonance on single defect centers. Sci New York NY 276:2012–2014
Hartmann M, Betz P, Sun Y et al (2012) Saccharide-modified nanodiamond conjugates for the efficient detection and removal of pathogenic bacteria. Chem A Eur J 18:6485–6492
Hawelek L, Brodka A, Dore JC et al (2008) Structural studies of nanodiamond by high-energy X-ray diffraction. Diam Relat Mater 17:1186–1193
Hens SC, Cunningham G, Tyler T et al (2008) Nanodiamond bioconjugate probes and their collection by electrophoresis. Diam Relat Mater 17:1858–1866
Ho D, Wang CK, Chow EK (2015) Nanodiamonds: the intersection of nanotechnology, drug development, and personalized medicine. Biomed Eng 1:1–14
Houshyar S, Kumar GS, Rifai A et al (2019) Nanodiamond/poly-ε-caprolactone nanofibrous scaffold for wound management. Mater Sci Eng C 100:378–387
Huang LL, Chang H (2004) Adsorption and immobilization of cytochrome c on nanodiamonds. Langumir 20(14):5879–5884
Huang H, Pierstorff E, Osawa E et al (2007) Active nanodiamond hydrogels for chemotherapeutic delivery. Nano Lett 7:3305–3314
Huang H, Dai L, Wang DH et al (2008) Large-scale self-assembly of dispersed nanodiamonds. J Mater Chem 18:1347–1352
Huang H, Liu M, Tuo X et al (2017) One-step fabrication of PEGylated fluorescent nanodiamonds through the thiol-ene click reaction and their potential for biological imaging. Appl Surf Sci 439:1143–1151
Huang H, Liu M, Jiang R et al (2018) Water-dispersible fluorescent nanodiamonds for biological imaging prepared by thiol-ene click chemistry. J Taiwan Inst Chem Eng 95:481–486
Hui YY, Zhang B, Chang Y et al (2010) Two-photon fluorescence correlation spectroscopy of lipid-encapsulated fluorescent nanodiamonds in living cells. Opt Express 18:2135–2143
Janaszczyk A, Brzecka P, Chuchracki M et al (2017) Use of nano-diamond in cosmetic microdermabrasion. Eng Biomater 20(143):87
Jaque D, Vetrone F (2012) Luminescence nanothermometry. Nanoscale 4:4301–4326
Jarre G, Heyer S, Memmel E et al (2014) Synthesis of nanodiamond derivatives carrying amino functions and quantification by a modified Kaiser test. Beilstein J Org Chem 10:2729–2737
Ji S, Jiang T, Xu K et al (1998) FTIR study of the adsorption of water on ultradispersed diamond powder surface. Appl Surf Sci 133:231–238
Khachatryan AK, Aloyan SG, May PW et al (2008) Graphite-to-diamond transformation induced by ultrasound cavitation. Diam Relat Mater 17:931–936
Kondo T, Okada N, Tatsuo A et al (2015) Boron-doped nanodiamond powder prepared by solid-state diffusion method. Chem Lett 44:627–629
Konopa I, Beranova J (2012) Antibacterial behavior of diamond nanoparticles against Escherichia coli. Basic Solid State Phys 2584:2581–2584
Kru A, Kataoka F, Ozawa M et al (2005) Unusually tight aggregation in detonation nanodiamond: identification and disintegration. Carbon N Y 43:1722–1730
Krueger BA (2008) Diamond nanoparticles: jewels for chemistry and physics. Adv Mater 20:2445–2449
Krutko OB, Kosel PB, Wu RLC et al (2000) P-type polycrystalline diamond layers by rapid thermal diffusion of boron P-type polycrystalline diamond layers by rapid thermal diffusion of boron. Appl Phys Lett 76:849–850

Kucsko G, Maurer PC, Yao NY et al (2013) Nanometre-scale thermometry in a living cell. Nature 500:54–59

Kumar A, Lin PA, Xue A et al (2013) Formation of nanodiamonds at near-ambient conditions via microplasma dissociation of ethanol vapour. Nat Commun 4:1–8

Laan KJ, Der V, Hasani M et al (2018) Nanodiamonds for in vivo applications. Small 14:1703838

Lai H, Stenzel MH, Xiao P (2019) Surface engineering and applications of nanodiamonds in cancer treatment and imaging. Int Mater Rev 65(4):189–225

Lam R, Chen M, Pierstorff E et al (2008) Nanodiamond-embedded microfilm elution. ACS Nano 2:2095–2102

Lee D, Kim SV, Limansubroto AN et al (2015) Nanodiamond À Gutta Percha composite biomaterials for root canal therapy. ACS Nano 9:11490–11501

Li L, Tian L, Zhao W et al (2016) Integrative biology acetate ions enhance load and stability of doxorubicin onto PEGylated nanodiamond for selective tumor intracellular controlled release and therapy. Integr Biol 8:956–967

Li D, Chen X, Wang H et al (2017) Cetuximab-conjugated nanodiamonds drug delivery system for enhanced targeting therapy and 3D Raman imaging. J Biophotonics 11:1–12

Lim DG, Jung JH, Ko HW et al (2016) Paclitaxel-nanodiamond nanocomplexes enhance aqueous dispersibility and drug retention in cells. ACS Appl Mater Interfaces 8:23558–23567

Lin Y, Su D (2014) Fabrication of nitrogen-modified annealed nanodiamond with improved catalytic activity. ACS Nano 8:7823–7833

Lin B, Chen C, Kunuku S et al (2018) Fe doped magnetic nanodiamonds made by ion implantation as contrast agent for MRI. Sci Rep 8:1–6

Liu Y, Klep V, Zdyrko B et al (2004) Polymer grafting via ATRP initiated from macroinitiator synthesized on surface. Langumir 20:8716–8724

Liu K, Cheng C, Chang C (2007) Biocompatible and detectable carboxylated nanodiamond on human cell. Nanotechnology 18:325102–325110

Liu K, Wang C, Cheng C et al (2009) Biomaterials endocytic carboxylated nanodiamond for the labeling and tracking of cell division and differentiation in cancer and stem cells. Biomaterials 30:4249–4259

Marchand AP (2012) Diamondoid hydrocarbons—delving into nature's bounty. Science 299:96–99

Master AM, Sen Gupta A (2012) EGF receptor-targeted nanocarriers for enhanced cancer treatment review. Nanomedicine 7:1895–1906

Maze JR, Stanwix PL, Hodges JS et al (2008) Nanoscale magnetic sensing with an individual electronic spin in diamond. Nature 455:644–647

Merson TD, Castelletto S, Aharonovich I et al (2013) Nanodiamonds with silicon vacancy defects for nontoxic photostable fluorescent labeling of neural precursor cells. Opt Lett 38:4170–4173

Miguel M, Jacobs David SB (2012) Powering the green economy: the feed-in tariff handbook. Ecol Econ 73:228–229

Mitev D, Dimitrova R, Spassova M et al (2007) Surface peculiarities of detonation nanodiamonds in dependence of fabrication and purification methods. Diam Relat Mater 16:776–780

Mitev DP, Townsend AT, Paull B et al (2014) Microwave-assisted purification of detonation nanodiamond. Diam Relat Mater 48:37–46

Mochalin VN, Gogotsi Y (2009) Wet chemistry route to hydrophobic blue fluorescent nanodiamond. J Am Chem Soc 131:4594–4595

Mochalin VN, Shenderova O, Ho D et al (2011) The properties and applications of nanodiamonds. Nat Nanotechnol 7:11–23

Mogilnaya O, Bondar V (2012) Antibacterial properties of lysozyme immobilized on nanodiamonds. Micro Nano Syst 4:41–47

Namdar R, Nafisi S (2018) Nanodiamond applications in skin preparations. Drug Discov Today 23:1152–1158

Neu E, Arend C, Gross E et al (2013) Narrowband fluorescent nanodiamonds produced from chemical vapor deposition films narrowband fluorescent nanodiamonds produced from chemical vapor. Appl Phys Lett 98:3107

Ntziachristos V (2010) Going deeper than microscopy: the optical imaging frontier in biology. Nat Publ Gr 7:603–614
Osswald S, Yushin G, Mochalin V et al (2006) Control of sp2/sp3 carbon ratio and surface chemistry of nanodiamond powders by selective oxidation in air. J Am Chem Soc 128:11635–11642
Ozawa BM, Inaguma M, Takahashi M et al (2007) Preparation and behavior of brownish, clear nanodiamond colloids. Adv Mater 19:1201–1206
Paget V, Sergent JA, Grall R et al (2014) Carboxylated nanodiamonds are neither cytotoxic nor genotoxic on liver, kidney, intestine and lung human cell lines. Nanotoxicology 8:46–56
Peng W, Mahfouz R, Pan J et al (2013) Gram-scale fractionation of nanodiamonds by density gradient ultracentrifugation. Nanoscale 5:5017–5026
Pentecost A, Gour S, Mochalin V et al (2010) Deaggregation of nanodiamond powders using salt- and sugar-assisted milling. ACS Appl Mater Interfaces 2:3289–3294
Pentecost AE, Witherel CE, Gogotsia Y et al (2017) Anti-inflammatory effects of octadecylamine-functionalized nanodiamond on primary human macrophages. Biomater Sci 5:2131–2143
Perevedentseva E (2013) Nanodiamond internalization in cells and the cell uptake mechanism. J Nanopart Res 15:1883418–1883430
Petit T, Puskar L (2018) FTIR spectroscopy of nanodiamonds: methods and interpretation. Daim Relat Mater 89:52–66
Picollo F, Mino L, Battiato A et al (2019) Synthesis and characterization of porphyrin functionalized nanodiamonds. Diam Relat Mater 91:22–28
Prabhakar N, Rosenholm JM (2019) ScienceDirect nanodiamonds for advanced optical bioimaging and beyond. Curr Opin Colloid Interface Sci 39:220–231
Prabhakar N, Khan MH, Peurla M et al (2017) Intracellular trafficking of fluorescent nanodiamonds and regulation of their cellular toxicity. ACS Omega 2:2689–2693
Prelas M, Ghosh TK, Tompson R et al (2014) Diffusion of boron into polycrystalline diamond films using the electric field enhanced diffusion (EFED) technique. J Wide Bandgap 10:15–27
Puzyr AP, Neshumayev DA, Tarskikh SV et al (2004) Destruction of human blood cells in interaction with detonation nanodiamonds in experiments in vitro. Diam Relat Mater 13:2020–2023
Rojas S, Gispert JD, Martin R et al (2011) Biodistribution of nanoparticles. In vivo studies based on 18 F radionuclide emission. ACS Nano 5:5552–5559
Schrand AM, Huang H, Carlson C et al (2006) Are diamond nanoparticles cytotoxic? J Phys Chem B 111:2–7
Schrand AM, Dai L, Schlager JJ et al (2007) Differential biocompatibility of carbon nanotubes and nanodiamonds. Diam Relat Mater 16:2118–2123
Schrand AM, Hens SAC, Shenderova OA et al (2009) Critical reviews in solid state and materials sciences nanodiamond particles: properties and perspectives for bioapplications nanodiamond particles: properties and perspectives for bioapplications. Crit Rev Solid State Mater Sci 34:18–74
Shenderova OA, Grichko VP (2016) Nanodiamond UV protectant formulation. US 9,283,155
Shenderova O, Nunn N (2017) Chapter 2. Production and purification of nanodiamonds. In: Arnault JC (eds) Nanodiamonds, 6th edn. Elsevier Inc, pp 25–56
Shenderova O, Koscheev A, Zaripov N et al (2011a) Surface chemistry and properties of ozone-purified detonation nanodiamonds. J Phys Chem 115:9827–9837
Shenderova OA, Vlasov II, Turner S et al (2011b) Nitrogen control in nanodiamond produced by detonation shock-wave-assisted synthesis. J Phys Chem 115:14014–14024
Shimkunas RA, Robinson E, Lam R et al (2009) Biomaterials nanodiamond—insulin complexes as pH-dependent protein delivery vehicles. Biomaterials 30:5720–5728
Silva AR, Ribeiro C, Carabineiro SAC et al (2017) Nanodiamonds/poly(vinylidene fluoride) composites for tissue engineering applications. Compos Part B 111:37–44
Spitsyn BV, Davidson JL, Gradoboev MN et al (2006) Inroad to modification of detonation nanodiamond. Diam Relat Mater 15:296–299

Su L, Fang C, Chang Y (2013) Creation of high density ensembles of nitrogen-vacancy centers in nitrogen-rich type Ib nanodiamonds. Nanotechnology 24:315702–315710
Sukhova AA, Gofman IV, Skorik YA (2019) Preparation and properties of chitosan–nanodiamond dispersions and composite films. Diam Relat Mater 98:107483
Tatsii VF, Bochko AV, Oleinik GS (2009) Structure and properties of Dalan detonation diamonds. Combust Explos Shock Waves 45:95–103
Tiainen T, Myllymäki TTT, Hatanpää T et al (2019) Polyelectrolyte stabilized nanodiamond dispersions. Diam Relat Mater 95:185–194
Ting CC, Young TF, Jwo CS (2007) Fabrication of diamond nanopowder using microwave plasma torch technique. Int J Adv Manuf Technol 34:316–322
Tinwala H, Wairkar S (2019) Production, surface modification and biomedical applications of nanodiamonds: a sparkling tool for theranostics. Mater Sci Eng C 97:913–931
Torelli MD, Rickard AG, Backer M et al (2019) Targeting fluorescent nanodiamonds to vascular endothelial growth factor receptors in tumor. Bioconjug Chem 30:604–613
Turcheniuk V, Raks V, Issa R et al (2014) Antimicrobial activity of menthol modified nanodiamond particles. Diam Relat Mater 57:2–8
Tzeng Y, Faklaris O, Chang B et al (2011) Superresolution imaging of albumin-conjugated fluorescent nanodiamonds in cells by stimulated emission depletion. Angew Chemie Int Ed English 50:2262–2265
Uchida T, Hamano A, Kawashima N et al (2007) Disaggregation and surface modification of nano-size diamond by ultrasound exposure: relationships among acoustic intensity, disaggregation, and surface modification. Electron Commun Japan 90:93–100
Vaijayanthimala V, Lee DK, Kim SV et al (2015) Nanodiamond-mediated drug delivery and imaging: challenges and opportunities. Expert Opin Drug Deliv 12:735–749
Van Thiel M, Ree FH, Van Thiel M et al (1987) Transition. J Appl Phys 62:1761–1767
Vlasov BII, Barnard AS, Ralchenko VG et al (2009) Nanodiamond Photoemitters based on strong narrow-band luminescence from silicon-vacancy defects. Adv Mater 21:808–812
Wahab Z, Marsh ZM, Tessema A et al (2017) Effect of nanodiamond (ND) surface functionalization on the properties of ND/PEEK composites. IEEE Trans Components, Packag Manuf Technol, pp 1–13
Wang LV, Hu S (2013) Photoacoustic tomography: in vivo imaging from organelles to organs. Sci New York NY 335:1458–1462
Wang C, Kurtsiefer C, Weinfurter H (2006) Single photon emission from SiV centres in diamond. J Phys B Mol Opt Phys 39:37–41
Wehling J, Dringen R, Zare RN et al (2014) Bactericidal activity of partially oxidized nanodiamonds. ACS Nano 8:6475–6483
Weng M, Chiang S, Wang N et al (2009) Fluorescent nanodiamonds for specifically targeted bioimaging: application to the interaction of transferrin with transferrin receptor. Diam Relat Mater 18:587–591
Whitehead KA, Langer R, Anderson DG (2009) Knocking down barriers: advances in siRNA delivery. Nat Rev Drug Discov 8:129–138
Wiethoff CM, Middaugh CR (2003) Barriers to nonviral gene delivery. J Pharm Sci 92:203–217
Wu Y, Ermakova A, Liu W et al (2015) Programmable biopolymers for advancing biomedical applications of fluorescent nanodiamonds. Adv Funct Mater 25:6576–6585
Xing Y, Dai L (2009) Nanodiamonds for nanomedicine review. Nanomedicine 4:207–218
Xing Y, Xiong W, Zhu L et al (2011) DNA damage in embryonic stem cells caused by nanodiamonds. ACS Nano 5:2376–2384
Xu D, Liu M, Zhang Q et al (2018) Preparation of water dispersible and biocompatible nanodiamond-poly (amino acid) composites through the ring-opening polymerization. Mater Sci Eng C 91:496–501
Yang L, May PW, Yin L et al (2007) Growth of diamond nanocrystals by pulsed laser ablation of graphite in liquid. Diam Relat Mater 16:725–729

Yoichi M, Takimoto T, Yamanaka H et al (2008) A facile and scalable process for size- controllable separation of nanodiamond particles as small as 4 nm. Nano Micro Small 4:2154–2157
Yu S, Kang M, Chang H et al (2005) Bright fluorescent nanodiamonds: no photobleaching and low cytotoxicity. J Am Chem Soc 127:17604–17605
Yu Y, Yang X, Liu M et al (2019) Anticancer drug delivery to cancer cells using alkyl amine-functionalized nanodiamond supraparticles. Nanoscale Adv 1:3291–3772
Yuan Y, Chen Y, Liu J et al (2009) Biodistribution and fate of nanodiamonds in vivo. Diam Relat Mater 18:95–100
Zhang X, Chen M, Lam R et al (2009) Polymer-functionalized nanodiamond platforms as vehicles for gene delivery. ACS Nano 3:2609–2616
Zhang Q, Mochalin VN, Neitzel I et al (2011) Biomaterials fluorescent PLLA-nanodiamond composites for bone tissue engineering. Biomaterials 32:87–94
Zhang X, Fu C, Feng L et al (2012a) PEGylation and polyPEGylation of nanodiamond. Polymer (Guildf) 53:3178–3184
Zhang X, Hu W, Li J et al (2012b) A comparative study of cellular uptake and cytotoxicity of multi-walled. Toxicol Res (Camb) 1:62–68
Zhang H, Aharonovich I, Glenn DR et al (2014) Silicon-vacancy color centers in nanodiamonds: cathodoluminescence imaging markers in the near infrared. Nano Micro Small 10:1908–1913
Zhang F, Song Q, Huang X et al (2015) A novel high mechanical property PLGA composite matrix loaded with nanodiamond-phospholipid compound for bone tissue engineering. ACS Appl Mater Interfaces 8:1087–1097
Zousman B, Levinson O (2014) Pure nanodiamonds produced by laser-assisted technique. In: Willims OA (ed) Nanodiamonds. Royal Society of Chemistry, Cambridge, pp 112–127

Evaluation of Nanotoxicity Using Zebrafish: Preclinical Model

7

Rajesh A. Maheshwari, Dhanya B. Sen, Aarti S. Zanwar, and Ashim Kumar Sen

Abstract

Throughout the globe, nanotechnology has emerged as a segment which produces a multitrillion-dollar business opportunity that covers a wide range of industries such as medicine, electronics, and chemistry. Due to the rapid development of application-oriented nanoparticles, from targeted drug delivery to diagnostics, in vivo toxicological examinations for assessing the potential hazardous effects of nanoparticles on natural and human safety are in urgent need. Therefore, it is essential to assess their toxicity and possible hazards to humans and ecosystem. Zebrafish is considered as the "gold standard" among animal models for assessment of several metal and metal oxide nanoparticle toxicity due to its cost-effectiveness, high fecundity, optical transparency, short life cycle, well-characterized developmental stages, etc. The chapter emphasizes on how zebrafish (*Danio rerio*) can be utilized to assess nanotoxicity at different levels, including genotoxicity, developmental toxicity, immunotoxicity, cardiovascular toxicity, teratogenicity, neurotoxicity, reproductive toxicity, hepatotoxicity, as well as change in behavior and disruption of gill, skin, and endocrine system. The harmful impacts of chosen metal and metal oxide nanoparticles are also reviewed. The advantages, drawbacks, and future aspects of utilization of zebrafish model in nanotoxicity studies are also argued. Overall, zebrafish is projected to fulfill as a high-throughput screening platform for drug delivery assessment and nanotoxicity, which may help in designing safe and more effective nanomedicines.

R. A. Maheshwari (✉) · D. B. Sen · A. S. Zanwar · A. K. Sen
Department of Pharmacy, Sumandeep Vidyapeeth Deemed to be University, Vadodara, Gujarat, India
e-mail: rmaheshwari.dop@sumandeepvidyapeethdu.edu.in; dhanyasen.dop@sumandeepvidyapeethdu.edu.in; aartizanwar.dop@sumandeepvidyapeethdu.edu.in; ashimsen.dop@sumandeepvidyapeethdu.edu.in

N. Shah (ed.), *Nanocarriers: Drug Delivery System*,
https://doi.org/10.1007/978-981-33-4497-6_7

Keywords

Nanoparticles · Toxicity · Zebrafish · Teratogenicity · Genotoxicity

7.1 Introduction

7.1.1 Overview of Nanoparticles

Nanotechnology is the engineering of functional systems at the atomic, molecular, and supramolecular scale. In recent times, nanoparticle (NP)-based research achieved overwhelming attention of scientific community due to its widespread area of applications. In nanotechnology, a particle is termed as a small entity which acts as an entire unit due to its unique properties and transportation capabilities. A particle having reduced dimension (1–100 nm) is described as nanoparticles (NPs) by agencies like "International Organization for Standardization," "American Society of Testing and Materials," and "National Institute of Occupational Safety and Health" (Horikoshi and Serpone 2013). Over the years, exponential growth in preparation, characterization, and innovative application of NPs has been observed (De Crozals et al. 2016). Extensive research on NPs resulted in engineered nanosized particles like various metal and metal oxide NPs, nanopolymers, fullerenes, carbon nanotubes (CNT), and crystalline materials, which possess numerous properties and are labeled as multifunctional NPs (Seaton et al. 2010; McNamara and Tofail 2013).

7.1.2 Applications of Nanoparticles

NPs are used, or are being evaluated for usefulness, in many fields due to its widespread area of applications. NPs possess diverse properties and are useful in industrial manufacturing as chemically inert additives, anticaking agents, pigments, and fillers and more prominently to generate functional surfaces/membranes which exert UV protection, antimicrobial property, catalytic function, filtration, etc. (Stark et al. 2015). Newer areas like nanomedicine have evolved as a cumulative outcome of well-known subjects like medicine, physics, and chemistry which are the driving force behind various biomedical applications. Characteristic electrochemical, piezoelectric, optical, and photoluminescence properties of NP are the basis of making biosensors for drugs, proteins, pathogens, nucleic acids, metabolites, cancer cells, etc. (Stark et al. 2015; Das et al. 2013). Most of the time NPs are designed to act as a delivery system where containment of surface characteristic and dimension is a prerequisite for drug-release pattern to exert site-specific action of the drug at an optimal rate and dose (McNamara and Tofail 2013; Das et al. 2013; McNamara and Tofail 2015). Site-specific distributions of NPs are possible due to their unique physicochemical properties when compared with fine particles (FPs). NPs are available in various forms and compositions like metallic based and carbon-based nanomaterials, polymeric particulate materials, and quantum dots (Wang and Tang 2018; Wu and Tang 2018). Among all, metal NPs and metal oxide NPs contribute majority of them in terms of manufacturing output and application (Djurisic et al. 2015). Specific metal NPs like silver (AgNPs), gold (AuNPs), nickel (NiNPs), copper oxide (CuNPs), and metal oxide NPs (titanium dioxide [TiO_2], zinc oxide [ZnO], iron oxides [Fe_2O_3, Fe_3O_4]) are produced in large quantity and supplied in various fields of healthcare, medicine, transportation, construction, energy, defense, etc. along with engineered nanoparticles (ENPs) as major components or as additives for performance enhancement (Kessler 2011; Rudramurthy and Swamy 2018). Researchers are exploring the possible anticancer activity of biologically synthesized AgNPs, AuNPs, and platinum NPs (PtNPs) (Bendale et al. 2017; Ning et al. 2017; Yamada et al. 2015; Zhang et al. 2016), whereas manufacturers of sunscreen products are using TiO_2 and ZnO NPs in the formulation due to their capability to block ultraviolet radiation. Research on drug delivery uses iron oxide NPs (IONPs), including Fe_3O_4 and γ-Fe_2O_3, and magnetic resonance imaging uses superparamagnetic IONPs widely (Ding and Guo 2013; Namvar et al. 2014). However, assessment of adverse impact on the environment and human health has explored a new area of research.

7.1.3 Measurement of Nanotoxicity

Nontoxicity is a prerequisite for NPs used in biomedical field. However, environmental exposure of toxic NPs used in manufacturing and other applications is a major concern (De Crozals et al. 2016; Friedman et al. 2013). Metal and metal oxide NPs possess good dispersibility and stability in the presence of organic matter present in water and thus can pollute aquatic environment by means of direct discharge and waste discharge and during routine use. Metal and metal oxide NPs entering into the aquatic environment can reach and accumulate in the human body through food chain while drinking water and eating vegetables, fish, and livestock and can be a threat to human health (Xing et al. 2016; Nowack and Bucheli 2007; Wang and Wang 2014). Researchers have been working on developing newer assessment or evaluation methodologies to check exposure levels and toxicity of specific nanomaterials. Largely, toxicological assessment of NPs is carried out using in vitro and in vivo models starting with in vitro cell culture assays to basic model organisms, such as algae, protozoa, zooplankton, and advanced higher vertebrate models, such as rodents, rabbits, and nonhuman primates (Li and Chen 2011; Schrand et al. 2010). Cellular level toxicity and genotoxicity can be assessed using simple organism and cell lines, whereas complex physiological interactions can be assessed only on higher vertebrates. However, rodents and rabbits have a drawback as an animal model due to their ethical concerns, cost, slower and inaccessible embryo development, and amount of testing material required (as per animal size), whereas primate model shows similar issues with greater extent (Gad 2006). Therefore, zebrafish can be used as a compelling alternative model for the evaluation of in vivo nanotoxicity due to its efficiency, cost-effectiveness, and smaller size (Chakraborty et al. 2016).

7.2 Zebrafish: Preclinical Model

7.2.1 Outline

Zebrafish (*Danio rerio*) has been a well-established vertebrate model since 1960s (Kalueff et al. 2014) and is being used extensively in preclinical and toxicity studies due to the number of favorable traits available (Strähle et al. 2012; Chakraborty and Agoramoorthy 2010). In recent times, zebrafish (*Danio rerio*) had drawn much attention as an in vivo model as it carries unique features like lower cost, high fecundity, embryonic transparency, rapid and well-characterized growth, shorter reproduction time, and gene manipulation accessibility. Ecotoxicology research recognizes zebrafish to assess embryo toxicity and it is used as one of the standard methods for evaluating toxicity due to single chemical entity as per the guidelines of national standards organizations (Fako and Furgeson 2009). A fully grown adult fish shown in Fig. 7.1 can be used for studying a large number of testing materials due to their advantage of having lower size. They possess high fertility rate where a single female can produce about 300 eggs, which proves the completeness as a model (Westerfield 1995; Ribas and Piferrer 2014). The genome of zebrafish and humans shows ~70% resemblance (Howe et al. 2013; Kettleborough et al. 2013). Majority of investigations using *Danio rerio* concentrate on teratogenic and developmental effects of materials on the larvae and on the fry. *Danio rerio* is commonly used to assess the potential toxic effects of NPs due to its capability of rapid reproduction, ease of breeding, availability of embryos round the year, and transparency of the larva body.

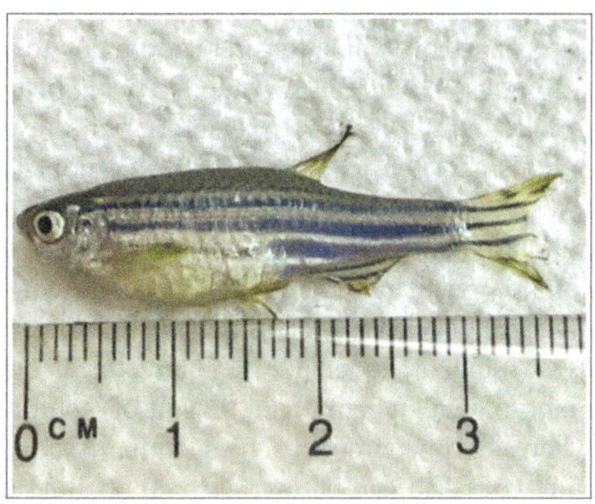

Fig. 7.1 Medium-size adult zebrafish (*Danio rerio*) (De León et al. 2019)

7.2.2 Advantages of Zebrafish in Nanotoxicity Research

In recent times, utilization of the zebrafish model has become popular in the screening of toxicants (Chun et al. 2017; Da et al. 2018; Sangabathuni et al. 2017; Vicario-Pares et al. 2018). Various attributes make zebrafish a substitute model for toxicological investigations of nanomaterials as follows.

Firstly, as a multicellular entity, zebrafish can provide more comprehensive data regarding kinetic, passage, and transformation of nanomaterials against in vitro cell culture analysis, despite the fact that in vitro analysis is mostly used to assess toxicological impacts of nanomaterials and is recognized as a successful model for toxicity studies even at the cellular level (Gad 2015).

Secondly, due to the smaller size, ease of cultivation, shorter life cycle, and higher fecundity compared to rodents, zebrafish became the most accessible model for the vast majority of research facilities around the world. They achieve mature reproductive system in laboratories within a short span of time (3–6 months) postfertilization under optimum temperature, food supply, and rearing densities (Spence et al. 2008). A fully grown female fish can yield about 100–300 embryos per day, and therefore may be useful in high-throughput analysis which improves the statistical power of experiments (Castranova et al. 2011; Spence and Smith 2005).

Third, the rapid embryogenesis and developmental processes in zebrafish compared with other animals make it a superior model for evaluating developmental toxicity (Haffter et al. 1996; Kimmel et al. 1995; Westerfield 2007; Lin et al. 2013). Toxicological effects like lethality, reproductive toxicity, and teratogenicity can be observed easily due to their transparency during embryo stages (Choi et al. 2016; Ma et al. 2018; Mesquita et al. 2017; Pecoraro et al. 2017).

Fourth, information gathered post-gene sequencing elaborates that zebrafish have 26,206 protein-coding genes and around 85% of these are similar to their human counterparts, making zebrafish a popular model for investigating genotoxicity and developmental toxicity (Collins et al. 2012; Howe et al. 2013; Renier et al. 2007; Garcia et al. 2016; Rizzo et al. 2013; Sarmah and Marrs 2016; Zhu et al. 2014).

7.2.3 Developmental Stages of Zebrafish

Eggs of zebrafish are robust in nature and grow externally, so it is possible to engineer them easily for high-throughput applications. In addition, optical transparency of zebrafish permits impeccable visual examination, including fluorescent and different markers (Stainier and Fishman 1994; Dooley and Zon 2000). The basic development of zebrafish requires only 24 h postfertilization (hpf), whereas embryogenesis completes by 72 hpf, and it takes 96 hpf to develop organs and around 3 months to arrive at adulthood (Stainier and Fishman 1994). There are a number of screening methods to study the developmental stages of zebrafish to measure toxicological responses to metal and metal oxide NPs, in terms of developmental toxicity, neurotoxicity, hepatotoxicity, genotoxicity, immunotoxicity, cardiovascular toxicity, reproductive toxicity, etc. (Fig. 7.2).

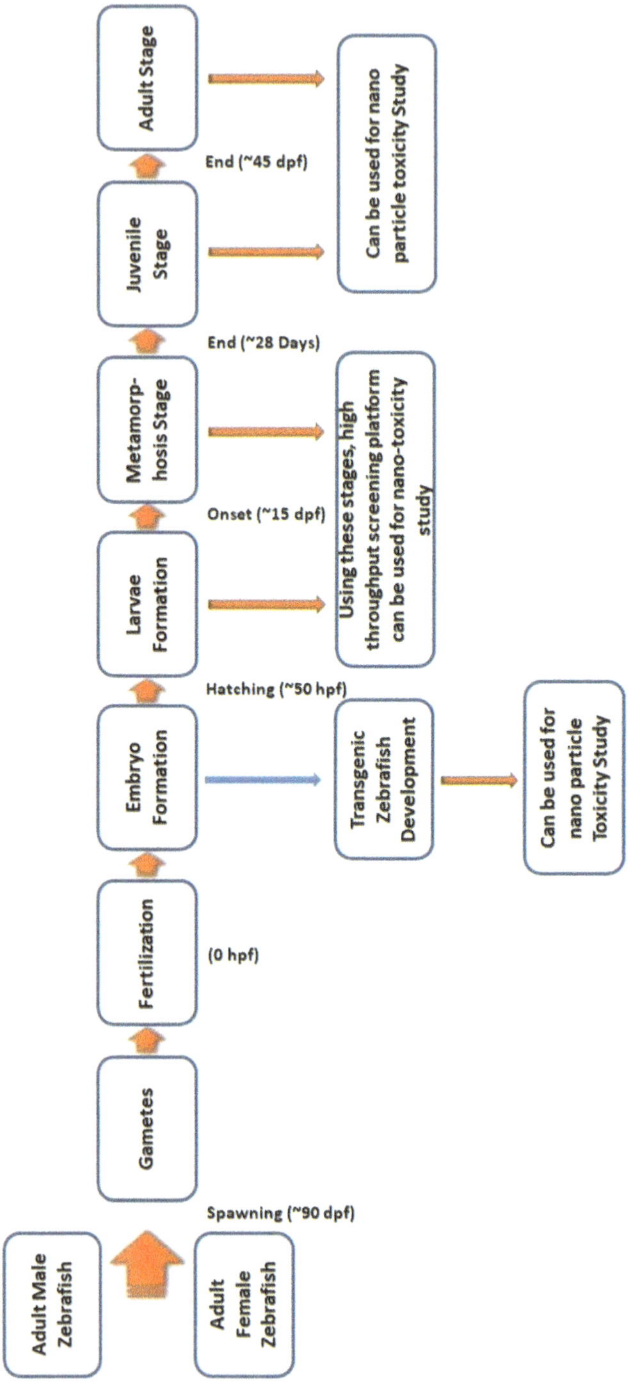

Fig. 7.2 Schematic diagram of different stages of zebrafish development and their relevance to nanotoxicity study (Chakraborty et al. 2016)

7.3 Various Methods to Assess Nanotoxicity

7.3.1 Developmental Toxicity

Teratogenicity, mortality, and hatching rate are the developmental toxicity parameters of nanomaterials in zebrafish. NP assessment on zebrafish used for the evaluation of developmental toxicity of embryos has been found to be mature than the toxicity evaluation of target organs or other systems. It is also appropriate for image-based detection and is capable of recording a range of teratogenic indicators like cell movement throughout intestinal phase, blood circulation, brain formation, and heartbeat due to its in vitro fertilization and lucidity during embryo stages. Embryonic development events are capable of being utilized as endpoints for toxicological assessment (Fig. 7.3). Additionally, the embryo teratogenic test cycle in zebrafish is short and appropriate for gene mutant screening and analysis in large scale.

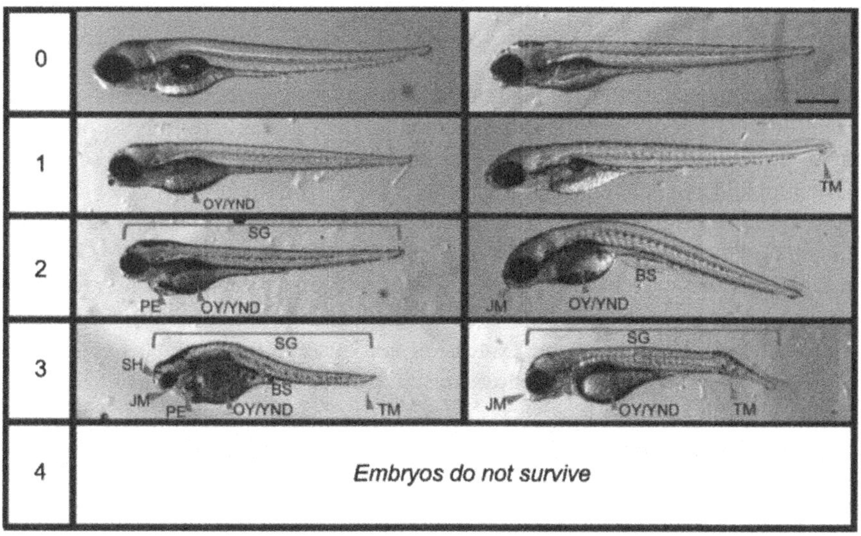

Fig. 7.3 The scoring spectrum utilized for screening nanoparticle-induced toxicity is portrayed by representative micrographs of 120 hpf zebrafish embryos that were exposed to different toxicants. This screening method was used as a semiquantitative analysis for scoring at 4, 24, 48, 96, and 120 hpf time points. Embryos were scored for severity of morphological defects, survival, and toxic adverse effects. Scores range from 0 to 4, with 0 indicating no visible deleterious effects and 4 signifying death. The intervening numbers correspond to various degrees and quantities of morphological anomalies (i.e., 1 = one to two minor toxic effects; 2 = one moderate or three to four minor toxic effects; and 3 = one (or more) severe or more than four minor toxic effects). Scores were used to yield a mean cumulative toxicity score for each treatment group at each time point to evaluate toxicity. Most of the sublethal endpoints included in the studies are depicted in the figure: bent spine (BS), jaw malformation (JM), opaque yolk (OY), pericardial edema (PE), stunted growth (SG), small head (SH), tail malformations (TM), and nondepleted yolk (YND). Scale bar = 0.5 mm (Bar-Ilan et al. 2009)

AgNP-treated embryos showed mortality and hatching delay. Furthermore, developmental toxicity like pericardial edema, slow blood flow, arrhythmia, twisted notochord, and body axis abnormality were the outcomes of AgNP treatment (Asharani et al. 2008; Shaw et al. 2016). Zebrafish embryos, when exposed to gold nanorods coated with cetyltrimethyl ammonium bromide (CTAB), were shown to induce delayed embryonic developments such as delayed eye, head and tail elongation development, pericardial edema, and tail deformities. These embryos were also found to induce mortality when exposed to CTAB (Mesquita et al. 2017). It has been observed that metal oxide NPs are also capable of inducing developmental and acute toxicity in zebrafish. Abnormal phenotypes like delayed epiboly and smaller head and eyes in zebrafish can be observed as a result of copper oxide NP exposure (Xu et al. 2017). Another metal oxide, namely ZnONPs, can cause toxic effects such as skin ulceration, hatching delay, and high mortality in zebrafish (Zhu et al. 2008). Toxicity to zebrafish embryos due to TiO_2NPs was also evaluated and found to affect the hatching time of embryos (Samaee et al. 2015).

7.3.2 Immunotoxicity

Application of zebrafish in the field of immunology has gained momentum in recent years. It has been observed that the immune system is sensitive to NPs, predominantly inducing an inflammatory response in addition to accumulation and activation of neutrophils and macrophages (Johnston et al. 2018). The process in which toxic substances destroy the function of immune system is known as immunotoxicity (Giannakou et al. 2016; Selgrade 2007; Xu et al. 2015; Jin et al. 2011) (Fig. 7.4). For instance, AuNPs have been proved to disrupt inflammatory and immune response pathways (Truong et al. 2013). In another study, an adult zebrafish was exposed to AgNPs and subsequently a gene expression study was performed in its liver tissues. The study proved that AgNP exposure resulted in immunotoxicity in adult zebrafish because of oxidative stress (Krishnaraj et al. 2016). ZnONP exposure also resulted in transcriptional changes of pro-inflammatory cytokines, interleukin (IL)-1β, and tumor necrosis factor-α and a significant upregulation in eleuthero embryos and a downregulation in zebrafish embryos. Therefore, ZnONPs have been proved to cause modulation of pro-inflammatory reactions (Brun et al. 2014).

7.3.3 Neurotoxicity

Zebrafish model has emerged as a sensitive and useful animal model for the assessment of neurotoxicity induced by NPs. The damage of nervous tissue and subsequent irregular activity of nervous system, when exposed to toxic substances, is called neurotoxicity and these toxic substances are known as neurotoxins (Segura-Aguilar and Kostrzewa 2006). A variety of NPs can activate free radical actions at their surfaces, thus generating oxidative stress at particle deposition and translocation site (Sato et al. 1998; Dellinger et al. 2001; Li et al. 2003). Specific behavioral

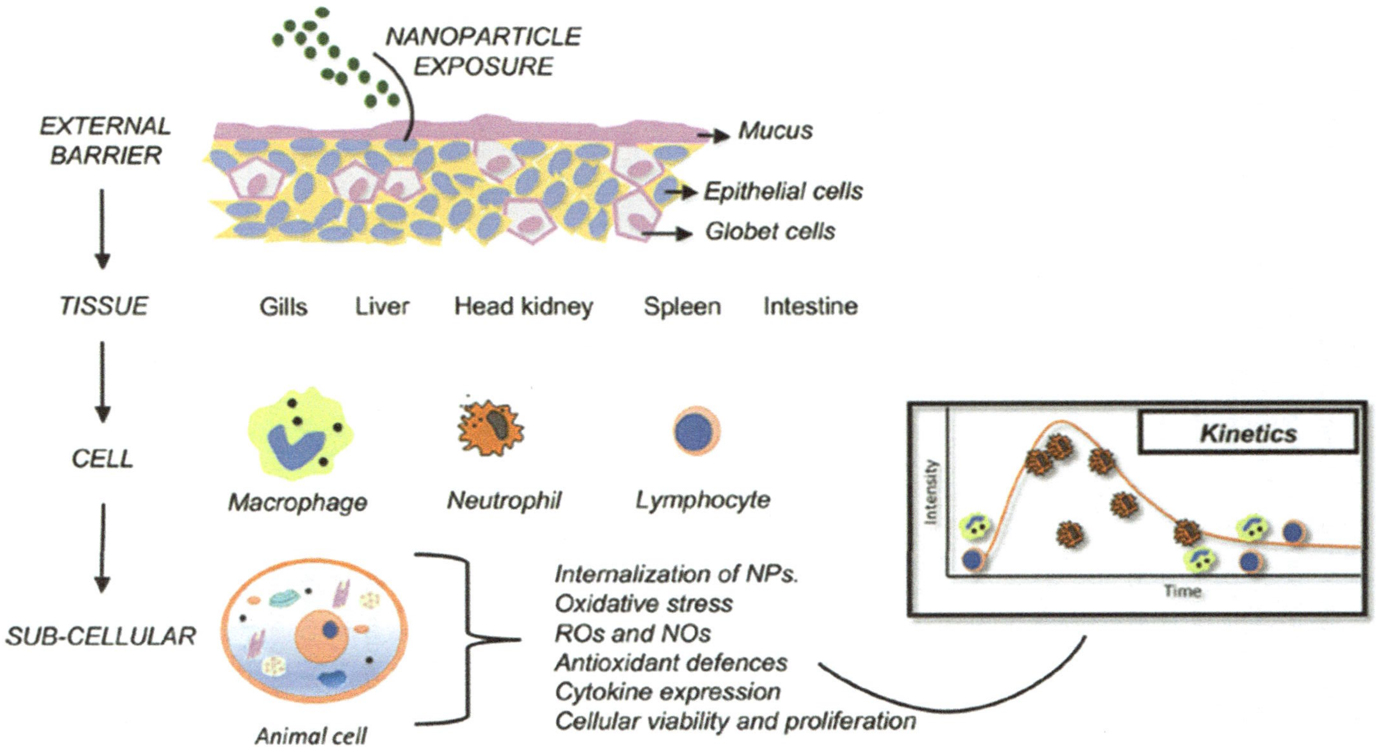

Fig. 7.4 Innate immunity as a bioindicator of health for teleost fish exposed to nanoparticles. Following NP exposure, fish immunity is evident at different levels (external barrier, tissue, cellular, and subcellular). Each provides unique insights into changes to homeostasis and, thus, can be used to detect nanoparticle-induced immunotoxicity (Torrealba et al. 2019)

effects for particular NPs are also seen. The brain tissues of juvenile zebrafish after 5 days of fertilization have been differentiated into telencephalon, diencephalon, midbrain, hindbrain, and rhomboidal ganglia. Behavioral toxicity of NPs such as learning, motion, and memory ability can also be evaluated using well-differentiated brain tissues of juvenile zebrafish. Furthermore, neurotoxicity of NPs to zebrafish embryos can also be evaluated using apoptosis of neurons, necrosis, morphological changes, and biochemical indicators. Neurotoxicity can be seen commonly in NPs that are capable of reaching brain and causing neurodegeneration (Win-Shwe and Fujimaki 2011; Chakraborty et al. 2009). Combustion-derived NPs have been proved urotoxic from in vivo and in vitro studies, due to the incidence of NP aggregation (Morimoto et al. 2010). For instance, silicon dioxide NPs resulted in altered color preferences (Li et al. 2014), whereas cadmium telluride quantum dots affected locomotor activity (Zhang et al. 2012). A size-dependent effect was observed on zebrafish due to polyvinyl pyrrolidone-coated AgNPs-PVP. The smaller AgNP-PVP sized 10 nm resulted in decreased locomotor activity, while hyperactivity was caused by the larger one (50 nm) under specific light conditions (Powers et al. 2011). Earlier studies have shown that TiO_2NP activates expression of genes like BDNF C-fos and C-jun. On the contrary, these NPs suppress the expression of genes such as NGF, p38, and CRE causing brain damage of zebrafish (Sheng et al. 2016). Alteration of neurotransmission and subsequent increase in brain acetylcholine esterase activity were caused by AuNP exposure (Dedeh et al. 2015). A delay in retinal neurodifferentiation with subsequent reduced locomotor activity was caused by CuONP exposure at high doses ($\geq$12.5 mg/L) (Sun et al. 2016). Exposure of FeONPs coated with dextran was also found to be neurotoxic to zebrafish. The toxic effects included higher levels of ferric iron in the brain, reduction in the exploratory performance, decreased acetylcholine esterase activity, and induction of casp8, casp9, and jun genes (De Oliveira et al. 2014).

7.3.4 Genotoxicity

Genotoxicity is the damage of genetic information inside a cell because of chemical agents which cause DNA damage, gene mutation, and chromosomal alteration (Bolognesi 2003). Genotoxicity is a major risk factor for carcinogenesis. Zebrafish model can be used to study various chemical-induced genotoxicities with the help of different techniques. Genotoxicity can be evaluated in embryos, larvae, or adult tissues and various techniques such as quantitative RAPD-PCR methodology for demonstrating dose-dependent genotoxicity of TiO_2NPs (Rocco et al. 2015) and comet assays for checking the effect of ferric oxide (Fe_2O_3) NPs can be used (Villacis et al. 2017). Moreover, RAPD-based methodology was used to assess genotoxic effects of gold NP on zebrafish (Dedeh et al. 2015; Geffroy et al. 2012). However, only fewer studies have been reported on the assessment of genotoxicity of NPs on zebrafish; hence, this area has to be studied extensively.

7.3.5 Cardiovascular Toxicity

Cardiac toxicological evaluation of NPs can be successfully performed using zebrafish embryos. Resemblance of zebrafish heart to human embryonic heart and direct observation of shape and rhythm of heart like heartbeats, cell activity in blood vessels, and blood vessel morphology using a microscope have greatly enabled efficient toxicity evaluation and toxicological research of NPs. Regular heartbeats in zebrafish commence at 36 h after fertilization. Monitoring and quantitative evaluation of cardiovascular damage on exposure to specific NPs have been effectively established using transgenic zebrafish lines. A study using transgenic zebrafish Tg (nacre/fli1: EGFP) revealed that CuONPs inhibit vasculogenesis through induction of apoptosis and reduction of vascular endothelial growth factor expression (Chang et al. 2015). The hematopoietic system of zebrafish is regulated by molecular pathways that are quite conventional. Particularly, the early development of cardiovascular system resembles that of humans. Therefore, AgNP toxicity in hematopoiesis was studied using a zebrafish model. Transcriptional responses of zebrafish embryos to AgNPs were revealed using microarray analysis. This analysis was performed at 24 h after fertilization. Gene ontology analysis revealed that AgNPs were responsible for downregulation of hemoglobin genes. It was also studied that erythrogenesis inhibition caused by AgNPs was cell specific and developmental stage specific. Further, it was found that this inhibition was caused mostly by AgNPs compared to their releasing ions (Cui et al. 2016).

7.3.6 Hepatotoxicity

The liver performs many important functions of body as it is the main metabolic organ of human body. Toxic effects of various chemicals can cause functional damage to liver and this may affect the normal functioning of body. The way in which the liver of zebrafish in its early developmental stages responds to toxic chemicals is similar to that of humans. Therefore, zebrafish model is ideal for studying NP-induced hepatotoxicity. Earlier studies have shown that when zebrafish embryos and larvae are exposed to CuONPs at high doses for a short period of time, hepatotoxicity and neurotoxicity, displaying as hepatic hypoplasia and delayed retinal neurodifferentiation coupled with decreased locomotor capacity, can be observed (Sun et al. 2016). Another study on the effects of oxidative stress and ZnOP damage on intestine, gill, and liver of zebrafish revealed that liver tissues were mainly targeted by oxidative damage. It was shown in the further study that ZnOPs produced higher OH radicals. The malondialdehyde, which is one of the biomarkers of oxidative stress, was increased in gills and liver of zebrafish (Xiong et al. 2011).

7.3.7 Reproductive Toxicity

Partial or whole life cycle tests of zebrafish can be used for testing reproductive toxicity of NPs. For instance, AgNP exposure resulted in oxidative stress, followed by germ cell apoptosis through mitochondrial dependent pathway. This finally led to damage of reproductive ability of zebrafish (Ma et al. 2018). In another study, AuNP (10–50 nm) exposure to adult female zebrafish gave rise to strand breaks in ovarian cells due to the ability of AuNPs to enter zebrafish ovaries (Dayal et al. 2016). Reproductive toxicity to zebrafish testis on exposure to TiO_2NPs was also studied. TiO_2NPs in higher doses were found to induce autophagy and necrosis in Sertoli cells and thus had a negative impact on testicular morphology and spermatogenic cells of zebrafish. It gave rise to mitochondrial degeneration with swelling and crista loss (Kotil et al. 2017).

7.3.8 Disruption of Gill, Skin, and Endocrine System

NP-induced toxicity also interrupts gills, skin, and endocrine system. Waterborne NPs mainly target gills of zebrafish. Silver ions (Ag+) produced by AgNPs show acute toxicity as they interact with the gills. Osmoregulation is affected in the gills, due to inhibition of Na^+/K^+-ATPase action and enzymes related to Na^+ and Cl^- uptake by Ag+ ions (Bury et al. 1999; Wood et al. 1999). Insoluble forms of CuNPs were also found to be very toxic and their suspensions may cause damage to gill lamellae (Griffitt et al. 2007). Moreover, NPs such as Ag-BSA enter embryo skin via diffusion or endocytosis, get deposited on the epidermis layer of larvae, and lead to skin abnormalities through apoptosis (Asharani et al. 2008). It was also suggested that TiO_2NPs cause an increase in the bioconcentration of lead, and lead to interruption of thyroid endocrine system in zebrafish larvae (Miao et al. 2015).

7.4 Nanotoxicology in Zebrafish

Nanotechnology has emerged as an interdisciplinary field which is linked to various subjects like physics, chemistry, biology, medicine, and toxicology (Weiss and Diabate 2011; Donaldson et al. 2004). Nanotechnology research primarily requires animal models to check nanotoxicity and zebrafish has the potential for the same as notable advancement has been made in the mentioned field using zebrafish (Jang et al. 2014). This section emphasizes on some recent studies and available data related to toxicity of NPs using zebrafish model.

7.4.1 Metal Nanoparticles

7.4.1.1 Gold

Unique properties AuNPs make it a preferred choice for various fields like cellular labeling, drug delivery, imaging and diagnostics for cancer, diabetes, and Alzheimer's disease (Li and Chen 2011). However, AuNPs may cause cytotoxicity in humans (Goodman et al. 2004; Gerber et al. 2013). Therefore, zebrafish has become a popular in vivo model for the assessment of toxicity caused by most commonly studied NPs (AuNPs) at present. Among all the engineered nanomaterials, AuNPs have the least empiric proof of adverse impacts on organisms, yet fewer number of investigations have been carried out to assess in vivo toxicity (Caballero-Diaz and Valcarcel 2014). In vitro assessment postulates some mechanisms such as genotoxicity, apoptosis, generation of ROS, leakage of toxic materials, interactions with lipids and proteins, mitochondrial damage, endocrine disruption, cellular morphology changes, and altered gene expression (Caballero-Diaz and Valcarcel 2014). There are a number of reported studies where embryos are exposed to 100 μL/mL of gold nanoclusters, but none showed toxic impact on mortality, gene expression, heart rate, hatching rate, and malformations (Chandirasekar et al. 2016). However, toxic impact was observed at relatively higher concentration, which does not have environmental importance. At 300 mg/mL, AuNPs showed 100% embryo mortality as an anticancer agent (Ramachandran et al. 2017). AuNPs were turned out to be less toxic toward embryos or adult zebrafish compared to other NPs such as Ag, Pt, and Cu (Ramachandran et al. 2018; Browning et al. 2019; Bar-Ilan et al. 2009; Asharani et al. 2010). But some studies reported toxic effect of AuNPs on zebrafish which may end up with embryonic lethality, neurotoxicity, developmental toxicity, and immunotoxicity (Truong et al. 2012; Kim et al. 2013). Presence of AuNPs (12 and 50 nm) in food leads to a variety of cellular malfunctions and genome modifications in adult zebrafish depending on size, exposure time, and concentration (Geffroy et al. 2012). Genome alteration in various adult tissues was observed when zebrafish was exposed to sediment containing 14 nm AuNPs for a longer period of time, which may be due to increase in oxidative stress (Dedeh et al. 2015). AuNPs were found to have more potential toxic effects than ionic Au if accumulated in tissues. Another work confirmed that 0–50 nm AuNPs could induce strand breaks in zebrafish ovaries (Dayal et al. 2017).

7.4.1.2 Silver

AgNPs are one of the most extensively studied NPs used as therapeutic agents, antimicrobials, and biosensors, in various cosmetic products and drug delivery systems (Czupryna and Tsourkas 2006; Yoon et al. 2007; Jin and Ye 2007; Prow et al. 2006; Perugini et al. 2002). AgNPs exert size-based toxicity which indicates that the dimension of NPs plays a crucial role in their toxicity profiling. A previous study established this fact by performing in vivo quantitative study in zebrafish to verify size-dependent transport and toxicity of AgNPs (Lee et al. 2012). In the abovementioned study, it was found that AgNPs having 30–72 nm diameter were capable to diffuse into the zebrafish embryos through chorionic pores due to random Brownian motion and may produce more potent toxic effect. However, different size (3, 10, 50, and 200 nm) of AgNPs (synthesized) showed 100% mortality rate after 120 hpf when administered to zebrafish embryos irrespective of size (Bar-Ilan et al. 2009). Hence, size-dependent toxicity profile of AgNPs is conclusive. A number of toxicities were observed including damage to neuromast hair cells, reduction in heart rate, teratogenicity, and mortality when AgNPs were exposed to zebrafish during early development (Yoo et al. 2016). But another study concluded that low concentrations of 10–20 nm AgNPs (<5 mg/L) do not have much impact on normal embryonic development, but higher concentrations showed significant impact on the growth of ectodermal and mesodermal tissues, probably due to delayed or inhibited cell division (Xia et al. 2016). Immunotoxicity and oxidative stress were observed due to the localization of AgNPs in the gills and liver when an adult zebrafish was exposed to it (Krishnaraj et al. 2016). A number of AgNPs possess different shapes and are known to induce oxidative stress, but plate-shaped AgNPs were more prone to show toxic effect than spherical and wire-shaped forms (George et al. 2012; Abramenko et al. 2018). Interestingly, these effects were associated with the presence of surface defects rather than Ag shedding (George et al. 2012). However, reductions in oxidative stress in embryos or adults were observed when AgNPs were coated with cysteine (George et al. 2012) or sulfidation (Devi et al. 2015). Increase in embryonic toxicity of AgNPs was detected after exposure to simulated solar light (George et al. 2014). Collectively, this suggests complex interplay of factors, where a range of physiochemical properties underpin biocompatibility.

7.4.2 Metal Oxide Nanoparticles

7.4.2.1 Titanium Dioxide

Among all, TiO_2NPs are one of the most extensively manufactured and commercially applied nanomaterials due to its area of application from colorants in sunscreens to excipients of toothpastes, shampoos, soaps, etc. which projects enormous growth potential; presently global annual production stands at around 10,000 Tm (Noman et al. 2018; Drobne 2018). Low-dose TiO_2NP does not show major developmental abnormalities in zebrafish when embryos are exposed to it (Wang et al. 2014). But various studies have reported their capability to trigger premature hatching in a dose-dependent manner (Samaee et al. 2015; Clemente et al. 2014). As per some studies, higher dose of TiO_2NPs may trigger embryonic malformation and death (Chakraborty et al. 2009). Another study reveals that the capability of TiO_2NPs to absorb photons may trigger production of electron-hole pairs which can interact with water and oxygen molecules to produce reactive oxygen species that are poisonous to zebrafish larvae (Bar-Ilan et al. 2012). Prolonged exposure of adult zebrafish to TiO_2NPs for 6 months at low concentrations (<4 mg/L) was also linked with low toxicity, judged by mortality rate. However, higher concentration leads to accumulation of NPs in various parts of the fish, including the heart, liver, gill, and brain (Chen et al. 2011a, b) and exhibits genotoxic effects (Rocco et al.

2015). Exposure of zebrafish embryos to TiO_2NPs starting from fertilization to the free-swimming phase does affect hatchability, survival, and malformation rate. However, larval swimming parameters such as average velocity and maximum velocity were considerably altered, indicating that the behavioral endpoints were far more sensitive than other parameters like hatchability and survival (Bar-Ilan et al. 2012; Chen et al. 2011a, b). However, the foremost consequence of TiO_2NP exposure is neurotoxicity. Even low level of TiO_2NPs may damage brain by crossing the blood-brain barrier, causing neuronal differentiation and neurogenesis (Wang et al. 2014; Chakraborty et al. 2009). Long-term low-dose exposure of TiO_2NPs to adult zebrafish for 45 days showed alteration in behavior and histopathological variations in the zebrafish brain due to the reduction in neurotransmitter level which were linked to dose-dependent elevation in nitric oxide levels (Sheng et al. 2016).

7.4.2.2 Copper and Copper Oxide

Utilization of copper has seen an upward trend over the years due to its considerable demand in various sectors like electronics, petroleum lubricants, catalysis, sintering active agents, consumer products of the pharmaceutical industry, adsorbents for water purification, and biomedical industries (Adeleye et al. 2016; Dankovich and Smith 2014; Lee et al. 2016; Liu and Astruc 2018; Goel et al. 2014). Copper and its oxides have been utilized in many areas including biosensing (Mao et al. 2015), energy storage (Dar et al. 2015), and development of antibacterial agents (Chatterjee et al. 2014). However, these materials can simply discharge copper particles which are exhibited to initiate cellular damage by prompting oxidative stress. Assessment of toxic impacts of Cu-based nanomaterials is far more difficult, as the toxicity is not only caused by the dissolved copper ions. One examination uncovered that CuNP introduction on zebrafish embryos indicated that CuNP creates ROS in a concentration-dependent manner (Denluck et al. 2018). CuNPs deferred embryo hatching time and produced teratogenicity of larvae. Dose-dependent mortality in zebrafish embryos was observed when CuNPs were exposed to it, whereas higher concentration leads to death of gastrula-stage zebrafish embryos (Bai et al. 2010). A previous study revealed that CuNPs cause acute toxicity to zebrafish embryos followed by gill injury (Griffitt et al. 2007). A new work reported on earlier report further disclosed that CuNPs (25 nm, 1 mg/L) might induce significant transcriptional changes in the pro-inflammatory linked genes in the skin and intestine and raise the movement of neutrophils in the tail of zebrafish embryos (Brun et al. 2018). Mentioned statement revealed CuNP-induced dermal and intestinal inborn immune responses, which may indicate the possible adverse events of CuNPs at higher levels of biological organization. CuONPs are vastly used in numerous fields like batteries, gas sensors, high-temperature superconductors, agricultural biocides, photocatalysts, energy transfer fluids, and antimicrobial agents (Batley et al. 2013; Hou et al. 2017; Kim et al. 2012; Llorens et al. 2012). Therefore, extensive use and its production may cause possible threats to individual organisms and ecosystem too. The outcome of the potential toxicity assessment of CuONPs in zebrafish embryos and larvae (Bai et al. 2010) exposed that CuONPs have the capability to interfere in embryo hatching in a dose-oriented way and produced amplified expression of the heat-shock protein 70 in zebrafish larvae when a higher dose was given (Lin et al. 2011). Additionally, administration of CuONPs in zebrafish embryos was discovered as a source of oxidative stress-mediated teratogenicity and this observation was primarily attributed to the particles themselves rather than dissolved Cu. Reactive oxygen species (ROS) may be generated due to the accumulation of CuONPs in embryos, which may further lead to cell apoptosis followed by production of deformed embryos (Ganesan et al. 2016).

7.4.2.3 Zinc Oxide

ZnONPs are considered as one of the most promising nanomaterials with widespread biomedical applications (e.g., anticancer and antibacterial therapy) (Mishra et al. 2017; Sirelkhatim et al. 2015) and possess characteristic properties like transparency, biocompatibility, high isoelectric point, and photocatalytic efficiency; they are frequently used in cosmetics, sunscreens, ceramics, photonics, and electrical appliances (Mirzaei and Darroudi 2017). Unfortunately, ZnONPs are categorized as "extremely toxic" and may cause severe threat to the environment and ecosystem (Kahru and Dubourguier 2010). Usually, the ZnONP-caused toxicity is primarily due to the dissolution of Zn^{2+} which can trigger various biological effects starting from lysosomal damage, mitochondrial perturbation, generation of ROS, initiation of pro-inflammatory responses, and lastly cell death (George et al. 2010; Xia et al. 2008, 2011). Zebrafish embryos and larvae show toxic effects such as retarded hatching, tail malformations, reduction in body length of the larvae, and tissue damage when they are exposed to ZnONPs at lower concentrations, but higher concentrations may lead to embryonic mortality (Zhu et al. 2008, 2009; Kteeba et al. 2017). Shape of the particle and surface coating play a significant role in experiencing ZnONP toxicity. Polymer-coated ZnONPs were considered to be more biocompatible compared to spherical ZnO, whereas leaf-shaped ZnONPs show extreme influence on hatching (Ong et al. 2014). Another research work on shape-based toxicity study of ZnONPs showed that nanospheres and cuboidal submicron particles were found to be less toxic than nanosticks in terms of hatching and overall mortality (Hua et al. 2014). Zhao et al. (2013) predicted the fundamental mechanism of ZnONP exposure-induced developmental toxicity which is linked to cellular oxidative stress, DNA damage, and altered actions of several critical defense enzymes (i.e., catalase, glutathione peroxidase, and superoxide dismutase).

7.4.2.4 Magnesium Oxide

MgONPs are commonly utilized in medicine, manufacturing, and anticancer therapy and as an antibacterial agent in the food industries. Extensive use of these NPs in our everyday lives results in unavoidable discharge and environmental exposure. Many researches have revealed variable toxicity of other metal oxide NPs. Exposure of MgONPs initiated increased mortality in zebrafish (Kovrižnych et al. 2013). Many researches evidenced concentration-dependent MgONP-induced cellular apoptosis and ROS. Dose-dependent alteration in hatching rate, malformations, and survival of zebrafish embryos were observed due to the exposure of MgONPs (20 nm) (Ghobadian et al. 2015).

7.4.2.5 Aluminum and Aluminum Oxide

Other NPs like aluminum nanoparticles (AlNPs) and Al_2O_3NPs have been broadly utilized in the drug delivery systems, optoelectronics industry, electronics, and biomedical products. Al_2O_3NPs and Al_2O_3 bulk showed very little acute toxicity to zebrafish embryos and larvae (Griffitt et al. 2008, 2011).

7.5 Limitations of Zebrafish Model for Nanotoxicity Study

Zebrafish as an in vivo model for toxicity profile of nanomaterials is a well-accepted phenomenon. The extent of toxicity of these NPs was evaluated by noticing the functional defects and malformations in zebrafish. However, literature survey reveals numerous lacunae in assaying nanomaterial-based immunotoxicity. Moreover, it is very challenging to assess embryo-based nanotoxicity assays systematically due to the fast developmental stages witnessed in zebrafish. However, advanced technologies along with automation help in screening nanotoxicity using zebrafish embryos. A number of nanomaterials are used for the purpose of therapeutic intervention in the area of antimicrobial therapy and drug delivery. Therefore, it is necessary to figure out the pharmacokinetic profiling of these nanomaterials. However, it is a bit challenging to perform ADME assay in zebrafish model after nanodrug delivery.

7.6 Future Prospects

Zebrafish as an in vivo model for toxicity profiling of nanomaterials has shown enormous potential. At present, several advance molecular biology techniques and zebrafish model transgenic lines are available for this purpose. Several zebrafish microarrays along with huge genomic resources are currently accessible for the purpose of nanotoxicity evaluation. These extremely advance resources make zebrafish a flexible system for toxicogenomic studies of nanomaterial in the coming days. Evaluation of nanomaterial toxicity on zebrafish development with the help of proteins and gene expression studies has enormous potential. Although zebrafish as a high-throughput screening system utilizing larval stages was previously explored for the purpose of evaluating nanomaterial toxicity, still huge scope persists for nanomaterial toxicity assays.

7.7 Conclusion

At present, zebrafish presents itself as a smart vertebrate model for testing NP toxicity and biocompatibility. Furthermore, this animal model has been much cheaper, faster, easy to conserve, and able to test agents efficiently via various routes for more than a decade. Additionally, definite physiological influences can be evaluated at multiple developmental stages. With the help of advance and up-to-date technology, zebrafish can become a meaningful alternative than other mammalian models for evaluating toxicity of nanomaterial in the coming days.

References

Abramenko NB, Demidova TB, Abkhalimov EV et al (2018) Eco toxicity of different-shaped silver nanoparticles: case of zebrafish embryos. J Hazard Mater 347:89–94

Adeleye AS, Oranu EA, Tao M et al (2016) Release and detection of nanosized copper from a commercial antifouling paint. Water Res 102:374–382

Asharani PV, Lian Wu Y, Gong Z et al (2008) Toxicity of silver nanoparticles in zebrafish models. Nanotechnology 19:255102

Asharani PV, Lianwu Y, Gong Z et al (2010) Comparison of the toxicity of silver, gold and platinum nanoparticles in developing zebrafish embryos. Nanotoxicology 5:43–54

Bai W, Tian W, Zhang Z et al (2010) Effects of copper nanoparticles on the development of zebrafish embryos. J Nanosci Nanotechnol 10(12):8670–8676

Bar-Ilan O, Albrecht RM, Fako VE et al (2009) Toxicity assessments of multi sized gold and silver nanoparticles in zebrafish embryos. Small 5:1897–1910

Bar-Ilan O, Louis KM, Yang SP et al (2012) Titanium dioxide nanoparticles produce phototoxicity in the developing zebrafish. Nanotoxicology 6:670–679

Batley GE, Kirby JK, McLaughlin MJ (2013) Fate and risks of nanomaterials in aquatic and terrestrial environments. Acc Chem Res 46(3):854–862

Bendale Y, Bendale V, Paul S (2017) Evaluation of cytotoxic activity of platinum nanoparticles against normal and cancer cells and its anticancer potential through induction of apoptosis. Integr Med Res 6(2):141–148

Bolognesi C (2003) Genotoxicity of pesticides: a review of human bio monitoring studies. Mutat Res 543:251–272

Browning LM, Lee KJ, Huang T et al (2019) Random walk of single gold nanoparticles in zebrafish embryos leading to stochastic toxic effects on embryonic developments. Nanoscale 1 (1):138–152

Brun NR, Lenz M, Wehrli B et al (2014) Comparative effects of zinc oxide nanoparticles and dissolved zinc on zebrafish embryos and eleuthero-embryos: importance of zinc ions. Sci Total Environ 476–477:657–666

Brun NR, Koch BVE, Varela M et al (2018) Nanoparticles induce dermal and intestinal innate immune system responses in zebrafish embryos. Environ Sci Nano 5:904–916

Bury NR, Grosell M, Grover AK et al (1999) ATP-dependent silver transport across the basolateral membrane of rainbow trout gills. Toxicol Appl Pharmacol 159:1–8

Caballero-Diaz E, Valcarcel M (2014) Toxicity of gold nanoparticles. In: Varcarcel M, Lopez-Lorente A (eds) Comprehensive analytical chemistry. Elsevier, Oxford, pp 207–254

Castranova D, Lawton A, Lawrence C et al (2011) The effect of stocking densities on reproductive performance in laboratory zebrafish (*Danio rerio*). Zebrafish 8(3):141–146

Chakraborty C, Agoramoorthy G (2010) Why zebrafish? Riv Biol 103:25

Chakraborty C, Sarkar B, Hsu C et al (2009) Future prospects of nanoparticles on brain targeted drug delivery. J Neurooncol 93:285–286

Chakraborty C, Sharma AR, Sharma G et al (2016) Zebrafish: a complete animal model to enumerate the nanoparticle toxicity. J Nanobiotechnol 14(1):1–13
Chandirasekar S, Chandrasekaran C, Muthukumarasamyvel T et al (2016) Biosurfactant templated quantum sized fluorescent gold nanoclusters for *in vivo* bioimaging in zebrafish embryos. Colloids Surf B Biointerfaces 143:472–480
Chang J, Ichihara G, Shimada Y et al (2015) Copper oxide nanoparticles reduce vasculogenesis in transgenic zebrafish through down-regulation of vascular endothelial growth factor expression and induction of apoptosis. J Nanosci Nanotechnol 15(3):2140–2147
Chatterjee AK, Chakraborty R, Basu T (2014) Mechanism of antibacterial activity of copper nanoparticles. Nanotechnology 25:135101
Chen J, Dong X, Xin Y et al (2011a) Effects of titanium dioxide nano-particles on growth and some histological parameters of zebrafish (*Danio rerio*) after a long-term exposure. Aquat Toxicol 101:493–499
Chen TH, Lin CY, Tseng MC (2011b) Behavioral effects of titanium dioxide nanoparticles on larval zebrafish (*Danio rerio*). Mar Pollut Bull 63(5–12):303–308
Choi JS, Kim RO, Yoon S et al (2016) Developmental toxicity of zinc oxide nanoparticles to zebrafish (*Danio rerio*): a transcriptomic analysis. PLoS One 11(8):e160763
Chun HS, Park D, Eun L et al (2017) Two zinc-aminoclays' in-vitro cytotoxicity assessment in HeLa cells and in-vivo embryotoxicity assay in zebrafish. Ecotoxicol Environ Safe 137:103–112
Clemente Z, Castro V, Moura M et al (2014) Toxicity assessment of TiO2 nanoparticles in zebrafish embryos under different exposure conditions. Aquat Toxicol 147:129–139
Collins JE, White S, Searle SMJ et al (2012) Incorporating RNA-seq data into the zebrafish Ensembl genebuild. Genome Res 22(10):2067–2078
Cui B, Ren L, Xu QH et al (2016) Silver nanoparticles inhibited erythrogenesis during zebrafish embryogenesis. Aquat Toxicol 177:295–305
Czupryna J, Tsourkas A (2006) Suicide gene delivery by calcium phosphate nanoparticles: a novel method of targeted therapy for gastric cancer. Cancer Biol Ther 5:1691–1692
Da SG, Clemente Z, Khan LU et al (2018) Toxicity assessment of TiO2-MWCNT nanohybrid material with enhanced photocatalytic activity on *Danio rerio* (zebrafish) embryos. Ecotoxicol Environ Safe 165:136–143
Dankovich TA, Smith JA (2014) Incorporation of copper nanoparticles into paper for point-of-use water purification. Water Res 63:245–251
Dar RA, Naikoo GA, Kalambate PK et al (2015) Enhancement of the energy storage properties of super capacitors using graphene nano sheets dispersed with macro-structured porous copper oxide. Electrochim Acta 163:196–203
Das S, Mitra S, Khurana SMP et al (2013) Nanomaterials for biomedical applications. Front Life Sci 7:90–98
Dayal N, Thakur M, Patil P et al (2016) Histological and genotoxic evaluation of gold nanoparticles in ovarian cells of zebrafish (*Danio rerio*). J Nanopart Res 18:291
Dayal N, Singh D, Patil P et al (2017) Effect of bioaccumulation of gold nanoparticles on ovarian morphology of female zebrafish (*Danio rerio*). World J Pathol 6:1
De Crozals G, Bonnet R, Farre C et al (2016) Nanoparticles with multiple properties for biomedical applications: a strategic guide. Nano Today 11(4):435–463
De León J, Cotto M, Márquez F (2019) Toxicology of nanomaterials on zebrafish. Am J Eng Appl Sci 12:193–203
De Oliveira GM, Kist LW, Pereira TC et al (2014) Transient modulation of acetylcholinesterase activity caused by exposure to dextran-coated iron oxide nanoparticles in brain of adult zebrafish. Comp Biochem Phys Part C 162:77–84
Dedeh A, Ciutat A, Treguer-Delapierre M et al (2015) Impact of gold nanoparticles on zebrafish exposed to a spiked sediment. Nanotoxicology 9:71–80
Dellinger B, Pryor WA, Cueto R et al (2001) Role of free radicals in the toxicity of airborne fine particulate matter. Chem Res Toxicol 14:1371–1377

Denluck L, Wu F, Crandon LE et al (2018) Reactive oxygen species generation is likely a driver of copper-based nanomaterial toxicity. Environ Sci 5(6):1473–1481
Devi GP, Ahmed KBA, Varsha MS et al (2015) Sulfidation of silver nanoparticle reduces its toxicity in zebrafish. Aquat Toxicol 158:149–156
Ding W, Guo L (2013) Immobilized transferrin Fe3O4@ SiO2 nanoparticle with high doxorubicin loading for dual-targeted tumor drug delivery. Int J Nanomedicine 8:4631–4639
Djurisic AB, Leung YH, Ng AM et al (2015) Toxicity of metal oxide nanoparticles: mechanisms, characterization, and avoiding experimental artefacts. Small 11(1):26–44
Donaldson K, Stone V, Tran CL et al (2004) Nanotoxicology. Occup Environ Med 61:727–728
Dooley K, Zon LI (2000) Zebrafish: a model system for the study of human disease. Curr Opin Genet Dev 10:252–256
Drobne D (2018) Spotlighting CLH report for TiO2: Nano-safety perspective. Chem Eng J 340:192–195
Fako VE, Furgeson DY (2009) Zebrafish as a correlative and predictive model for assessing biomaterial nanotoxicity. Adv Drug Deliv Rev 61(6):478–486
Friedman AD, Claypool SE, Liu R (2013) The smart targeting of nanoparticles. Curr Pharm Des 19:6315–6329
Gad SC (2006) Animal models in toxicology. CRC Press, Boca Raton
Gad SC (2015) Animal models in toxicology, 3rd edn. CRC Press, Boca Raton
Ganesan S, Thirumurthi NA, Raghunath A et al (2016) Acute and sub-lethal exposure to copper oxide nanoparticles causes oxidative stress and teratogenicity in zebrafish embryos. J Appl Toxicol 36:554–567
Garcia GR, Noyes PD, Tanguay RL (2016) Advancements in zebrafish applications for 21st century toxicology. Pharmacol Ther 161:11–21
Geffroy B, Ladhar C, Cambier S et al (2012) Impact of dietary gold nanoparticles in zebrafish at very low contamination pressure: the role of size, concentration and exposure time. Nanotoxicology 6:144–160
George S, Pokhrel S, Xia T et al (2010) Use of a rapid cytotoxicity screening approach to engineer a safer zinc oxide nanoparticle through iron doping. ACS Nano 4:15–29
George S, Lin S, Ji Z et al (2012) Surface defects on plate-shaped silver nanoparticles contributes to its hazard potential in a fish gill cell line and zebrafish embryos. ACS Nano 6:3745–3759
George S, Gardner H, Seng EK et al (2014) Differential effect of solar light in increasing the toxicity of silver and titanium dioxide nanoparticles to a fish cell line and zebrafish embryos. Environ Sci Technol 48:6374–6382
Gerber A, Bundschuh M, Klingelhofer D et al (2013) Gold nanoparticles: recent aspects for human toxicology. J Occup Med Toxicol 8:32
Ghobadian M, Nabiuni M, Parivar K et al (2015) Toxic effects of magnesium oxide nanoparticles on early developmental and larval stages of zebrafish (*Danio rerio*). Ecotoxicol Environ Safe 122:260–267
Giannakou C, Park MV, de Jong WH et al (2016) A comparison of immunotoxic effects of nanomedicinal products with regulatory immunotoxicity testing requirements. Int J Nanomedicine 11:2935–2952
Goel S, Chen F, Cai WB (2014) Synthesis and biomedical applications of copper sulfide nanoparticles: from sensors to theranostics. Small 10:631–645
Goodman CM, McCusker CD, Yilmaz T et al (2004) Toxicity of gold nanoparticles functionalized with cationic and anionic side chains. Bioconjug Chem 15:897–900
Griffitt RJ, Weil R, Hyndman KA et al (2007) Exposure to copper nanoparticles causes gill injury and acute lethality in zebrafish (*Danio rerio*). Environ Sci Technol 41:8178–8186
Griffitt RJ, Luo J, Gao J et al (2008) Effects of particle composition and species on toxicity of metallic nanomaterials in aquatic organisms. Environ Toxicol Chem 27(9):1972–1978
Griffitt RJ, Feswick A, Weil R et al (2011) Investigation of acute nanoparticulate aluminum toxicity in zebrafish. Environ Toxicol 26(5):541–551

Haffter P, Granato M, Brand M et al (1996) The identification of genes with unique and essential functions in the development of the zebrafish, *Danio rerio*. Development 123(1):1–36

Horikoshi S, Serpone N (eds) (2013) Microwaves in nanoparticle synthesis: fundamentals and applications. Wiley, Hoboken

Hou J, Wang X, Hayat T et al (2017) Ecotoxicological effects and mechanism of CuO nanoparticles to individual organisms. Environ Pollut 221:209–217

Howe K, Clark MD, Torroja CF (2013) The zebrafish reference genome sequence and its relationship to the human genome. Nature 496:498–503

Hua J, Vijver MG, Richardson MK et al (2014) Particle-specific toxic effects of differently shaped zinc oxide nanoparticles to zebrafish embryos (*Danio rerio*). Environ Toxicol Chem 33:2859–2868

Jang GH, Hwang MP, Kim SY et al (2014) A systematic *in-vivo* toxicity evaluation of nanophosphor particles via zebrafish models. Biomaterials 35:440–449

Jin S, Ye K (2007) Nanoparticle-mediated drug delivery and gene therapy. Biotechnol Prog 23:32–41

Jin Y, Zheng S, Fu Z (2011) Embryonic exposure to cypermethrin induces apoptosis and immunotoxicity in zebrafish (*Danio rerio*). Fish Shellfish Immunol 30:1049–1054

Johnston HJ, Verdon R, Gillies S et al (2018) Adoption of *in vitro* systems and zebrafish embryos as alternative models for reducing rodent use in assessments of immunological and oxidative stress responses to nanomaterials. Crit Rev Toxicol 48:252–271

Kahru A, Dubourguier HC (2010) From ecotoxicology to nanoecotoxicology. Toxicology 269:105–119

Kalueff AV, Stewart AM, Gerlai R (2014) Zebrafish as an emerging model for studying complex brain disorders. Trends Pharmacol Sci 35:63–75

Kessler R (2011) Engineered nanoparticles in consumer products: understanding a new ingredient. Environ Health Perspect 119(3):120–125

Kettleborough RN, Busch-Nentwich EM, Harvey SA et al (2013) A systematic genome-wide analysis of zebrafish protein-coding gene function. Nature 496:494–497

Kim S, Lee S, Lee I (2012) Alteration of phytotoxicity and oxidant stress potential by metal oxide nanoparticles in *Cucumis sativus*. Water Air Soil Pollut 223(5):2799–2806

Kim KT, Zaikova T, Hutchison JE et al (2013) Gold nanoparticles disrupt zebrafish eye development and pigmentation. Toxicol Sci 133(2):275–288

Kimmel CB, Ballard WW, Kimmel SR et al (1995) Stages of embryonic development of the zebrafish. Dev Dynam 203(3):253–310

Kotil T, Akbulut C, Yon ND (2017) The effects of titanium dioxide nanoparticles on ultra-structure of zebrafish testis (*Danio rerio*). Micron 100(6):38–44

Kovrižnych JA, Sotniková R, Zeljenková D et al (2013) Acute toxicity of 31 different nanoparticles to zebrafish (*Danio rerio*) tested in adulthood and in early life stage comparative study. Interdiscip Toxicol 6(2):67–73

Krishnaraj C, Harper SL, Yun SI (2016) *In vivo* toxicological assessment of biologically synthesized silver nanoparticles in adult zebrafish (*Danio rerio*). J Hazard Mater 301:480–491

Kteeba SM, El-Adawi HI, El-Rayis OA et al (2017) Zinc oxide nanoparticle toxicity in embryonic zebrafish: mitigation with different natural organic matter. Environ Pollut 230:1125–1140

Lee KJ, Browning LM, Nallathamby PD et al (2012) *In vivo* quantitative study of sized-dependent transport and toxicity of single silver nanoparticles using zebrafish embryos. Chem Res Toxicol 25:1029–1046

Lee IC, Ko JW, Park SH et al (2016) Comparative toxicity and bio-distribution of copper nanoparticles and cupric ions in rats. Int J Nanomedicine 11:2883–2900

Li YF, Chen C (2011) Fate and toxicity of metallic and metal containing nanoparticles for biomedical applications. Small 7(21):2965–2980

Li N, Sioutas C, Cho A et al (2003) Ultrafine particulate pollutants induce oxidative stress and mitochondrial damage. Environ Health Perspect 111:455–460

Li X, Liu B, Li XL et al (2014) SiO2 nanoparticles change colour preference and cause Parkinson's-like behaviour in zebrafish. Sci Rep 4:3810
Lin S, Zhao Y, Xia T et al (2011) High content screening in zebrafish speeds up hazard ranking of transition metal oxide nanoparticles. ACS Nano 5(9):7284–7295
Lin S, Zhao Y, Nel AE et al (2013) Zebrafish: an *in vivo* model for nano EHS studies. Small 9 (9–10):1608–1618
Liu X, Astruc D (2018) Atomically precise copper nanoclusters and their applications. Coord Chem Rev 359:112–126
Llorens A, Lloret E, Picouet PA et al (2012) Metallic-based micro and nanocomposites in food contact materials and active food packaging. Trends Food Sci Tech 24(1):19–29
Ma YB, Lu CJ, Junaid M et al (2018) Potential adverse outcome pathway (AOP) of silver nanoparticles mediated reproductive toxicity in zebrafish. Chemosphere 207:320–328
Mao ZG, Qing ZH, Qing TP et al (2015) Poly(thymine)-templated copper nanoparticles as a fluorescent indicator for hydrogen peroxide and oxidase-based biosensing. Anal Chem 87:7454–7460
McNamara K, Tofail SA (2013) Biomedical applications of nanoalloys. In: Nanoalloys: from fundamentals to emergent applications. Elsevier Inc., Amsterdam, pp 345–371
McNamara K, Tofail SA (2015) Nanosystems: the use of nanoalloys, metallic, bimetallic, and magnetic nanoparticles in biomedical applications. Phys Chem Chem Phys 17:27981–27995
Mesquita B, Lopes I, Silva S et al (2017) Gold nanorods induce early embryonic developmental delay and lethality in zebrafish (*Danio rerio*). J Toxicol Environ Health Part A 80 (13–15):672–687
Miao W, Zhu B, Xiao X et al (2015) Effects of titanium dioxide nanoparticles on lead bioconcentration and toxicity on thyroid endocrine system and neuronal development in zebrafish larvae. Aquat Toxicol 161:117–126
Mirzaei H, Darroudi M (2017) Zinc oxide nanoparticles: biological synthesis and biomedical applications. Ceram Int 43(1):907–914
Mishra PK, Mishra H, Ekielski A et al (2017) Zinc oxide nanoparticles: a promising nanomaterial for biomedical applications. Drug Discov Today 22:1825–1834
Morimoto Y, Kobayashi N, Shinohara N et al (2010) Hazard assessments of manufactured nanomaterials. J Occup Health 52:325–334
Namvar F, Rahman HS, Mohamad R et al (2014) Cytotoxic effect of magnetic iron oxide nanoparticles synthesized via seaweed aqueous extract. Int J Nanomedicine 9:2479–2488
Ning L, Zhu B, Gao T (2017) Gold nanoparticles: promising agent to improve the diagnosis and therapy of cancer. Curr Drug Metab 18(11):1055–1067
Noman MT, Ashraf MA, Ali A (2018) Synthesis and applications of nano-TiO2: a review. Environ Sci Pollut Res 26:3262–3291
Nowack B, Bucheli TD (2007) Occurrence, behavior and effects of nanoparticles in the environment. Environ Pollut 150(1):5–22
Ong KJ, Zhao X, Thistle ME et al (2014) Mechanistic insights into the effect of nanoparticles on zebrafish hatch. Nanotoxicology 8:295–304
Pecoraro R, Salvaggio A, Marino F et al (2017) Metallic nano-composite toxicity evaluation by zebrafish embryo toxicity test with identification of specific exposure biomarkers. Curr Protoc Toxicol 74:1–14
Perugini P, Simeoni S, Scalia S et al (2002) Effect of nanoparticle encapsulation on the photostability of the sunscreen agent, 2-ethylhexyl-p-methoxycinnamate. Int J Pharm 246:37–45
Powers CM, Slotkin TA, Seidler FJ et al (2011) Silver nanoparticles alter zebrafish development and larval behavior: distinct roles for particle size, coating and composition. Neurotoxicol Teratol 33(6):708–714
Prow T, Grebe R, Merges C et al (2006) Nanoparticle tethered biosensors for auto regulated gene therapy in hyperoxic endothelium. Nanomed Nanotechnol Biol Med 2(4):276

Ramachandran R, Krishnaraj C, Sivakumar AS et al (2017) Anti-cancer activity of biologically synthesized silver and gold nanoparticles on mouse myoblast cancer cells and their toxicity against embryonic zebrafish. Mater Sci Eng 73:674–683
Ramachandran R, Krishnaraj C, Kumar VA et al (2018) *In vivo* toxicity evaluation of biologically synthesized silver nanoparticles and gold nanoparticles on adult zebrafish: a comparative study. Biotech 8(10):441
Renier C, Faraco JH, Bourgin P et al (2007) Genomic and functional conservation of sedative-hypnotic targets in the zebrafish. Pharmacogenet Genom 17(4):237–253
Ribas L, Piferrer F (2014) The zebrafish (*Danio rerio*) as a model organism, with emphasis on applications for finfish aquaculture research. Rev Aquac 6:209–240
Rizzo LY, Golombek SK, Mertens ME et al (2013) *In vivo* nanotoxicity testing using the zebrafish embryo assay. J Mater Chem B 1(32):3918–3925
Rocco L, Santonastaso M, Mottola F et al (2015) Genotoxicity assessment of TiO2 nanoparticles in the teleost *Danio rerio*. Ecotoxicol Environ Safe 113:223–230
Rudramurthy GR, Swamy MK (2018) Potential applications of engineered nanoparticles in medicine and biology: an update. J Biol Inorg Chem 23(8):1185–1204
Samaee SM, Rabbani S, Jovanovic B et al (2015) Efficacy of the hatching event in assessing the embryo toxicity of the nano-sized TiO2 particles in zebrafish: a comparison between two different classes of hatching-derived variables. Ecotoxicol Environ Safe 116:121–128
Sangabathuni S, Murthy RV, Chaudhary PM et al (2017) Mapping the glyco-gold nanoparticles of different shapes toxicity, biodistribution and sequestration in adult zebrafish. Sci Rep 7(1):1–7
Sarmah S, Marrs J (2016) Zebrafish as a vertebrate model system to evaluate effects of environmental toxicants on cardiac development and function. Int J Mol Sci 17(12):2123
Sato Q, Zhang Y, Kusaka K et al (1998) Differences in the extent of inflammation caused by intratracheal exposure to three ultrafine metals: role of free radicals. J Toxicol Environ Health Part A 53:423–438
Schrand AM, Rahman MF, Hussain SM et al (2010) Metal-based nanoparticles and their toxicity assessment. Wiley Interdiscip Rev Nanomed Nanobiotech 2(5):544–568
Seaton A, Tran L, Aitken R et al (2010) Nanoparticles, human health hazard and regulation. J R Soc Interface 7:S119–S129
Segura-Aguilar J, Kostrzewa RM (2006) Neurotoxins and neurotoxicity mechanisms. An overview. Neurotox Res 10:263–287
Selgrade MK (2007) Immunotoxicity: the risk is real. Toxicol Sci 100(2):328–332
Shaw BJ, Liddle CC, Windeatt KM et al (2016) A critical evaluation of the fish early-life stage toxicity test for engineered nanomaterials: experimental modifications and recommendations. Arch Toxicol 90(9):2077–2107
Sheng L, Wang L, Su M et al (2016) Mechanism of TiO2 nanoparticle-induced neurotoxicity in zebrafish (*Danio rerio*). Environ Toxicol 31:163–175
Sirelkhatim A, Mahmud S, Seeni A et al (2015) Review on zinc oxide nanoparticles: antibacterial activity and toxicity mechanism. Nano Micro Lett 7:219–242
Spence R, Smith C (2005) Male territoriality mediates density and sex ratio effects on oviposition in the zebrafish, *Danio rerio*. Anim Behav 69(6):1317–1323
Spence R, Gerlach G, Lawrence C et al (2008) The behavior and ecology of the zebrafish, *Danio rerio*. Biol Rev Camb Philos Soc 83(1):13–34
Stainier DY, Fishman MC (1994) The zebrafish as a model system to study cardiovascular development. Trends Cardiovasc Med 4:207–212
Stark WJ, Stoessel PR, Wohlleben W et al (2015) Industrial applications of nanoparticles. Chem Soc Rev 44:5793–5805
Strähle U, Scholz S, Geisler R (2012) Zebrafish embryos as an alternative to animal experiments—a commentary on the definition of the onset of protected life stages in animal welfare regulations. Reprod Toxicol 33:128–132
Sun Y, Zhang G, He Z et al (2016) Effects of copper oxide nanoparticles on developing zebrafish embryos and larvae. Int J Nanomedicine 11:905–918

Torrealba D, More-Bayona JA, Wakaruk J et al (2019) Innate immunity provides biomarkers of health for teleosts exposed to nanoparticles. Front Immunol 9:3074
Truong L, Saili KS, Miller JM et al (2012) Persistent adult zebrafish behavioral deficits results from acute embryonic exposure to gold nanoparticles. Comp Biochem Phys C 155(2):269–274
Truong L, Tilton SC, Zaikova T et al (2013) Surface functionalities of gold nanoparticles impact embryonic gene expression responses. Nanotoxicology 7:192–201
Vicario-Pares U, Lacave JM, Reip P et al (2018) Cellular and molecular responses of adult zebrafish after exposure to CuO nanoparticles or ionic copper. Ecotoxicology 27(1):89–101
Villacis RAR, Filho JS, Pina B et al (2017) Integrated assessment of toxic effects of maghemite (g-Fe2O3) nanoparticles in zebrafish. Aquat Toxicol 191:219–225
Wang Y, Tang M (2018) Review of *in vitro* toxicological research of quantum dot and potentially involved mechanisms. Sci Total Environ 625:940–962
Wang J, Wang W (2014) Significance of physicochemical and uptake kinetics in controlling the toxicity of metallic nanomaterials to aquatic organisms. J Zhejiang Univ SC A 15:573592
Wang YJ, He ZZ, Fang YW et al (2014) Effect of titanium dioxide nanoparticles on zebrafish embryos and developing retina. Int J Ophthalmol 7:917–923
Weiss C, Diabate S (2011) A special issue on nanotoxicology. Arch Toxicol 85:705–706
Westerfield M (1995) The Zebrafish book: a guide for the laboratory use of Zebrafish (*Brachy Danio rerio*). University of Oregon Press, Eugene
Westerfield M (2007) The Zebrafish book: a guide for the laboratory use of zebrafish (*Danio rerio*). University of Oregon Press, Eugene
Win-Shwe TT, Fujimaki H (2011) Nanoparticles and neurotoxicity. Int J Mol Sci 12:6267–6280
Wood CM, Playle RC, Hogstrand C (1999) Physiology and modeling of mechanisms of silver uptake and toxicity in fish. Environ Toxicol Chem 18:71–83
Wu T, Tang M (2018) Review of the effects of manufactured nanoparticles on mammalian target organs. J Appl Toxicol 38(1):25–40
Xia T, Kovochich M, Liong M et al (2008) Comparison of the mechanism of toxicity of zinc oxide and cerium oxide nanoparticles based on dissolution and oxidative stress properties. ACS Nano 2:2121–2134
Xia T, Zhao Y, Sager T et al (2011) Decreased dissolution of ZnO by iron doping yields nanoparticles with reduced toxicity in the rodent lung and zebrafish embryos. ACS Nano 5:1223–1235
Xia G, Liu T, Wang Z et al (2016) The effect of silver nanoparticles on zebrafish embryonic development and toxicology. Artif Cells Nanomed Biotechnol 44:1116–1121
Xing B, Vecitis CD, Senesi N (eds) (2016) Engineered nanoparticles and the environment: biophysicochemical processes and toxicity. Wiley, Hoboken
Xiong D, Fang T, Yu L et al (2011) Effects of nano-scale TiO2, ZnO and their bulk counterparts on zebrafish: acute toxicity, oxidative stress and oxidative damage. Sci Total Environ 409 (8):1444–1452
Xu H, Dong X, Zhang Z et al (2015) Assessment of immunotoxicity of dibutyl phthalate using live zebrafish embryos. Fish Shellfish Immunol 45:286–292
Xu J, Zhang R, Zhang T et al (2017) Copper impairs zebrafish swim bladder development by down regulating Wnt signaling. Aquat Toxicol 192:155–164
Yamada M, Foote M, Prow TW (2015) Therapeutic gold, silver, and platinum nanoparticles. Wiley Interdiscip Rev Nanomed Nanobiotechnol 7(3):428–445
Yoo MH, Rah YC, Choi J et al (2016) Embryo toxicity and hair cell toxicity of silver nanoparticles in zebrafish embryos. Int J Pediatr Otorhinolaryngol 83:168–174
Yoon KY, HoonByeon J, Park JH et al (2007) Susceptibility constants of *Escherichia coli* and *Bacillus subtilis* to silver and copper nanoparticles. Sci Total Environ 373:572–575
Zhang W, Sun X, Chen L et al (2012) Toxicological effect of joint cadmium selenium quantum dots and copper ion exposure on zebrafish. Environ Toxicol Chem 31:2117–2123
Zhang XF, Liu ZG, Shen W et al (2016) Silver nanoparticles: synthesis, characterization, properties, applications, and therapeutic approaches. Int J Mol Sci 17(9):1534

Zhao XS, Wang ST, Wu Y et al (2013) Acute ZnO nanoparticles exposure induces developmental toxicity, oxidative stress and DNA damage in embryo-larval zebrafish. Aquat Toxicol 136:49–59

Zhu X, Zhu L, Duan Z et al (2008) Comparative toxicity of several metal oxide nanoparticle aqueous suspensions to Zebrafish (*Danio rerio*) early developmental stage. J Environ Sci Health Part A 43(3):278–284

Zhu X, Wang J, Zhang X et al (2009) The impact of ZnO nanoparticle aggregates on the embryonic development of zebrafish (*Danio rerio*). Nanotechnology 20(19):195103

Zhu JJ, Xu YQ, He JH et al (2014) Human cardiotoxic drugs delivered by soaking and microinjection induce cardiovascular toxicity in zebrafish. J Appl Toxicol 34(2):139–148

Part III

Nanovesicular Drug Delivery Carriers

Niosomes: A Novel Nanometric Vesicular System for Drug Delivery

8

Pranav Shah, Rinkal Jariwala, Swinal Kapadiya, Vidya P. Sabale, Priya Patel, and Pallavi M. Chaudhari

Abstract

Vesicular systems like niosomes, liposomes, transferosomes, pharmacosomes, and ethosomes have been employed as an all-rounder tool to upgrade the delivery of drug, since the past decade. Niosomes are submicron-sized vesicular drug delivery systems including nonionic surfactants, which are perishable, inexpensive, steadier, and relatively nontoxic in nature. There are various methods, which are meant for the preparation of niosomes. Niosomes are capable of delivering the drug in accordance with a predetermined rate and release therapeutically effective amount of drug at the targeted site. Various nonionic surfactants have been employed in the development of bilayer structure of niosomes. They have been known to possess the merits of drug targeting in a controlled or sustained manner, long shelf life, and high stability. In recent years, the potential of niosomes, as a carrier for delivery of therapeutic agents, diagnostic agents, and biotechnology-based products like DNA and RNA, has been widely studied. This review acknowledges us regarding the techniques of formulation, characterization process, and recent utilization of niosomes in the delivery of drugs.

P. Shah (✉) · R. Jariwala · S. Kapadiya
Maliba Pharmacy College, Uka Tarsadia University, Gopal Vidyanagar, Surat, Gujarat, India
e-mail: pranav.shah@utu.ac.in

V. P. Sabale
Dadasaheb Balpande College of Pharmacy, Besa, Nagpur, Maharashtra, India

P. Patel
Department of Pharmaceutical Sciences, Saurashtra University, Rajkot, Gujarat, India
e-mail: priyapatel@sauuni.ac.in

P. M. Chaudhari
Department of Pharmaceutics, Dr. D. Y. Patil College of Pharmacy, Pune, Maharashtra, India
e-mail: pallavichaudhari@dyppharmaakurdi.ac.in

N. Shah (ed.), *Nanocarriers: Drug Delivery System*,
https://doi.org/10.1007/978-981-33-4497-6_8

Keywords

Niosomes · Vesicular systems · Drug delivery · Drug targeting · Nonionic surfactants · Bilayer structure

8.1 Introduction

In recent years, delivery of drug at controlled rate and targeted rate has gained wide acceptance. The field of nanotechnology has helped to develop a number of drug carriers that are effective in loading of the drug. The application of nanocarriers has provided advantages like protection of drug from degradation, targeted delivery, low toxicity, and controlled release. Niosomes (nonionic vesicles) are basically unilamellar or multilamellar vesicles made of lipophilic and aqueous components. These vesicles are formulated by hydrated self-assembly of surfactant monomers. They have the capability of encapsulating hydrophilic and lipophilic drugs. They were previously used in the cosmetic industry, and the first was devised by L'Oréal, but now its importance is recognized in the pharmaceutical field too. Niosomes are more stable due to their constituents like nonionic surfactants which are more stable than lipids in terms of physical and chemical stability. As both hydrophilic and lipophilic drugs can be incorporated in niosomes, better targeting can be obtained by use of niosomes and it can be treated as a potential carrier for delivery of drugs effectively. Thus, due to these advantages of niosomes, it serves as a potential carrier for different drug targeting like ophthalmic, parenteral, and topical.

8.2 Components of Niosomes

Materials which are used in the formulation of niosomes are depicted in Table 8.1.

8.2.1 Nonionic Surfactants

Nonionic surfactants do not have any charge in their hydrophilic heads (Jiao 2008). Surfactant should be selected based on HLB value and critical packing parameter. Nonionic surfactants are amphiphilic molecules containing two different regions including hydrophilic (water soluble) and hydrophobic (organic soluble) and among them hydrophilic surfactants have high aqueous solubility.

8.2.2 Cholesterol

Steroids can affect bilayer fluidity and permeability on cell membrane. Cholesterol forms hydrogen bonds along with hydrophilic head of a surfactant in the bilayer form of niosomes (Mandal et al. 2013; Moghassemi and Hadjizadeh 2014). Physical

Table 8.1 Components of niosomes

Nonionic surfactant	Example	References
Alkyl ether 1. Alkyl glyceryl ether 2. Polyoxyethylene glycolAlkyl ethers	Hexadecyl diglycerol ether Brij 78, Brij 52	Arunothayanun et al. (2000), Pardakhty et al. (2007), Manconi et al. (2003), Bayindir and Yuksel (2010)
Alkyl ester	Span 20, Tween 20, Span 60, Tween 60, Span 80, Tween 80	Yoshioka et al. (1994), Okore et al. (2011), Akhilesh et al. (2012), Jain and Vyas (1995), Marianecci et al. (2012)
Fatty acids	Palmitic acid and stearic acid	Bandyopadhyay and Johnson (2007)
Lipidic components	Cholesterol	Bragagni et al. (2014)
Charged molecule	Diacetyl phosphate, phosphatidic acid, stearylamine, stearyl pyridinium chloride	Sankhyan and Pawar (2012), Junyaprasert et al. (2012)

properties such as stability, drug release, and entrapment efficiency are dependable on cholesterol content of niosomes (Akhilesh et al. 2012; Nasseri 2005). Drug entrapment efficiency shows major impact on the amount of cholesterol. In one study it is proved that cholesterol improves the stability and entrapment efficiency of enoxacin-loaded niosomes (Agarwal et al. 2004). Depending on the physicochemical characteristic of loaded drug and surfactants, cholesterol amount and its type need to be optimized.

8.2.3 Charged Molecule

For increased stability, charged molecules are added to niosomes in the bilayer vesicles. They prevent vesicle aggregation by increased surface charge density. Phosphatidic acid is a negatively charged molecule and stearyl pyridinium chloride is a positively charged molecule which are used in niosomal preparations.

8.3 Advantages and Limitations of Niosomes

8.3.1 Advantages

- It offers more patient compliance, as it is a water-based vesicular system in contrast to oily system.
- As the drug is entrapped into the vesicle, it enhances the steadiness of drug.
- They are stable in osmolality and active.
- They can be dispensed through oral, parenteral, as well as topical route.

- The therapeutic efficacy of drug molecules can be improved by delaying the clearance from the circulation; also it protects the drug from biological environment and provides targeted action.
- There is no specific requirement of handling and storage of surfactant as they are biocompatible, decomposable, and unsusceptible in nature.
- The oral bioavailability of the poorly absorbed drugs is increased by niosomes.
- They alleviate the penetration of drugs into the skin.
- The drug molecules can be accommodated into a wide range of solubility because niosomes are incorporated by the hydrophilic, lipophilic, and amphiphilic moieties.
- Controlled release of drug can be achieved by niosomes.
- They are physically stable than liposomes.

8.3.2 Limitations

- The physical stability and sterilization hinder niosomes from being used as potential drug delivery system.
- There are chances of leakage of the entrapped drugs.
- The aqueous suspension of niosomes may manifest fusion, aggregation, leaching, or hydrolysis of the entrapped drugs concluding in lowering of shelf life.

8.4 Differentiating Facts Between Niosomes and Liposomes

Table 8.2 shows differentiating important parameters of niosomes and liposomes.

8.5 Types of Niosomes

Niosomes are classified based on their function, method of preparation, and size of the vesicle (Gharbavi et al. 2018; Mayer et al. 1985) and are represented in Fig. 8.1.

8.5.1 Small Unilamellar Vesicles (SUVs)

Small unilamellar vesicles are prepared using various methods such as sonication, high-pressure homogenization, and extrusion methods.

8.5.2 Multilamellar Vesicles (MLVs)

MLVs (Fig. 8.1) are formed from some bilayers nearby to the aqueous lipid section separately. The diameter of these vesicles ranges from 100 to 1000 nm.

Table 8.2 Difference between niosomes and liposomes (Lotfabadi 2019; Lanka 2019)

Liposomes	Niosomes
Liposomes consist of phospholipid bilayers	Niosomes are constructed from nonionic surfactant vesicles
Size ranges from 10 to 3000 nm	Size ranges from 10 to 100 nm
Special storage conditions are required	No such conditions are required
Phospholipids are unstable	Nonionic surfactants are stable
Comparatively toxic	Less toxic
Comparatively expensive	Inexpensive
Phospholipid molecules contain two tails	Nonionic surfactants contain single tail
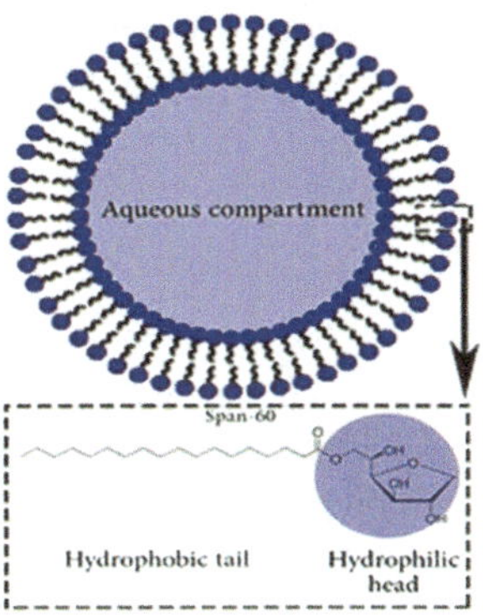 Niosomes using Span 60 as surfactant	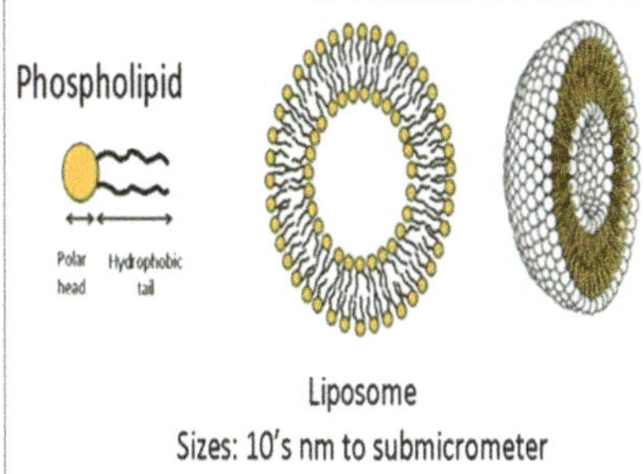

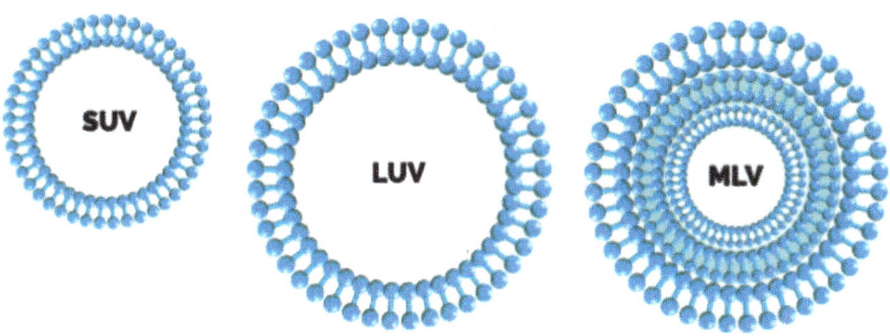

Small unilamellar vesicles | Large unilamellar vesicles | Multi lamellar vesicles

Fig. 8.1 Schematic diagram showing typical vesicle sizes of niosomes

8.5.3 Large Unilamellar Vesicles (LUVs)

LUVs hold high aqueous part compared to lipid part and due to this the bioactive resources can be taken up by membrane lipids. The approximate diameter of these vesicles ranges from 100 to 250 nm.

8.5.4 Innovative Niosomes of Alpha, Omega-Hexadecyl-Bis-(1-Aza-18-Crown-6) (Bola)

In these, two hydrophilic heads can be linked by one or two long lipophilic spacers. The surfactant used in bola surfactant-containing niosomes is omega hexadecyl-bis-(1-aza-18 crown-6) (bola surfactant):Span 80:cholesterol in 2:3:1 percentage.

8.5.5 Proniosomes

Proniosomes (Fig. 8.2) comprise water-soluble carriers and surfactants which are dehydrated and which would be hydrated for earlier usage. Problems of niosomes such as aggregation, fusion, and leakage of medication can be reduced with the use of proniosomes.

8.5.6 Aspasomes

Cholesterol, ascorbyl palmitate, and highly charged lipid such as dihexadecyl phosphate (DCP) are hydrated by water solvent and sonicated to produce the aspasome. Improved transdermal drug delivery systems are promising with the use of aspasome.

8.5.7 Discosomes

These are large disk-shaped structures having low cholesterol concentration. Discosomes act at the ocular site as potential drug delivery carriers for sustained-release system.

8.5.8 Elastic Niosomes (Ethoniosomes)

These are deformable niosomes made up of nonionic surfactants, ethanol, and water. These could better penetrate the intact skin compared to the conventional vesicles

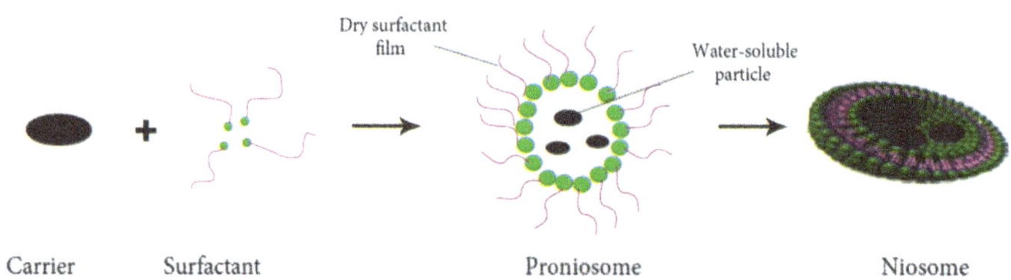

Fig. 8.2 Schematic representation showing proniosome and niosome formation process

since they can squeeze through small pores in stratum corneum which are smaller in size. Ethoniosomes prolong the drug release and represent a better biological activity by delivering the drugs or compounds at low and high molecular weight.

8.5.9 Polyhedral Niosomes

In the absence of cholesterol, polyhedral niosomes are formed by hexadecyl diglycerol ether (C16G2) and a series of polyoxyethylene alkyl ethers. Water-soluble particles are entrapped in these vesicles.

8.5.10 Niosomes (Vesicles) in Water-in-Oil System (V/W/O)

Vesicle-in-water-in-oil (v/w/o) system is formed from an aqueous dispersion of niosomes (nonionic surfactant vesicles) emulsified in an external oil phase which permits the delivery of vesicles in a nonaqueous vehicle. These types of niosomes may be used for protein drug delivery and protection from enzymatic degradation after oral administration and controlled release.

8.5.11 Niosomes in Carbopol Gel (Niosomal Gel)

In this system, prepared niosomes from the drug, nonionic surfactant, and cholesterol are incorporated in Carbopol-934 gel (1% w/w).

8.6 Factors Affecting the Formulations of Niosomes

8.6.1 Nature of Surfactant

HLB is commonly used to describe the vesicle-forming capacity of any surfactant. Encapsulation efficiency, stability, and toxicity of niosomes mainly depend on the nature of surfactant. The vesicle size of niosomes increased proportionally with increased HLB of the surfactants such as Span 60 having HLB value of 4.7 and Span 40 having HLB value of 6.7 to Span 20 having HLB value of 8.6 due to decreased surface free energy.

8.6.2 Surfactant Hydrophobicity

The particle size of niosomes increases with more HLB value of surfactant because of reduction in the surface free energy with increase in hydrophobicity.

8.6.3 Surfactant Lipid Ratio

Surfactant lipid ratio is an important parameter which affects physical stability of niosomes. Ideally the surfactant lipid ratio is 10–30 mm.

8.6.4 Effect of Cholesterol

Cholesterol is mainly responsible for particle size and encapsulation efficiency of niosomes along with its structure stability. Cholesterol mainly acts by two ways by increasing chain order of liquid-state bilayers and by increasing chain order of gel-state bilayers. Cholesterol is mainly responsible for stabilizing the system against aggregation due to the repulsive and electrostatic effects.

8.6.5 Hydration Temperature

Hydration temperature mainly affects the size and shape of the niosomes. Vesicle shape modification are mainly due to hydration temperature. Hydration time and volume of hydration are modified according to the structure requirement. Drug leakage problems mainly occur due to the improper hydration time and temperature.

8.6.6 Hydration Time

For stable structure hydration time should be above the gel-to-liquid phase transition temperature of system as it mainly influences the shape and size of the niosome. Both hydration volume and time are crucial factors for niosomes. Formation of fragile niosomes or production of drug leakage problems may result if these factors are not chosen properly (Shakya and Bansal 2014).

8.7 Methods for Preparation of Niosomes

There are some standard methods for the preparation of niosomes. It should begin with the hydration of a lipid mixture and surfactant at elevated required temperatures, followed by optional niosome size reduction in order to obtain a colloidal suspension (Sahin 2007). Hand shaking, ether injection, sonication, and microfluidization methods are a few examples (Kumar and Rajeshwarrao 2011; Kazi et al. 2010; Keservani et al. 2011). Centrifugation, gel filtration, and dialysis techniques are commonly used to separate unentrapped drug from the entrapped drug (Sahin 2007).

8.7.1 Ether Injection Method

In this method the surfactant is dissolved in diethyl ether followed by injection of solution through a 14-gauge needle into an aqueous solution of drug maintained at 60 °C (Akhilesh et al. 2012). Finally as a result, niosomes are formed in the form of large multilamellar vesicles (LMVs) (Baillie et al. 1986; Vyas and Khar 2002).

8.7.2 Hand-Shaking Method

This method is also known as thin-film hydration technique. In this method a mixture of surfactant and cholesterol is dissolved in an organic solvent in a round-bottom flask of rotary evaporator. Organic solvent is removed by low pressure at room temperature. The resultant dry surfactant film is hydrated by agitation at 50–60 °C and multilamellar vesicles (MLVs) are formed (Varshosaz et al. 2003).

8.7.3 Reverse-Phase Evaporation

Szoka and Papahadjopoulos described reverse-phase evaporation technique for liposome preparation (Szoka and Papahadjopoulos 1978). In this method water-in-oil emulsion is formed from an organic solution of surfactants and lipids. Both phases are emulsified by sonication. Final product of niosomes is obtained after heating the mixture at 60 °C for 10 min (Diljyot 2012; Kazi et al. 2010; Naresh et al. 1994). The organic solvent is removed at 40 °C under a low pressure, followed by dilution of remaining suspension with phosphate-buffered saline. The paste-like consistency is observed after removal of organic solvent and it contains an aqueous suspension of large unilamellar vesicles (Vemuri and Rhodes 1995).

8.7.4 Bubbling of Nitrogen

This method does not require any organic solvents. All the components required for niosomes are dispersed in the appropriate aqueous solutions and then homogenized using a homogenizer to obtain a homogeneous dispersion. After homogeneous dispersion, it is placed in a round-bottom flask with three necks attached to a water-cooled reflux along with thermometer and nitrogen supply and immersed in a water bath. A continuous stream of nitrogen gas bubbles is generated and introduced through the dispersion. Niosome dispersions are formed with mean particle size between 0.2 and 0.5 μm (Talsma et al. 1994).

8.7.5 Method of Handjani-Vila

Homogenous lamellar film is formed by addition of equivalent amount of synthetic nonionic lipids and aqueous solution of drug by shaking. The resultant mixture is homogenized with controlled temperature employing ultracentrifugation and agitation (Handjani-vila et al. 1982).

8.7.6 Sonication

The aqueous phase-containing drug is added into the mixture of surfactant and cholesterol in a scintillation vial. Resulting solution is homogenized using a sonication probe at 60 °C for 5 min. By this method, niosomes with control particle size can be produced (Baillie et al. 1985; Kazi et al. 2010).

8.7.7 Microfluidization

Microfluidization is a commonly used technique to maintain this size uniformity. Fluidized streams move forward through a precisely defined microchannel, and with ultrahigh velocity these streams interact with each other and produce the vesicles.

8.8 Pharmaceutical Applications of Niosomes for Oral Drug Delivery

Niosomes are very promising carriers for the delivery of numerous pharmacological and diagnostic agents due to their nonionic nature, excellent biocompatibility, and low toxicity. Due to the structure of niosomes they allow the development of effective novel drug delivery systems with the ability of loading both hydrophilic and lipophilic drugs. Hydrophilic drugs are entrapped into an aqueous core and lipophilic drugs are entrapped into the membrane bilayer of niosomes.

Oral drug delivery faces problems such as poor absorption and degradation of drug with acids and digestive enzymes in the stomach and hence there is a need to improve the bioavailability of drug through novel drug delivery systems such as niosomes (Azmin et al. 1985). Niosomes are an alternative drug delivery system for drug delivery to liposomes and polymersomes. Niosomes have some unique advantages when compared to liposomes and polymersomes such as biocompatibility, low toxicity, and biodegradability and are suitable for both hydrophilic cavity and hydrophobic shell.

8.8.1 Antineoplastic Agents

Several ligand-targeting agents were used for brain delivery such as insulin receptor and transferrin receptor by modification of the carrier structure, molecular size, and surface properties. Several studies have been functionalized for transport of vasoactive intestinal peptide (VIP) through glucose-targeted niosomes (Dufes et al. 2004). Another study with glucose derivative developed a probable transporter for brain-targeted delivery (Kazi et al. 2010). For CNS drug delivery, folic acid-targeted niosome was used as a carrier (Dufes et al. 2004).

In cancer chemotherapy, three targeting transporter systems can be used such as passive targeting, physical targeting, and active targeting.

1. Passive targeting: In passive targeting, nanoparticles are deposited in the surrounding of tumor cell through enhanced permeation and retention effect (EPR), due to particular properties inherent to the tumor cell which are not normally present in healthy tissues (Dufes et al. 2004).
2. Physical targeting: In this delivery system drug is released only when it is exposed to a specific microenvironment such as a change in pH or temperature by using an external magnetic field.
3. Active targeting: Nanoparticles can reach the targeted site in the tumor cells by active targeting agent. It can work with the versatile molecules or active targeting agent to identify tumor tissue targets (Dufes et al. 2004).

Conventional cancer chemotherapy is having numerous side effects to healthy tissue and produces lesser therapeutic efficiency. Negatively charged niosomes of paclitaxel showed efficient oral delivery with reduced side effects. Effect of 5-fluorouracil (5-FU) in nonmelanoma skin cancer was enhanced by delivery through niosomes (Kazi et al. 2010). Doxorubicin-loaded Span 60-containing niosomes have provided prolonged-release effect on tumor activity, with increased metabolic rate in liver and reduced clearance.

8.8.2 Antileishmanial Agents

Leishmania parasite invades cells of spleen and liver damaging the heart, liver, and kidney. One study shows that sodium stibogluconate in niosomal form was more effective compared to free drug in murine visceral leishmaniasis (Baillie et al. 1986). Paromomycin in the niosome form has efficient activity in the liver parasites without any sign of bone marrow parasites (Williams et al. 1998). Also amarogentin niosomes possess good applicability in leishmaniasis compared to the free drug (Medda et al. 1999). Most of the potent drugs which are used in the treatment of leishmaniasis show maximum toxicity; hence potent drugs such as quercetin increase its efficacy along with reduced side effects as proven in the hamster model (Sarkar et al. 2002).

8.8.3 Gene Delivery

Gene therapy is an emerging powerful tool nowadays for the treatment of diseases, but delivery remains a challenging task for clinical applications. Polymer- and lipid-based approaches are employed for the administration of gene materials as a nonviral gene carrier (Mahato et al. 1997; Mintzer and Simanek 2009). Antisense oligonucleotides showed positive cellular uptake in the form of niosomes (Huang et al. 2005). A new carrier may be developed by auto-coacervation through electronic effect-modified surface of cationic liposomes, which shows a positive effect on gene expression (Huang et al. 2006).

8.8.4 Proteins

Oral administration is not suitable for protein delivery due to the proteolytic enzymes, pH, and epithelial permeability. Pardakhty et al. (2007) had proved that the insulin given by oral route in a niosomal form composed of polyoxyethylene alkyl ethers and nonionic surfactants was stable with proteolytic action of pepsin and trypsin (Pardakhty et al. 2007). Niosomes can also be used for oral sustained-release delivery of protein and peptide molecules (Khaksa et al. 2000).

8.8.5 Antifungal Agents

Griseofulvin is commonly used as an antifungal agent but it suffers from the issue of poor and variable oral bioavailability; hence griseofulvin-loaded niosomes are prepared using different methods by using cholesterol and nonionic surfactant. Among them Span 60 shows promising results of entrapment efficiency and sustained-release effect (Jadon et al. 2009). Parenteral administration of niosomal nystatin is more effective with less toxicity of an antifungal agent. Nystatin in the form of niosomes shows higher amount of drug in vital organ with less hepatotoxicity and nephrotoxicity (Abdelbary et al. 2011). One study also proved that clotrimazole for vaginal administration shows controlled release for vaginal therapy in the form of niosomes (Ning et al. 2005). Fluconazole-loaded niosomes also show the sustained-release effect with prominently enhanced cutaneous retention of the drug (Gupta et al. 2011).

8.8.6 Antibiotics

For the ophthalmic delivery, an antibiotic in the form of niosomes with nonionic surfactant shows promising effect for the administration (Abdelbary and El-Gendy 2008). One study shows that cefpodoxime proxetil-loaded niosomes showed sustained release of 66% drug for 24 h, reducing the dose-dumping effect (Sambathkumar et al. 2011).

8.8.7 Antitubercular Drugs

Isoniazid-loaded niosomal formulation achieves effective treatment of tuberculosis with 62% of cellular uptake by macrophage cells. It shows reduced toxicity, reduced dosage frequency, and improved patient compliance with effective treatment of tuberculosis (Singh et al. 2011). Isoniazid in niosomal form also showed sustained-release pattern with lesser toxicity up to 48 h (Karki et al. 2008). Encapsulation of pyrazinamide in niosome form also delivered the maximum amount of pyrazinamide to the targeted site with reduced adverse effects and toxicity along with incredulous problem related to drug resistance (El-Ridy et al. 2011).

8.8.8 Antibacterial Drugs

Enoxacin-loaded nonionic surfactant modulated the drug delivery system without any sign of toxicity (Jia-You et al. 2001). Niacin-loaded niosomes also improved the performance of drug molecule. Thus, niosome is an alternative approach for the delivery of antibacterial drug compared to liposomes (Kopermsuba et al. 2011).

8.8.9 Hormones

Luteinizing releasing hormone (LRH)-loaded niosome formulation showed sustained-release effect and remained stable in plasma (Arunothayanun et al. 2000).

8.8.10 Niosomes as Carriers for Hemoglobin

Niosomes can also be used as a transporter of hemoglobin inside the blood (Radha et al. 2013).

8.9 Pharmaceutical Applications of Niosomes in Parenterals

The niosomes in submicron size are generally used for parenterals. They are administered via intramuscular or intraperitoneal route, with a size up to 10 μm. As per the reports of Florence and Cable, who prepared the 59Fe-deferroxamine trioxyethylene cholesterol vesicles, these were administered by i.v. route and they found that the distribution of vesicles depended on the size, with larger distribution in liver and spleen (Reddy et al. 2012).

Another study described the results by investigating the administration of doxorubicin niosomes prepared using C16G2 (a hexadecyl-diglycerol ether), Solulan C24, and cholesterol to healthy, nontumor mice, by the intraperitoneal route and they showed that there was no lung toxicity in mice dosed with niosomes of doxorubicin (Uchegbu et al. 1994).

In one study the prepared niosomes with large diameters of 800–900 nm using CI6 triglyceryl ether, with and without cholesterol and drug doxorubicin (Adriamycin), were administered to SI 80 tumor-bearing NMRI mice through bolus injection. The results showed that release was delayed for the one that contained cholesterol, but in vivo result showed little difference in plasma levels. Compared to the free solution half-life was found to be prolonged. Encapsulated doxorubicin noisome did not affect the liver uptake, but there was aggregation of vesicles that led to accumulation in the lung. When cholesterol niosomes were administered, it had a good effect on tumor, by reducing it (Rogerson et al. 1988).

As per one study, niosomes were prepared by the use of cholesterol, nonionic surfactant, and dicetyl phosphate with methotrexate (MTX) as the drug that was administered in mice. When the niosomes were administered intravenously levels of MTX in the blood were found to be prolonged and majority of the drug was taken up by the liver. Due to the permeability to BBB, drug may enter into the brain. When the drug was given orally, absorption was found to be increased at a few doses. Thus, from their study, they concluded that on administration of niosomes, there was a possibility of change in metabolic patterns of the drug (Azmin and Florence 1987).

One study reported the use of niosomes of glucose as a brain-targeted delivery system for the vasoactive intestinal peptide (VIP). These niosomes were intravenously injected to mice. Use of ^{125}I-labeled radioactivity determined the uptake of brain. HPLC was utilized for brain extract analysis. Presence of intact VIP in brain was confirmed by HPLC after administering the niosomes, but not after the solution of VIP was administered. The niosomes showed higher uptake of brain, when compared to control niosomes (Christine et al. 2004).

Saravanan and Popli (1998) had studied the effect of metronidazole-loaded niosomes in rats. The tissue distribution of metronidazole after intravenous administration comparing free drug solution and niosomes was verified in rat liver. When it was compared with the conventional delivery, niosomes were readily taken up by the reticuloendothelial system, so it was concluded that hepatic localization of niosomes was important for the treatment of hepatic amebiasis. When the drug was entrapped in niosomes, higher concentrations of drug were found in liver. Thus, niosomes can be of great potential for hepatic localization of metronidazole (Saravanan and Popli 1998).

One literature reports the use of cisplatin-loaded niosomes (CP-NMs) for anticancer treatment, when injected in rabbits with VX2 sarcoma. The niosomes gave superior results in inhibiting the tumor, with low mortality and not much change in body weight of rabbits, when it was compared with the solution of drug. Thus, they concluded that niosomes, when administered by parenteral route, showed good anticancer activity with less toxicity (Yang et al. 2013).

Thus, from the literature and studies reported, niosomes play an active role in the delivery of drugs by parenteral route.

8.10 Niosomes for Cancer

Cancer rules the main cause of death in the world, irrespective of the financial status of countries (Prager et al. 2018). In recent years, the rate of episodes of cancer has been increasing worldwide (Thompson 2010; Wiseman 2019). The major etiology for the development of cancer may be contributed by various factors like improper and poor diet, smoking, changes in reproduction, environmental conditions, and inactiveness in physical activities (Torre et al. 2015). As the expectations of longer life are increased, more occurrence of cancer has also been reported. Thus, various chemotherapies, in association with radiation and surgery, have been used to treat cancer.

Niosomes are bilayer-structured nanoparticles, by combination of cholesterol and nonionic surfactants (Fig. 8.3). These have the property of incorporating both hydrophilic and hydrophobic drugs. It is one of the promising drug deliveries for the treatment of cancer due to the modification on the surfaces that can be done for targeting only cancerous cells (Ahmad 2016).

Nowadays, the treatment for cancer relies on chemotherapy, but the poor penetration of anticancer drugs limits their therapeutic efficacy to cancer cells and due to this it leads to side effects on healthy cells. Nanotechnology is a novel field which may bring solution for overcoming all side effects of anticancer drugs and act as reservoir for drugs. There are a number of formulations for the treatment of cancer but a class of nanocarriers like niosomes possess a number of advantages over the conventional dosage forms that can become promising carriers for the treatment of cancer (Seleci et al. 2016).

8.10.1 Applications of Niosomes for Cancer

The niosomes due to their properties have been used for targeting cancer cells (Shi et al. 2005; Ruckmani et al. 2000; Abdelkader et al. 2012a, 2012b, 2012c; Abd-Elbary et al. 2008; Agarwal et al. 2001).

Cancer is a disease that involves uncontrolled growth of cells, invasion (meaning destruction of tissues that are adjacent), and even metastasis (that spreads through

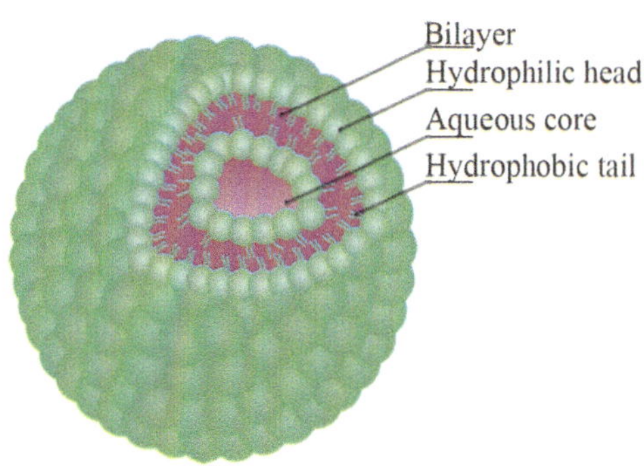

Fig. 8.3 Structure of niosome

blood/lymph to other parts of the body). It is one of the deadliest diseases that may affect people of all ages, and the incidence may increase with age. The diagnosis, treatment, and ways to prevent cancer are called as oncology. There are a number of causes of cancer, as mentioned above, and some other reasons include errors in DNA replication and genetic abnormalities (Mensah et al. 2018). Niosome can be one of the steps towards cancer treatment through targeted drug delivery system. This niosome potential in cancer treatment can be explored through its numerous applications. The conventional formulations possessed various side effects, so niosomes have been a choice for researchers (Kalra and Jeyabalan 2017).

The disadvantages of anticancer drugs are that they possess dose-dependent toxicity, neurotoxicity, ototoxicity, nephrotoxicity, depression in bone marrow, etc. When these drugs are administered orally, they get rapidly absorbed and there is a need for frequent dosing of the drug in large dose for the treatment. Thus, to avoid this frequent administration and avoid the limitations of anticancer drugs, niosome possesses the necessary properties for targeting cancer cells by using these anticancer drugs.

A good example of anticancer broad-spectrum anthracycline drug doxorubicin showed dose-dependent cardiotoxicity (Gandhi et al. 2012). But the same drug when given in the form of niosome showed increase in life span. The reason for this may be contributed to the property of niosomes of encapsulating the drug, without altering the metabolism of the drug and thereby causing prolonged circulation.

Another drug used for chemotherapy is methotrexate, individually or in combination. The niosome preparation of this drug in parenteral form was shown to weaken the tumors, enhance drug encapsulation, and offer high plasma drug level concentration when studied in mice having tumor (Udupa et al. 1993).

The drug tamoxifen citrate is used for breast cancer and when encapsulated in niosome and evaluated, the limitations of the drug were minimized against cancer cell line of breast (Shaker et al. 2015). Similarly another drug used for breast cancer is tocotrienol, which when encapsulated in niosomes showed higher cytotoxic effect to kill cancer cells of breast and there was drastic increase in the uptake of the drug too.

Paolino et al. had formulated 5-fluorouracil-loaded niosomes for treatment of skin cancer. The niosomal formulation showed reduction in toxicity of the drug and antitumor activity was found to increase when niosomes were administered (Paolino et al. 2008).

Curcumin possesses anticancer properties and hence niosome of this has also been formulated that leads to maximum encapsulation of about 92.3%. The formulation was used for ovarian cancer A2780 cells, where the results showed that the niosome showed controlled release (Xu et al. 2016). Study of curcumin- and doxorubicin hydrochloride-loaded niosomes proved to be very effective for the treatment of cervical cancer (HeLa) cells (Sharma et al. 2015).

Artemisinin is generally used for chills and fever, but it also possesses anticancer property. Niosome of artemisinin was developed for melanoma cells and results showed its significant effect on the treatment of melanoma (Dwivedi et al. 2015).

Dalia et al. had formulated tamoxifen citrate niosomes to study cytotoxicity effect on breast cancer cells. The niosomes showed prolonged drug release, increased cellular uptake, and higher entrapment efficiency (Dalia et al. 2015).

8.10.2 Target Specificity

Targeted drug delivery is to deliver the drug to the target by expansion of therapeutic window of the respective drug (Sudimack and Lee 2000). The accumulation of the drug at target site should not depend on the method utilized and route (Torchilin 2000). Effective target specificity needs four basic necessities of retaining, evading, targeting, and releasing of the drug (Mills and Needham 1999). The aim of this concept is to increase the efficiency by reduction of the side effects and target only the tumor cells without having any effect on the normal cells. There are two mechanisms by which targeting of the drug is done either by active or by passive targeting.

8.10.2.1 Active Targeting

It refers to the interaction with drug and target by ligand and receptor (Beduneau et al. 2007; Hong et al. 2009). Thus, localization of active targeting involves ligand–receptor interaction, and this can be done by PEGylation. This modifies the surface of the carrier and improves permeation and retention to the tumor site.

8.10.2.2 Passive Targeting

In this, the pathological condition of target can be utilized for accumulation of the drug for targeting the tumor cells. It can be utilized for tumor in the form of nanocarriers (Danhier et al. 2010). These nanocarriers take advantage of accumulation of drug for tumor cells and due to the surface charge they help to treat the tumor cells by use of EPR effect (Byrne et al. 2008).

8.10.3 Advances/Modifications in Niosomes for Delivery to Cancer

A number of techniques and modification agents are utilized for alteration of the side effects of niosomes, one of them being the use of polyethylene glycol (PEG). The coatings by PEG to niosomes help their retention in bloodstream for a longer time, due to their non-detection by the immune system, which leads to avoidance in engulfment by RES (Laouini et al. 2011). The surface modification of niosomes improves the target specificity for cancer treatment (Ahmad 2016). The surface modification of niosomes with folic receptor is considered the best for breast cancer (Gaber 2003). The various studies for the delivery of anticancer drug through niosomes are shown in Table 8.3.

Table 8.4 shows various studies of niosomes in different drug delivery systems.

Table 8.3 Summary of niosomal formulations for delivery of anticancer drugs

Type of study	Drug	References
Cancer and atherosclerosis treatment	Doxorubicin	Pawar and Vavia (2016)
Intravenous administration	Paclitaxel	Bayindir et al. (2015)
Breast cancer	Doxorubicin	Tavano et al. (2013a, 2013b)
Enhanced antitumor activity against sarcoma	Vincristine	Parthasarathi et al. (1994)
Enhanced cytotoxicity against cell lines of ovary and breast	Mitoxantrone	Tila et al. (2015)
Increase in cytotoxicity against breast cancer cell lines	Cisplatin	Kanaani et al. (2017)
Skin cancer	5-Fluorouracil (5-FU)	Abdelbary et al. (2015)
Cytotoxic studies	Paclitaxel	Timmakondu et al. (2011)
Synergistic antitumor efficacy	Paclitaxel and curcumin	Alemi et al. (2018)
Treatment of various cancers	Capecitabine	Vanani et al. (2019)
Colon cancer	Capecitabine	Anbarasan et al. (2013)
Colon cancer	5-Fluorouracil and leucovorin	Karthick and Kumaran (2016)
Breast cancer	Carum-loaded niosomes	Barani et al. (2019)
MCF-7 and HeLa cell lines	Lycopene	Sharma et al. (2016)
Breast cancer tumor and antitumor activity	Theranostic	Nowroozi et al. (2018)
Cancer treatment	Ciclopirox olamine	Shaikh et al. (2012)
PEGylated niosomes used for targeting to tumor	Hydroxycamptothecin	Hong et al. (2009)

Table 8.4 Various studies of niosomes in different drug delivery systems

Drug	Composition	Experimental study	Year	References
Candesartan cilexetil	Cholesterol, Span 60, maltodextrin, dicetyl phosphate	Pharmacokinetic analysis of proniosomal tablet	2016	Yuksel et al. (2016)
Naproxen	Tween 80, Tween 20, cholesterol	Preformulation study and release behavior	2016	Shah (2016)
Moxifloxacin	Tween 60, cholesterol	Release behavior and antimicrobial activity	2016	Sohrabi et al. (2016)
Paclitaxel	Span 40, cholesterol, dicetyl phosphate	Pharmacokinetic and tissue distribution studies	2015	Bayindir et al. (2015)
Nevirapine	Tyloxapol, cholesterol	In vitro release study and diffusion kinetics of drug	2014	Mehta and Jindal (2014)

8.11 Patents

The available patents for niosomal delivery systems are displayed in Table 8.5 (Manpreet and Kumar 2018).

8.12 Toxicity of Niosomes

Niosome formulation reported less toxicity issue. In one study the effects of surfactant in terms of molecular structure and addition of cholesterol on cell proliferation were checked. Polyoxyethylene chain lengths of surfactants with different hydrocarbon were also examined and it was seen that both show a minor effect on cell proliferation. As compared to ester-type surfactants ether-type surfactants inhibited 50% of cell proliferation (Hofland et al. 1991).

Ocular toxicity in terms of corneal irritation in niosomal formulation has been studied with Span 60-containing different bilayer material additives such as dicetyl phosphate and sodium cholate. The study proved that it had less ocular irritations with good ocular tolerability (Abdelkader et al. 2012a, 2012b, 2012c).

8.13 Marketed Formulations

Niosome-loaded drugs are also available in the market as shown in Table 8.6 (Kaur and Kumar 2018).

Table 8.5 List of patents related to niosomes

Cited patent	Applicant	Title
FR2571963B1	L'Oréal	Cosmetic and pharmaceutical compositions containing niosome
1892/MUM/2007	Murthy R. S.R.	PEGylated liposomal drug delivery system for doxorubicin hydrochloride

Table 8.6 Marketed formulations of niosomes

Brand	Name of the product
Lancome—foundation and complexation	Flash retouch brush on concealer
Britney Spears—Curious	Curious Coffret: Edp spray 100 mL + Dual-Ended Parfum and Pink Lip Gloss + body soufflé 100 mL
Loris Azzaro—Chrome	Chrome Eau De Toilette Spray
Orlane—lip color and lipstick	Lip gloss

8.14 Strengths, Limitations, and Future Direction of Niosome Drug Delivery System

As a potential drug delivery system stability of niosomes and sterilization are biggest issues. For improving the physical stability, dispersing liposomes in a viscous gel may become one way to reduce the leakage of drug from the formulation (Meisnera and Mezeib 1995).

Additionally lyophilization and spray drying approach can be used to enhance the physical stability of vesicles by converting the final product into the powder form (Ingvarsson et al. 2011). Meanwhile the oxidative stability issue of oxidizable drug molecules can also be reduced by reducing the formation of hydroxyl free radicals (Uchegbu and Vyas 1998). But main drawbacks are high manufacture cost, more production time, and many complications in formulations. In future research, many techniques and strategies are needed to be developed for improving the stability of niosomes.

Heat sterilization is not suitable for lipid-based drug delivery system; hence sterilization is again a challenging task and this area needs further research for commercial niosomal formulations (Zuidam et al. 1993). Drug leakage from bilayer vesicles may occur particularly with dry heat and steam sterilization process. Preparation under aseptic condition might be useful because minimum heat is generated during the process. Membrane filtration is also not feasible for particulate drug delivery system having a pore size larger than 0.22 μm (Hathout et al. 2007). Gamma radiation can be used for thermolabile drug and packaged drug as they contain high penetration power. In future, these methods put a good potential value for sterile niosomes (Waterman et al. 2002).

Niosomes show the greatest chemical stability and are more stable against oxidation and degradation during the storage compared to liposomes (Uchegbu and Vyas 1998). Commonly biodegradable and biocompatible surfactants are used for niosome preparation (Waddad et al. 2013). Stability along with size, lamellarity, and surface charge of niosomes can be controlled by surfactant, content of cholesterol, and type of method (Verma et al. 2010). During the storage of niosomes it shows risk of aggregation, leakage, and hydrolysis of drug.

Nasal route is considered as a potential route to achieve higher level and faster drug adsorption due to more permeability compared to gastrointestinal tract. It also has lack of various enzymatic activities with nasal mucosal pH (Kisan et al. 2007). As compared to the oral route it shows faster onset of action with avoidance of hepatic first-pass metabolism. Nasal mucosal area has less absorption surface area compared to gastrointestinal area and there are chances of damage to cilia on nasal mucosa and side effects; achieving an ideal concentrated dosing volume at the site is challenging (Chajed et al. 2011). Priprem et al. (2012) prepared melatonin-loaded niosomes by extrusion technique through polycarbonate membrane and it is found that intranasally administered niosomes distribute a significant amount of melatonin to the liver and testis of male rats (Priprem et al. 2012).

References

Abdelbary G, El-Gendy N (2008) Niosome-encapsulated gentamicin for ophthalmic controlled delivery. AAPS PharmSciTech 9:740–747

Abdelbary A, Essam T, Abd El-Salam RM et al (2011) Niosomes as a potential drug delivery system for increasing the efficacy and safety of nystatin (antifungal). Drug Dev Ind Pharm 37:149–508

Abdelbary A, Salem F, Khallaf R (2015) Niosomal 5-fluorouracil gel for effective treatment of skin cancer; in-vitro and in-vivo evaluation. Int J Drug Deliv 7(4):223–232

Abd-Elbary A, El-laithy H, Tadros M (2008) Sucrose stearate-based proniosome derived niosomes for the nebulisable delivery of cromolyn sodium. Int J Pharm 357(1–2):189–198

Abdelkader H, Ismail S, Kamal A et al (2012a) Conjunctival and corneal tolerability assessment of ocular naltrexone niosomes and their ingredients on the hen's egg chorioallantoic membrane and excised bovine cornea models. Int J Pharm 432:1–10

Abdelkader H, Ismail S, Hussein A et al (2012b) Conjunctival and corneal tolerability assessment of ocular naltrexone niosomes and their ingredients on the hen's egg chorioallantoic membrane and excised bovine cornea models. Int J Pharm 432(1–2):1–10

Abdelkader H, Wu Z, Al-Kassas R et al (2012c) Niosomes and discosomes for ocular delivery of naltrexone hydrochloride: morphological, rheological, spreading properties and photo-protective effects. Int J Pharm 433(1–2):142–148

Agarwal R, Katare O, Vyas S (2001) Preparation and in vitro evaluation of liposomal/niosomal delivery systems for antipsoriatic drug dithranol. Int J Pharm 228(1):43–52

Agarwal S, Bakshi V, Vitta P et al (2004) Effect of cholesterol content and surfactant HLB on vesicle properties of niosomes. Indian J Pharm Sci 66(1):121–123

Ahmad S (2016) Current status and future prospects of application specific engineered nanocurcumin compounds. Int J Pharm Pharm Sci 9:3391–3451

Akhilesh D, Bini KB, Kamath JV (2012) Review on span-60 based non-ionic surfactant vesicles (niosomes) as novel drug delivery. Int J Pharm Biomed Res 3:6–12

Alemi A, Zavar R, Haghiralsadat F et al (2018) Paclitaxel and curcumin coadministration in novel cationic PEGylated niosomal formulations exhibit enhanced synergistic antitumor efficacy. J Nanobiotechnology 16(1):28

Anbarasan B, Rekha S, Elango K et al (2013) Optimization of the formulation and *in vitro* evaluation of capecitabine niosomes for the treatment of colon cancer. Int J Pharm Sci Res 37:1504–1513

Arunothayanun P, Bernard MS, Craig DQ et al (2000) The effect of processing variables on the physical characteristics of nonionic surfactant vesicles (niosomes) formed from hexadecyl diglycerol ether. Int J Pharm 201:7–14

Azmin M, Florence A (1987) The effect of non-ionic surfactant vesicle (niosome) entrapment on the absorption and distribution of methotrexate in mice. J Microencapsul 4(4):321–328

Azmin MN, Florence AT, Handjani-Vila RM et al (1985) The effect of non-ionic surfactant vesicle (niosome) entrapment on the absorption and distribution of methotrexate in mice. J Pharm Pharmacol 37:237–242

Baillie AJ, Florence AT, Hume LR et al (1985) The preparation and properties of niosomes-non-ionic surfactant vesicles. J Pharm Pharmacol 37:863–868

Baillie AJ, Coomb GH, Dolan TF et al (1986) Non-ionic surfactant vesicles, niosomes, as a delivery system for the anti-leishmanial drug, sodium stibogluconate. J Pharm Pharmacol 38:502–505

Bandyopadhyay P, Johnson M (2007) Fatty alcohols or fatty acids as niosomal hybrid carrier: effect on vesicle size, encapsulation efficiency and in vitro dye release. Colloids Surf B Biointerfaces 58(1):68–71

Barani M, Mirzaei M, Mahani M et al (2019) Evaluation of carum-loaded niosomes on breast cancer cells: physicochemical properties, in vitro cytotoxicity, flow cytometric, DNA fragmentation and cell migration assay. Sci Rep 9:7139

Bayindir ZS, Yuksel N (2010) Characterization of niosomes prepared with various nonionic surfactants for paclitaxel oral delivery. J Pharm Sci 99(4):2049–2060

Bayindir Z, Besikci A, Yuksel N (2015) Paclitaxel-loaded niosomes for intravenous administration: pharmacokinetics and tissue distribution in rats. Turk J Med Sci 45(6):1403–1412

Beduneau A, Saulnier P, Hindre F et al (2007) Design of targeted lipid nanocapsules by conjugation of whole antibodies and antibody Fab' fragments. Biomaterials 28:4978–4990

Bragagni M, Mennini N, Furlanetto S et al (2014) Development and characterization of functionalized niosomes for brain targeting of dynorphin-B. Eur J Pharm Biopharm 87(1):73–79

Byrne J, Betancourt T, Brannon L (2008) Active targeting schemes for nanoparticle systems in cancer therapeutics. Adv Drug Deliv Rev 60(15):1615–1626

Chajed S, Sangle S, Barhate S (2011) Advantageous nasal drug delivery system: a review. Int J Pharm Sci Res 2(6):1322–1336

Christine D, Frederic B, Gaillard C (2004) Glucose-targeted niosomes deliver vasoactive intestinal peptide (VIP) to the brain. Int J Pharm 285:77–85

Dalia S, Shaker A, Hanafy M (2015) Cellular uptake, cytotoxicity and in-vivo evaluation of Tamoxifen citrate loaded niosomes. Int J Pharm 493(1–2):285–294

Danhier F, Feron O, Préat V et al (2010) To exploit the tumor microenvironment: passive and active tumor targeting of nanocarriers for anti-cancer drug delivery. J Control Release 148(2):135–146

Diljyot K (2012) Niosomes: a new approach to targeted drug delivery. Int J Pharm Phytopharm Res 2(1):53–59

Dufes C, Gaillard F, Uchegbu IF et al (2004) Glucose-targeted niosomes deliver vasoactive intestinal peptide (VIP) to the brain. Int J Pharm 285(1–2):77–85

Dwivedi A, Mazumder A, Plessis L et al (2015) In vitro anti-cancer effects of artemisone nano-vesicular formulations on melanoma cells. Nanomedicine 11:2041–2050

El-Ridy MS, Abdelbary A, Nasr EA et al (2011) Niosomal encapsulation of the antitubercular drug, pyrazinamide. Drug Dev Ind Pharm 37:1110–1118

Gaber M (2003) Enhanced cell killing by methotrexate encapsulated in folate targeted thermosensitive liposomes. Rom J Biophys 13:31–41

Gandhi A, Sen S, Paul A (2012) Current trend in niosome as vesicular drug delivery system. Asian J Pharm Life Sci 2:339–353

Gharbavi M, Jafar A, Kheiri-Manjili H et al (2018) Niosome: a promising nanocarrier for natural drug delivery through blood-brain barrier. Adv Pharmacol Sci 2018:1–15

Gupta M, Vaidya B, Mishra N et al (2011) Effect of surfactants on the characteristics of fluconazole niosomes for enhanced cutaneous delivery. Artif Cells Blood Substit Immobil Biotechnol 39:376–384

Handjani-vila RM, Ribier A, Vanlerberghe G (1982) Les liposomes. In: Lavoisier (ed) Les liposomes. Paris, pp 297–313

Hathout RM, Mansour S, Mortada ND, Guinedi AS (2007) Liposomes as an ocular delivery system for acetazolamide: in vitro and in vivo studies. AAPS PharmSciTech 8:E1–E12

Hofland HE, Bouwstra JA, Ponec M et al (1991) Interactions of non-ionic surfactant vesicles with cultured keratinocytes and human skin in vitro: a survey of toxicological aspects and ultra-structural changes in stratum corneum. J Control Release 16(1–2):55–167

Hong M, Zhu S, Jiang Y et al (2009) Efficient tumor targeting of hydroxycamptothecin loaded PEGylated niosomes modified with transferrin. J Control Release 133:96–102

Huang YZ, Han G, Wang H et al (2005) Cationic niosomes as gene carriers: preparation and cellular uptake in vitro. Pharmazie 60:473–474

Huang YZ, Gao JQ, Chen JL et al (2006) Cationic liposomes modified with non-ionic surfactants as effective non-viral carrier for gene transfer. Colloids Surf B Biointerfaces 49:158–164

Ingvarsson PT, Yang M, Nielsen HM et al (2011) Stabilization of liposomes during drying. Expert Opin Drug Deliv 8:375–388

Jadon PS, Gajbhiye V, Jadon RS et al (2009) Enhanced oral bioavailability of griseofulvin via niosomes. AAPS PharmSciTech 10(4):1186–1192

Jain CP, Vyas SP (1995) Preparation and characterization of niosomes containing rifampicin for lung targeting. J Microencapsul 12(4):401–407
Jiao J (2008) Poly oxy ethylated nonionic surfactants and their applications in topical ocular drug delivery. Adv Drug Deliv Rev 60(15):1663–1673
Jia-You F, Chi-Tzong H, Wen-Ta C et al (2001) Effect of liposomes and niosomes on skin permeation of enoxacin. Int J Pharm 219:61–72
Junyaprasert VB, Teeranachaideekul V, Supaperm T (2012) Effect of charged and non-ionic membrane additives on physicochemical properties and stability of niosomes. AAPS PharmSciTech 9(3):851–859
Kanaani L, Tabrizi MM, Khiyavi AA et al (2017) Improving the efficacy of cisplatin using niosome nanoparticlesAgainst human breast cancer cell line BT-20: An In Vitro Study. APJCB 2(2):27-29
Kalra N, Jeyabalan G (2017) Formulation and in-vitro evaluation of niosomal drug delivery in cancer chemotherapy. Indian J Pharm Biol Res 5(4):29–33
Karki R, Mamatha GC, Subramanya G et al (2008) Preparation, characterization and tissue disposition of niosomes containing isoniazid. Rasayan J Chem 1:224–227
Karthick K, Kumaran K (2016) Formulation and preclinical evaluation of niosomes co-loaded with 5-fluorouracil and leucovorin. Int J Res Pharm Nano Sci 5(5):283–292
Kaur D, Kumar S (2018) Niosomes: present scenario and future aspects. J Drug Deliv Therap 8 (5):35–43
Kazi KM, Mandal AS, Biswas N et al (2010) Niosome: a future of targeted drug delivery systems. J Adv Pharm Tech Res 1:374–380
Keservani RK, Sharma AK, Ayaz MD (2011) Novel drug delivery system for the vesicular delivery of drug by the niosomes. Int J Res Control Release 1:1–8
Khaksa G, D'Souza R, Lewis S et al (2000) Pharmacokinetic study of niosome encapsulated insulin. Indian J Exp Biol 38:901–905
Kisan RJ, Manoj NG, Ishaque MS et al (2007) Nasal drug delivery system-factors affecting and applications. Curr Drug Ther 2:27–38
Kopermsuba P, Mayena V, Warin C (2011) Potential use of niosomes for encapsulation of nisin and EDTA and their antibacterial activity enhancement. Food Res Int 44:605–612
Kumar GP, Rajeshwarrao P (2011) Nonionic surfactant vesicular systems for effective drug delivery-an overview. Acta Pharm Sin B 1:208–219
Lanka (2019) What is the difference between liposomes and niosomes? https://pediaa.com. Accessed 27 June 2019
Laouini A, Jaafar C, Sfar S et al (2011) Liposome preparation using a hollow fiber membrane contactor application to spironolactone encapsulation. Int J Pharm 415:53–61
Lotfabadi AS (2019) Difference between liposomes and niosomes. https://www.researchgate.net/post. Accessed 18 June 2019
Mahato RI, Rolland A, Tomlinson E (1997) Cationic lipid-based gene delivery systems: pharmaceutical perspectives. Pharm Res 14:853–859
Manconi M, Valenti D, Sinico C et al (2003) Niosomes as carriers for tretinoin: II. Influence of vesicular incorporation on tretinoin photostability. Int J Pharm 260(2):261–272
Mandal S, Banerjee C, Ghosh S et al (2013) Modulation of the photophysical properties of curcumin in nonionic surfactant (Tween-20) forming micelles and niosomes: a comparative study of different microenvironments. J Phys Chem B 117(23):6957–6968
Manpreet K, Kumar S (2018) Progress in the field of niosomes as novel drug delivery system. Indo Am J Pharm Sci 5(5):3417–3424
Marianecci C, Rinaldi F, Mastriotae M et al (2012) Anti-inflammatory activity of novel ammonium glycyrrhizinate/niosomes delivery system: human and murine models. J Control Release 164 (1):17–25
Mayer LD, Bally MB, Hope MJ (1985) Uptake of antineoplastic agents into large unilamellar vesicles in response to a membrane potential. Biochim Biophys Acta 816(2):294–302

Medda S, Mukhopadhyay S, Basu MK (1999) Evaluation of the in vivo activity and toxicity of amarogentin, an antileishmanial agent, in both liposomal and niosomal forms. J Antimicrob Chemother 44:791–794
Mehta SK, Jindal N (2014) Tyloxapol niosomes as prospective drug delivery module for antiretroviral drug nevirapine. AAPS PharmSciTech 16(1):67–75
Meisnera D, Mezeib M (1995) Liposome ocular delivery systems. Adv Drug Deliv Rev 16:75–93
Mensah K, Oosthuizen F, Bonsu A (2018) Cancer awareness among community pharmacist: a systematic review. BMC Cancer 18:299
Mills J, Needham D (1999) Targeted drug delivery. Expert Opin Ther Patents 9:1499–1513
Mintzer MA, Simanek EE (2009) Nonviral vectors for gene delivery. Chem Rev 109:259–302
Moghassemi S, Hadjizadeh A (2014) Nano-niosomes as nano scale drug delivery systems: an illustrated review. J Control Release 185(1):22–36
Naresh RA, Chandrashekhar G, Pillai GK et al (1994) Anti-inflammatory activity of niosome encapsulated diclofenac sodium in arthritic rats. Indian J Pharmacol 26(1):46–48
Nasseri B (2005) Effect of cholesterol and temperature on the elastic properties of niosomal membranes. Int J Pharm 300(1):95–101
Ning M, Guo Y, Pan H et al (2005) Preparation, in vitro and in vivo evaluation of liposomal/niosomal gel delivery systems for clotrimazole. Drug Dev Ind Pharm 31:375–383
Nowroozi F, Dadashzadeh S, Soleimanjahi H et al (2018) Theranostic niosomes for direct intratumoral injection: marked enhancement in tumor retention and anticancer efficacy. Nanomedicine 13(17):2201–2219
Okore VC, Attama AA, Ofokansi KC et al (2011) Formulation and evaluation of niosomes. Indian J Pharm Sci 73(3):323–328
Paolino D, Cosco D, Muzzalupo R et al (2008) Innovative bola-surfactant niosomes as topical delivery systems of 5-fluorouracil for the treatment of skin cancer. Int J Pharm 353:233–242
Pardakhty A, Varshosaz J, Rouholamini A (2007) In vitro study of poly oxy ethylene alkyl ether niosomes for delivery of insulin. Int J Pharm 328(2):130–141
Parthasarathi G, Udupa N, Umadevi P et al (1994) Niosome encapsulated of vincristine sulfate: improved anticancer activity with reduced toxicity in mice. J Drug Target 2(2):173–182
Pawar S, Vavia P (2016) Glucosamine anchored cancer targeted nano-vesicular drug delivery system of doxorubicin. J Drug Target 24(1):68–79
Prager GW, Braga S, Bystricky B et al (2018) Global cancer control: responding to the growing burden, rising costs and inequalities in access. ESMO Open 3(2):e000285
Priprem A, Limphirat W, Limsitthichaikoon S et al (2012) Intranasal delivery of nanosized melatonin encapsulated niosomes in rats. Open Access Sci Rep 1(4):232–237
Radha GV, Rani TS, Sarvani B (2013) A review on proniosomal drug delivery system for targeted drug action. J Basic Clin Pharm 4(2):42–48
Reddy B, Padman J, Voruganti S (2012) Niosomes as nanocarrier systems: a review. Int J Pharm Sci Res 3(6):1560–1568
Rogerson A, Cummings J, Willmott N et al (1988) The distribution of doxorubicin in mice following administration in niosomes. J Pharm Pharmacol 40:337–342
Ruckmani K, Jayakar B, Ghosal S (2000) Nonionic surfactant vesicles (niosomes) of cytarabine hydrochloride for effective treatment of leukemias: encapsulation, storage, and in vitro release. Drug Dev Ind Pharm 26(2):217–222
Sahin NO (2007) Niosomes as nano carrier systems. In: Mozafari MR (ed) Nanomaterials and nanosystems for biomedical applications. Springer, Dordrecht, pp 67–82
Sambathkumar R, Sekharbabu V, Perumal P et al (2011) Development and evaluation of cefpodoxime proxetil niosomes using various sorbitan esters. Res J Pharm Biol Chem Sci 2:213–219
Sankhyan A, Pawar P (2012) Recent trends in niosome as vesicular drug delivery system. J Appl Pharm 2(6):20–32
Saravanan D, Popli H (1998) Preparation and evaluation of metronidazole-loaded niosomes in rats. Pharm Pharmacol Commun 4:485–487

Sarkar S, Mandal S, Sinha J et al (2002) Quercetin: critical evaluation as an antileishmanial agent in vivo in hamsters using different vesicular delivery modes. J Drug Target 10:573–578
Seleci D, Seleci M, Walter J et al (2016) Niosomes as nanoparticular drug carriers: fundamentals and recent applications. J Nanomater 2016:1–13
Shah N (2016) Characterization, optimization and formulation of niosome containing naproxen. J Biomed Pharm Res 5(1):1–6
Shaikh K, Pawar A, Aphale S et al (2012) Effect of vesicular encapsulation on in-vitro cytotoxicity of ciclopirox olamine. Int J Drug Deliv 4(2):139–146
Shaker D, Shaker M, Hanafy M (2015) Cellular uptake, cytotoxicity and in-vivo evaluation of Tamoxifen citrate loaded niosomes. Int J Pharm 493:285–294
Shakya V, Bansal BK (2014) Niosomes: a novel trend in drug delivery. Int J Res Dev Pharm Life Sci 3(4):1036–1041
Sharma V, Anandhakumar S, Sasidharan M (2015) Self-degrading niosomes for encapsulation of hydrophilic and hydrophobic drugs: an efficient carrier for cancer multi-drug delivery. Mater Sci Eng C Mater Biol Appl 56:393–400
Sharma P, Saxena P, Jaswanth A et al (2016) Novel encapsulation of lycopene in niosomes and assessment of its anticancer activity. J Bioequivalence Biovailab 8:224–232
Shi B, Fang C, You M et al (2005) Influence of PEG chain length on in vitro drug release and in vivo pharmacokinetics of hydroxycamptothecin (HCPT) loaded PEG-PHDCA niosomes. J Chin Pharm 40(21):1643–1646
Singh G, Dwivedi H, Saraf SK et al (2011) Niosomal delivery of isoniazid—development and characterization. Trop J Pharm Res 10:203–210
Sohrabi S, Haeri A, Mahboubi A et al (2016) Chitosan gel-embedded moxifloxacin niosomes: an efficient antimicrobial hybrid system for burn infection. Int J Biol Macromol 85:625–633
Sudimack J, Lee R (2000) Targeted drug delivery via the folate receptor. Adv Drug Deliv Rev 41 (2):147–162
Szoka J, Papahadjopoulos D (1978) Procedure for preparation of liposomes with large internal aqueous space and high capture by reverse phase evaporation. Proc Natl Acad Sci U S A 75 (9):4194–4198
Talsma H, Van MJ, Steenbergen JC et al (1994) A novel technique for the one-step preparation of liposomes and nonionic surfactant vesicles without the use of organic solvents. Liposome formation in a continuous gas stream: the "bubble" method. J Pharm Sci 83(3):276–280
Tavano L, Gentile L, Oliviero Rossi C et al (2013a) Novel gel-niosome formulations as multicomponent systems for transdermal drug delivery. Colloids Surf B Biointerfaces 110:281–288
Tavano L, Muzzalupo R, Mauro L et al (2013b) Transferrin-conjugated pluronic niosomes as a new drug delivery system for anticancer therapy. Langmuir 29(41):12638–12646
Thompson R (2010) Preventing cancer: the role of food, nutrition and physical activity. J Fam Health Care 20(3):100–102
Tila D, Narjes S, Saeed G et al (2015) pH-sensitive, polymer modified, plasma stable niosomes: promising carriers for anti-cancer drugs. EXCLI J 14:21–32
Timmakondu S, Parthiban S, Prabu S et al (2011) Formulation and evaluation of paclitaxel niosome for its improved anti-cancer activity. Acta Pharm Sci 53(3):469–475
Torchilin VP (2000) Drug targeting. Eur J Pharm Sci 11:S81–S91
Torre L, Bray F, Siegel R et al (2015) Global cancer statistics. J Clin 65(2):87–108
Uchegbu IF, Vyas SP (1998) Non-ionic surfactant based vesicles (niosomes) in drug delivery. Int J Pharm 172(1–2):33–70
Uchegbu I, Turton J, Double J et al (1994) Drug distribution and a pulmonary adverse effect of intraperitoneally administered doxorubicin niosomes in the mouse. Biopharm Drug Dispos 15 (8):691–707
Udupa N, Chandraprakash K, Umadevi P et al (1993) Formulation and evaluation of methotrexate niosomes. Drug Dev Ind Pharm 19:1331–1342
Vanani R, Karimian K, Azarpira N et al (2019) Capecitabine-loaded nanoniosomes and evaluation of anticancer efficacy. Artif Cell Nanomed B 47(1):420–426

Varshosaz J, Pardakhty A, Hajhashemi V et al (2003) Development and physical characterization of sorbitan monoester niosomes for insulin oral delivery. Drug Deliv 10:251–262
Vemuri S, Rhodes CT (1995) Preparation and characterization of liposomes as therapeutic delivery systems: a review. Pharm Acta Helv 70:95–111
Verma S, Singh S, Navneet S et al (2010) Nanoparticle vesicular systems: a versatile tool for drug delivery. J Chem Pharm Res 2(2):496–509
Vyas SP, Khar RK (2002) Controlled drug delivery—concepts and advances, 1st edn. Vallabh Prakashan, New Delhi, pp 38–50
Waddad AY, Abbad S, Yu F et al (2013) Formulation, characterization and pharmacokinetics of Morin hydrate niosomes prepared from various non-ionic surfactants. Int J Pharm 456 (2):446–458
Waterman KC, Adami RC, Alsante KM et al (2002) Stabilization of pharmaceuticals to oxidative degradation. Pharm Dev Technol 7:1–32
Williams D, Mullen AB, Baillie AJ et al (1998) Comparison of the efficacy of free and non-ionic-surfactant vesicular formulations of paromomycin in a murine model of visceral leishmaniasis. J Pharm Pharmacol 50:1351–1356
Wiseman M (2019) Nutrition and cancer: prevention and survival. Br J Nutr 122(5):481–487
Xu Y, Chen W, Tsosie J et al (2016) Niosome encapsulation of Curcumin: characterization and cytotoxic effect on ovarian cancer cells. J Nanomater 2:1–9
Yang H, Deng A, Jingqing Z (2013) Preparation, characterization and anticancer therapeutic efficacy of cisplatin-loaded niosomes. J Microencapsul 30(3):237–244
Yoshioka T, Sternberg B, Florence AT (1994) Preparation and properties of vesicles (niosomes) of sorbitan monoesters (Span 20, 40, 60 and 80) and a sorbitan triester (Span 85). Int J Pharm 105 (1):1–6
Yuksel N, Bayindir ZS, Aksakal E et al (2016) In situ niosome forming maltodextrin proniosomes of candesartan cilexetil: in vitro and in vivo evaluations. Int J Biol Macromol 82:453–463
Zuidam NJ, Lee SS, Crommelin DJA (1993) Sterilization of liposomes by heat treatment. Pharm Res 10:1591–1596

9 Cubosomes: Novel Nanocarriers for Drug Delivery

Brijesh Patel and Hetal P. Thakkar

Abstract

Cubosomes are nanostructured systems which are made up of various amphiphilic lipids and stabilizers in a definite proportion. Cubosomes have a curved bicontinuous lipid bilayer which is organized in a three-dimensional space in such a way that gives it a honeycomb-like structure. In general, cubosomes are the colloidal dispersion of a bicontinuous cubic liquid phase in a solution of suitable stabilizers like poloxamers. Cubosomes are associated with a number of advantages compared to vesicular structures like liposomes and niosomes. They can incorporate a wide range of drugs like hydrophilic, lipophilic, and amphiphilic and moreover they are more thermodynamically stable than liposomes, niosomes, and other vesicular nanocarriers. They are also bioadhesive in nature and provide controlled release of a drug over longer period of time. Due to its unique properties, cubosomes are proving to be promising drug delivery systems. This chapter focuses on various aspects of cubosomes such as the mechanism of their formation, advantages and limitations, methods of preparation and characterization, applications in drug delivery, and safety and toxicity concerns. The various research works reported for therapeutic applications of cubosomes and the related patents have also been included in this chapter.

Keywords

Cubosomes · Bicontinuous cubic liquid crystalline phase · Drug delivery · Nanocarriers

B. Patel · H. P. Thakkar (✉)
Faculty of Pharmacy, The Maharaja Sayajirao University of Baroda, Vadodara, Gujarat, India
e-mail: hetal.thakkar-pharmacy@msubaroda.ac.in

N. Shah (ed.), *Nanocarriers: Drug Delivery System*,
https://doi.org/10.1007/978-981-33-4497-6_9

9.1 Introduction

Nanotechnology has received significant consideration of the researchers working in the area of drug delivery due to the novel and versatile applications associated with it. Various nanocarriers such as nanoparticles are made up of polymers, lipids, etc. and vesicular systems like liposomes, niosomes, and cubosomes have been widely studied for their ability in improving the safety and efficacy of drugs. Nanocarriers are colloidal dispersions having the drug either entrapped or coated on its surface via adsorption or conjugation. The nanocarriers of various drugs are developed for targeted or site-specific delivery, enhanced bioavailability, controlled release, reduced toxicity, improved stability, etc. Varieties of materials are explored for preparation of nanocarriers and depending on that they belong to different classes such as polymeric nanoparticles, lipid nanoparticles, metallic nanoparticles, vesicular structures, micelles, and dendrimers. Depending upon the material used for their preparation, nanocarriers can efficiently entrap hydrophilic, lipophilic, or amphiphilic drug molecules. Amphiphilic lipid molecules are used for formulating vesicular structures and can entrap both hydrophilic and lipophilic drugs (Mishra et al. 2019; Lombardo et al. 2019).

Amphiphilic molecules have a very unique property of self-assembly leading to formation of well-defined structures having higher thermodynamic stability known as liquid crystals (Karami and Hamidi 2016). Liquid crystals have intermediate properties between the liquid and solid crystals. Two types of liquid crystals are identified, i.e., thermotropic and lyotropic. In thermotropic liquid crystals, there is a change in the order of molecules with a change in temperature. Thus, their formation depends on the processing temperature. The preparation of lyotropic liquid crystals, however, requires the addition of immiscible solvent systems. Thus, amphiphilic molecules and solvent both are required to form any of these types of liquid crystals. In addition to this, the packing parameter, type of surfactant, and interfacial curvature energy are important factors to form lyotropic liquid crystals (Garti et al. 2012; Gin et al. 2007). Among the different types of amphiphilic molecules, lipids have been widely investigated as drug carriers due to their property of self-assembling leading to formation of lamellar, hexagonal, and bicontinuous cubic phases. Figure 9.1 depicts the ternary diagram showing the composition and temperature required to form different types of liquid crystals.

Vesicular carriers like liposomes, niosomes, and ethosomes are lamellar while cubosomes are bicontinuous lyotropic liquid crystals (Karami and Hamidi 2016; Garti et al. 2012). These vesicular systems are associated with a number of advantages like the ability of controlled or sustained release of drug with their ordered 3D mesoporous internal structure, site-specific release of therapeutic moiety which ultimately results in reduction of side effects, and increased lipid fraction per particle due to which the larger lipophilic area is available for poorly water-soluble lipophilic moieties (Chong et al. 2015). They also have the capability to increase the bioavailability of drugs having low water solubility. Moreover, the amphiphilic lipids used in the formulations are nontoxic and biodegradable and some of these

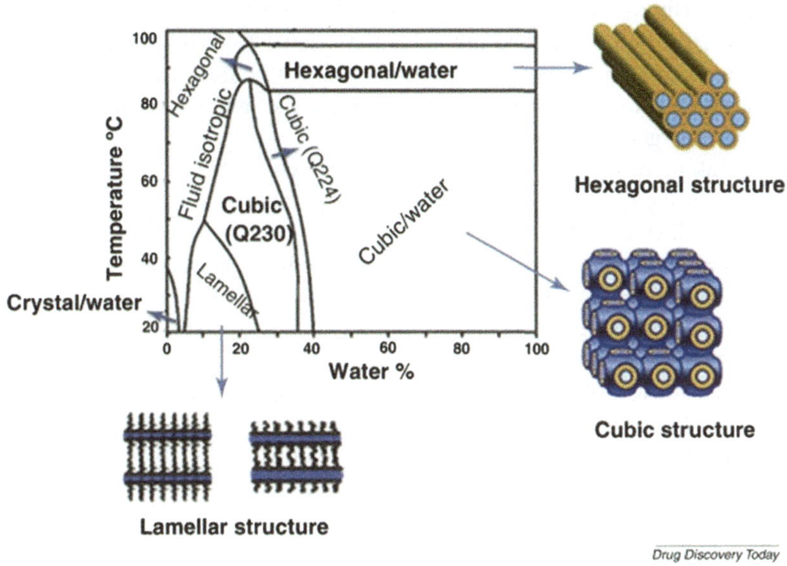

Fig. 9.1 Ternary phase diagram of the GMO: water system (Karami and Hamidi 2016)

have bioadhesive property (Guo et al. 2010; Yaghmur and Glatter 2009; Karami and Hamidi 2016; Shah et al. 2001; Chong et al. 2015).

Cubosomes are defined as "discrete, submicron, nanostructured particles of the bicontinuous cubic liquid crystalline phase." The term "bicontinuous" indicates two separate hydrophilic areas divided by the bilayer. Bicontinuous cubic crystalline materials have been an interesting research area as their structure lends itself well to controlled-release applications. Cubosomes are the colloidal dispersion (size ranging from 100 to 300 nm) made by dispersing the bicontinuous cubic liquid crystalline structures in aqueous medium having surface active agents (Karami and Hamidi 2016; Garg et al. 2007). The recently increasing research on cubosomes as drug nanocarriers indicates its large number of applications in drug delivery. Cubosomes have the ability to encapsulate a variety of drug molecules falling into the hydrophilic, lipophilic, and amphiphilic classes. Total eight types of cubic phases have been identified yet: *Ia3d* (Q^{230}), *Pn3m* (Q^{224}), *Im3m* (Q^{229}), *Fm3m* (Q^{225}), *Pm3n* (Q^{223}), *Fd3m* (Q^{227}), $P6_3Immc$, and $P4_332$. All these cubic phases are bicontinuous in nature except Q^{223} and Q^{227}, which are discrete or discontinuous in nature (Chong et al. 2015).

There are various differences between cubosomes and the most common vesicular carrier—liposomes. The differences and advantages of cubosomes over liposomes are listed below:

- Cubosomes are bicontinuous cubic liquid crystalline structures while liposomes and other vesicular structures have a spherical structure.

- Materials (lipids) used to prepare cubosomes are comparatively cheaper than those used in liposomes and they also have better tolerability than lipids used for the preparation of liposomes (Bansal et al. 2012; Anbarasan et al. 2015).
- Compared to the liposome, the ratio between the area of bilayer and particle volume is larger in cubosomes.
- The breaking resistance is higher in cubosomes compared to liposomes.
- Manufacturing processes are quite simpler than those used for liposome preparation (Bansal et al. 2012).
- The stability of cubosomes is much better than other vesicular systems and it also has better drug entrapment and loading (Bansal et al. 2012).
- Cubosomes can carry amphiphilic, lipophilic, and hydrophilic drugs (Bansal et al. 2012; Anbarasan et al. 2015).
- Sustained release of an entrapped drug molecule is possible due to the large interfacial area.
- The lipid component of the formulation is bioadhesive, biocompatible, and digestible (Rarokar and Khedekar 2018).

However, manufacturing of cubosomes on large scale is challenging due to high viscosity of the bulk phase (Bansal et al. 2012). Generally, the bulk gel phase involves viscous, clear, and semisolid gel which is similar to the cross-linked polymer hydrogels in appearance and viscosity (Spicer et al. 2001; Guo et al. 2010). The high viscosity is the main barrier in its application due to difficulty in handling and irritation to the biological epithelia (Rosen 2005; Guo et al. 2010). The cubic bulk phase is converted into a nanodispersion to overcome the main drawback of the bulk gel phase. This is done by dispersing the bulk phase into the water with the help of suitable surfactant to form small particles known as cubosomes (Gustafsson et al. 1996, 1997; Barauskas et al. 2005a).

The structure of the bicontinuous liquid crystals is unique and consists of a curved bicontinuous lipid bilayer protruding in three dimensions and two interpenetrating, however distinct, nano-channels having high interfacial area of 400 m^2/g. The lipid bilayer is approximately 3 nm thick whereas the estimated diameter of the fully swollen aqueous nano-channels is 5 nm (Drummond and Fong 1999; Spicer et al. 2001; Yaghmur and Glatter 2009). Glyceryl monooleate and phytantriol are widely investigated amphiphilic lipids for this purpose (Kim et al. 2015; Garg et al. 2007; Karami and Hamidi 2016). This compartmental structure in cubosomes is able to entrap hydrophilic, lipophilic, and amphiphilic guest molecules. In this cubic phase, hydrophilic molecules are located near the polar portion of emulsifier or in the water channels, while lipophilic drugs entangle into the lipid bilayer whereas amphiphilic molecules are situated at the interface (Sagalowicz et al. 2006; Leser et al. 2003).

9.2 Mechanism for the Formation of the Cubic Structure

The information of mechanism by which cubosomes are formed is not much clear. However, Patrick T. Spicer et al. conducted an experiment to understand the formation of cubosomes. They studied it by diluting monoolein-ethanol solution with poloxamer 407 solution in water. Upon dilution, creation of immediate interfacial turbulence with spontaneous emulsification took place resulting in the formation of numerous submicron vesicles. During the formation of vesicular structure, intermediate myelinic particle morphologies are formed. The diffusion of poloxamer 407 into the droplets and ethanol out of the droplets takes place. Figure 9.1 represents the phase diagram of GMO and water and also suggests the favorable condition for the formation of cubosome.

It represents the aqueous-phase behavior of the GMO-water system with temperature and also a schematic illustration of the lamellar, cubic, and hexagonal structure. From the phase diagram, it is clearly seen that by heating or by increasing the water concentration, a transition from lamellar to cubic to hexagonal is possible, and thus all phases are interconvertible by adjusting the manufacturing temperature and the water content (Karami and Hamidi 2016).

The X-ray crystallographic studies indicated only three discrete reversed bicontinuous cubic phases from the above-listed seven cubic phases: the body-centered cubic phase (Im3m, Q_{229}), the double-diamond lattice (Pn3m, Q_{224}), and the gyroid lattice (Ia3d, Q_{230}) (Larsson 2000; Shah et al. 2001; Tardieu and Luzzati 1970; Tenchov et al. 1998; Chong et al. 2015). It is also reported that, due to swelling at high water volumes, the Ia3d surface can be transformed into Pn3m surface as shown in the phase diagram in Fig. 9.1. On the other hand, the formation of Im3m surface requires the addition of the third component to the systems which may be caseins or amphiphilic block copolymer, which means that raising the water level is not sufficient to produce Im2m surface (Buchheim and Larsson 1987).

9.3 The Main Structural Components of Cubosomes

9.3.1 Amphiphilic Molecules

Amphiphilic materials which are reported to form a cubic phase are cationic and anionic soaps, nonionic and zwitterionic surfactants, and materials from the biological origin such as sphingolipids, monoglycerides, phospholipids, galactolipids, tetra-ether lipids, and glycolipids (Fontell 1990). The most investigated amphiphilic lipids used for the preparation of cubic crystalline phase are glyceryl monooleate (GMO) and phytantriol (Murgia et al. 2015; Demurtas et al. 2015; Larsson 1989; Azmi et al. 2013; Akhlaghi and Loh 2017; Azhari et al. 2016). GMO has varied applications in the pharmaceutical and food industry due to its amphiphilic nature. GMO is insoluble in water (10^{-6} M); however, its polar region forms hydrogen bond with water molecules. Thus, GMO can self-assemble into a different type of liquid crystalline structure under the influence of parameters like temperature and solvent concentration (water) resulting in the formation of cubic phase. Moreover, GMO is biodegradable, nontoxic, biocompatible, and generally recognized as safe (GRAS) (Ganem-Quintanar et al. 2000; Kulkarni et al. 2011; Lutton 1965; Larsson 1989; Hyde and Andersson 1984). Phytantriol is generally used as an active ingredient in the cosmetic products for hair and skin care. It has a phytanyl backbone structure which is chemically different from that of monoglycerides (Boyd et al. 2006). Fatty acid materials like GMO have a severe disadvantage of being hydrolyzed by the esterase, which reduces the stability of the GMO-based formulations or products; instead of this phytantriol provides more stability than GMO against esterase hydrolysis (Boyd et al. 2006; Rizwan et al. 2007). Moreover, phytantriol shows similar phase transitions as GMO upon increased water content and temperature. Thus, it is proposed as the best suited alternative to GMO for the preparation of bicontinuous cubic phase (Barauskas and Landh 2003). Phytantriol is biocompatible, biodegradable, and nontoxic in nature and it is insoluble in water (Boyd et al. 2006).

9.3.2 Stabilizers

As discussed above, amphiphilic lipids generally self-assemble in water leading to formation of lyotropic liquid crystalline phases. However, they possess very high viscosity which renders them inappropriate for direct ingestion or intravenous and nasal administration. In spite of the thermodynamically stable internal structure of these liquid crystalline particles, they have greater tendency of aggregation than a regular emulsion in aqueous solutions. To resolve these problems, the lyotropic liquid crystalline phase is fragmented into the solvent systems by using the stabilizer which provides colloidal stability to the liquid crystalline phase (Guillot et al. 2010; Larsson 1989). An ideal stabilizer is the one which prevents undesirable interaction between the lipid domains without disturbing the internal structure of the cubosomes. This may be achieved by the formation of an electro-repulsive and/or steric barrier between the approaching cubosomal particles. Thus, stabilizers are the most important component of cubosomal preparation (Tilley et al. 2013). However, it is reported that the negatively charged surfactant molecules disturb the internal structure of the cubosomes (Lindell et al. 1998). Moreover, it is also well known that negatively charged liposomes are rapidly removed from the systemic circulation compared to the neutrally charged liposome and positively charged liposomes are more toxic (Nishikawa et al. 1990; Senior 1987; Chonn et al. 1991; Kudchodkar et al. 1983; Senior et al. 1991). Thus, it can be inferred that cubosomes with neutral charge are more stable in systemic circulation and also more biocompatible than charged bicontinuous reverse cubic liquid phase.

Steric and stealth barrier is provided by a variety of polymers which are reported to bestow force of repulsion to the surfaces. They possess properties such as high hydrophilicity which prevents opsonization and subsequent uptake by reticuloendothelial system (Wattendorf and Merkle 2008; Ostuni et al. 2001). Numerous steric stabilizers are reported for the preparation of cubosomes which are classified into four different groups (Chong et al. 2015): (1) amphiphilic block copolymer like poloxamer (Rattanapak et al. 2012; Kwon et al. 2012; Hong et al. 2012a; Kwon and Kim 2014) and poloxamine (Boyd et al. 2009); (2) PEGylated lipids like vitamin E TPGS (Barauskas et al. 2005b), GMO-PEG (Johnsson et al. 2005), DOPE-PEG (Angelov et al. 2012), DSPE-PEG (1,2-distearoyl phosphatidyl ethanolamine-PEG) (Johnsson and Edwards 2001), and DMPE-PEG (1,2-dimyristoyl-sn-glycero-3-phosphoethanolamine-PEG) (Koynova et al. 1999; Koynova et al. 1997); (3) customized/designer lipid and various copolymer series such as P(ODA)-b-P (PEGA-OMe) (poly(octadecyl acrylate)-block-poly(PEG methyl ether acrylate) series); and (4) alternative steric stabilizers such as various proteins (Zhai et al. 2011; Boyd et al. 2009), polysaccharides (Uyama et al. 2009; Spicer et al. 2002), and bile salts (Gustafsson et al. 1999).

Most widely used and suitable steric stabilizer for the formulation of cubosomes is poloxamer 407 which is also known as Pluronic® F127. Poloxamer 407 is a nonionic triblock copolymer mostly used in pharmaceutical formulation and in personal care products and is composed of polypropylene oxide (PPO) and polyethylene glycol (PEG): $PEG_{100}PPO_{65}PEG_{100}$. It has an approximate molecular weight of 12,600 Da (Chong et al. 2015; Karami and Hamidi 2016). It acts by incorporating or absorbing its hydrophobic PPO block on the surface of nanostructured particle while exposing the PEO chain to the surrounding aqueous phase. In case of PPO, it acts as an anchor to the particle while the PEO chain extends covering the surface and providing steric stabilization to the particle in colloidal dispersion (Alexandridis 1996). Poloxamer 407 is most frequently used to stabilize bicontinuous liquid cubic phase, where GMO, glycerol monolinoleate, and phytantriol are used as bulk lipid phase. Among these the GMO-based formulation is the most extensively investigated lipid (Chong et al. 2015). Low concentration of (>4%) poloxamer 407 compared to a GMO forms cubosomes having a Q^{224} (*Pn3m*) surface while higher concentration (7.4 or 10 or more) forms a Q^{229} (*Im3m*) surface. However, low concentration of poloxamer 407 results in the product having overall poor quality with visual aggregates. Thus, a higher concentration of a stabilizer is used (approximately 10%) compared to a GMO for preparation of the bicontinuous cubic phase, as they give a product which is aggregate free (Wörle et al. 2007; Gustafsson et al. 1997; Landh 1994).

9.4 Methods of Preparation

The size of prepared nanostructured formulation is a very important characteristic for its desired pharmaceutical application. The method of preparation has a significant effect on the size of nanoformulation and thus proper selection of a method for preparation of the cubosomes is very important. There are two methods which can be used to prepare cubosome formulation: (1) "bottom-up" and (2) "top-down." In both methods, poloxamer 407 is the most commonly used stabilizer as described above.

9.4.1 Top-Down Method

This is the most widely used method for preparation of a cubosomal formulation. It involves two steps: (1) preparation of viscous bulk cubic phase by simply mixing the lipid(s) and the stabilizer(s) and (2) dispersion of the prepared bulk cubic phase into water using high-energy techniques like sonication or high-pressure homogenization. This results in the formation of cubosomes which are stable against aggregation for a period of at least 1 year (Karami and Hamidi 2016; Akhlaghi et al. 2016). However, this method has a major disadvantage that it involves the formation of highly viscous bulk cubic phase, difficult to handle during large-scale production. Moreover, the dispersion of viscous phase requires high energy which may produce heat and thus limits the incorporation of the temperature-sensitive materials into it. Furthermore, this high energy consumption enhances the cost of production. Moreover, the cubosomal formulation prepared by this method always coexists with the lamellar crystalline phase like liposomes (Rosen 2005; Karami and Hamidi 2016). According to the research study, the stable colloidal formulation of a cubic particle could be achieved only between 40 and 60 °C. Thus, if the homogenization temperature is increased to about 80 °C, coexistence of H_2 (hexagonal phase) and L_2 (inverse micellar phase) occurs which will decrease the overall quality of the formulation because coexisting phase affects the particle size, PDI (polydispersity index), entrapment of drug, etc. (Wörle et al. 2007; Muller et al. 2010). The top-down method for the preparation of cubosomes is depicted in Fig. 9.2.

9.4.2 Bottom-Up Method

In this method, the cubic dispersion is prepared by crystallization from the precursors and hence it is also known as liquid precursor or solvent dilution method (Spicer et al. 2001). Two solutions are prepared: (1) solution of lipid(s) in suitable hydrotrope and (2) stabilizer(s) in water. The first solution is added to the second solution dropwise with continuous stirring at 1000 rpm (Akhlaghi et al. 2016; Boge et al. 2019). The hydrotrope is added to the system to avoid the formation of a viscous liquid phase at high concentration (Um et al. 2003). Hydrotropes used in the preparation of cubosomes are ethanol, polyethylene glycol, and propylene glycol, among which ethanol is most widely used (Rizwan et al. 2011). The advantages of the liquid precursor method over the top-down method are as follows: (1) It can be prepared at low temperature, making it suitable for the temperature-sensitive materials. (2) It requires low energy input to prepare the cubosomes. (3) It has long-term stability due to the homogenous distribution of the stabilizer(s) onto the surface of the cubosomes. (4) The use of the hydrotrope simplifies the preparation process of the cubosomes and yields the product which has better or similar characteristics to those which are prepared by top-down approach. Due to all of the above advantages, the bottom-up method is preferable for scale-up to industrial manufacturing (Karami and Hamidi 2016). However, the use of hydrotrope has also some limitations such as the possible occurrence of allergic reaction and irritation

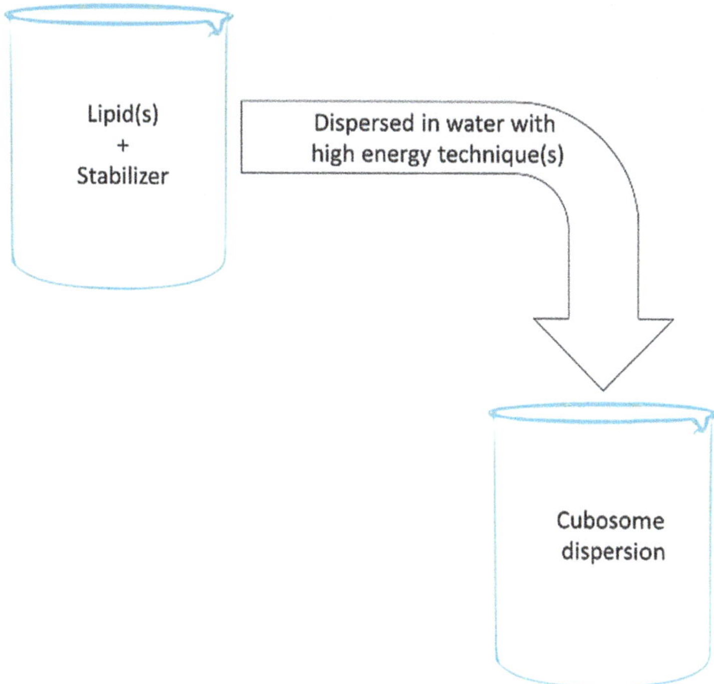

Fig. 9.2 Flowchart of the top-down method for the preparation of cubosome

when the cubosomal formulation is administered. Moreover, the formation of vesicular systems cannot be avoided during the preparation process. Thus, some new approaches are developed to overcome these limitations of the bottom-up approach, in which phosphate-buffered saline is used in a dual-lipid system of phytantriol and dodecyldimethylammonium bromide (DDAB) as a charged lipid. The added PBS in this system restores the bicontinuous cubic phase by creating the charged shield on DDAB which alters the bilayer curvature. Similar response is produced with the anionic lipid, 1,2-dipalmitoyl phosphatidylserine (DPPS) and phytantriol mixture (Hartnett et al. 2014).

During the preparation of cubosomes, some amount of drug does not get entrapped and remains as free drug in the formulation, which needs to be removed. The separation can be achieved using ultracentrifugation or passing the dispersion through Sephadex column. Centrifugation technique can be used to separate the free hydrophobic drug by rotating it at around 3000–5000 rpm, while in case of hydrophilic drug, Sephadex column is used. The bottom-up method for preparation of cubosomes is depicted in Fig. 9.3.

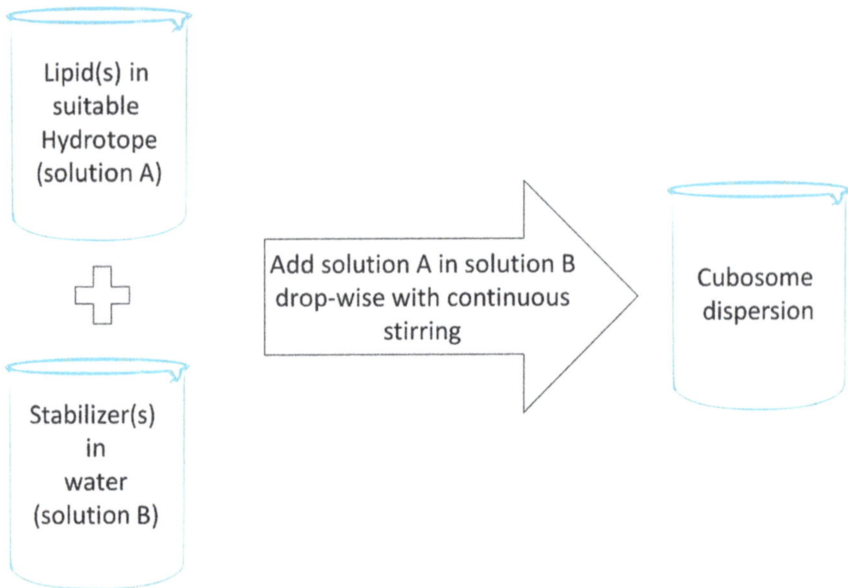

Fig. 9.3 Flowchart of the bottom-up method for preparation of cubosome

9.5 Characterization of Cubosomes

9.5.1 Particle Size Analysis and Zeta Potential

Dynamic laser light scattering technique is used to measure the particle size and distribution of nanoformulations such as cubosomes. Particle size analyzer and zeta potential analyzer are the instruments working on this principle. The sample is diluted with double-ultrapure water and filled in a cuvette which has an electrode fitted into it to measure the zeta potential of the formulation. Then, the suitable light scattering intensity is adjusted (around 300 Hz) and size distribution is measured at 25 °C. The measurements are done in triplicate and the mean is reported. The data collection is done for 8 min and average volume-weight size is reported.

The particle size of cubosomes is affected by variables such as concentration of lipid and stabilizer (Bei et al. 2009a), method of preparation (Karami and Hamidi 2016), and homogenization speed (Bei et al. 2009b). As the concentration of lipid and stabilizer increases, the particle size of a formulation increases (Bei et al. 2009a). Out of the two methods of preparation described above, top-down method gives formulation which has a lower particle size compared to bottom-up method (Karami and Hamidi 2016). An increase in homogenization speed results in decreased particle size of cubosomes (Bei et al. 2009b).

For measurement of zeta potential, the mode of the instrument is changed from particle size to zeta potential. Due to the electrode fitted inside the cuvette, potential measurement is possible. Each measurement is done in triplicate and mean is reported (Li et al. 2015). Zeta potential of cubosomes depends on the formulation components used for the preparation. Olof Svensson et al. prepared the cubosomal formulation using the glyceryl monooleate as a lipid and Lutrol F127 as a stabilizing polymer. In this study they found that the prepared formulation has a slightly negative charge ranging from −3 to −9 mV at pH range of 3–8. According to them this was possible due to the presence of a free oleic acid in a lipid. They also added that the adsorption of a hydroxyl ion at the oil-lipid interface was also responsible for the negative charge produced by the cubosomes. Each measurement is done in triplicate and mean is reported (Svensson et al. 2008).

9.5.1.1 Polarizing Light Microscopy

The cubosomal dispersion is subjected to polarizing light microscope employing a $\lambda/4$ compensator to analyze the existence of birefringence. The examination is done at a magnification of around 400X and photomicrographs are captured (Pitzalis et al. 2000; Wu et al. 2012). The identification of liquid crystalline phases is done as per the classification provided by Rosevear (1954).

9.5.2 Encapsulation Efficiency and Loading Capacity

The cubosomes are dissolved in solvents such as methanol or ethanol and the amount of drug entrapped is determined by the appropriate analytical method. The degree of encapsulation and loading efficiency depend on the type of the materials, i.e., lipid(s), stabilizer(s), hydrotrope(s), etc., and on its concentration. Moreover, they are also affected by the speed of homogenization (Bei et al. 2009a; Wu et al. 2012). The calculation of encapsulation efficiency (EE) and loading capacity (DLC) is done using Eqs. 9.1 and 9.2, respectively (Wu et al. 2012):

$$\mathrm{EE}(\%) = \frac{\text{Encapsulated drug in cubosomes}}{\text{Total drug added}} \times 100 \tag{9.1}$$

$$\mathrm{DLC}(\%) = \frac{\text{Amount of drug in cubosomes}}{\text{Amount of lipid added}} \times 100 \tag{9.2}$$

9.5.3 Cryo-Transmission Electron Microscopy (Cryo-TEM)

A controlled atmosphere is necessary to prepare samples to be analyzed by cryo-TEM. To avoid evaporation from the sample during preparation, the temperature of the chamber is set to 25–28 °C and the relative humidity is adjusted near to saturation, i.e., 95% or more. A drop of the sample is positioned on a holey film which is coated with carbon and supported by copper grid. Then, it is moderately blotted using filter paper to achieve a thin layer of the sample on the copper grid. The

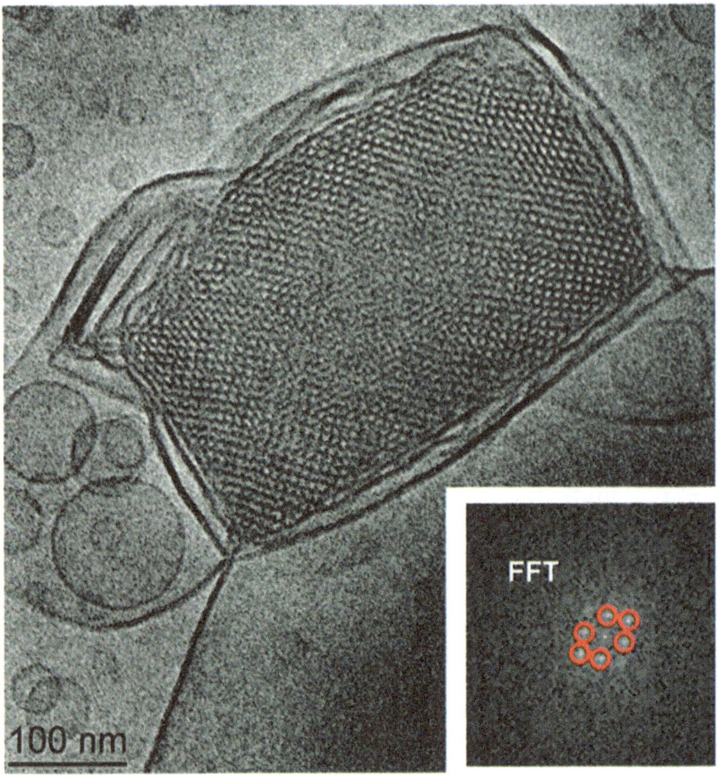

Fig. 9.4 Cryo-TEM image of the cubosome dispersion (Angelov et al. 2015)

prepared grid is placed on a project area of the cryo-plunger and promptly propelled in liquid ethane at around 90 K and excess liquid is eliminated. After that, the leftover ethane is removed from the same grid and the obtained implanted sample is shifted to a cryoelectron microscope. The sample is analyzed at a steady temperature of 90–100 K and voltage acceleration close to 200 kV under low-dose conditions. The pictures of the sample are taken with the help of a charge-coupled device (CCD) camera (Wu et al. 2012). Figure 9.4 depicts the cryo-TEM image of a cubosomal dispersion.

9.5.4 In Vitro Drug Release

It is reported that cubosomes can incorporate a variety of drug molecules and control its release (Anbarasan et al. 2015). A cubic phase follows Higuchi-type diffusion-controlled drug release kinetic (Higuchi 1967). Diffusion of the drug from the matrix shows a linear relationship with a square root of time as per Eq. 9.3 (Higuchi 1967):

$$Q = [D_m D_d(2A - C_d)]^{1/2} \quad (9.3)$$

where Q = amount of drug release per unit area of the matrix, D_m = diffusion coefficient of drug, C_d = solubility of a drug in matrix, A = primary quantity of drug in unit volume of matrix, and t = time.

Various factors like physicochemical properties of the drug, interaction between drug and lipid, swelling capacity, structure, and composition of the lipid bilayers affect the release profile of the drug (Boyd 2003). This can be explained in detail by the following examples. Lai et al. studied the release of simvastatin from cubosomes. Simvastatin is a lipophilic drug, so theoretically it has some interaction with the lipid bilayers of the cubosomes. In case of simvastatin crystal, a rapid release of around 90% in 1 h in simulated gastric fluid and in fasted-state simulated intestinal fluid was observed. However, only 3% of drug released from cubosomes after 10 h indicated that simvastatin has a high affinity towards the hydrophobic portion of the cubic matrix which retards the release of the drug from the formulation (Lai et al. 2009). Lee et al. compared the release rates of hydrophilic drugs from cubosomes prepared using GMO and phytantriol. They reported that the release rate of a drug decreased as the matrix was changed from Q_H GMO to Q_H phytantriol (Lee et al. 2009). Clogston et al. reported the effect of molecular size of an incorporated entity in the cubosomes on its release rate. They found that in case of smaller molecules like rubipy and tryptophan, the release of a drug from the cubic phase occurred up to 1–3 days. The midsized molecules such as cytochrome, lysozyme, and myoglobin were released up to a period of around 1 week. The macromolecules such as ovalbumin, conalbumin, calf thymus DNA, and apoferritin showed very slow release of up to 12% in 3 weeks (Clogston and Caffrey 2005).

In vitro release profile of drug from the cubosomes is studied using a glass-stopped tube. The cubosomal dispersion is filled into a dialysis bag having appropriate molecular weight cutoff (12,000–14,000 Da), which is tied at both the ends and taken in a glass tube containing the diffusion medium. The glass tube is incubated using a rotary shaker incubator at 37 °C. Sampling is done from the tube at predetermined time periods followed by replacement of equal volume of fresh diffusion medium. The concentration of the released drug is measured by HPLC or any other appropriate analytical method (Wu et al. 2012).

9.5.5 Stability Studies

9.5.5.1 Thermal Stability

Stability of the cubosomal dispersion is conducted at 4 and 25 °C for 28 days. At day 7, 14, 21, and 28 samples are checked by visual inspection and entrapment efficiency, particle size, and zeta potential are measured (Abdel-Bar and el Basset Sanad 2017).

9.5.5.2 Freeze-Thaw Stability

The cubosomal dispersion is subjected to three freeze-thaw cycles, each comprising freezing at −80 °C for 2 days and then storage at 25 °C for 2 days. The samples are then evaluated for entrapment efficiency, particle size, and zeta potential (Abdel-Bar and el Basset Sanad 2017).

9.6 Cubosomes for Various Therapeutic Applications

9.6.1 As Theranostic Drug Delivery System

Recently, cubosomes as theranostic agents are being investigated by many research groups. Caltagirone et al. developed cubosomes loaded with camptothecin, an anticancer drug, and squaraine-based NIR (near-infrared)-emitting fluorescent probe. Moreover, to improve the targeting ability of the cubosomes they used a mixture of Pluronic F108 and folate-conjugated Pluronic F108 in appropriate ratios. The superior ability of this nanocarrier in targeted delivery of drug was investigated by the lipid droplet alteration in HeLa cells (Caltagirone et al. 2014). Murgia et al. developed a cubosome formulation containing quercetin, an anticancer drug stabilized by the dansyl-conjugated Pluronic F108. Here, they attached the fluorescent imaging agent dansyl to the terminal ethylene oxide moieties of the block copolymer Pluronic F108 and used it in combination with nonconjugated Pluronic F108 in the preparation of cubosome in a ratio of 65/35. The fluorescent cubosome formulation was then investigated for its cytotoxicity feature in HeLa cell culture and the optical image of the cell line highlights the liquid droplet accumulation (Murgia et al. 2015). Yong Tian et al. investigated the etoposide cubosome for its targeted delivery by using rhodamine B. They prepared the targeted drug delivery system by using folic acid, studied the cellular uptake of the rhodamine B-loaded cubosome-folic acid by human breast carcinoma MCF-7 cell line, and found a marked increase in cellular accumulation of rhodamine B-loaded folic acid compared to rhodamine B and rhodamine B-loaded cubosomes (Tian et al. 2017). Sergio Murgia et al. developed quercetin, an anticancer drug-loaded fluorescent cubosome. They used two fluorescent probes, fluorescein and dansyl, modified with a hydrocarbon chain to enhance their encapsulation efficiency with the cubic phase. They demonstrated the cellular uptake of monoolein-based cubosomes doped with fluorescein and dansyl using the 3T3 fibroblasts and concluded that fluorescein and dansyl penetrate into the cell via the cubosome carrier based on the fluorescence detected in the cytosol and particularly in the perinuclear region of living cells (Murgia et al. 2013). The tumor fluorescence images are shown in Fig. 9.5.

9.6.2 Delivery of Peptide and Proteins

Due to the recent advancement in the field of pharmaceutical biotechnology, large-scale production of the protein or peptide drugs is now possible. However, drugs

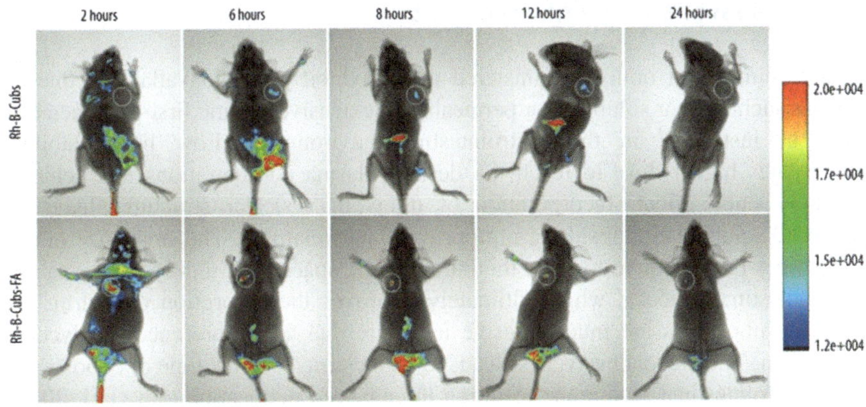

Fig. 9.5 Whole-body and tumor fluorescence images (white circles indicate the inoculated tumor) in MCF-7 tumor-bearing mice after intravenous injection of RhB-cubosomes and RhB-cubosome-folic acid (Tian et al. 2017)

from this class can easily be inactivated by precipitation, denaturation, aggregation, surface adsorption, etc. Moreover, instability in the gastric and intestinal environment makes oral delivery of these drugs difficult. Thus, the therapeutic application of these drugs is challenging and specialized drug delivery systems are required to overcome the limitations mentioned above. Cubosomes have been investigated for their capability to deliver the protein and peptide drugs by many research groups. These proteins and peptides are encapsulated in the core of the cubosomes making them resistant to precipitation, denaturation, enzymatic degradation, etc. Moreover, cubosomes also enhance the absorption of these proteins and peptides leading to enhancement of their bioavailability (S Duttagupta et al. 2016). Borislav Angelov et al. used synchrotron radiation SAXS and cryo-TEM imaging and identified a large aqueous channel in cationic PEGylated cubosomes which can encapsulate larger amount of siRNA for cancer treatment of cancer and also give tumor-targeted delivery (Angelov et al. 2015). Rizwan et al. investigated cubosomes for its ability to deliver the protein vaccine. They encapsulated oval albumin in cubosomes prepared using GMO or phytantriol and poloxamer F127, among which phytantriol cubosomes were able to give high entrapment and slow release of the oval albumin compared to the GMO cubosomes (Rizwan et al. 2011). Liu H. developed cubosomes to deliver the earthworm fibrinolytic enzyme (EFC) to the inner ear in order to enhance the inner ear bioavailability of the drug. They also performed the cytotoxicity study of the prepared formulation and found that the viability of the L929 cell declined by enhancing the concentration of cubosomes (Liu et al. 2013). L.H. Nielsen et al. formulated cubosomes containing the model antigen ovalbumin by spray drying to investigate its potential for oral vaccine delivery. The prepared formulation gave sustained release of the antigen up to 47.9 ± 2.8% in 96 h in a buffer of pH 6.8 (Nielsen et al. 2017).

9.6.3 Bioavailability Enhancement

A large number of orally administered drugs exhibit poor bioavailability due to reasons such as poor solubility or permeability, extensive hepatic first-pass metabolism, or instability in the gastrointestinal environment. Low bioavailability necessitates higher dose to reach the desired plasma concentrations and achieve the therapeutic effect. Incorporation of drugs in vesicular structures leads to enhanced solubilization resulting in improved bioavailability. The small size of the cubosomes permits them to enter the intravascular spaces and tightly adhere to the gastrointestinal mucosa, which ultimately improves the absorption of a drug. In general, cubosomes are mucoadhesive in nature due to the presence of glyceryl monoolein. The mucoadhesiveness leads to an increased contact time of cubosomes with gastrointestinal membrane resulting in enhanced permeation (Ali et al. 2017; Nguyen et al. 2010). Moreover, the lipids of cubosomes stimulate the release of bile salts from the gallbladder into the small intestine. These bile salts interact with cubosomes and form mixed micelles, which could be absorbed along with the drug into the systemic circulation (Das and Chaudhury 2011; Nguyen et al. 2010). Furthermore, cubosomes also favor lymphatic passage from a small intestine in a similar manner to liposomes (Kim et al. 2013; Nguyen et al. 2010). Entrapment of a drug in cubosomes also increases the gastrointestinal stability of drugs which are susceptible to degradation. Zhiwen Yang et al. incorporated amphotericin B in cubosomes by using SolEmuls technology. The relative bioavailability of amphotericin B cubosomes was 285% compared to Fungizone® with minimal nephrotoxicity (Yang et al. 2012). Mohamed Nasr developed sorbitol-based powder precursor of cubosomes to improve the oral bioavailability of poorly water-soluble drugs. Tamoxifen citrate was used as a prototype drug and oral bioavailability of a drug was improved to $152.50 \pm 32.67\%$ compared to the plain tamoxifen citrate powder (Nasr and Dawoud 2016).

9.6.4 Enhancement of Transdermal Permeation

Transdermal delivery is limited by the presence of tough stratum corneum barrier which prevents the permeation of majority of drugs. Enhancement in transdermal permeation of drugs is necessary to achieve therapeutic plasma concentrations. Several mechanisms of enhancement of transdermal permeation using cubosomes have been proposed by different research groups. Firstly, glyceryl monooleate (GMO) and poloxamer, the major building blocks of cubosomes, are known penetration enhancers. GMO increases the fluidity by modifying the intercellular ordered structure of lipid bilayer in the stratum corneum resulting in enhancement of transdermal permeation of a drug across the stratum corneum (Kwon et al. 2012). Poloxamer used in the preparation of cubosomes has hydrophilic ethylene oxide and hydrophobic propylene oxide in its structure. This allows poloxamer to partition between the lipophilic skin component and the hydrophilic protein component, resulting in a deeper penetration of cubosomes through the stratum corneum. It also disrupts the lipid arrangement of skin and increases its fluidity, ultimately resulting in enhanced transdermal permeation (Yapar and Ýnal 2012). Secondly, the structural arrangement of cubosomes is similar to that of skin which allows them to be pressed through the pores of stratum corneum, resulting in enhanced transdermal permeation (Esposito et al. 2005). Salwa Salah et al. developed etodolac transdermal cubosomes for the treatment of rheumatoid arthritis and they were able to encapsulate 100% of drug with sustained release up to 24 h. The relative bioavailability of the etodolac cubosome was 266.11% compared to the oral capsule (Salah et al. 2017). Xinsheng Peng et al. produced phytantriol- and GMO-based cubosomal formulation of capsaicin for sustained and targeted transdermal delivery of a drug. The drug permeation and retention study using Franz diffusion cell indicated that in case of cubosome-loaded drug, skin retention was significantly higher than that in case of capsaicin cream (Peng et al. 2015). Sana Khan et al. formulated a cubosomal dispersion of erythromycin for topical delivery to treat acne.

9.6.5 Brain Targeting via Intranasal Route

Intranasal administration has gained significant attention of researchers because of its noninvasive nature for bypassing blood-brain barrier (BBB) and delivering the drug directly to the brain (Mittal et al. 2014). Moreover, the nasal mucosa is a highly permeable and vascularized site for drug delivery, which offers the rapid absorption of nasally administered drug (Mittal et al. 2014). Drug delivery from nose to brain is facilitated through the olfactory and/or trigeminal nerve pathways, placed at the roof of the nasal cavity as its neuroepithelium is the only part of the CNS that is directly open to the external environment. Thus, the olfactory region is known to be the gateway for a therapeutic molecule to enter from nose to brain following nasal absorption (Crowe et al. 2018). Hence, transport across the olfactory epithelium is the principal concern for brain-targeted intranasal delivery. Despite this advantage, the nasal route of drug administration has some limitations such as rapid removal of the formulation by mucociliary clearance, low permeability of the nasal epithelium, and enzymatic degradation (Crowe et al. 2018; Mittal et al. 2014). Cubosomes are beneficial in these limitations due to its mucoadhesive property and penetration enhancement property of its building components. Moreover, they also protect the therapeutic moiety from the enzymatic degradation as it is incorporated into the core of cubosomes, so that the drug does come in direct contact with the enzymes (Kwon et al. 2012; Ali et al. 2017; Yapar and Ýnal 2012; Nguyen et al. 2010). Hongbing Wu developed Odorrana lectin-bearing cubosomes for the targeted nose-to-brain delivery and performed the biodistribution study of the prepared formulation by using a fluorescent marker, coumarin-6, and found that the relative uptake of the coumarin Odorrana lectin cubosomes was 1.66–3.46-fold in brain tissues than the normal cubosomal preparation. Moreover, they incorporated Gly14-Humanin as a model peptide drug in the above-prepared cubosomes and also evaluated it for its pharmacodynamics activity in Alzheimer's disease on rats following the intranasal administration by Morris water maze and acetylcholinesterase activity test. In the end, they concluded that Odorrana lectin functionalization increased the therapeutic activity of Gly14-Humanin-loaded cubosomes in Alzheimer's disease (Wu et al. 2012). Mayuri Ahirrao et al. prepared resveratrol cubosomes with 83.08% entrapment for brain targeting by using the nasal route of drug administration. They showed that cubosomes gave significantly higher transnasal permeation and better distribution to brain compared to the i.m. drug solution and oral drug solution (Ahirrao and Shrotriya 2017).

9.7 Latest Research Work Done

Owing to its superior advantages, it has undergone lot of researches as shown in Table 9.1.

9.8 Patents Reported for Cubosomes

The patents reported for cubosomes are shown in Table 9.2.

9.9 Safety Concerns of Cubosomes

The most important concerns for any delivery system are safety and efficacy. The increasing research work on cubosomes as a delivery system necessitates the evaluation of its safety and toxicity. Toxicity study of cubosomes is performed using suitable cell lines. Hinton et al. carried out the cytotoxic effects of cubosomes using Chinese hamster ovary cells (CHO-GFP) and human alveolar basal epithelial cells (A549) and compared them in terms of their toxicity. They found that GMO-based cubosomes were nontoxic at a concentration of 100 μg/mL in A549 cells while there was 40% toxicity at 75 and 100 μg/mL in case of CHO-GFP cells. Moreover, they did not find any toxicity in either cell line at 50 μg/mL concentration. In the case of phytantriol cubosomes, the formulation was nontoxic at a concentration of 25 μg/mL for both CHO-GFP and A549 cells, but highly toxic at higher concentration than this. This finding suggests that phytantriol-based cubosomes are more toxic to these cell lines compared to the GMO-based cubosomes. The possible reason behind more toxicity of phytantriol-based cubosomes is that phytantriol cubosomes show a higher degree of membrane disruption than GMO-based cubosomes. Moreover, phytantriol-based cubosomes result in a significant sustained inflammatory response compared to GMO cubosomes. They also performed another study to find out a maximum tolerated dose (MTD) of phytantriol cubosomes in rats via bolus delivery and found that MTD of phytantriol cubosomes was 350 mg/kg, while for GMO cubosomes this study was not performed due to high in vivo tolerance (Hinton et al. 2014).

Table 9.1 Research work reported on cubosomes as a drug delivery system

Drug used	Lipid used	Stabilizer used	Method of preparation	Purpose of the study
Dapsone	GMO	Poloxamer 407	Top-down	To develop a transdermal drug delivery system (Nithya et al. 2018)
Antimicrobial peptide LL-37	GMO	Poloxamer 407	Top-down, bottom-up, post-loading	To develop the topical drug delivery system (Boge et al. 2019)
Resveratrol	GMO	Poloxamer 407	Top-down	A cytotoxic study using human hepatoma cells (Abdel-Bar and el Basset Sanad 2017)
Palmitoyl peptides (GHKcube, GQPRcube)	Phytantriol	Pluronic F127	Top-down	Interaction and release of palmitoyl peptides (Akhlaghi and Loh 2017)
Chlorin e6, meso-tetraphenylporphine-Mn(III) chloride	GMO, phospholipid	Propylene glycol	Top-down	To develop a transdermal delivery system (Bazylińska et al. 2018)
Antimicrobial peptides (AP114, DPK-060, LL-37)	GMO	Kolliphor® P407	Bottom-up	To enhance the bactericidal effect and proteolytic stability of the antimicrobial peptides (Boge et al. 2017)
Ovalbumin, Quil-A	Dimodan MO 90/D (GMO)	Dextran as a carrier	Spray drying	Spray-dried cubosomes for subcutaneous and oral delivery (von Halling et al. 2018)
Gramicidin A, melittin, alamethicin	GMO	Pluronic F127	Bottom-up	To incorporate these antimicrobial peptides in cubosomes (Meikle et al. 2017)
Sertaconazole nitrate	GMO	Pluronic F127, Pluronic F108, Tween 80, PVA, Brij 58	Top-down	To develop the corneal targeted drug delivery system (Younes et al. 2018)
Triclosan	GMO	Pluronic F127	Top-down	To study the skin permeation of a drug

(continued)

Table 9.1 (continued)

Drug used	Lipid used	Stabilizer used	Method of preparation	Purpose of the study
Gambogenic acid	GMO	Pluronic F127	Top-down, bottom-up	To develop intraperitoneal drug delivery system and study cytotoxicity and intracellular uptake
Clonazepam	GMO	Pluronic F127	Top-down	To develop a transdermal patch for the treatment of epilepsy in children
Silver sulfadiazine	GMO	Poloxamer 407, PVA	Top-down	To develop topical cubosomal gel for the treatment of burns (Morsi et al. 2014)
Annexin V	Phytantriol, DPPS sodium salt (1,2-dihexadecanoyl-sn-glycero-3-phospho-L-serine)	Pluronic F127	Bottom-up	Targeted detection of phosphatidylserine in biomimetic membrane (Shen et al. 2013)
Odorrana lectin	GMO, maleimide-PEG-oleate	Pluronic F127	Bottom-up	Brain delivery using the intranasal route
Polysaccharide from *Ulva fasciata*	GMO	Poloxamer 407	Top-down	To develop the cubic liquid crystalline nanoparticles of polysaccharide (Matloub et al. 2018)
Ondansetron	GMO, glyceryl-monolinoleate, polyglycerol 3-dioleate (Plurol®), soya PC, dipalmitoyl-phosphatidylcholine (DPPC), dipalmitoyl-phosphatidyl-glycerol (DPPG)	Poloxamer 407	Bottom-up	To develop polyglycerol-dioleate-based novel cubosome with tailored physical characteristic (Mansour et al. 2017)
Efavirenz	Phytantriol	Poloxamer 407	Bottom-up, spray drying	To enhance the oral bioavailability and to provide sustained release of efavirenz (Avachat and Parpani 2015)
Plectasin derivative AP114	GMO	Lutrol® F127, disaccharide	Top-down	To develop freeze-dried and rehydrated cubosome stabilized using disaccharide (Boge et al. 2018)

α-Lipoic acid	GMO	Poloxamer 407	Bottom-up and top-down	To study the cosmeceutical application of cubosome containing α-lipoic acid (Sherif et al. 2014)
Dexamethasone	GMO	Poloxamer 407	Top-down	To improve the preocular retention and enhance the ocular bioavailability (Gan et al. 2010)
NIL	Selachyl alcohol	Pluronic F127, Tween 80	Bottom-up	To study the effect of Tween 80 and Pluronic F127 on aqueous behavior of an endogenous lipid, selachyl alcohol (Younus et al. 2018)
Iron oxide	GMO	Pluronic F127	Top-down	To prepare iron oxide nanoparticles within the monoolein cubic phase (Hong et al. 2012b)
20(S)-Protopanaxadiol	GMO	Poloxamer 407	Top-down	To improve oral bioavailability
Risperidone	GMO	Pluronic F127, Tween 80	Top-down	Brain targeting of drug using the intranasal route of drug delivery (Abdelrahman et al. 2015)
Nil	GMO, dipalmitoyl phosphatidylserine (DPPS)	Pluronic F127	Top-down	To investigate the effect of DPPS on phase behavior and cellular response of lyotropic liquid crystalline dispersion (Shen et al. 2010)
Progesterone	GMO	Poloxamer 407	Top-down, bottom-up	To prepare a transdermal delivery system for progesterone (Elgindy et al. 2016)
Nil	GMO	Pluronic F127	Top-down	To prepare magnetocubosomes for targeted and control release (Montis et al. 2015)
Cinnarizine	GMO, phytantriol	Pluronic F127	Top-down	To provide a sustained-release effect to the poorly soluble drug after oral administration (Nguyen et al. 2011)
Fenofibrate	GMO	Poloxamer 407	Spray drying	To improve oral bioavailability (Wei et al. 2017)

Table 9.2 Patents reported on cubosomes as a drug delivery system

Application number	Title	Status
WO2005077336A1	Composition and formulation of colloidal system comprising biocompatible aqueous-soluble polymer, and preparation method thereof (Chung et al. 2005)	Pending
US7008646B2	Cubic liquid crystalline compositions and methods for their preparation (Spicer et al. 2006)	Abandoned
US5531925A	Particles, method of preparing said particles and uses thereof (Landh and Larsson 1996)	Expired
US5230895A	The sustained-release delivery system for use in the periodontal pocket (Czarnecki and Williams 1993)	Expired

9.10 Future Aspect

Blood plasma has some amount of lipase enzyme which can degrade glyceryl monooleate (GMO), a major component of cubosomes. Thus, the total plasma circulation time of the formulation in blood can decrease after intravenous administration leading to rapid termination of therapeutic action. Lipase is also present in the gastrointestinal tract leading to possible degradation of GMO-based cubosomes after oral administration. Thus, extended research on overcoming this limitation is needed to improve the in vivo stability of cubosomes.

References

Abdel-Bar HM, el Basset Sanad RA (2017) Endocytic pathways of optimized resveratrol cubosomes capturing into human hepatoma cells. Biomed Pharmacother 93:561–569

Abdelrahman FE, Elsayed I, Gad MK et al (2015) Investigating the cubosomal ability for transnasal brain targeting: in vitro optimization, ex vivo permeation and in vivo biodistribution. Int J Pharm 490(1–2):281–291

Ahirrao M, Shrotriya S (2017) In vitro and in vivo evaluation of cubosomal in situ nasal gel containing resveratrol for brain targeting. Drug Dev Ind Pharm 43(10):1686–1693

Akhlaghi SP, Loh W (2017) Interactions and release of two palmitoyl peptides from phytantriol cubosomes. Eur J Pharm Biopharm 117:60–67

Akhlaghi SP, Ribeiro IR, Boyd BJ et al (2016) Impact of preparation method and variables on the internal structure, morphology, and presence of liposomes in phytantriol-Pluronic® F127 cubosomes. Colloids Surf B Biointerfaces 145:845–853

Alexandridis P (1996) Amphiphilic copolymers and their applications. Curr Opin Colloid Interface Sci 1(4):490–501

Ali MA, Kataoka N, Ranneh A-H et al (2017) Enhancing the solubility and oral bioavailability of poorly water-soluble drugs using monoolein cubosomes. Chem Pharm Bull 65(1):42–48

Anbarasan B, Grace X, Shanmuganathan S (2015) An overview of cubosomes—smart drug delivery system. Sri Ramachandra J Med 8:1–4

Angelov B, Angelova A, Garamus VM et al (2012) Earliest stage of the tetrahedral nanochannel formation in cubosome particles from unilamellar nanovesicles. Langmuir 28(48):16647–16655

Angelov B, Angelova A, Drechsler M et al (2015) Identification of large channels in cationic PEGylated cubosome nanoparticles by synchrotron radiation SAXS and Cryo-TEM imaging. Soft Matter 11(18):3686–3692

Avachat AM, Parpani SS (2015) Formulation and development of bicontinuous nanostructured liquid crystalline particles of efavirenz. Colloids Surf B Biointerfaces 126:87–97

Azhari H, Strauss M, Hook S et al (2016) Stabilising cubosomes with Tween 80 as a step towards targeting lipid nanocarriers to the blood–brain barrier. Eur J Pharm Biopharm 104:148–155

Azmi IM, Nilsson C, Stürup S et al (2013) Characterization of cisplatin-loaded cubosomes and hexosomes: effect of mixing with human plasma. J Geriatr Oncol 4:S62

Bansal S, Kashyap CP, Aggarwal G et al (2012) A comparative review on vesicular drug delivery system and stability issues. Int J Res Pharm Chem 2(3):704–713

Barauskas J, Landh T (2003) Phase behavior of the phytantriol/water system. Langmuir 19 (23):9562–9565

Barauskas J, Johnsson M, Joabsson F et al (2005a) Cubic phase nanoparticles (cubosome): principles for controlling size, structure, and stability. Langmuir 21(6):2569–2577

Barauskas J, Johnsson M, Tiberg F (2005b) Self-assembled lipid superstructures: beyond vesicles and liposomes. Nano Lett 5(8):1615–1619

Bazylińska U, Kulbacka J, Schmidt J et al (2018) Polymer-free cubosomes for simultaneous bioimaging and photodynamic action of photosensitizers in melanoma skin cancer cells. J Colloid Interface Sci 522:163–173

Bei D, Marszalek J, Youan BBC (2009a) Formulation of dacarbazine-loaded cubosomes—part I: influence of formulation variables. AAPS PharmSciTech 10(3):1032

Bei D, Marszalek J, Youan BBC (2009b) Formulation of dacarbazine-loaded cubosomes—part II: influence of process parameters. AAPS PharmSciTech 10(3):1040

Boge L, Umerska A, Matougui N et al (2017) Cubosomes post-loaded with antimicrobial peptides: characterization, bactericidal effect and proteolytic stability. Int J Pharm 526(1–2):400–412

Boge L, Västberg A, Umerska A et al (2018) Freeze-dried and re-hydrated liquid crystalline nanoparticles stabilized with disaccharides for drug-delivery of the plectasin derivative AP114 antimicrobial peptide. J Colloid Interface Sci 522:126–135

Boge L, Hallstensson K, Ringstad L et al (2019) Cubosomes for topical delivery of the antimicrobial peptide LL-37. Eur J Pharm Biopharm 134:60–67

Boyd BJ (2003) Characterisation of drug release from cubosomes using the pressure ultrafiltration method. Int J Pharm 260(2):239–247

Boyd BJ, Whittaker DV, Khoo SM et al (2006) Lyotropic liquid crystalline phases formed from glycerate surfactants as sustained release drug delivery systems. Int J Pharm 309(1–2):218–226

Boyd BJ, Dong YD, Rades T (2009) Nonlamellar liquid crystalline nanostructured particles: advances in materials and structure determination. J Liposome Res 19(1):12–28

Buchheim W, Larsson K (1987) Cubic lipid-protein-water phases. J Colloid Interface Sci 117:582–583

Caltagirone C, Falchi AM, Lampis S et al (2014) Cancer-cell-targeted theranostic cubosomes. Langmuir 30(21):6228–6236

Chong JY, Mulet X, Boyd BJ et al (2015) Steric stabilizers for cubic phase lyotropic liquid crystal nanodispersions (cubosomes). In: Advances in planar lipid bilayers and liposomes, vol. 21. Elsevier, pp 131–187

Chonn A, Cullis P, Devine D (1991) The role of surface charge in the activation of the classical and alternative pathways of complement by liposomes. J Immunol 146(12):4234–4241

Chung H, Jeong SY, Kwon IC et al (2005) Composition and formulation of colloidal system comprising biocompatible aqueous-soluble polymer, and preparation method thereof. Google Patents

Clogston J, Caffrey M (2005) Controlling release from the lipidic cubic phase. Amino acids, peptides, proteins and nucleic acids. J Control Release 107(1):97–111

Crowe TP, Greenlee MHW, Kanthasamy AG et al (2018) Mechanism of intranasal drug delivery directly to the brain. Life Sci 195:44–52

Czarnecki RF, Williams DL (1993) Sustained released delivery system for use in the periodontal pocket. Google Patents

Das S, Chaudhury A (2011) Recent advances in lipid nanoparticle formulations with solid matrix for oral drug delivery. AAPS PharmSciTech 12(1):62–76

Demurtas D, Guichard P, Martiel I et al (2015) Direct visualization of dispersed lipid bicontinuous cubic phases by cryo-electron tomography. Nat Commun 6:8915

Drummond CJ, Fong C (1999) Surfactant self-assembly objects as novel drug delivery vehicles. Curr Opin Colloid Interface Sci 4(6):449–456

Duttagupta AS, Chaudhary HM, Jadhav KR et al (2016) Cubosomes: innovative nanostructures for drug delivery. Curr Drug Deliv 13(4):482–493

Elgindy NA, Mehanna MM, Mohyeldin SM (2016) Self-assembled nano-architecture liquid crystalline particles as a promising carrier for progesterone transdermal delivery. Int J Pharm 501 (1–2):167–179

Esposito E, Cortesi R, Drechsler M et al (2005) Cubosome dispersions as delivery systems for percutaneous administration of indomethacin. Pharm Res 22(12):2163–2173

Fontell K (1990) Cubic phases in surfactant and surfactant-like lipid systems. Colloid Polym Sci 268(3):264–285

Gan L, Han S, Shen J et al (2010) Self-assembled liquid crystalline nanoparticles as a novel ophthalmic delivery system for dexamethasone: improving preocular retention and ocular bioavailability. Int J Pharm 396(1–2):179–187

Ganem-Quintanar A, Quintanar-Guerrero D, Buri P (2000) Monoolein: a review of the pharmaceutical applications. Drug Dev Ind Pharm 26(8):809–820

Garg G, Saraf S, Saraf S (2007) Cubosomes: an overview. Biol Pharm Bull 30(2):350–353

Garti N, Libster D, Aserin A (2012) Lipid polymorphism in lyotropic liquid crystals for triggered release of bioactives. Food Funct 3(7):700–713

Gin DL, Pecinovsky CS, Bara JE et al (2007) Functional lyotropic liquid crystal materials. In: Liquid crystalline functional assemblies and their supramolecular structures. Springer, pp 181–222

Guillot S, Salentinig S, Chemelli A et al (2010) Influence of the stabilizer concentration on the internal liquid crystalline order and the size of oil-loaded monolinolein-based dispersions. Langmuir 26(9):6222–6229

Guo C, Wang J, Cao F et al (2010) Lyotropic liquid crystal systems in drug delivery. Drug Discov Today 15(23–24):1032–1040

Gustafsson J, Ljusberg-Wahren H, Almgren M et al (1996) Cubic lipid–water phase dispersed into submicron particles. Langmuir 12(20):4611–4613

Gustafsson J, Ljusberg-Wahren H, Almgren M et al (1997) Submicron particles of reversed lipid phases in water stabilized by a nonionic amphiphilic polymer. Langmuir 13(26):6964–6971

Gustafsson J, Nylander T, Almgren M et al (1999) Phase behavior and aggregate structure in aqueous mixtures of sodium cholate and glycerol monooleate. J Colloid Interface Sci 211 (2):326–335

Hartnett TE, Ladewig K, O'Connor AJ et al (2014) Size and phase control of cubic lyotropic liquid crystal nanoparticles. J Phys Chem B 118(26):7430–7439

Higuchi WI (1967) Diffusional models useful in biopharmaceutics: drug release rate processes. J Pharm Sci 56(3):315–324

Hinton TM, Grusche F, Acharya D et al (2014) Bicontinuous cubic phase nanoparticle lipid chemistry affects toxicity in cultured cells. Toxicol Res 3(1):11–22

Hong SK, Ma JY, Kim JC (2012a) In vitro skin permeation enhancement of KIOM-MA-128 by monoolein cubosomes. J Dispers Sci Technol 33(10):1503–1508

Hong SK, Ma JY, Kim JC (2012b) Preparation of iron oxide nanoparticles within monoolein cubic phase. J Ind Eng Chem 18(6):1977–1982

Hyde S, Andersson S (1984) A cubic structure consisting of a lipid bilayer forming an infinite periodic minimum surface of the gyroid type in the glycerol monooleat-water system. Zeitschrift für Kristallographie Crystal Mater 168(1–4):213–220

Johnsson M, Edwards K (2001) Phase behavior and aggregate structure in mixtures of dioleoyl phosphatidylethanolamine and poly (ethylene glycol)-lipids. Biophys J 80(1):313–323

Johnsson M, Barauskas J, Tiberg F (2005) Cubic phases and cubic phase dispersions in a phospholipid-based system. J Am Chem Soc 127(4):1076–1077

Karami Z, Hamidi M (2016) Cubosomes: remarkable drug delivery potential. Drug Discov Today 21(5):789–801

Kim H, Kim Y, Lee J (2013) Liposomal formulations for enhanced lymphatic drug delivery. Asian J Pharm Sci 8(2):96–103

Kim D-H, Jahn A, Cho S-J et al (2015) Lyotropic liquid crystal systems in drug delivery: a review. J Pharm Invest 45(1):1–11

Koynova R, Tenchov B, Rapp G (1997) Low amounts of PEG-lipid induce cubic phase in phosphatidylethanolamine dispersions. Biochim Biophys Acta Biomembr 1326(2):167–170

Koynova R, Tenchov B, Rapp G (1999) Effect of PEG-lipid conjugates on the phase behavior of phosphatidylethanolamine dispersions. Colloids Surf A Physicochem Eng Asp 149 (1–3):571–575

Kudchodkar BJ, Albers JJ, Bierman EL (1983) Effect of positively charged sphingomyelin liposomes on cholesterol metabolism of cells in culture. Atherosclerosis 46(3):353–367

Kulkarni CV, Wachter W, Iglesias-Salto G et al (2011) Monoolein: a magic lipid? Phys Chem Chem Phys 13(8):3004–3021

Kwon TK, Kim J-C (2014) In vitro skin permeation and anti-atopic efficacy of lipid nanocarriers containing water soluble extracts of Houttuynia cordata. Drug Dev Ind Pharm 40 (10):1350–1357

Kwon TK, Hong SK, Kim J-C (2012) In vitro skin permeation of cubosomes containing triclosan. J Ind Eng Chem 18(1):563–567

Lai J, Chen J, Lu Y et al (2009) Glyceryl monooleate/poloxamer 407 cubic nanoparticles as oral drug delivery systems: I. In vitro evaluation and enhanced oral bioavailability of the poorly water-soluble drug simvastatin. AAPS PharmSciTech 10(3):960

Landh T (1994) Phase behavior in the system pine needle oil monoglycerides-Poloxamer 407-water at 20 degree. J Phys Chem 98(34):8453–8467

Landh T, Larsson K (1996) Particles, method of preparing said particles and uses thereof. Google patents

Larsson K (1989) Cubic lipid-water phases: structures and biomembrane aspects. J Phys Chem 93 (21):7304–7314

Larsson K (2000) Aqueous dispersions of cubic lipid–water phases. Curr Opin Colloid Interface Sci 5(1–2):64–69

Lee KW, Nguyen T-H, Hanley T et al (2009) Nanostructure of liquid crystalline matrix determines in vitro sustained release and in vivo oral absorption kinetics for hydrophilic model drugs. Int J Pharm 365(1–2):190–199

Leser ME, Michel M, Watzke HJ (2003) Food goes nano—new horizons for food structure research. Food Colloids Biopolym Mater 284:3–13

Li J-C, Zhu N, Zhu J-X et al (2015) Self-assembled cubic liquid crystalline nanoparticles for transdermal delivery of paeonol. Med Sci Monitor 21:3298

Lindell K, Engblom J, Engström S et al (1998) Influence of a charged phospholipid on the release pattern of timolol maleate from cubic liquid crystalline phases. In: The colloid science of lipids. Springer, pp 111–118

Liu H, Wang Y, Wang Q et al (2013) Protein-bearing cubosomes prepared by liquid precursor dilution: inner ear delivery and pharmacokinetic study following intratympanic administration. J Biomed Nanotechnol 9(10):1784–1793

Lombardo D, Kiselev MA, Caccamo MT (2019) Smart nanoparticles for drug delivery application: development of versatile nanocarrier platforms in biotechnology and nanomedicine. J Nanomater 2019:3702518

Lutton E (1965) Phase behavior of aqueous systems of monoglycerides. J Am Oil Chem Soc 42 (12):1068–1070

Mansour M, Kamel A, Mansour S et al (2017) Novel polyglycerol-dioleate based cubosomal dispersion with tailored physical characteristics for controlled delivery of ondansetron. Colloids Surf B Biointerfaces 156:44–54

Matloub AA, AbouSamra MM, Salama AH et al (2018) Cubic liquid crystalline nanoparticles containing a polysaccharide from Ulva fasciata with potent antihyperlipidaemic activity. Saudi Pharm J 26(2):224–231

Meikle TG, Zabara A, Waddington LJ et al (2017) Incorporation of antimicrobial peptides in nanostructured lipid membrane mimetic bilayer cubosomes. Colloids Surf B Biointerfaces 152:143–151

Mishra RK, Tiwari SK, Mohapatra S et al (2019) Efficient Nanocarriers for drug-delivery systems: types and fabrication. In: Nanocarriers drug delivery. Elsevier, pp 1–41

Mittal D, Ali A, Md S et al (2014) Insights into direct nose to brain delivery: current status and future perspective. Drug Deliv 21(2):75–86

Montis C, Castroflorio B, Mendozza M et al (2015) Magnetocubosomes for the delivery and controlled release of therapeutics. J Colloid Interface Sci 449:317–326

Morsi NM, Abdelbary GA, Ahmed MA (2014) Silver sulfadiazine based cubosome hydrogels for topical treatment of burns: development and in vitro/in vivo characterization. Eur J Pharm Biopharm 86(2):178–189

Muller F, Salonen A, Glatter O (2010) Monoglyceride-based cubosomes stabilized by Laponite: separating the effects of stabilizer, pH and temperature. Colloids Surf A Physicochem Eng Asp 358(1–3):50–56

Murgia S, Bonacchi S, Falchi AM et al (2013) Drug-loaded fluorescent cubosomes: versatile nanoparticles for potential theranostic applications. Langmuir 29(22):6673–6679

Murgia S, Falchi AM, Meli V et al (2015) Cubosome formulations stabilized by a dansyl-conjugated block copolymer for possible nanomedicine applications. Colloids Surf B Biointerfaces 129:87–94

Nasr M, Dawoud M (2016) Sorbitol based powder precursor of cubosomes as an oral delivery system for improved bioavailability of poorly water soluble drugs. J Drug Deliv Sci Technol 35:106–113

Nguyen TH, Hanley T, Porter CJ et al (2010) Phytantriol and glyceryl monooleate cubic liquid crystalline phases as sustained-release oral drug delivery systems for poorly water soluble drugs I. Phase behaviour in physiologically-relevant media. J Pharm Pharmacol 62(7):844–855

Nguyen T-H, Hanley T, Porter CJ et al (2011) Nanostructured liquid crystalline particles provide long duration sustained-release effect for a poorly water soluble drug after oral administration. J Control Release 153(2):180–186

Nielsen LH, Rades T, Boyd B et al (2017) Microcontainers as an oral delivery system for spray dried cubosomes containing ovalbumin. Eur J Pharm Biopharm 118:13–20

Nishikawa K, Arai H, Inoue K (1990) Scavenger receptor-mediated uptake and metabolism of lipid vesicles containing acidic phospholipids by mouse peritoneal macrophages. J Biol Chem 265 (9):5226–5231

Nithya R, Jerold P, Siram K (2018) Cubosomes of dapsone enhanced permeation across the skin. J Drug Deliv Sci Technol 48:75–81

Ostuni E, Chapman RG, Holmlin RE et al (2001) A survey of structure−property relationships of surfaces that resist the adsorption of protein. Langmuir 17(18):5605–5620

Peng X, Zhou Y, Han K et al (2015) Characterization of cubosomes as a targeted and sustained transdermal delivery system for capsaicin. Drug Des Dev Ther 9:4209

Pitzalis P, Monduzzi M, Krog N et al (2000) Characterization of the liquid−crystalline phases in the glycerol monooleate/diglycerol monooleate/water system. Langmuir 16(15):6358–6365

Rarokar NR, Khedekar PB (2018) Cubosomes: a vehicle for delivery of various therapeutic agents. MOJ Toxicol 4(1):19–21

Rattanapak T, Young K, Rades T et al (2012) Comparative study of liposomes, transfersomes, ethosomes and cubosomes for transcutaneous immunisation: characterisation and in vitro skin penetration. J Pharm Pharmacol 64(11):1560–1569

Rizwan S, Dong Y-D, Boyd B et al (2007) Characterisation of bicontinuous cubic liquid crystalline systems of phytantriol and water using cryo-field emission scanning electron microscopy (cryo-FESEM). Micron 38(5):478–485
Rizwan S, Assmus D, Boehnke A et al (2011) Preparation of phytantriol cubosomes by solvent precursor dilution for the delivery of protein vaccines. Eur J Pharm Biopharm 79(1):15–22
Rosen M (2005) Delivery system handbook for personal care and cosmetic products: technology, applications and formulations. William Andrew
Rosevear F (1954) The microscopy of the liquid crystalline neat and middle phases of soaps and synthetic detergents. J Am Oil Chem Soc 31(12):628–639
Sagalowicz L, Leser M, Watzke H et al (2006) Monoglyceride self-assembly structures as delivery vehicles. Trends Food Sci Technol 17(5):204–214
Salah S, Mahmoud AA, Kamel AO (2017) Etodolac transdermal cubosomes for the treatment of rheumatoid arthritis: ex vivo permeation and in vivo pharmacokinetic studies. Drug Deliv 24 (1):846–856
Senior J (1987) Fate and behavior of liposomes in vivo: a review of controlling factors. Crit Rev Ther Drug Carrier Syst 3(2):123–193
Senior JH, Trimble KR, Maskiewicz R (1991) Interaction of positively-charged liposomes with blood: implications for their application in vivo. Biochim Biophys Acta Biomemb 1070 (1):173–179
Shah JC, Sadhale Y, Chilukuri DM (2001) Cubic phase gels as drug delivery systems. Adv Drug Deliv Rev 47(2–3):229–250
Shen H-H, Crowston JG, Huber F et al (2010) The influence of dipalmitoyl phosphatidylserine on phase behaviour of and cellular response to lyotropic liquid crystalline dispersions. Biomaterials 31(36):9473–9481
Shen H-H, Lake V, Le Brun AP et al (2013) Targeted detection of phosphatidylserine in biomimetic membranes and in vitro cell systems using annexin V-containing cubosomes. Biomaterials 34 (33):8361–8369
Sherif S, Bendas ER, Badawy S (2014) The clinical efficacy of cosmeceutical application of liquid crystalline nanostructured dispersions of alpha lipoic acid as anti-wrinkle. Eur J Pharm Biopharm 86(2):251–259
Spicer PT, Hayden KL, Lynch ML et al (2001) Novel process for producing cubic liquid crystalline nanoparticles (cubosomes). Langmuir 17(19):5748–5756
Spicer PT, Small WB, Lynch ML et al (2002) Dry powder precursors of cubic liquid crystalline nanoparticles (cubosomes). J Nanopart Res 4(4):297–311
Spicer PT, William BSI, Lynch ML (2006) Cubic liquid crystalline compositions and methods for their preparation. Google patents
Svensson O, Thuresson K, Arnebrant T (2008) Interactions between drug delivery particles and mucin in solution and at interfaces. Langmuir 24(6):2573–2579
Tardieu A, Luzzati V (1970) A novel cubic phase—a cage-like network of rods with enclosed spherical micelles. Biochim Biophys Acta Biomemb 219(1):11–17
Tenchov B, Koynova R, Rapp G (1998) Accelerated formation of cubic phases in phosphatidylethanolamine dispersions. Biophys J 75(2):853–866
Tian Y, Li JC, Zhu JX et al (2017) Folic acid-targeted etoposide cubosomes for theranostic application of cancer cell imaging and therapy. Med Sci Monitor 23:2426
Tilley AJ, Drummond CJ, Boyd BJ (2013) Disposition and association of the steric stabilizer Pluronic® F127 in lyotropic liquid crystalline nanostructured particle dispersions. J Colloid Interface Sci 392:288–296
Um JY, Chung H, Kim KS et al (2003) In vitro cellular interaction and absorption of dispersed cubic particles. Int J Pharm 253(1–2):71–80
Uyama M, Nakano M, Yamashita J et al (2009) Useful modified cellulose polymers as new emulsifiers of cubosomes. Langmuir 25(8):4336–4338
von Halling LC, Gibson B, van de Weert M et al (2018) Spray dried cubosomes with ovalbumin and Quil-A as a nanoparticulate dry powder vaccine formulation. Int J Pharm 550(1–2):35–44

Wattendorf U, Merkle HP (2008) PEGylation as a tool for the biomedical engineering of surface modified microparticles. Int J Pharm 97(11):4655–4669
Wei S, Ren J, Li N et al (2017) Preparation and pharmacokinetic study of fenofibrate cubic liquid crystalline. Asian J Pharm Sci 12(6):580–585
Wörle G, Drechsler M, Koch M et al (2007) Influence of composition and preparation parameters on the properties of aqueous monoolein dispersions. Int J Pharm 329(1–2):150–157
Wu H, Li J, Zhang Q et al (2012) A novel small Odorrana lectin-bearing cubosomes: preparation, brain delivery and pharmacodynamic study on amyloid-β25–35-treated rats following intranasal administration. Eur J Pharm Biopharm 80(2):368–378
Yaghmur A, Glatter O (2009) Characterization and potential applications of nanostructured aqueous dispersions. Adv Colloid Interf Sci 147:333–342
Yang Z, Tan Y, Chen M et al (2012) Development of amphotericin B-loaded cubosomes through the SolEmuls technology for enhancing the oral bioavailability. AAPS PharmSciTech 13 (4):1483–1491
Yapar EA, Ýnal Ö (2012) Poly (ethylene oxide)–poly (propylene oxide)-based copolymers for transdermal drug delivery: an overview. Trop J Pharm Res 11(5):855–866
Younes NF, Abdel-Halim SA, Elassasy AI (2018) Corneal targeted sertaconazole nitrate loaded cubosomes: preparation, statistical optimization, in vitro characterization, ex vivo permeation and in vivo studies. Int J Pharm 553(1–2):386–397
Younus M, Hawley A, Boyd BJ et al (2018) Bulk and dispersed aqueous behaviour of an endogenous lipid, selachyl alcohol: effect of Tween 80 and Pluronic F127 on nanostructure. Colloids Surf B Biointerfaces 169:135–142
Zhai J, Waddington L, Wooster TJ et al (2011) Revisiting β-casein as a stabilizer for lipid liquid crystalline nanostructured particles. Langmuir 27(24):14757–14766

Self-Nanoemulsifying Drug Delivery Systems (SNEDDS): An Innovative Approach to Improve Oral Bioavailability

10

Girish U. Sailor

Abstract

In recent years, drug discovery programme has been utilizing a high-throughput approach for screening of molecules along with combinatorial chemistry which resulted in increasing amount of diverse new chemical entities (NCEs). However, it has been estimated that nearly 40–70% of these NCEs are found to be poorly water soluble. Lipid-based formulation, particularly self-nanoemulsifying drug delivery systems (SNEDDS), presents viable means for the augmentation of bioavailability of highly lipophilic, poorly water-soluble drug by various mechanisms. This is a homogenous liquid system made up of lipid, emulsifier and co-emulsifier. The system is spontaneously emulsified to form nanoemulsion, with moderate stirring when diluted with water. In this chapter, a comprehensive overview of different literature related to self-nanoemulsifying formulations is provided and important aspects like formulation attributes, mechanism, evaluation parameters and different forms of SNEDDS are discussed.

Keywords

Self-nanoemulsification · Nanoemulsion · Lipid-based formulation · SNEDDS

10.1 Introduction

Administration of drugs by oral route is the easiest and convenient way by most consumers. However, several factors underlie the poor or unpredictable oral bioavailability of many drug molecules: poor water solubility, permeability, extensive presystemic metabolism and intestinal P-glycoprotein (PGP) efflux or presystemic

G. U. Sailor (✉)
Bhagwan Mahavir College of Pharmacy, Bhagwan Mahavir University, Vesu, Surat, Gujarat, India

N. Shah (ed.), *Nanocarriers: Drug Delivery System*,
https://doi.org/10.1007/978-981-33-4497-6_10

elimination of the drug (Li 2011; Lipinski 2002). In addition, orally administered drugs face significant challenges in retaining its original chemical structure due to the varied environment of GI tract (Li 2011).

Currently various formulation strategies are available to overcome these problems, such as the use of permeation enhancers (Perng et al. 1998), chemical modification (salt formation and prodrug), complexation with cyclodextrins (Aungst et al. 1988), solid dispersion (Serajuddin et al. 1988), solubilization in surfactant systems (Miyako et al. 2010), liposomes (Schwendener and Schott 1996), nanosuspension (NS), lipid-based delivery system (Pouton 2000) and polymeric nanoparticles (Desai et al. 2012). However, lipid-based formulation has gained much attention with more focus on drug delivery system with self-emulsifying characteristics (Humberstone and Charman 1997). This formulation is used to enhance the absorption of orally administered drug with poor solubility and/or permeability by increasing the dissolution and lymphatic transport, and evading P-gp efflux and hepatic enzyme inhibition (Pouton 1985; Singh et al. 2009; Cherniakov et al. 2015; Pouton 2000). The increase in absorption leads to enhancement in bioavailability which improves the efficacy of drug. The efficacy is also improved by protecting the drug against physical and chemical degradation and also via active or passive targeting (Cherniakov et al. 2015). These formulation techniques solve the complexities associated with commercial production of emulsion system including stability and manufacturing problem. The emergence of the spontaneous nanoemulsification process has re-established the interest of formulation scientist for oral administration of drug-loaded nanoemulsion.

10.2 Self-Nanoemulsifying Drug Delivery System

Self-emulsifying systems for drug delivery have arisen as a prospective formulation with enormous potential in augmenting the oral bioavailability of low-aqueous-solubility molecules, which unless otherwise may not be delivered orally. These systems are anhydrous isotropic mixtures in which drugs are dissolved in the oil phase in the presence of surfactants and co-surfactants which upon dilution with water or gastrointestinal fluid form fine O/W emulsions or microemulsion (Fig. 10.1). This emulsification process requires little agitation provided by motility of the stomach and small intestine (Charman et al. 1992). This system avoids the dissolution step and upon dispersion it forms micro/nanosized globules of oil giving large interfacial area for quick absorption which improves the bioavailability of drug.

The improved drug absorption may be offered by solubilization of drug in the intestinal colloidal dispersion formed by bile salt, lipid-digested products, cholesterol and phospholipid (Pouton 1997, 2000; Neslihan Gursoy and Benita 2004; Gershanik and Benita 2000). Throughout this absorption process, either direct or mixed micelle transport system, the drug remains in solubilized state (Pouton 2000).

Self-emulsifying system by nature is a thermodynamically stable emulsion compared to unstable regular emulsions and is also able to solubilize more lipophilic

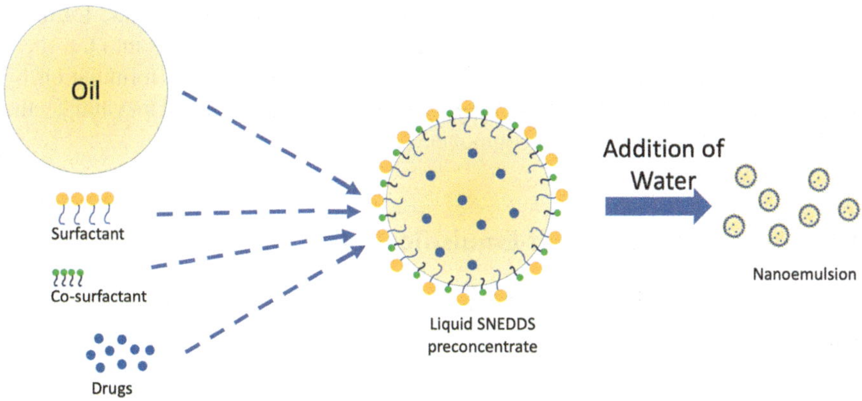

Fig. 10.1 Schematic flowchart of preparation of SNEDDS formulation with subsequent formation of nanoemulsion

Table 10.1 Types of self-emulsifying formulation

Criteria	Self-emulsifying system	Self-microemulsifying system	Self-nanoemulsifying system
Size of droplet (nm)	>300	100–250	<10
Formulation appearance	Turbid	Optically clear	Optically clear
HLB value—surfactant	<12	>12	>12
Types as per lipid formulation classification	II	IIIB	IIIB
Quantity of lipid (%)	40–80	>20	>20
Quantity of emulsifier (%)	30–40	40–80	40–80

HLB hydrophilic-lipophilic balance, *LFCS* lipid formulation classification system

drug. Their stability is thought to be dependent on their relatively small dispersed oil droplet size and narrow range of droplet size distribution (Leal-Calderon et al. 2007). As per the droplet size formed during in situ self-emulsification process, the formulations are categorized as given in Table 10.1 (Dokania and Joshi 2015). As compared to SMEDDS, SNEDDS offers greater advantage for drug absorption and transport due to its higher surface area after dilution. These liquid self-emulsifying systems can be formulated in soft gelatin or hard gelatin capsules for easy oral administration. Although liquid formulation perhaps presents certain drawbacks, it can be adsorbed on certain solid carriers to make it convenient for oral administration in the form of solid pellets, granules, tablets, solid dispersion, etc. with no or moderate effects on the in vivo behavior. (Singh et al. 2014; Schwarz 2003).

However, development of a precise in vitro method which can predict the in vivo performance of self-emulsifying drug delivery systems is the major challenge. In addition to this, degradation of drug in GIT as well as irritation caused due to excessive amount of surfactant (~30–60%) in formulation is also a concern.

Furthermore, drugs in hard or soft gelatin capsule of convention SNEDDS formulation can also form precipitate due to migration of volatile co-solvents into the shell. The drug molecule also tends to precipitate due to the dilution of formulation by hydrophilic solvent. At the same time, it becomes more difficult to validate the formulations containing several components.

10.3 Mechanism of Self-Emulsification

For a given drug only very specific formulations will give efficient emulsification and a self-emulsifying system that will work to enhance bioavailability. Efficiency of SNEDDS therefore is determined by concentration of emulsifier, ratio of lipid/emulsifier, emulsion polarity, charge and size of the droplets (Gursoy and Benita 2004). However, the mechanism that governs self-emulsification has not yet been fully understood, but it can be predicted using Gibbs free energy concept. Self-emulsification takes place when energy requirements to increase the surface area of dispersion are below the entropy change (Singh et al. 2009; Porter et al. 2008). The system works on the surroundings in a spontaneous manner if free energy for the system must be negative. Thus, it is essential to keep the energy required for the formation of droplet as low as possible to form thermodynamic stable emulsion which can be determined by using Eq. 10.1 at fixed temperature:

$$\Delta G = \gamma_{\mathrm{LL}} \Delta A - T\Delta S \tag{10.1}$$

where γ_{LL} is the interfacial surface tension, ΔA interfacial surface area change and ΔS entropy of emulsification.

In addition, Reiss has also explained about the need of energy to produce new surface area in conventional emulsion using Eq. 10.2:

$$\Delta G(\text{Free energy}) = \sum\nolimits_i N_i 4\pi r_i^2 \sigma \tag{10.2}$$

In this equation, ΔG is the free energy of process, r_i droplet radius, N_i no. of droplets and σ interfacial energy.

In nanoemulsion, the work done to increase the interfacial surface is not sufficient to compensate for entropy of the system which makes nanoemulsion thermodynamically unstable with respect to separate phases, conversely to the microemulsion (Denton and Rostron 2013). Hence, to reduce the free energy two phases tend to break up after a certain period of time to minimize the surface area of interface. This increase in the interfacial tension between oil and water can be decreased by addition of more surfactant which is absorbed on the surface which consequently yields delayed coalescence of droplets by electrostatic and steric repulsion (Rosen and Kunjappu 2012; Becher 1965).

The free energy available to develop spontaneous emulsion is quite less for a self-emulsifying system. It is suggested that water penetrates through the gel and liquid crystal (LC) phases that occur at the surface of the droplets. Then, water is solubilized in oil phase until the solubilization limits. After the limit is reached, dispersion of LC phase is formed and this depends on the proportion of oil, surfactant and available water (Craig et al. 1995). With this formation, self-emulsifying system becomes resistant to coalescence. However with the addition of high amounts of drug, which is a common case for potential oral dosage forms, it is harder to have stabilized emulsions. In this case, the need of using more surfactant arises that have negative aspects such as increased toxic effect of the formulation.

10.4 Formulation Components

Composition of excipients plays an important role in the formulation of SNEDDS. The various excipients commonly used in the formulation of SNEDDS are listed in Table 10.2.

10.4.1 Lipid Phase

Lipid is the major component of formulation which facilitates solubilization of lipophilic drugs. It is essential for self-emulsification process as well as to augment the lymphatic transport of hydrophobic drug which ultimately improves the absorption of drug from the GIT (Gursoy et al. 2003; Gershanik and Benita 2000; Holm et al. 2002; Tenjarla 1999). In spite of higher potential of lipid excipients, insufficient data in terms of physical chemistry of lipids and also physicochemical stability of drug limits the formulation to reach the pharma market.

Additionally, it is also important to consider the interaction of formulation containing lipids with GI milieu which can affect the absorption of drug. The saturation state of lipids and its chain length can be able to affect the capacity of solvent and also the formulation digestibility (Rahman et al. 2013; Hauss 2007). It was reported that both types of triglycerides (long and medium chain) have been used in SNEDDS formulations (Constantinides 1995). However, self-emulsification process is greatly influenced by the chain length of hydrocarbon (Deckelbaum et al. 1990). Generally, long-chain triglycerides have profound self-emulsifying property as compared to medium-chain triglycerides. The MCT also has greater solubility due to its low molecular size. All these properties lead to improvement of drug absorption due to rapid hydrolysis of low-molecular-size lipid and greater mobility at O/W interface (Alander and Warnheim 1989; Grove et al. 2005).

In addition, it is interesting to know that LCT and MCT follow different absorption mechanisms as given in Fig. 10.6. LCT enhances lymphatic transport of drug via formation of chylomicron, thereby bypassing the extensive first-pass metabolism with regard to MCT (Khoo et al. 2003). On the other hand, MCT are more water soluble (Pouton and Porter 2008) and transported directly to systemic circulation via portal blood (Sek et al. 2002).

Table 10.2 List of commonly used excipients in SNEDDS formulation

Excipient class	Example
Lipid phase	
Long-chain triglycerides (LCT)	Castor oil, soybean oil
Medium-chain triglycerides (MCT)	Miglyol, Labrafac, Crodamol, Captex
Medium and/or long chain—mono- and diglycerides or mixed glycerides	Medium chain: Capmul, Imwitor, Alkoline, Labrasol, Gleurice, Acconon Long chain: Maisine 35-1, Compritol, Peceol, Geleol, Capmul GMO, Labrafil
Propylene glycol fatty acid esters	Capryol, Capmul PG-8, Sefsol 218, Lauroglycol
Fatty acid esters	Isopropyl myristate, isopropyl palmitate, ethyl oleate
Fatty acid	Oleic acid, caprylic acid
Surfactants	
Polysorbates	Tween 20, Tween 80, Crillet 1, Crillet 4
Sorbitan esters	Span 20, Sapn 80, Crill 1, Crill 4
PEO-PPO-block copolymers	Pluronic/Lutrol F68, Pluronic/Lutrol F127
Polyoxyethylene alkyl ethers	Brij, Cremophor A
Polyoxyethylene castor oil derivatives	Cremophor (RH 40, RH 60, EL), Eumulgin RO, Nikkol CO 40 TX, Etocas 35 HV, HCO-40, HCO-60
Polyoxyethylene stearate	Solutol HS 15
Polyoxylglycerides	Labrafil 2125 CS, Labrafil 1944 CS, Labrasol, Gleurice 44/14, Acconon MC-8 Acconon C44
Polyoxyethylene vitamin E	Vitamin E TPGS
Phospholipids	Soybean lecithin
Co-solvents	
Ethylene glycol and its derivatives	PEG 400, Transcutol
Short-chain alcohols	Ethanol, benzyl alcohol
Alkane diols and triols	Propylene glycol, glycerol

The droplet size of nanoemulsion is directly affected by lipophilicity and amount of oil phase. Additionally, it is important to note that the same oil should solubilize the maximum amount of drug. It was also known that digestible lipids can be able to increase the absorption of drugs with poor solubility as compared to the lipid which is nondigestible (Nanjwade et al. 2011; Palin et al. 1982; Gallo-Torres et al. 1978). Therefore, it is important to choose the lipid phase which can solubilize the drug and at the same time be able to form nanoemulsion with required properties. Few examples of lipid components used in self-emulsifying formulation are listed in Table 10.2.

10.4.2 Surfactants

It is also considered as a critical component for emulsifying properties of SNEDDS formulation. The amphiphilic nature of surfactant can dissolve high amounts of hydrophobic drug compounds. Several properties of surfactant like affinity for lipid phase, cloud point and HLB are known to effect the droplet size, self-emulsification area and emulsification process (Becher 1965; Wang et al. 2009).

In addition, the type of surfactant also affects the emulsification property. Self-emulsifying formulation can be prepared by using natural and/or synthetic surfactants. However, natural surfactant owing to its poor emulsifying characteristics has not been used in the formulation in spite of its high safety profile. Conversely, synthetic surfactant is chosen over natural surfactant due to its non-ionic characteristic and high HLB value. Surfactants with high HLB instantly emulsify to form o/w droplets and are quickly dispersed into aqueous media (Pouton 2000; Constantinides 1995) and non-ionic properties give rise to enhancement of absorption of drug by increasing the intestinal permeability without affecting the cellular viability (Tenjarla 1999). The surfactant also protects the drug from precipitation in GI fluid after dilution (Constantinides 1995; Shah et al. 1994; Attwood and Florence 1983). Therefore, the selection of surfatant with suitable properties is critically important.

Moreover, the surfactant concentration also play important rome in the formulation. It has been reported that a surfactant concentration above 25% (w/w) is required for self-emulsification and quick dispersion of SNEDDS. However, surfactant concentration of 50–60% (w/w) forms a crystalline gel-like liquid at the boundary of O/W which could destroy the progression of emulsification (Becher 1965). Additionally, high surfactant concentration may also cause gastric irritation and could even be toxic (Gursoy and Benita 2004). Nevertheless, the irritation potential can be reduced due to rapid stomach emptying of small lipid droplet size of formulation and extensive distribution of formulation all over GIT which reduce the concentration of drug at local sight.

10.4.3 Co-solvents

The co-solvent is amphiphilic in nature and plays an important role in the formulation of SNEDDS. It facilitates the dissolution of high amount of drug and/or hydrophilic surfactant in lipid phase and supports self-nanoemulsification (Constantinides 1995; Basalious et al. 2010; Ren et al. 2009). It significantly enhances the solubilization capacity of solvent at high concentration which has its own disadvantages. After dilution, solvent capacity of co-solvents is lost which may cause drug precipitation. High amount of co-solvents can be immiscible with the oil components and low-molecular-weight co-solvents could be incompatible with the capsule shells. The solubility of many drugs is near to logarithmic relationship with the amount of co-solvent (Cole et al. 2008). Another purpose to include the co-solvents is to produce very fine dispersion under gentle agitation and also to increase the area of self-emulsification in ternary diagram.

10.4.4 Aqueous Phase

The SNEDDS on oral administration goes to GIT and then gets absorbed into blood and distributed throughout the body where it comes in contact with diverse pH range; physiological milieu ranges from pH 1.2 to 7.4. Moreover, physiological milieu contains various ions which can also influence properties of nanoemulsion created from SNEDDS. Hence, it is important to consider aqueous-phase pH and ionic content while formulating SNEDDS as it affects the droplet size and stability of nanoemulsion (Moreira de Morais et al. 2006; Date and Nagarsenker 2007).

10.4.5 Drug

Generally, bioavailability is the major concern for the drug due to poor solubility and permeability, P-gp efflux, presystemic metabolism and metabolism by hepatic enzyme. Lipid-based formulations are widely used to solve these problems of drug (Singh et al. 2013). The concentration of drug as well as its physicochemical properties like molecular structure, molecular weight, log p, pKa and ionizable group significantly affects the various properties of SNEDDS, like droplet size and phase behaviour (Date and Nagarsenker 2007; Wang et al. 2009). Some studies have also observed the effect of pH-dependent solubility of drug on nanoemulsification region (Date and Nagarsenker 2007).

According to Biopharmaceutical Classification System (BCS), drug molecules are classified into classes I–IV as per their solubility and permeability properties. The SNEDDS could be able to combat the issue related to all categories of BCS class drugs (Fig. 10.2). In addition, Lipinski's rule of five is a widely used qualitative prediction tool for oral absorption trend. According to this rule drug should have no more than one violation and if so then the drug is not suitable as an orally active drug in humans. However, Lipinski's rule does not take account of efflux of substrate drug. Prediction of in vivo behaviour of drug is also one of the challenges as individual drugs with same properties behave differently in similar vehicles (Kohli et al. 2010).

10.4.6 Other Additives

10.4.6.1 Antioxidants

Antioxidants that are soluble in lipid like propyl gallate, vitamin E and carotenoids may also be incorporated to prevent the oxidation of excipients especially of lipids (Schaich et al. 2013; Bowtle 2007).

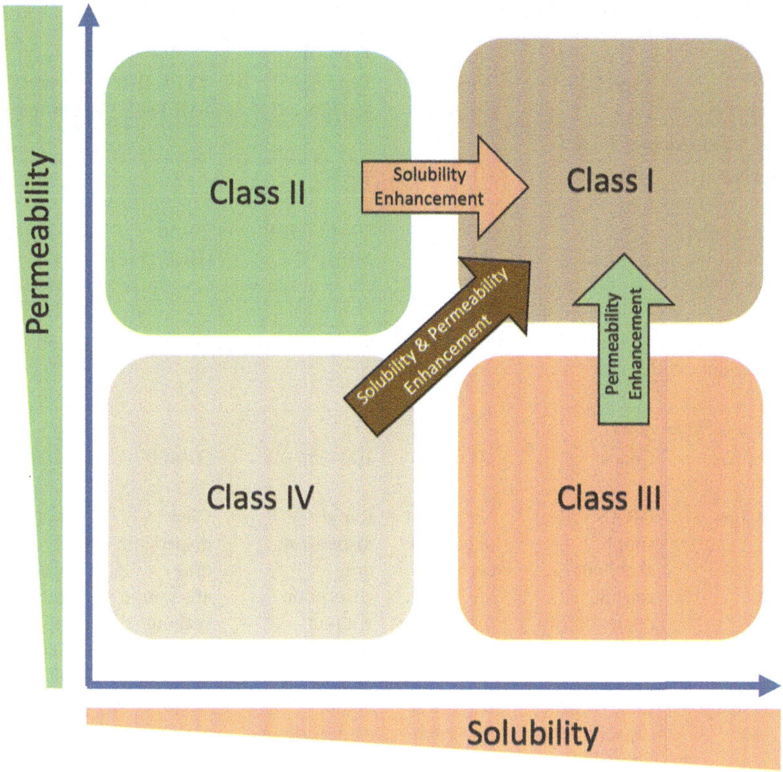

Fig. 10.2 Role of SNEDDS to enhance oral bioavailability of BCS class II to IV drugs

10.4.6.2 Polymers

In many cases, the solvent capacity of SEDDS is lost which causes drug precipitation. The use of supersaturated systems that maintain drug solubilization beyond equilibrium phase of solubility and do not allow drug precipitation for a long time is a new approach to enhance drug absorption (Vithani et al. 2018; Jo et al. 2019). Generally, the precipitation inhibitors are hydrophilic polymers such as polyethylene glycol (PEG) 4000, hydroxypropyl methylcellulose (HPMC) and polyvinyl pyrrolidone (PVP) which facilitate the formation of supersaturable systems. These inhibitors could inhibit the crystallization of drug by adsorbing on the surface of the drug and prevent the precipitation by hydrogen bonding (Chen et al. 2012).

Table 10.3 Types of lipid formulations

Formulation	Type I Oil	Type II SEDDS	Type III Type IIIA SEDDS	TYPE IIIB SMEDDS	Type IV Oil free
Typical composition (%w/w)					
Lipid phase (%)	100	40–80	40–80	<20	–
Surfactant (%)		20–60 (HLB <12)	20–40 (HLB >12)	20–50 (HLB >12)	0–20 (HLB <12) 30–80 (HLB >12)
Water-soluble co-surfactant (%)			0–40	20–50	0–50
Characteristic features					
Droplet size (nm)	Coarse	100–250	100–250	50–100	<100
Advantages	GRAS status; simple; excellent capsule compatibility, digestible	No impact of solvent capacity on dispersion	Rapid dispersion; drug absorption without digestion	Clear dispersion; drug absorption without digestion	Good solvent capacity for many drugs; disperses to micellar solution
Disadvantages	Poor solvent capacity with limited dispersion unless drug is highly lipophilic	Digestion likely but not crucial	Upon dispersion or digestion, solvent capacity probably lost; may be digestible	Upon dispersion, possible loss of solvent ability	Upon dispersion, possible loss of solvent ability; not digestible

10.5 Formulation Steps of Self-Nanoemulsifying Drug Delivery System

10.5.1 Screening of Components

SNEDDS are composed of a defined mixture of several lipid excipients including oils, surfactants and co-surfactants. There are large varieties of liquid or waxy excipients available which can be used to formulate the drug-loaded emulsions. However, identification of appropriate excipient or excipient combination plays an important role in the formulation which solubilizes the drug dose in unit volume suitable for oral consumption. To assist in selecting the optimum composition of formulation components, such as lipid, surfactant and co-solvents or co-emulsifier,

Pouton (2006) has classified these formulations into five types as illustrated in Table 10.3.

Generally, excipients' screening study involves drug excipient testing for compatibility, solubility and stability. Out of these, solubility of drug in excipients is the most prominent characteristic and can be performed by flask shaking techniques. Excess amount of drug is added to fixed volume of excipients and subjected to vortexing. Then, the mixture is allowed to shake continuously for 72 h at ambient room temperature (~25 °C) and then centrifuged. The supernatant is taken and filtered for any visible impurities and drug content is quantitatively determined using UV spectrophotometer.

Since formulation has multiple components, it is essential to find the drug solubility in a complete system rather than rely on drug solubility in individual components. Maximum drug loading, minimum droplet size and self-emulsification time in gastric environment and protection of drug against GI degradation or avoidance of metabolism are the main objectives of selection process (Pouton 2000; Kararli et al. 1992). To simplify this process Williams and co-workers (Williams et al. 2013) have suggested a flowchart for general guidance for excipient selection to formulate lipid-based formulation (Fig. 10.3).

In addition, selection of component also depends on the formulation route of administration. SNEDDS are mostly delivered via oral route (Kim et al. 2015; Parikh and Sawant 2019; Quan et al. 2013); however few studies related to parenteral (Shete et al. 2015) and transdermal route (El Maghraby 2010) administration have also been reported.

Additionally, excipient toxicity is a major concern for regulatory agencies. The United States Food and Drug Administration (USFDA) has listed certain excipients and inactive ingredients in its publication as "generally recognized as safe" (GRAS). However, additive toxicological effect should also be looked upon for the formulation with more than one excipient. Lastly, other factors such as capsule compatibility, purity, excipient miscibility, melting temperature and cost of goods also affect the performance of lipid-based formulation.

10.5.2 Construction of Ternary Phase Diagrams

Optimum amount of lipid, surfactant and co-surfactant is determined by constructing a diagram for pseudo-ternary phase which serves as a main foundation for the development of emulsion formulations. Figure 10.4 illustrates typical phase diagram with ternary system containing lipid, surfactant and water Garti and Aserin 2012). The corner of triangle signifies 100% of the phase mentioned. Firstly, the phase diagram is constructed by mixing different excipients in different quantities to identify the region with homogenous mixture. After identification of this region, the next step will be to identify which compositions form nano- or microemulsion with desired properties. To identify this, the pre-concentrate is titrated with water until change in the phase is observed from gels to microemulsion. During this process the viscosity of system is increased initially until the formation of continuous

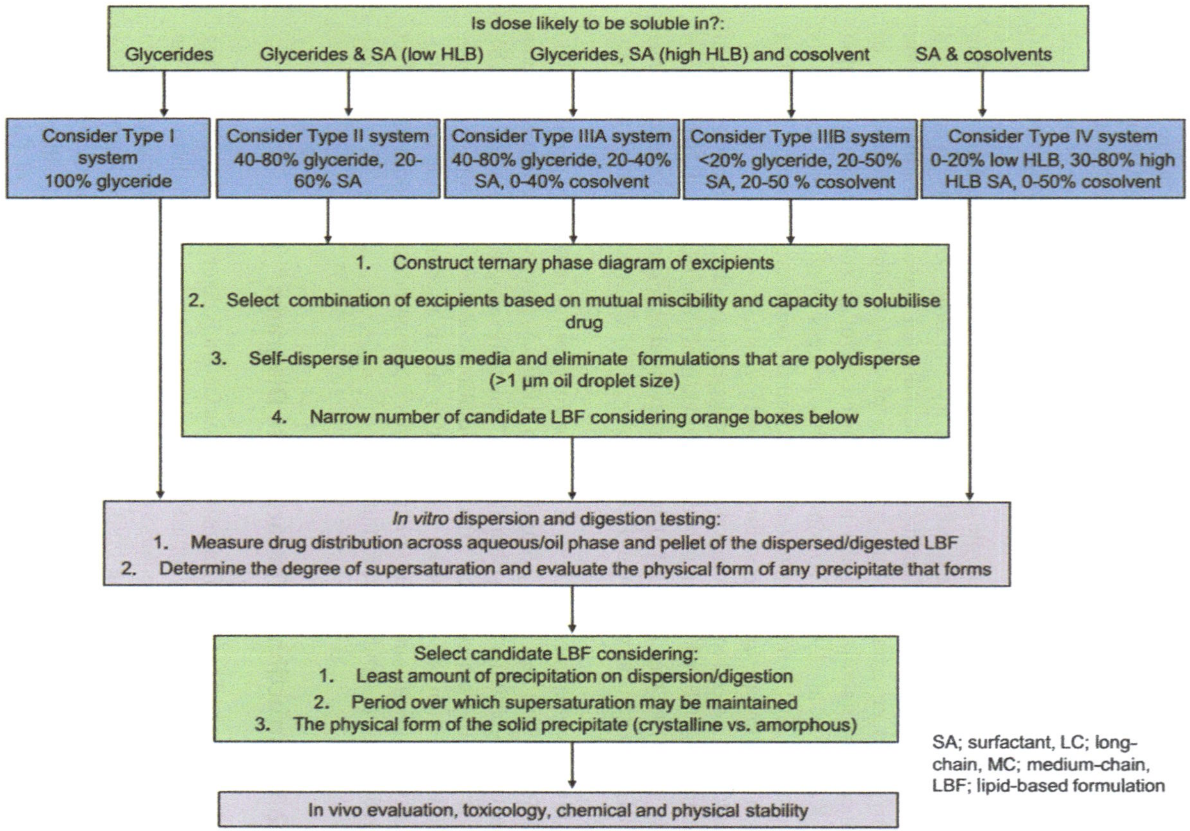

Fig. 10.3 A flowchart suggesting a general guide to lipid-based formulation design (reproduced from Williams et al. with permission from American Society for Pharmacology and Experimental Therapeutics, Williams et al. 2013)

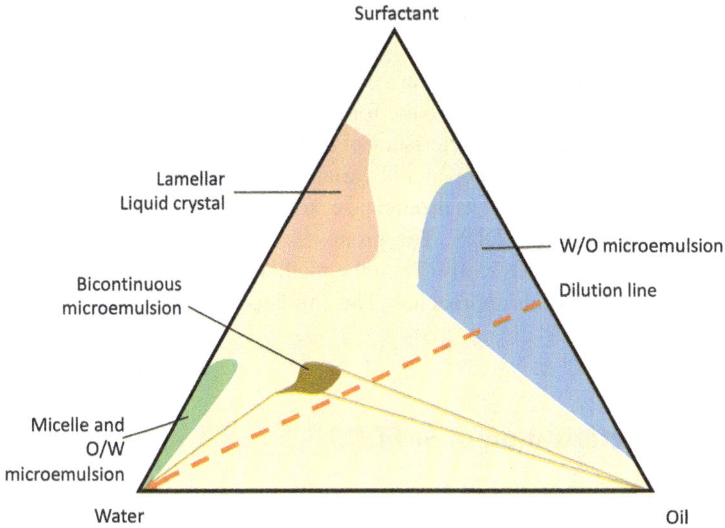

Fig. 10.4 Typical ternary phase diagram of oil, surfactant and water system

water phase, and then there is a decrease in viscosity close to water in dispersed SNEDDS. These mesomorphic structures can be determined by continuous monitoring of viscosity (Gradzielski 1999). All these details such as amount of individual phase required for optimized formulation along with the amount of water added are used to construct the diagram using triplot or chemix school software (Mu et al. 2013).

10.5.3 Self-Emulsification Characteristics

The self-nanoemulsification region can be identified by ternary phase diagram after observing spontaneous nanoemulsion region and phase behaviour. The different composition of ternary phase was prepared as per the diagram and diluted with constant amount of water. Then, droplet size of nanoemulsion is determined to identify the self-nanoemulsification region. It is also imperative to study the effect of physicochemical characteristics of drug on spontaneous self-emulsification region in ternary diagram for successful formulation (Date et al. 2010).

10.5.4 Preparation of SNEDDS

Figure 10.1 represents the flowchart for the preparation of SNEDDS formulation. The formulation is diluted in gastric milieu after the oral administration with subsequent formation of nanoemulsion.

10.5.4.1 Optimization Techniques

There are various techniques available to optimize the formulation. Response surface methodology is advantageous over other methods in terms of number of experiments. This tool can optimize the formulation with minimum number of trials without compromising the characteristics of final product. This methodology can be used to study the influence of dependent variables such as the amount of component on independent variables such as droplet size, solubility and self-nanoemulsification time characteristics of SNEDDS. The proposed design is constructed using these variables and analysed statistically. Then the mathematical correlation is established and validated using various approaches. The validated design is used to optimize the formulation with desired characteristics.

10.6 Characterization of SNEDDS

The final SNEDDS formulation is evaluated using various in vitro, in vivo and ex vivo parameters (Singh et al. 2009). Various techniques have been utilized to evaluate these parameters to determine the practicality of formulation process, formulation stability and in vivo behaviour as given in Table 10.4.

10.7 Potential of SNEDDS in Bioavailability Enhancement

Self-emulsifying drug delivery systems have gained much popularity as oral drug delivery system due to its ability to enhance the absorption of orally administered drug with poor solubility and/or permeability by increasing the dissolution and lymphatic transport, and evading P-gp efflux and hepatic enzyme inhibition. In addition, SNEDDS can also be useful to enhance oral bioavailability of BCS class II–IV drugs (Fig. 10.2) by enhancing the solubility of hydrophobic drug, enhancing the permeability of poorly permeable drugs, modifying the pharmacokinetic behaviour and improving the stability of drug in gastric environment (Fig. 10.5).

10.7.1 Drug Transport Mechanism

The SNEDDS formulation follows three pathways after oral administration: digestive, absorptive and circulatory phases same as the all lipid-based formulation. It is well known that drug properties and lipid phase also affect the absorption mechanism and can be illustrated by in vitro lipolysis model. This model predicts several mechanisms for drug absorption (Fig. 10.6). SNEDDS after oral administration undergoes digestive phase in which it is dispersed to form emulsion on contact with GI milieu. The digestion of lipids is initiated by gastric lipase. The peristalsis movement of the stomach along with lipid-digested product (fatty acid and diglyceride) facilitates the development of crude emulsion. Then crude emulsion is

Table 10.4 Characterization parameters for SNEDDS

Parameters	Significance
In vitro tests	
Visual characterization	Identification of emulsion system
Solubility and pseudo-ternary phase diagram study	To identify the region of nanoemulsion
Dissolution	To understand the release of drug
Phase behaviour and optical property	To identify the stability
Dispersion behaviour	Determination of self-emulsification ability
Dye tests	Determination of self-emulsification ability of solid SNEDDS
Cloud point	To determine the stability of self-emulsification properties to temperature
Time of liquefaction	To predict the time required for melting of solid SNEDDS
Dilution test	For the formulation stability on dilutions
Lipolysis model	Study of lipid digestion and estimation of lipolysis rate
Freeze-thaw cycles	Study of thermodynamic stability
Centrifugation test	Determination of phase separation behaviour
Characterization by instrument	
Nanosizer/zetasizer	To determine the droplet size and its distribution, and zeta potential
Rheometer	To predict the existence of globule with spherical structure to dilution
Refractometer	Study of isotropic behaviour of diluted SNEDDS
Differential scanning calorimetry	To identify the thermal behaviour of SNEDDS
Nephelometry and turbidimetry	To determine cloudy and turbidity characteristics
Infrared spectroscopy (FTIR)	For compatibility study of drug with excipient
Transmission electron microscopy	To study the shape and morphological characteristics of SNEDDS
Scanning electron microscope	Study of surface characteristics of solid SNEDDS
Infrared spectroscopy (NIR)	Study of turbidity and size of particle
Diffusing-wave spectroscopy	Study of flow characteristics
NMR and its modified techniques	To judge the shape, diffusion property of SNEDDS component and formulation behaviour after dilution
Ultrasonic technology	For self-emulsification ability and stability of droplet
X-ray scattering—small angle	To determine the structure of lipid
Neutron scattering, small angle	To determine the droplet size and shape
Ex vivo study	
Franz diffusion cell studies	To predict the permeability
Simulation study (in vitro)	
Caco-2 cells, epithelial cell, isolated cell, vesicle at the membrane of brush border	Study of drug transport and its rate

(continued)

Table 10.4 (continued)

Parameters	Significance
In situ study	
Perfusion study	To determine the mechanism of drug absorption Study the effect of efflux transporter on oral bioavailability
In vivo studies	
Pharmacokinetic	Determine the amount of drug in biological fluids
Pharmacodynamic	For the determination of pharmacological, physiological and biochemical effect of drug

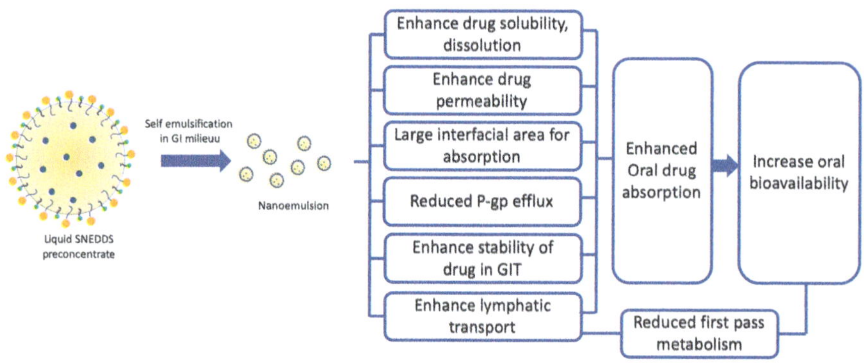

Fig. 10.5 Overview of potential mechanism of SNEDDS to enhance bioavailability of drugs

completely digested into monoglyceride by pancreatic lipase and its cofactor co-lipase into small intestine. In case of drug-loaded SNEDDS, gastric lipase as well as pancreatic lipase also digest the lipids of SNEDDS and the drug is released and redistributed (Ye et al. 2019). In the presence of exogenous lipid, bile salt is also secreted. This raised bile salt and lipid digestion products are eventually converted into mixed micelles and vesicle. This environment of intestine increases the solubilization capacity of drug and lipid-digested products. The remaining drug which is not solubilized by this mechanism is transported after its dispersion in the aqueous media via carrier-mediated uptake in ionized state or by passive diffusion (Porter et al. 2007).

After completion of digestive phase, these mixed micelles and vesicles are transported through enterocyte membrane via collisional transfer and binding followed by endocytosis or uptake to the lymph (Porter et al. 2007; Trevaskis et al. 2015) or blood, respectively. On the other hand, the upper lipid layer contained digested product of MCT or LCT. LCT enhanced lymphatic transport of drug (log $P > 5$) via formation of chylomicron as mentioned in Sect. 10.4.2. On the other hand, MCT was transported directly to enterocyte and formed vesicle of free drug (log $P < 5$) which was than absorbed into the portal system. The remaining drugs

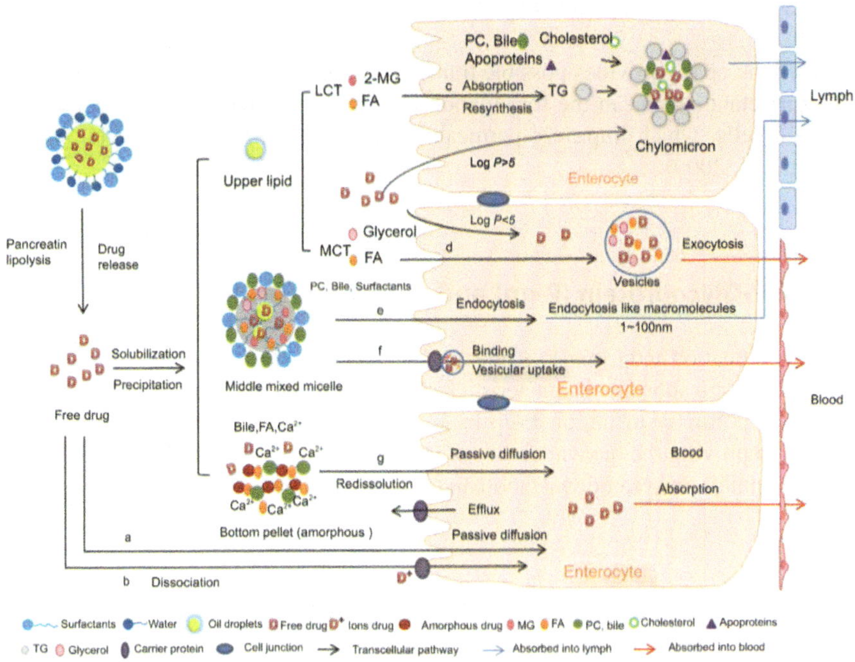

Fig. 10.6 Absorption routes of SNEDDS with different oil species (MCT/LCT) and drug properties (log P > 5/log P < 5) (reproduced from Jingyi et al. with permission from Dove Medical Press)

precipitated or precipitate of amorphous substance was transported via passive diffusion (Khan et al. 2016; Alskär et al. 2018; Tanaka et al. 2017).

Apart from this mechanism drug can also be transported via paracellular and transcellular absorption, inhibition of P-gp, inhibition of CYP450 and also resynthesis of lipoprotein chylomicron.

10.7.2 Transcellular Pathway

Transport of chylomicron across enterocyte is also done by transcellular route (Porter et al. 2007) but their mechanism is not fully understood. However, it is believed to result in disorientation of lipid structure by surfactant, thus resulting in enhanced drug permeation.

10.7.3 Paracellular Pathway

Intercellular space provides possible transport channel for drug through enterocyte. The surfactant present in the formulation may cause opening of the tight junction between cells, which improves permeability of drug with high molecular weight (Singh et al. 2009).

10.7.4 P-Glycoprotein (P-gp) and CYP450 Inhibition

BCS classification and Lipinski's rule for drug candidate selection have failed to take account of P-gp substrate characteristics of drug. SNEDDS formulation can be able to give opportunity to inhibit P-gp by excipients such as Cremophor and TPGS, thereby improving the bioavailability of such drugs (Collnot et al. 2007). SNEDDS can also inhibit presystemic metabolism of drug via inhibition of CYP450 enzymes.

10.8 Types of SNEDDS for Oral Delivery

SNEDDS formulations are classified on the basis of its physical state like solid, semisolid and liquid. Additionally, solid and semi-solid SNEDDS can also be subdivided on the basis of excipients and technology (Bansal et al. 2008; Sato 2001).

10.8.1 Liquid Formulation

The formulation contains lipids, surfactant and co-surfactant and is optically stable. These formulations, owing to their liquid nature, can be filled into soft/hard gelatin capsule shell.

10.8.2 Semi-solid Formulation

This formulation is more or less similar to liquid SNEDDS but semi-solid at room temperature. The stability and handling issue of formulation during the production can be solved due to its semi-solid nature. Additionally, the in vivo performance of formulation is also not hampered due to the semi-solid nature because of high mobility when dispersed in aqueous media at body temperature. However, self-emulsification properties get compromised and it is a major challenge for formulation scientist (Singh et al. 2013; Beg et al. 2015).

10.8.3 Solid Formulation

Liquid SNEDDS can be converted to solid SNEDDS to overcome various limitations associated with liquid state of formulation. Various approaches like adsorption, melt granulation and extrusion spheronization have been widely used to produce this formulation. Various techniques, such as adsorption on the inert carriers, spray drying, extrusion spheronization and melt granulation, have been explored for the formulation of S-SNEDDS (Tan et al. 2013).

SNEDDS formulation with extended-release profile (Patil et al. 2009) and eutectic characteristics (Nazzal and Khan 2006) can also be prepared in tablet dosage form. In addition to this formulation, various novel SNEDDS formulation approaches such as gastroretentive (Wang et al. 2011), mucoadhesive (Whittle and Guy 2001), osmotic pumps (Zhang et al. 2013; Dong et al. 2002), phospholipid-based system (Zhang et al. 2015), cationic (Gershanik and Benita 2000), polar lipid-based system (Bhalani and Patel 2012), polymeric (Holmberg and Siekmann 2010) and self-double-emulsifying (Qi et al. 2011) can also be used on the basis of the merits and applications of formulations.

10.9 Advantage of SNEDDS in Oral Drug Delivery

SNEDDS offers the following advantages:

1. Protection of drugs which are prone to be destroyed or metabolized by enzymes in GI tract.
2. Reduction of gastric irritancy occurring due to prolonged exposure of drug to the inner surface of GI tract.
3. Generally fed state and fasted state offer variable bioavailability in individuals which can be overcome by SNEDDS.
4. Enhanced absorption of orally administered drug via various mechanisms illustrated in Fig. 10.6, and production of rapid onset of action.
5. Tiny droplet formed after self-emulsification process in GI tract substantially increases the interfacial surface for drug partition in two phases.
6. The emulsifying agents like Tween 80, SPAN, TPGS and Cremophor EL are used for the SNEDDS formulation exhibiting efflux transporter inhibitory effect to improve the bioavailability of substrate drug.
7. The emulsifying agent with high HLB value can also be able to relax the tight junction between cells which improves penetration of drug.
8. It can be stored easily due to high thermodynamic stability.
9. The manufacturing and scale-up are easy.

10.10 Limitation of SNEDDS

There are few issues associated with SNEDDS as follows:

1. Precipitation of drug molecule during dilution in GI milieu. The most essential requirements for the self-emulsifying formulation are that the drug should be in dissolved state.
2. Some conventional SNEDDS formulations utilize volatile co-solvents (ethanol and propylene glycol), which cause precipitation of hydrophobic drugs due to the migration of this co-solvent into the shell of hard or soft gelatin capsule.
3. Incompatibility of liquid SNEDDS component with capsule shell on long-term storage. In addition, storage of liquid SNEDDS at low temperature is also one of the major concerns. All this can be overcome by formulation of solid SNEDDS.
4. The SNEDDS cannot be suitable for hydrophobic drugs that are susceptible to be catalysed at acidic pH.
5. Lack of predictive in vitro model to mimic the digestion of lipid before absorption of drug to assess the formulation.
6. Some of the lipids used in the preparation of SNEDDS are unsaturated in nature which is liable to oxidation. High degree of unsaturation will lead to high rate of oxidation of lipid.
7. Normally, lymphatic transport of drug through SNEDDS requires high log P and high triglyceride solubility (Caliph et al. 2000). However, the amount of drug uptake via lymphatic system suffers from high variability and also depends upon the physiological nature of the individual. Therefore, lipophilicity and triglyceride solubility of the drug with respect to lymphatic transport need in-depth study which requires a robust predictive in vitro or in vivo model (Trevaskis et al. 2008).
8. The dispersion of SNEDDS in physiological or aqueous media is not affected by polymorphic changes. However, it can notably affect the release of drug, if matrix is slow or incapable of erosion in dissolution media (Khan and Craig 2003; Freitas and Müller 1999; Brubach et al. 2004).

10.11 Challenges in Formulations

The problem of drug precipitation in the GI tract can be overcome by designing supersaturable SNEDDS. This is the thermodynamically stable system with less amount of emulsifier and a precipitation inhibitor polymer. Generally, drugs in conventional SNEDDS are precipitated after dilution in GI milieu due to disturbance in supersaturated state. However, supersaturable SNEDDS can be able to prevent these conditions by producing and preserving the supersaturated state. Sometimes, drugs in the supersaturated formulation may also get crystallized during storage which can also be prevented by using supersaturable formulations. Methyl cellulose, hydroxypropyl methylcellulose (HPMC), sodium carboxymethyl and cellulose polyvinylpyrrolidone (PVP) are some of the most commonly used polymers to inhibit precipitation in SNEEDS (Xu and Dai 2013).

Generally, liquid SNEDDS formulations are filled in hard or soft gelatin capsule, but sometimes the shell of the capsule may soften or become brittle due to the leaching or interaction of formulation with shell. At lower temperature, drug may also precipitate. To solve these issues, liquid SNEDD are transformed into solid SNEDDS using several techniques (Tang et al. 2008). Thus, solid SNEDDS provides stability and robustness to the liquid formulation with better patient compliance and also lower manufacturing costs (Milović et al. 2012; Mu et al. 2013).

Lipid oxidation due to unsaturation or due to their architectures with oxidizable moiety can be minimized by addition of certain antioxidants. In addition, dissolution of lipid by heating the system at a temperature above 20 °C than the melting point of lipids could prevent the polymorphism of lipid. Moreover, solidification time can be increased by using the lipid excipients such as macrogol. This process ensures reduction and control of polymorphic change of lipid matrix. However, this adjustment process could negatively affect the formulation handling and also the packing of capsule (Sato 2001).

In respect to the lymphatic transport, there is an urgent need for adequate predictive model as lipid solubility of drug is not absolutely satisfactory for prediction. Currently, animal models like intestinal lymphatic duct cannulation are available to predict the transportation of drug via lymphatic route.

Another major concern is lack of precise in vitro model to predict the in vivo performance of self-emulsifying drug delivery systems which stops the formulations from entering into the market. To combat this situation, in vitro lipolysis as well as dynamic gastric model (DGM) is utilized to assess lipid-based formulations (Mercuri et al. 2011; Vardakou et al. 2011).

10.12 Conclusion

Currently, SNEDDS formulation technology gains much attention for drugs with poor solubility and permeability and several reports have appeared in the literature. However, a considerable gap exists between SNEDDS research in the laboratory and their marketed SNEDDS products. Furthermore, dearth of in vivo research data justify the relative results of similar formulations with different droplet sizes. Various efforts have been made to address the issues associated with SNEDDS. Additionally, handling and storage issue associated with liquid SMEDDS increases production cost. Therefore, conversion of liquid SNEDDS into solid could be an important alternative to combat the handling, production cost and stability-related issues. However, this strategy cannot be able to eliminate other issues related to drug degradation in most cases. Therefore, it is essential to develop a strategy which can modify the in vivo microenvironment to avoid drug degradation. Additionally, extensive investigation is required to develop a model to predict in vitro/in vivo correlation.

References

Alander J, Warnheim T (1989) Model microemulsions containing vegetable oils part 1: nonionic surfactant systems. J Am Oil Chem Soc 66(11):1656–1660

Alskär LC, Keemink J, Johannesson J et al (2018) Impact of drug physicochemical properties on lipolysis-triggered drug supersaturation and precipitation from lipid-based formulations. Mol Pharm 15(10):4733–4744

Attwood D, Florence AT (1983) Surfactant systems: their chemistry, pharmacy, and biology. Chapman and Hall, London

Aungst BJ, Rogers NJ, Shefter E (1988) Comparison of nasal, rectal, buccal, sublingual and intramuscular insulin efficacy and the effects of a bile salt absorption promoter. J Pharmacol Exp Ther 244(1):23–27

Bansal T, Mustafa G, Khan ZI et al (2008) Solid self-nanoemulsifying delivery systems as a platform technology for formulation of poorly soluble drugs. Crit Rev Ther Drug Carrier Syst 25(1):63–116

Basalious EB, Shawky N, Badr-Eldin SM (2010) SNEDDS containing bioenhancers for improvement of dissolution and oral absorption of lacidipine. I: development and optimization. Int J Pharm 391(1–2):203–211

Becher P (1965) Emulsions theory and practice. American Chemical Society monograph, 162, 2nd edn. Reinhold Publ. Corp.; Chapman & Hall, New York; London

Beg S, Sandhu PS, Batra RS et al (2015) QbD-based systematic development of novel optimized solid self-nanoemulsifying drug delivery systems (SNEDDS) of lovastatin with enhanced biopharmaceutical performance. Drug Deliv 22(6):765–784

Bhalani V, Patel S (2012) Pharmaceutical composition for lipophilic drugs. US Patent US8119157B2

Bowtle WJ (2007) Materials, process, and manufacturing considerations for lipid-based hard-capsule formats. In: Oral lipid-based formulations, Drug and the pharmaceutical sciences, vol 170, 1st edn. Informa Healthcare USA, New York, pp 79–106

Brubach JB, Ollivon M, Jannin V et al (2004) Structural and thermal characterization of mono- and diacyl polyoxyethylene glycol by infrared spectroscopy and X-ray diffraction coupled to differential calorimetry. J Phys Chem B 108(46):17721–17729

Caliph SM, Charman WN, Porter CJ (2000) Effect of short-, medium-, and long-chain fatty acid-based vehicles on the absolute oral bioavailability and intestinal lymphatic transport of halofantrine and assessment of mass balance in lymph-cannulated and non-cannulated rats. J Pharm Sci 89(8):1073–1084

Charman SA, Charman WN, Rogge MC et al (1992) Self-emulsifying drug delivery systems: formulation and biopharmaceutic evaluation of an investigational lipophilic compound. Pharm Res 9(1):87–93

Chen ZQ, Liu Y, Zhao JH et al (2012) Improved oral bioavailability of poorly water-soluble indirubin by a supersaturable self-microemulsifying drug delivery system. Int J Nanomedicine 7:1115–1125

Cherniakov I, Domb AJ, Hoffman A (2015) Self-nano-emulsifying drug delivery systems: an update of the biopharmaceutical aspects. Expert Opin Drug Deliv 12(7):1121–1133

Cole ET, Cadé D, Benameur H (2008) Challenges and opportunities in the encapsulation of liquid and semi-solid formulations into capsules for oral administration. Adv Drug Deliv Rev 60 (6):747–756

Collnot E-M, Baldes C, Wempe MF et al (2007) Mechanism of inhibition of P-glycoprotein mediated efflux by vitamin E TPGS: influence on ATPase activity and membrane fluidity. Mol Pharm 4(3):465–474

Constantinides PP (1995) Lipid microemulsions for improving drug dissolution and oral absorption: physical and biopharmaceutical aspects. Pharm Res 12(11):1561–1572

Craig DQM, Barker SA, Banning D et al (1995) An investigation into the mechanisms of self-emulsification using particle size analysis and low frequency dielectric spectroscopy. Int J Pharm 114(1):103–110

Date AA, Nagarsenker MS (2007) Design and evaluation of self-nanoemulsifying drug delivery systems (SNEDDS) for cefpodoxime proxetil. Int J Pharm 329(1–2):166–172

Date AA, Desai N, Dixit R et al (2010) Self-nanoemulsifying drug delivery systems: formulation insights, applications and advances. Nanomedicine (Lond) 5(10):1595–1616

Deckelbaum RJ, Hamilton JA, Moser A et al (1990) Medium-chain versus long-chain triacylglycerol emulsion hydrolysis by lipoprotein lipase and hepatic lipase: implications for the mechanisms of lipase action. Biochemistry 29(5):1136–1142

Denton P, Rostron C (2013) Pharmaceutics: the science of medicine design. Integrated foundations of pharmacy, 1st edn. Oxford University Press, Oxford

Desai PP, Date AA, Patravale VB (2012) Overcoming poor oral bioavailability using nanoparticle formulations—opportunities and limitations. Drug Discov Today Technol 9(2):e71–e174

Dokania S, Joshi AK (2015) Self-microemulsifying drug delivery system (SMEDDS)—challenges and road ahead. Drug Deliv 22(6):675–690

Dong L, Shafi K, Wong P, Wan J (2002) L-OROS® SOFTCAP™ for controlled release of non-aqueous liquid formulations. Drug Deliv Technol 2:1

El Maghraby GM (2010) Self-microemulsifying and microemulsion systems for transdermal delivery of indomethacin: effect of phase transition. Colloids Surf B Biointerfaces 75 (2):595–600

Freitas C, Müller RH (1999) Correlation between long-term stability of solid lipid nanoparticles (SLN) and crystallinity of the lipid phase. Eur J Pharm Biopharm 47(2):125–132

Gallo-Torres HE, Ludorf J, Brin M (1978) The effect of medium-chain triglycerides on the bioavailability of vitamin E. Int J Vitam Nutr Res 48(3):240–241

Garti N, Aserin A (2012) Micelles and microemulsions as food ingredient and nutraceutical delivery systems. In: Garti N, McClements DJ (eds) Encapsulation technologies and delivery systems for food ingredients and nutraceuticals, Food science, technology and nutrition, vol 239. Woodhead Publishing, Sawston, pp 211–251

Gershanik T, Benita S (2000) Self-dispersing lipid formulations for improving oral absorption of lipophilic drugs. Eur J Pharm Biopharm 50(1):179–188

Gradzielski MHH (1999) Rheological properties of microemulsion. In: Kumar P, Mittal KL (eds) Handbook of microemulsion science and technology. CRC press, New York, pp 357–386

Grove M, Pedersen GP, Nielsen JL et al (2005) Bioavailability of seocalcitol I: relating solubility in biorelevant media with oral bioavailability in rats-effect of medium and long chain triglycerides. J Pharm Sci 94(8):1830–1838

Gursoy RN, Benita S (2004) Self-emulsifying drug delivery systems (SEDDS) for improved oral delivery of lipophilic drugs. Biomed Pharmacother 58(3):173–182

Gursoy N, Garrigue JS, Razafindratsita A et al (2003) Excipient effects on in vitro cytotoxicity of a novel paclitaxel self-emulsifying drug delivery system. J Pharm Sci 92(12):2411–2418

Hauss DJ (2007) Oral lipid-based formulations. Adv Drug Deliv Rev 59(7):667–676

Holm R, Porter CJ, Mullertz A et al (2002) Structured triglyceride vehicles for oral delivery of halofantrine: examination of intestinal lymphatic transport and bioavailability in conscious rats. Pharm Res 19(9):1354–1361

Holmberg C, Siekmann B (2010) Self-emulsifying drug delivery system. US Patent US7815933B2

Humberstone AJ, Charman WN (1997) Lipid-based vehicles for the oral delivery of poorly water soluble drugs. Adv Drug Deliv Rev 25(1):103–128

Jo K, Kim H, Khadka P et al (2019) Enhanced intestinal lymphatic absorption of saquinavir through supersaturated self-microemulsifying drug delivery systems. Asian J Pharm Sci 15(3):336–346

Kararli TT, Needham TE, Griffin M et al (1992) Oral delivery of a renin inhibitor compound using emulsion formulations. Pharm Res 9(7):888–893

Khan N, Craig DQ (2003) The influence of drug incorporation on the structure and release properties of solid dispersions in lipid matrices. J Control Release 93(3):355–368

Khan J, Rades T, Boyd BJ (2016) Lipid-based formulations can enable the model poorly water-soluble weakly basic drug Cinnarizine to precipitate in an amorphous-salt form during in vitro digestion. Mol Pharm 13(11):3783–3793

Khoo SM, Shackleford DM, Porter CJ et al (2003) Intestinal lymphatic transport of halofantrine occurs after oral administration of a unit-dose lipid-based formulation to fasted dogs. Pharm Res 20(9):1460–1465

Kim MS, Ha ES, Choo GH et al (2015) Preparation and in vivo evaluation of a dutasteride-loaded solid-supersaturable self-microemulsifying drug delivery system. Int J Mol Sci 16 (5):10821–10833

Kohli K, Chopra S, Dhar D et al (2010) Self-emulsifying drug delivery systems: an approach to enhance oral bioavailability. Drug Discov Today 15(21–22):958–965

Leal-Calderon F, Schmitt V, Bibette J (2007) Emulsion science: basic principles, 2nd edn. Springer, New York

Li X (2011) Oral bioavailability: basic principles, advanced concepts, and applications, vol 16, 1st edn. Wiley, Hoboken

Lipinski C (2002) Poor aqueous solubility—an industry wide problem in drug discovery. Am Pharm Rev 5(3):82–85

Mercuri A, Passalacqua A, Wickham MS et al (2011) The effect of composition and gastric conditions on the self-emulsification process of ibuprofen-loaded self-emulsifying drug delivery systems: a microscopic and dynamic gastric model study. Pharm Res 28(7):1540–1551

Milović M, Djuriš J, Djekić L et al (2012) Characterization and evaluation of solid self-microemulsifying drug delivery systems with porous carriers as systems for improved carbamazepine release. Int J Pharm 436(1–2):58–65

Miyako Y, Khalef N, Matsuzaki K et al (2010) Solubility enhancement of hydrophobic compounds by cosolvents: role of solute hydrophobicity on the solubilization effect. Int J Pharm 393 (1–2):48–54

Moreira de Morais J, dos Santos ODH, Delicato T, Azzini Gonçalves R et al (2006) Physicochemical characterization of canola oil/water nano-emulsions obtained by determination of required HLB number and emulsion phase inversion methods. J Dispers Sci Technol 27(1):109–115

Mu H, Holm R, Müllertz A (2013) Lipid-based formulations for oral administration of poorly water-soluble drugs. Int J Pharm 453(1):215–224

Nanjwade BK, Patel DJ, Udhani RA, Manvi FV (2011) Functions of lipids for enhancement of oral bioavailability of poorly water-soluble drugs. Sci Pharm 79(4):705–727

Nazzal S, Khan MA (2006) Controlled release of a self-emulsifying formulation from a tablet dosage form: stability assessment and optimization of some processing parameters. Int J Pharm 315(1–2):110–121

Neslihan Gursoy R, Benita S (2004) Self-emulsifying drug delivery systems (SEDDS) for improved oral delivery of lipophilic drugs. Biomed Pharmacother 58(3):173–182

Palin KJ, Wilson CG, Davis SS et al (1982) The effect of oils on the lymphatic absorption of DDT. J Pharm Pharmacol 34(11):707–710

Parikh KJ, Sawant KK (2019) Solubilization of vardenafil HCl in lipid-based formulations enhances its oral bioavailability in vivo: a comparative study using Tween–20 and Cremophor–EL. J Mol Liq 277:189–199

Patil PR, Biradar SV, Paradkar AR (2009) Extended release felodipine self-nanoemulsifying system. AAPS PharmSciTech 10(2):515–523

Perng C-Y, Kearney AS, Patel K et al (1998) Investigation of formulation approaches to improve the dissolution of SB-210661, a poorly water soluble 5-lipoxygenase inhibitor. Int J Pharm 176 (1):31–38

Porter CJ, Trevaskis NL, Charman WN (2007) Lipids and lipid-based formulations: optimizing the oral delivery of lipophilic drugs. Nat Rev Drug Discov 6(3):231–248

Porter CJ, Wasan KM, Constantinides P (2008) Lipid-based systems for the enhanced delivery of poorly water soluble drugs. Adv Drug Deliv Rev 60(6):615–616

Pouton CW (1985) Self-emulsifying drug delivery systems: assessment of the efficiency of emulsification. Int J Pharm 27(2):335–348
Pouton CW (1997) Formulation of self-emulsifying drug delivery systems. Adv Drug Deliv Rev 25 (1):47–58
Pouton CW (2000) Lipid formulations for oral administration of drugs: non-emulsifying, self-emulsifying and 'self-microemulsifying' drug delivery systems. Eur J Pharm Sci 11(Suppl 2): S93–S98
Pouton CW (2006) Formulation of poorly water-soluble drugs for oral administration: physicochemical and physiological issues and the lipid formulation classification system. Eur J Pharm Sci 29(3–4):278–287
Pouton CW, Porter CJ (2008) Formulation of lipid-based delivery systems for oral administration: materials, methods and strategies. Adv Drug Deliv Rev 60(6):625–637
Qi X, Wang L, Zhu J et al (2011) Self-double-emulsifying drug delivery system (SDEDDS): a new way for oral delivery of drugs with high solubility and low permeability. Int J Pharm 409 (1–2):245–251
Quan Q, Kim DW, Marasini N et al (2013) Physicochemical characterization and in vivo evaluation of solid self-nanoemulsifying drug delivery system for oral administration of docetaxel. J Microencapsul 30(4):307–314
Rahman MA, Hussain A, Hussain MS et al (2013) Role of excipients in successful development of self-emulsifying/microemulsifying drug delivery system (SEDDS/SMEDDS). Drug Dev Ind Pharm 39(1):1–19
Ren F, Jing Q, Cui J et al (2009) Self-nanoemulsifying drug delivery system (SNEDDS) of anethole trithione by combined use of surfactants. J Dispers Sci Technol 30(5):664–670
Rosen MJ, Kunjappu JT (2012) Surfactants and interfacial phenomena, 4th edn. Wiley, Hoboken
Sato K (2001) Crystallization behaviour of fats and lipids—a review. Chem Eng Sci 56 (7):2255–2265
Schaich KM, Shahidi F, Zhong Y, Eskin NAM (2013) Lipid oxidation. In: Biochemistry of foods, vol 3. Academic, New York, pp 419–478
Schwarz J (2003) Solid self-emulsifying dosage form for improved delivery of poorly soluble hydrophobic compounds and the process for preparation thereof. US Patent 20030072798A1
Schwendener RA, Schott H (1996) Lipophilic 1-beta-D-arabinofuranosyl cytosine derivatives in liposomal formulations for oral and parenteral antileukemic therapy in the murine L1210 leukemia model. J Cancer Res Clin Oncol 122(12):723–726
Sek L, Porter CJ, Kaukonen AM et al (2002) Evaluation of the in-vitro digestion profiles of long and medium chain glycerides and the phase behaviour of their lipolytic products. J Pharm Pharmacol 54(1):29–41
Serajuddin AT, Sheen PC, Mufson D et al (1988) Effect of vehicle amphiphilicity on the dissolution and bioavailability of a poorly water-soluble drug from solid dispersions. J Pharm Sci 77 (5):414–417
Shah NH, Carvajal MT, Patel CI et al (1994) Self-emulsifying drug delivery systems (SEDDS) with polyglycolyzed glycerides for improving in vitro dissolution and oral absorption of lipophilic drugs. Int J Pharm 106(1):15–23
Shete H, Sable S, Tidke P et al (2015) Mono-guanidine heterolipid based SMEDDS: a promising tool for cytosolic delivery of antineoplastics. Biomaterials 57:116–132
Singh B, Bandopadhyay S, Kapil R et al (2009) Self-emulsifying drug delivery systems (SEDDS): formulation development, characterization, and applications. Crit Rev Ther Drug Carrier Syst 26(5):427–521
Singh B, Singh R, Bandyopadhyay S et al (2013) Optimized nanoemulsifying systems with enhanced bioavailability of carvedilol. Colloids Surf B Biointerfaces 101:465–474
Singh B, Beg S, Khurana RK et al (2014) Recent advances in self-emulsifying drug delivery systems (SEDDS). Crit Rev Ther Drug Carrier Syst 31(2):121–185

Tan A, Rao S, Prestidge CA (2013) Transforming lipid-based oral drug delivery systems into solid dosage forms: an overview of solid carriers, physicochemical properties, and biopharmaceutical performance. Pharm Res 30(12):2993–3017
Tanaka Y, Kawakami A, Nanimatsu A et al (2017) In vivo evaluation of supersaturation/precipitation/re-dissolution behavior of cinnarizine, a lipophilic weak base, in the gastrointestinal tract: the key process of oral absorption. Eur J Pharm Sci 96:464–471
Tang B, Cheng G, Gu JC et al (2008) Development of solid self-emulsifying drug delivery systems: preparation techniques and dosage forms. Drug Discov Today 13(13–14):606–612
Tenjarla S (1999) Microemulsions: an overview and pharmaceutical applications. Crit Rev Ther Drug Carrier Syst 16(5):461–521
Trevaskis NL, Charman WN, Porter CJ (2008) Lipid-based delivery systems and intestinal lymphatic drug transport: a mechanistic update. Adv Drug Deliv Rev 60(6):702–716
Trevaskis NL, Kaminskas LM, Porter CJ (2015) From sewer to saviour—targeting the lymphatic system to promote drug exposure and activity. Nat Rev Drug Discov 14(11):781–803
Vardakou M, Mercuri A, Naylor TA et al (2011) Predicting the human in vivo performance of different oral capsule shell types using a novel in vitro dynamic gastric model. Int J Pharm 419 (1–2):192–199
Vithani K, Hawley A, Jannin V et al (2018) Solubilisation behaviour of poorly water-soluble drugs during digestion of solid SMEDDS. Eur J Pharm Biopharm 130:236–246
Wang L, Dong J, Chen J et al (2009) Design and optimization of a new self-nanoemulsifying drug delivery system. J Colloid Interface Sci 330(2):443–448
Wang YP, Gan Y, Zhang XX (2011) Novel gastroretentive sustained-release tablet of tacrolimus based on self-microemulsifying mixture: in vitro evaluation and in vivo bioavailability test. Acta Pharmacol Sin 32(10):1294–1302
Whittle B, Guy G (2001) Pharmaceutical formulations. US Patent US20180042842A1
Williams HD, Trevaskis NL, Charman SA et al (2013) Strategies to address low drug solubility in discovery and development. Pharmacol Rev 65(1):315–499
Xu S, Dai WG (2013) Drug precipitation inhibitors in supersaturable formulations. Int J Pharm 453 (1):36–43
Ye J, Wu H, Huang C et al (2019) Comparisons of in vitro Fick's first law, lipolysis, and in vivo rat models for oral absorption on BCS II drugs in SNEDDS. Int J Nanomedicine 14:5623–5636
Zhang X, Yi Y, Qi J et al (2013) Controlled release of cyclosporine A self-nanoemulsifying systems from osmotic pump tablets: near zero-order release and pharmacokinetics in dogs. Int J Pharm 452(1–2):233–240
Zhang J, Li J, Ju Y et al (2015) Mechanism of enhanced oral absorption of morin by phospholipid complex based self-nanoemulsifying drug delivery system. Mol Pharm 12(2):504–513

Nanotechnological Approach for Design and Delivery of Phytopharmaceuticals

11

Ghanshyam R. Parmar and Girish U. Sailor

Abstract

Natural phytoconstituents play a vital role in the treatment of disease in humans. Problems associated with these natural compounds like solubility, absorption, stability, half-life, metabolism, and rapid excretion make it not suitable to dispense through the conventional dosage forms. Hence, researchers are in search for the development of novel drug delivery system of medicine to overcome aforementioned hindrances. Recently, research and development in nanotechnology has been combating these issues by developing nanocarriers for the efficient delivery of these phytoconstituents. In this chapter, we discuss the challenges associated with the development of novel formulations and focus on the design and development of various nanocarriers like phytosomes, liposomes, transfersomes, ethosomes, solid lipid nanoparticles, nanoemulsion, self-microemulsifying and nanoemulsifying system, and polymeric nanoparticle and microsphere and its future prospective.

Keywords

Phytoconstituents · Herbal extracts · Phytosomes · Liposomes · Nanoparticles · Nanoemulsion

G. R. Parmar (✉)
Department of Pharmacy, Sumandeep Vidyapeeth Deemed to be University, Vadodara, Gujarat, India
e-mail: ghanshyamparmar.dop@sumandeepvidyapeethdu.edu.in

G. U. Sailor
Bhagwan Mahavir College of Pharmacy, Bhagwan Mahavir University, VIP Road, Vesu, Surat, Gujarat, India

N. Shah (ed.), *Nanocarriers: Drug Delivery System*,
https://doi.org/10.1007/978-981-33-4497-6_11

11.1 Introduction

Since long history, human race has been acclimatized with nature and used medicinal plant-derived products to treat ailments. The foundation of present modern medicines is mainly on the utilization of herbs on the basis of traditional knowledge and practice. Around 80% of the world's population is using medicinal plants as remedies for the treatment of disease (Mohanty et al. 2017; Swamy and Sinniah 2016). In the current synthetic trend, the designing and discovery of new molecules fundamentally lie on the structural diversity of natural phytochemicals (Siddiqui et al. 2014a). Presently, majority of anticancer, antibacterial, immunosuppressive, and cardiovascular activities are derived from medicinal and aromatic plants. Additionally, 10% of volatile oils isolated from medicinal and aromatic plants are utilized with commercial importance in various sectors of pharmaceutical, cosmetic, agronomy, etc. Chemically, essential oils are majorly composed of terpenoids, aliphatic hydrocarbon derivatives. Hence these oils are therapeutically active (Ekor 2014).

Natural phytochemicals play a vital role in the treatment of major illnesses, i.e., metabolic syndrome, cancer, infectious disease, inflammatory bowel disease, diabetes, and dengue (Chaudhary et al. 2015; Farzaei et al. 2017; Graf et al. 2010; Nankar et al. 2017; Singh and Rawat 2017). These phytoconstituents bear ample advantages such as low toxicity profile, low adverse reactions, and low cost with therapeutic potential. However, in spite of numerous therapeutic benefits the major challenges with phytochemicals are poor absorption, poor water solubility, poor bioavailability, poor biocompatibility, poor stability, rapid metabolism in liver, and rapid excretion and toxicity (Ruan et al. 2010; Tang et al. 2012; Thanki et al. 2013) (Fig. 11.1). Hence, the novel drug delivery system (NDDS) can answer these challenges. Recently, nanotechnology has been playing a significant role in advance novel formulation, targeted delivery, and controlled drug release with high success rate.

The scenario of nanoparticle (NP) drug delivery system came into existence with the formulation of the first liposome by Bangham et al. (1965). Subsequently, an ample amount of polymer and lipids were invented for the formulation of NPs. In accordance with this several nanocarrier systems were developed such as solid lipid nanoparticles (SLNs), liposomes, polymeric nanoparticles, niosomes, and dendrimers. The literature stated that lipid-based NPs such as solid lipid nanoparticles (SLNs), microemulsions, nanostructured lipid carriers (NLCs), and self-microemulsifying drug delivery systems (SMEDDSs) are prevailing approaches to diminish the problem of poor solubility (Kalepu et al. 2013). Thermal and photostability of phytoconstituents or fractions improved by loading them into these carriers which significantly improved the stability of molecules or fractions. Furthermore, nanocarriers have effects on physicochemical properties of phytoconstituents and fractions, enhance the absorption and alter the rapid metabolism and elimination (Chen et al. 2012; Frozza et al. 2010; Wu et al. 2011; Zhang et al. 2014).

The NPs bear various advantages like specificity and selectivity of the target, low rate of degradation due to heat and light, low toxicity, enhancement of bioavailability and solubility, enhancement of therapeutic index, and sustained release of phytoconstituents. The basic methods involved in the formulation of NPs are solvent

Fig. 11.1 Factors that influence the absorption and bioavailability of phytoconstituents and fractions

evaporation, self-emulsion, high-pressure homogenization, precipitation, solvent displacement, supercritical fluid technology, and diffusion and emulsification.

11.2 Limitations of Conventional Formulations for Herbal Medicines

Herbal medicines have been used widely since thousands of years as remedies for many diseases. However, traditional medicines are not able to attract the modern scientist due to the lack of scientific data to support these medicines. Various issues associated with herbal medicines like standardization, adulteration, complexity of formulation, and high dose have raised objections about the safety and efficacy of these medicines. In addition, some studies have also raised serious concerns about high metal content in these medicines. Traditionally herbal medicines used in the form of churna, vati, gutika, rasayana, etc. have emerged as a new form of phytomedicine in which plant drugs are converted into a conventional formulation like tablet, capsule, and syrups. This form of medicine is accepted by not only patients but also physicians due to its convenient characteristics. However, some studies have shown that this medicine fails to reproduce the in vivo efficacy as compared to in vitro study. Scientists are continuously working to understand the mechanism of action, active constituents and their compatibility with other constituents, bioavailability, and also clinical efficacy of phytomedicine. Nowadays, due to the advancement of technology, isolation of medicinally important constituents is tested for the therapeutic effect instead of crude extract. Most of these biologically active constituents, such as tannins, flavonoids, and terpenoids, are highly water soluble, but their permeability is low which limits its absorption resulting in low bioavailability and less clinical efficacy. In addition, some active constituents have high molecular weight, poor solubility, degradation in the gastric environment, extensive metabolism, and reduced tissue target uptake particle size which seem to be the major limitations for its clinical use. Nanotechnology has drawn attention in terms of overcoming the oral bioavailability and efficacy problem associated with active constituents of plants by loading them into various nanocarrier systems.

This chapter summarizes the utilization of several NDDS like phytosomes, liposomes, transfersomes, ethosomes, SLNs, nanoemulsion, SMEDDSs, polymeric nanocarriers, dendrimers, and other novel nanocarriers.

11.3 Phytosomes

Phytosome is also termed as "herbosome." Phytosome is a combination of two words "phyto" and "some" which means plant- and cell-like, respectively. Phytosome is a novel nanocarrier system and liposome-like vesicles, loaded with the phytoconstituents or plant extracts in phospholipids to enhance the bioavailability and absorption and reduce the adverse effect of traditional herbal extracts as shown in Fig. 11.2 (Bhosale et al. 2016; Bombardelli et al. 1989; Chaudhary et al. 2015).

Mainly, phytosomes comprise phospholipid complex with polar phytoconstituents, e.g., phenolics, glycosides, and flavonoids, via formation of chemical bonding (Apostolova et al. 2015; Dai and Mumper 2010). This leads to the formation of a more stable complex which enhances the bioavailability of these water-soluble compounds. Natural phospholipids like soy phospholipids and phosphatidylcholine have been reported for the formulation of phytosomes. The major advantages of phytosomes are that it can easily cross the lipid membrane and

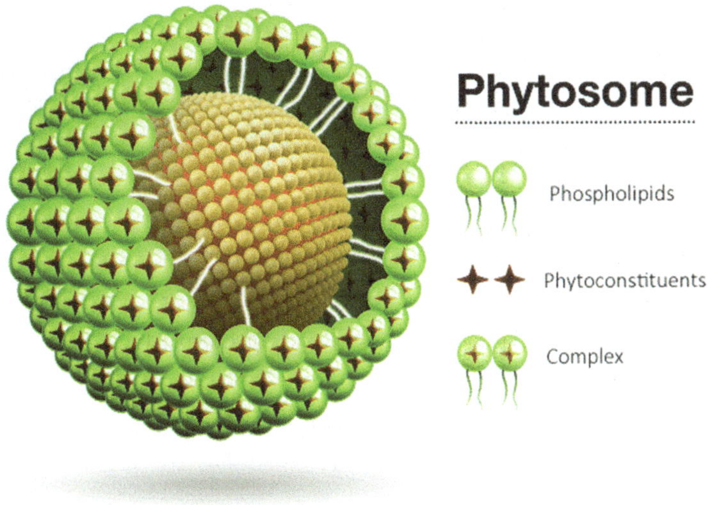

Fig. 11.2 Phytosomes: phytoconstituent-phospholipid complex

enhance the bioavailability of poorly lipid-soluble phytoconstituents via improvement in the absorption in GIT (Bhattacharya 2009; Kidd and Head 2005).

Phospholipids have greater affinity towards the polar compounds having a high number of free functional groups having the capability to form hydrogen bonding. This phospholipid complex theory is not limited to the polyphenols but extends to any phytoconstituent. In the present scenario with the advancement in lipid research, the stoichiometric ratio of solvent, phytoconstituents, and phospholipid is a vital factor for the formulation of phytosomes (Bombardelli et al. 1989). This potentiality of phytosomes may have better future prospect and applicability in the field of pharmaceuticals (Maiti et al. 2007; Yanyu et al. 2006).

There are many successful phytosome formulations marketed from the many popular herbal extracts, i.e., green tea, *Ginkgo biloba*, milk thistle, grape seed, ginseng, and hawthorn. The phytoconstituents like flavonoids, polyphenols, terpenoids, and herbal extracts form complex with the phosphatidylcholine directly to form stable formulations as shown in Table 11.1.

11.4 Liposomes

Liposomes are spherical nanosized (50–1000 nm) vesicles architected by arrangement of phospholipid bilayer in polar medium having the ability to incorporate the biological molecules (Fig. 11.3). Liposome is biodegradable and biocompatible, which make the liposomes a safer drug delivery system. Additionally, liposome structure entraps a wide range of molecular weight compounds. The liposomes have the advantage to provide protection against the enzymatic and chemical degradation

Table 11.1 Marketed formulations of phytosomes

Marketed name of phytosomes	Phytoconstituent	Application	Reference
Greenselect®	Epigallocatechin from *Thea sinensis*	Obesity	Di Pierro et al. (2009)
Virtiva®	Flavonoid from *Ginkgo biloba*	Memory improvement	Kennedy et al. (2007)
Silybin Phytosome®	Silybin from silymarin	Hepatoprotective	Khan et al. (2013)
Leucoselect®	Procyanidins	Anticancer	Vigna et al. (2003)
Oleaselect®	Polyphenols	Antihyperlipidemic	Cao et al. (2012)
Meriva®	Curcumin	Diabetes	Appendino et al. (2011)
18ß-Glycyrrhetinic Acid Phytosome®	18ß-Glycyrrhetinic acid	Anti-erythema activity	Bombardelli et al. (1994)

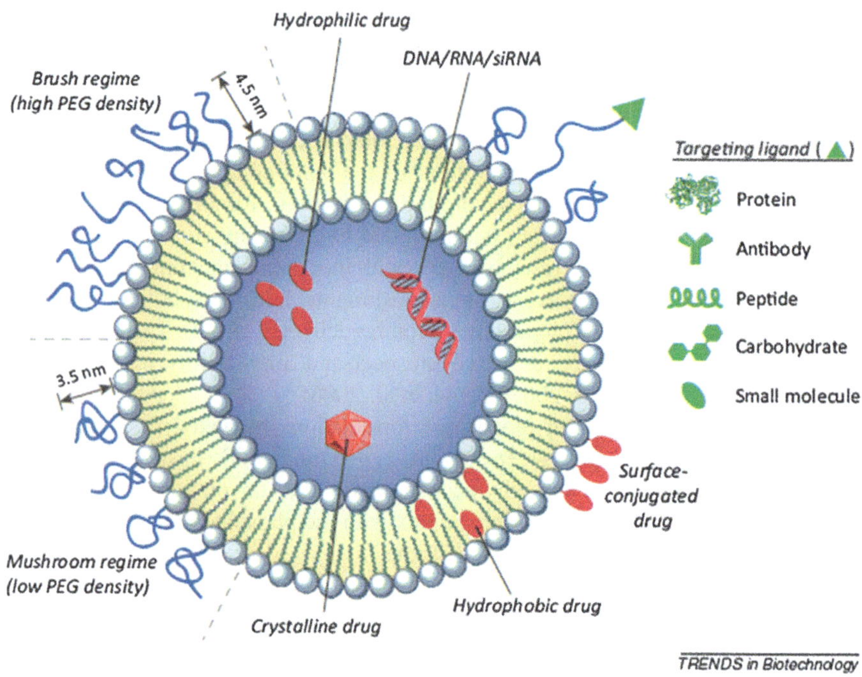

Fig. 11.3 Versatile liposomal formulation with possible modifications

in metabolic process. Liposomes deliver a tolerable amount of drug which helps in reduction of toxicity of drugs, improves pharmacokinetic profile, and enhances therapeutic efficacy (Almeida and Souto 2007; Bozzuto and Molinari 2015; Panahi et al. 2017).

Table 11.2 Liposomal formulation of phytoconstituents

Phytoconstituents/ extract	Type of liposome	Indication	Reference
Ursolic acid	pH-sensitive liposomes	Breast cancer	Caldeira de Araújo Lopes et al. (2013)
Vincristine	pH-gradient liposomes	Leukemia	Boman et al. (1995)
Berberine	Liposomes	Cancer	Sailor et al. (2015)
Thymoquinone	Liposomes	Breast cancer	Odeh et al. (2012)
Taxol	Liposomes	Breast cancer	Yan et al. (2013)
Celastrol	Liposomes	Cancer	Song et al. (2011)
Luteolin	Liposomes	Cancer	Li et al. (2016)
Ginkgo biloba extract	Liposomes	Cardioprotective	Panda and Naik (2008)
Silibinin	Liposomes	Hepatoprotective	Ochi et al. (2016)
Bombax ceiba extract	Liposomes	Hepatoprotective	Karole and Gupta (2019)
Essential oil of *Artemisia arborescens*	Liposomes	Antiviral	Sinico et al. (2005)
Apigenin	Liposomes	Antioxidant	Telange et al. (2017)
Naringenin	Liposomes	Antioxidant	Maiti et al. (2006)
Triptolide	Liposomes	Anti-inflammatory	Chen et al. (2015)
Quercetin	Liposomes	Anxiolytic and cognitive effects	Priprem et al. (2008)

Liposomes are mainly composed of phospholipids, phosphatidylcholine, and/or sphingomyelin. Liposome formulation method involves the mechanical dispersion method, solvent dispersion method, detergent removal method, and dilution method to incorporate the drug molecules for the therapeutic efficacy. Modification of the surface of the conventional liposomes with PVP, PA, cholesterol, and GA increases its circulation time in body and improves the site specificity of drug (Chadha et al. 2008; Mehrabi et al. 2016). The list of liposomal formulation of phytoconstituents is shown in Table 11.2.

11.5 Transfersomes and Ethosomes

Transfersomes resemble liposome vesicles formulated by phospholipids with surface modification. In distinction to liposomes, the transfersomes possess the elasticity which provides the penetration ability through the pores as same as in NPs (Duangjit et al. 2011; Rattanapak et al. 2012). Due to the ultra-flexibility transfersomes can facilitate the delivery of phytoconstituents in the deeper skin tissue but the mechanism is not understood well (Scognamiglio et al. 2013). In the previous reports, it was revealed that the topical absorption of capsaicin, colchicine, and vincristine was improved than the plain one (Xiao-Ying et al. 2006; Zheng et al. 2006). Ethosomes are deformable like transfersomes and enhance the permeability of drug through the skin via *stratum corneum*. It can administer both polar and nonpolar drug molecules with higher concentration in the deeper layer of the skin (Alvi et al. 2011). This can be achieved by using 30–45% of ethanol in ethosome formulation which may enhance the transdermal transport efficiency. Shen et al. (2014) have done an extensive study to enhance the skin deposition of apigenin using the ethosome formulation for anti-inflammatory action post-UV-B damage. Ethosomes are potential therapeutic drug carriers to deliver the poorly absorbed ligustrazine for the treatment of Alzheimer's disease to enhance memory. This designed ligustrazine-loaded ethosome attributed to the significant efficacy of transdermal administration in an in vivo model of amnesia via action on the oxidative system of brain (Shi et al. 2012).

11.6 Solid Lipid-Based Nanoparticles

Solid lipid nanoparticles (SLNs) have emerged as a safe and effective drug delivery system (Pardeike et al. 2009). SLNs are a colloidal system for drug delivery containing physiological lipids (solid at room temperature) that form the core which is stabilized by the emulsifier. Solid lipid nanoparticles typically have a particle size of 50–1000 nm. SLNs are able to overcome the temporal and in vivo stability issues that plague the emulsion and polymeric nanoparticle system (Martins et al. 2012). Additionally, solid state of drug provides better physicochemical stability to the formulation (Pardeike et al. 2009) by reducing the drug mobility which prevents the leakage of drug and also offers better release of drug in a controlled manner due to increase in resistance to mass transfer. Moreover, the formulation also offers other advantages like drug targeting, biocompatibility, prevention of drug degradation, being devoid of organic solvent, and accommodation of hydrophilic and lipophilic drug (Liu and Feng 2015). SLNs have been widely studied for delivery of drugs through dermal, peroral, parenteral, ocular, pulmonary, and rectal routes (Shah et al. 2015). SLNs were formulated to overcome the limitation associated with phytoconstituents with proven therapeutic potentials.

SLNs significantly augment the bioavailability of herbal drug constituents. For instance, the application of quercetin in such conditions, despite its extensive therapeutic effect in various diseases, is limited due to its low bioavailability. Several efforts have been made by researchers to overcome these limitations. Ahmad et al. reported improvement in relative bioavailability with significantly better osteoporotic activity (Ahmad et al. 2016). Similarly, Pandita et al. and Ban et al. have also prepared the solid lipid nanoparticle of resveratrol (Pandita et al. 2014) and curcumin (Ban et al. 2020), respectively, and found that phytoconstituent-loaded formulation produces significant enhancement in bioavailability and therapeutic potential as compared to drug suspension. Ahangarpour et al. formulated SLNs containing myricetin which release the drug in a sustained manner and presented superior in vitro and in vivo antidiabetic activity compared to metformin at low dose (Ahangarpour et al. 2018).

Additionally, lipids used in the SLN formulation like triglycerides, mixed triglycerides, or waxes are biocompatible. Silva et al. have studied the in vitro biocompatibility of SLNs using cytotoxicity assay on Vero and MDCK cell line (Silva et al. 2012a). They showed better biocompatible characteristics of SLNs as compared to metallic nanoparticles. However, these studies present several limitations and further comprehensive work is needed to establish the concrete evidence. Madureira et al. have extensively evaluated the toxicity profile of rosmarinic acid-loaded SLNs using various in vitro and in vivo approaches (Madureira et al. 2016). The formulation was prepared by taking WITEPSOL and carnauba wax as a lipid component. The drug-loaded formulation is devoid of any in vitro cytotoxicity and genotoxicity, and is also found to be safe in an in vivo model.

SLNs can also be able to protect the phytoconstituents which are liable to degrade in gastric environment by circumventing the direct contact and thereby enhancing the stability of drug-loaded formulation. Dwivedi et al. have developed the SLNs containing arteether, a derivative of artemisinin obtained from the herb *Artemisia annua* L. (Dwivedi et al. 2014). Degradation of epoxide ring of arteether in gastric environment restricted its use as an antimalarial drug. However, incorporation of drug in SLNs can be able to solve this problem which preserves the therapeutic potential of drug.

Previously it has been shown that loading of phytoconstituents in SLNs can augment the oral bioavailability. However, the formulation suffers from a problem of burst release in gastric environment which raises the need of another additional approach in which the formulation is protected by a coating layer (Venishetty et al. 2012). Generally, the surface of SLNs is modified by coating of chitosan owing to its biocompatibility, mucoadhesive, and high absorption properties. Recently, Baek and Cho have formulated chitosan (*N*-carboxymethyl chitosan)-coated SLNs and found that burst release of curcumin in gastric pH is inhibited while the formulation shows sustained release in intestinal milieu (Baek and Cho 2017). Likewise, silybin-loaded SLNs coated with chitosan also exhibit excellent properties such as stability, mucoadhesion, sustained release, higher absorption, and cellular uptake (Piazzini et al. 2019). In another investigation, Trotta and his co-worker have prepared resveratrol-loaded solid microparticle coated with chitosan for brain targeting (Trotta et al. 2018). The results show noticeable rise in the bioavailability in CSF without any systematic distribution which represents brain targeting property of formulation via nasal route.

11.7 Nanoemulsion

A nanoemulsion (NE) is a clear biphasic system comprising water and oil and is thermodynamically stable by an interfacial film of surfactant and co-surfactant (Fanun 2012). The nanoemulsion is of microdroplet size ranging from 10 to 100 nm as in Fig. 11.4 (Wu et al. 2009). The nanoemulsion is categorized into two: (1) W/O (water in oil) and (2) O/W (oil in water).

Generally, the natural or synthetic lipids and oils serve as oil phase along with the surfactant and co-surfactant like natural and synthetic PEG (polyethylene

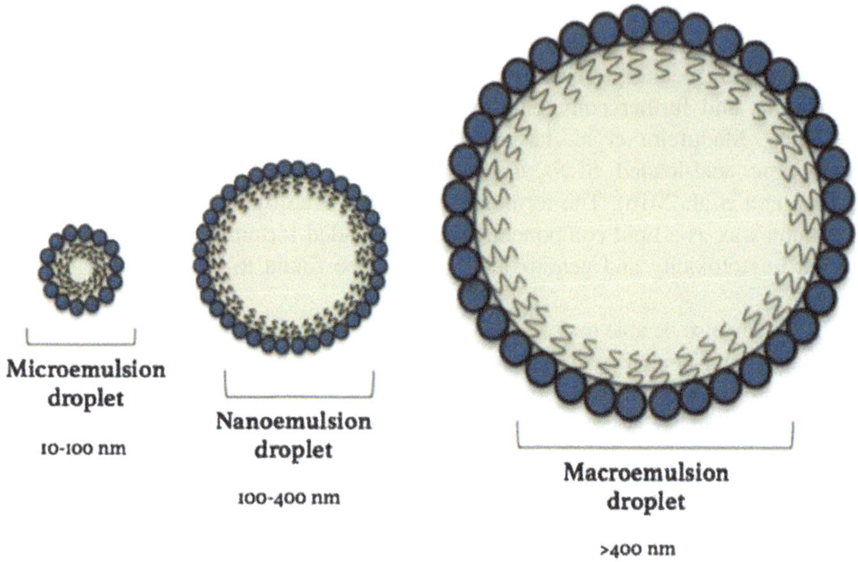

Fig. 11.4 Types and typical droplet diameter sizes of macro-, nano-, and microemulsions

glycol)-conjugated castor oil, lecithin, and triglycerides (Tiwari and Amiji 2006). The NE is an attractive nanocarrier in the pharmaceutical industry because of its transparency, low viscosity, and high kinetic stability (Abolmaali et al. 2011). The industry proposes numerous applications of NE in drug delivery of anticancer drugs, gene delivery, photosensitizer, and diagnosis (Tiwari and Amiji 2006). The NE preparation methods involve high-pressure homogenization, microfluidization, self-nanoemulsion, solvent evaporation technique, and solvent displacement (Mostafa et al. 2017).

The NE has the following advantages (Fanun 2008):

- Increase in the solubility of lipophilic molecules
- Improvement in the permeability and stability
- Enhancement in the bioavailability and absorption of drug
- Controlled release of water-soluble drugs
- Lesser pain during administration due to low viscosity
- Ease in formulation and development
- Versatile for many routes of administration like oral, intravenous, transdermal, ocular, parenteral, and mucosal
- Target and site specificity
- Reduced toxicity of drugs

The NE drug delivery is an emerging approach for the assimilation of phytoconstituents with many beneficial effects. Due to nanodroplet size phytoconstituents are immobilized in the lipid matrix which leads to minimization

Table 11.3 Nanoemulsions for enhancing the efficacy of phytonutrients

Name of species	Phytoconstituent	Application	Reference
Boswellia serrata	Boswellic acid	Anti-inflammatory	Mostafa et al. (2015b)
Glycyrrhiza glabra	Glycyrrhetinic acid	Antioxidant	Mostafa et al. (2014)
Curcuma longa	Curcumin	Anti-inflammatory	Liu et al. (2011)
Wheat bran oil	Tocopherols	Antioxidant	Rebolleda et al. (2015)
Foeniculum vulgare	Phenols and phenolic glycosides	Antidiabetic effect	Mostafa et al. (2015a)
Cuminum cyminum	Cuminaldehyde	Antioxidant and hepatoprotective	Mostafa et al. (2015c)

of separation of phase. It helps in the protection of phytoconstituents from the deterioration through oxidation of fat. It also enhances the permeability and bioavailability of phytoconstituents because of its subcellular size which manifests the enhancement in the passive transport mechanism (Silva et al. 2012b). Examples of NE of plant extract of liquorice, boswellia, essential oil, etc. are listed in Table 11.3.

11.8 Self-Microemulsifying and Self-Nanoemulsifying Drug Delivery Systems

Self-emulsifying systems for drug delivery have arisen as a prospective formulation with enormous potential in augmenting the oral bioavailability of low aqueous solubility molecules, which unless otherwise may not be delivered orally. These systems are anhydrous isotropic mixtures in which drugs are dissolved in the oil phase in the presence of surfactants and co-surfactants which upon dilution with water or gastrointestinal fluid form fine O/W emulsions or microemulsion. This emulsification process requires little agitation provided by motility of the stomach and small intestine (Charman et al. 1992). This system avoids the dissolution step and upon dispersion it forms micro/nanosized globules of oil that give large interfacial area for quick absorption which improves the bioavailability of drug. It is a thermodynamically stable emulsion system compared to unstable regular emulsions and is also able to solubilize more lipophilic drugs. This formulation can also be easily formulated into hard or soft gelatin capsules for easy oral administration.

Generally, the compound obtained from plant possessed well-known pharmacological activity. However, bioavailability is a major concern for the phytoconstituents due to poor solubility and permeability, P-gp efflux, presystemic metabolism, and metabolism by hepatic enzyme. Lipid-based formulations are widely used to solve these problems of drug (Singh et al. 2013). Therefore, drug delivery system plays an important role in phytoconstituent efficacy (Mukherjee 2015).

Self-emulsifying drug delivery systems have gained much popularity in oral drug delivery system due to its ability to overcome the problem associated with phytoconstituents. It can be effectively utilized to improve the solubility- as well as permeability-related issue and hence is applicable to all BCS class drug as well as phytoconstituents with similar properties. For instance, Liu et al. prepared baicalein-loaded SMEDDS to improve the dissolution of drug (Liu et al. 2012). The drug-loaded formulation shows 23% improvement in drug release as compared to baicalein suspension resulting in better absorption and bioavailability. In another study, Setthacheewakul and co-worker developed a self-microemulsifying formulation (liquid and pellet) of curcumin which could be able to improve the dissolution and absorption of curcumin compared to plain suspension of drug (Setthacheewakul et al. 2010). These studies suggest the importance of formulation in the enhancement of dissolution and absorption of orally administered phytoconstituents with poor solubility. Likewise, several studies have been carried out across the globe to augment the solubility, absorption, bioavailability, and efficacy of phytoconstituents, comprising naringenin (Khan et al. 2015), phillygenin (Wang et al. 2020), apigenin (Zhao et al. 2013), limonene (Zhu et al. 2019), phyllanthin (Duc Hanh et al. 2015), puerarin (Cheng et al. 2016), resveratrol (Balata et al. 2016), and curcumin (Shukla et al. 2017).

Furthermore, self-emulsifying formulation also improves the permeability which could be explained by in situ permeability study of piperine-loaded formulation (Shao et al. 2015). The study shows substantial improvement in the rate of effective permeability compared to nonformulated drug. Similarly, ex vivo permeability study of ellagic acid stabilized by the complexation with phospholipid indicates substantial enhancement in permeability across the stomach and intestinal membrane of rats (Avachat and Patel 2015).

The lipids used in the formulation of self-emulsifying formulations also facilitate the lymphatic transport of drug due to the formation of chylomicron by lipid-digested product in the enterocyte. Sato et al. have prepared self-microemulsifying formulation of lutein and evaluated the drug concentration in lymph as well as its distribution in tissue using thoracic lymph cannulation (Sato et al. 2018). The study found significant higher amount of lutein in lymph suggesting the augmentation of lymphatic transport of lutein as compared to plain drug. In another study, chylomicron flow blocking model was used to investigate the lymphatic transport of baicalein-incorporated SMEDDS. The study reported ~2-fold and 2.73-fold rise in the lymphatic transport of baicalein from conventional and modified SMEDDS, respectively, compared to free drug suspension.

In case of efflux problem related to P-gp substrate phytoconstituents, several attempts have been made to modulate P-glycoprotein activity to improve the bioavailability. One such approach is to use those excipients which can modulate the P-gp activity like peppermint oil (AboulFotouh et al. 2017), Miglyol 812, Peceol, Plurol oleique CC 497, Labrasol, Cremophor EL/RH40, Transcutol P, Tween 80, and vitamin E TPGS 1000 (Akhtar et al. 2011). Similarly, SMEDDS can also inhibit presystemic metabolism of drug via inhibition of CYP450 enzymes and also via lymphatic transport of drug. Additionally, excipients like Span 80, Tween

20, and Cremophor RH 40 also inhibit the CYP3A4 activity (Elgart et al. 2013) which may also augment the bioavailability of phytoconstituents metabolized extensively by hepatic enzyme. Lastly, self-emulsifying systems are of stable formulation which can be easily manufactured and scaled up. However there are certain limitations like precipitation of drug, oxidation of lipid, and dearth of predictive in vitro model associated with the formulation which provides scope for the researcher in further development of the formulation.

11.9 Polymeric Nanoparticles

Polymeric nanoparticles (PNPs) are polymeric solid nanosized spherical particles (ranging from 1 to 1000 nm) having the ability to encapsulate into the core or absorb the drug within matrix or conjugate on the surface of particles. The advantages of PNPs are sustained release of drugs, constant serum level concentration, stability, enhanced bioavailability, and low toxicity. The mechanism of release of drug through matrix is diffusion and erosion phenomena (Jain et al. 2013). The most widely used biocompatible and biodegradable polymer candidates are chitosan, dextran, gelatin, alginate, albumin, poly(lactic acid) (PLA), poly(glycolic acid) (PGA), and their copolymer poly(lactide-coglycolide) (PLGA) employed for the formulation of PNPs due to their versatile kinetic profile. The applications of PNPs are in controlled released, targeted delivery, enhancement of bioavailability, etc. (Avgoustakis 2004; Lavasanifar et al. 2002; Mitra et al. 2001; Singer et al. 2004). This versatility of PNPs makes it suitable to overcome the hurdles associated with the delivery of phytoconstituents or herbal extracts in BCS classification category. The PNPs enhance the solubility of drug to reach out at the target site in enough concentration compared to conventional formulations. Also PNPs protect the drug molecules against degradation by enzymes and chemicals and improve the absorption of phytoconstituents or extracts (Prabhu et al. 2010). Various nanoparticles of phytoconstituents or extracts are summarized in Table 11.4.

11.10 Microspheres

Microsphere is of globular particle size ranging from 1 to 300 μm. Microsphere is represented as the matrix of drug. Microsphere is represented as a matrix of drug which is dispersed in polymer. The mechanism of release of drug through the matrix follows the first-order kinetic. Initially, the matrix of polymer is diffuse which then solubilizes the entrapped drug molecules. In another mechanism the surface erosion of polymer leads to the release of drug molecules (Yim et al. 2010). This versatile feature makes it suitable for the drug administration via oral or parenteral route with controlled release and targeted drug delivery. The polymers widely used to fabricate microspheres are natural and synthetic such as dextran, albumin, gelatin, PLA, and PLG (Burgess and Hickey 1994). Various phytoconstituents, i.e., silymarin, ginsenoside, zedoary oil, quercetin, rutin, and camptothecin, have been formulated

Table 11.4 List of phytoconstituents or extracts loaded as PNP formulations

Phytoconstituent or extract	Polymer	Application	Reference
Silymarin	PLGA	Cytotoxicity	Snima et al. (2014)
Oridonin	PLA	Antitumor	Xu et al. (2012)
Polygala senega extract	PLGA	Anticancer	Paul et al. (2011)
Curcumin	PLGA	Anticancer	Yallapu et al. (2010)
Glycyrrhizic acid	PLGA	Anti-inflammatory	Zhang and Ye (2009)
Vincristine	PLGA	Breast cancer	Chen et al. (2014)
Apigenin	PLGA	Skin cancer	Das et al. (2013)
Paclitaxel	PLGA	Osteosarcoma	Wang et al. (2015)
Epigallocatechin 3-gallate	PEG-PLA	Cancer	Siddiqui et al. (2014b)
Etoposide	PLGA/P188	Glioblastoma	Callewaert et al. (2013)
Luteolin	PLA-PEG	Lung cancer	Majumdar et al. (2014)
Cuscuta chinensis extract	PF68	Antitumor	Yen et al. (2008)
Naringenin	PVA-Eudragit	Hepatoprotective	Yen et al. (2009)
Glycyrrhetinic acid	Chitosan	Hepatoprotective	Cheng et al. (2013)
Cucurbitacin	PLA	Anticancer	Zhang et al. (2007)

as microsphere (Garg and Gupta 2010; Liu et al. 2008; Natarajan et al. 2011; Xiao et al. 1994; You et al. 2006).

11.11 Future Prospects

There are significant efforts made for the delivery of phytoconstituents using nanoformulation to enhance the therapeutic potential and to overcome the various obstacles associated with it. In spite of these, significant challenges remain for the implementation of clinically viable therapies in this field. This is because of several obstacles like less reproducibility of nanoparticle formation, low entrapment efficiency, stability of nanoparticles, deformed release kinetics of phytoconstituents and ultimately high cost. In addition, interaction of nanomaterials with biological system, feasibility of scale-up processes, targeting efficiency, toxicity, and biocompatibility are some of the current challenges associated with nanoformulation which should be taken into consideration while developing phytoconstituent formulation.

11.12 Conclusion

Medicinal plants produce several phytoconstituents which are extremely effective for the treatment of numerous diseases. However, their use in medicine is challenging due to poor aqueous solubility, low permeability, poor pharmacokinetic parameters, and potential toxicity. Conventional formulations are unable to overcome these obstacles which lead to low therapeutic efficacy and undesirable toxicity. This chapter focused on the potential of various nanotechnological approaches for augmentation of solubility and bioavailability of phytochemicals which can improve its therapeutic potential. In addition, development of target-based therapy can be leveraged to minimize the toxicity. These nanoformulations have proven their potential as medicine but further development is necessary to tackle the problem associated with phytoconstituents to progress them from laboratory to bedside.

References

Abolmaali SS, Tamaddon AM, Farvadi FS et al (2011) Pharmaceutical nanoemulsions and their potential topical and transdermal applications. Iran J Pharm Res 7(3):139–150

AboulFotouh K, Allam AA, El-Badry M et al (2017) Development and in vitro/in vivo performance of self-nanoemulsifying drug delivery systems loaded with candesartan cilexetil. Eur J Pharm Sci 109:503–513

Ahangarpour A, Oroojan AA, Khorsandi L et al (2018) Solid lipid nanoparticles of myricitrin have antioxidant and antidiabetic effects on streptozotocin-nicotinamide-induced diabetic model and myotube cell of male mouse. Oxidative Med Cell Longev 2018:7496936

Ahmad N, Banala VT, Kushwaha P et al (2016) Quercetin-loaded solid lipid nanoparticles improve osteoprotective activity in an ovariectomized rat model: a preventive strategy for post-menopausal osteoporosis. RSC Adv 6(100):97613–97628

Akhtar N, Ahad A, Khar RK et al (2011) The emerging role of P-glycoprotein inhibitors in drug delivery: a patent review. Expert Opin Ther Pat 21(4):561–576

Almeida AJ, Souto E (2007) Solid lipid nanoparticles as a drug delivery system for peptides and proteins. Adv Drug Deliv Rev 59(6):478–490

Alvi IA, Madan J, Kaushik D et al (2011) Comparative study of transfersomes, liposomes, and niosomes for topical delivery of 5-fluorouracil to skin cancer cells: preparation, characterization, in-vitro release, and cytotoxicity analysis. Anti-Cancer Drugs 22(8):774–782

Apostolova E, Spaseska B, Crcarevska MS et al (2015) An overview of phytosomes as a novel herbal drug delivery system. In: International symposium at Faculty of Medical Sciences, 2015

Appendino G, Belcaro G, Cornelli U et al (2011) Potential role of curcumin phytosome (Meriva) in controlling the evolution of diabetic microangiopathy. A pilot study. Panminerva Med 53(3):43

Avachat AM, Patel VG (2015) Self nanoemulsifying drug delivery system of stabilized ellagic acid–phospholipid complex with improved dissolution and permeability. Saudi Pharm J 23 (3):276–289

Avgoustakis K (2004) Pegylated poly (lactide) and poly (lactide-co-glycolide) nanoparticles: preparation, properties and possible applications in drug delivery. Curr Drug Deliv 1 (4):321–333

Baek JS, Cho CW (2017) Surface modification of solid lipid nanoparticles for oral delivery of curcumin: improvement of bioavailability through enhanced cellular uptake, and lymphatic uptake. Eur J Pharm Biopharm 117:132–140

Balata GF, Essa EA, Shamardl HA et al (2016) Self-emulsifying drug delivery systems as a tool to improve solubility and bioavailability of resveratrol. Drug Des Devel Ther 10:117–128

Ban C, Jo M, Park YH et al (2020) Enhancing the oral bioavailability of curcumin using solid lipid nanoparticles. Food Chem 302:125328

Bangham AD, Standish MM, Watkins JC (1965) Diffusion of univalent ions across the lamellae of swollen phospholipids. J Mol Biol 13(1):238–252

Bhattacharya S (2009) Phytosomes: the new technology for enhancement of bioavailability of botanicals and nutraceuticals. Int J Health Res 2(3):225–232

Bhosale AP, Patil A, Swami M (2016) Herbosomes as a novel drug delivery system for absorption enhancement. World J Pharm Pharm Sci 5:345–355

Boman NL, Bally MB, Cullis PR et al (1995) Encapsulation of vincristine in liposomes reduces its toxicity and improves its anti-tumor efficacy. J Liposome Res 5(3):523–541

Bombardelli E, Curri SB, Della Loggia R et al (1989) Complexes between phospholipids and vegetal derivates of biological interest. Fitoterapia 60:1–9

Bombardelli E, Cristoni A, Morazzoni P (1994) Phytosome® s in functional cosmetics. Fitoterapia 65(5):387–401

Bozzuto G, Molinari A (2015) Liposomes as nanomedical devices. Int J Nanomedicine 10:975

Burgess DJ, Hickey AJ (1994) Microsphere technology and applications. Marcel Dekker, New York

Caldeira de Araújo Lopes S, Vinícius Melo Novais M, Salviano Teixeira C et al (2013) Preparation, physicochemical characterization, and cell viability evaluation of long-circulating and pH-sensitive liposomes containing ursolic acid. Biomed Res Int 2013:1–7

Callewaert M, Dukic S, Van Gulick L et al (2013) Etoposide encapsulation in surface-modified poly (lactide-co-glycolide) nanoparticles strongly enhances glioma antitumor efficiency. J Biomed Mater Res A 101(5):1319–1327

Cao F, Gao Y, Yin Z et al (2012) Enhanced oral bioavailability of oleanolic acid in rats with phospholipid complex. Lett Drug Des Discov 9(5):505–512

Chadha R, Kapoor VK, Thakur D et al (2008) Drug carrier systems for anticancer agents: a review. J Sci Ind Res 67:185–197

Charman SA, Charman WN, Rogge MC et al (1992) Self-emulsifying drug delivery systems: formulation and biopharmaceutic evaluation of an investigational lipophilic compound. Pharm Res 9(1):87–93

Chaudhary T, Chahar A, Sharma JK et al (2015) Phytomedicine in the treatment of cancer: a health technology assessment. J Clin Diagn Res 9(12):XC04–XC09

Chen Y, Yuan L, Zhou L et al (2012) Effect of cell-penetrating peptide-coated nanostructured lipid carriers on the oral absorption of tripterine. Int J Nanomedicine 7:4581–4591

Chen Y, Zheng XL, Fang DL et al (2014) Dual agent loaded PLGA nanoparticles enhanced antitumor activity in a multidrug-resistant breast tumor xenograft model. Int J Mol Sci 15 (2):2761–2772

Chen G, Hao B, Ju D et al (2015) Pharmacokinetic and pharmacodynamic study of triptolide-loaded liposome hydrogel patch under microneedles on rats with collagen-induced arthritis. Acta Pharm Sin B 5(6):569–576

Cheng M, Gao X, Wang Y et al (2013) Synthesis of glycyrrhetinic acid-modified chitosan 5-fluorouracil nanoparticles and its inhibition of liver cancer characteristics in vitro and in vivo. Mar Drugs 11(9):3517–3536

Cheng G, Hu R, Ye L et al (2016) Preparation and in vitro/in vivo evaluation of puerarin solid self-microemulsifying drug delivery system by spherical crystallization technique. AAPS PharmSciTech 17(6):1336–1346

Dai J, Mumper RJ (2010) Plant phenolics: extraction, analysis and their antioxidant and anticancer properties. Molecules 15(10):7313–7352

Das S, Das J, Samadder A et al (2013) Efficacy of PLGA-loaded apigenin nanoparticles in Benzo[a] pyrene and ultraviolet-B induced skin cancer of mice: mitochondria mediated apoptotic signalling cascades. Food Chem Toxicol 62:670–680

Di Pierro F, Menghi AB, Barreca A et al (2009) GreenSelect (R) phytosome as an adjunct to a low-calorie diet for treatment of obesity: a clinical trial. Altern Med Rev 14(2):154

Duangjit S, Opanasopit P, Rojanarata T et al (2011) Characterization and *in vitro* skin permeation of meloxicam-loaded liposomes versus transfersomes. J Drug Deliv 2011:1–9 (418316)

Duc Hanh N, Mitrevej A, Sathirakul K et al (2015) Development of phyllanthin-loaded self-microemulsifying drug delivery system for oral bioavailability enhancement. Drug Dev Ind Pharm 41(2):207–217

Dwivedi P, Khatik R, Khandelwal K et al (2014) Pharmacokinetics study of arteether loaded solid lipid nanoparticles: an improved oral bioavailability in rats. Int J Pharm 466(1–2):321–327

Ekor M (2014) The growing use of herbal medicines: issues relating to adverse reactions and challenges in monitoring safety. Front Pharmacol 4:1–10 (177)

Elgart A, Cherniakov I, Aldouby Y et al (2013) Improved oral bioavailability of BCS class 2 compounds by self nano-emulsifying drug delivery systems (SNEDDS): the underlying mechanisms for amiodarone and talinolol. Pharm Res 30(12):3029–3044

Fanun M (2008) Microemulsions: properties and applications. CRC Press, New York

Fanun M (2012) Microemulsions as delivery systems. Curr Opin Colloid Interface Sci 17 (5):306–313

Farzaei MH, Shahpiri Z, Bahramsoltani R et al (2017) Efficacy and tolerability of phytomedicines in multiple sclerosis patients: a review. CNS Drugs 31(10):867–889

Frozza RL, Bernardi A, Paese K et al (2010) Characterization of trans-resveratrol-loaded lipid-core nanocapsules and tissue distribution studies in rats. J Biomed Nanotechnol 6(6):694–703

Garg R, Gupta GD (2010) Gastroretentive floating microspheres of silymarin: preparation and in vitro evaluation. Trop J Pharm Res 9(1):59–66

Graf BL, Raskin I, Cefalu WT et al (2010) Plant-derived therapeutics for the treatment of metabolic syndrome. Curr Opin Investig Drugs 11(10):1107–1115

Jain AK, Thanki K, Jain S (2013) Co-encapsulation of tamoxifen and quercetin in polymeric nanoparticles: implications on oral bioavailability, antitumor efficacy, and drug-induced toxicity. Mol Pharm 10(9):3459–3474

Kalepu S, Manthina M, Padavala V (2013) Oral lipid-based drug delivery systems—an overview. Acta Pharm Sin B 3(6):361–372

Karole S, Gupta GKGS (2019) Preparation and evaluation of phytosomes containing ethanolic extract of leaves of Bombax ceiba for hepatoprotective activity. Evaluation 6(2):1–5

Kennedy DO, Haskell CF, Mauri PL et al (2007) Acute cognitive effects of standardised Ginkgo biloba extract complexed with phosphatidylserine. Hum Psychopharmacol 22(4):199–210

Khan J, Alexander A, Saraf S et al (2013) Recent advances and future prospects of phyto-phospholipid complexation technique for improving pharmacokinetic profile of plant actives. J Control Release 168(1):50–60

Khan AW, Kotta S, Ansari SH et al (2015) Self-nanoemulsifying drug delivery system (SNEDDS) of the poorly water-soluble grapefruit flavonoid naringenin: design, characterization, in vitro and in vivo evaluation. Drug Deliv 22(4):552–561

Kidd P, Head K (2005) A review of the bioavailability and clinical efficacy of milk thistle phytosome: a silybin-phosphatidylcholine complex (Siliphos). Altern Med Rev 10(3):193–203

Lavasanifar A, Samuel J, Kwon GS (2002) Poly (ethylene oxide)-block-poly (L-amino acid) micelles for drug delivery. Adv Drug Deliv Rev 54(2):169–190

Li J, Cheng X, Chen Y et al (2016) Vitamin E TPGS modified liposomes enhance cellular uptake and targeted delivery of luteolin: an in vivo/in vitro evaluation. Int J Pharm 512(1):262–272

Liu Y, Feng N (2015) Nanocarriers for the delivery of active ingredients and fractions extracted from natural products used in traditional Chinese medicine (TCM). Adv Colloid Interf Sci 221:60–76

Liu CB, Zhang D, Li DG et al (2008) Preparation and characterization of biodegradable polylactide (PLA) microspheres encapsulating ginsenoside Rg3. Chem Res Chin Univ 24(5):588–591

Liu CH, Chang FY, Hung DK (2011) Terpene microemulsions for transdermal curcumin delivery: effects of terpenes and cosurfactants. Colloids Surf B Biointerfaces 82(1):63–70

Liu W, Tian R, Hu W et al (2012) Preparation and evaluation of self-microemulsifying drug delivery system of baicalein. Fitoterapia 83(8):1532–1539

Madureira AR, Nunes S, Campos DA et al (2016) Safety profile of solid lipid nanoparticles loaded with rosmarinic acid for oral use: in vitro and animal approaches. Int J Nanomedicine 11:3621–3640

Maiti K, Mukherjee K, Gantait A et al (2006) Enhanced therapeutic potential of naringenin-phospholipid complex in rats. J Pharm Pharmacol 58(9):1227–1233

Maiti K, Mukherjee K, Gantait A et al (2007) Curcumin–phospholipid complex: preparation, therapeutic evaluation and pharmacokinetic study in rats. Int J Pharm 330(1–2):155–163

Majumdar D, Jung KH, Zhang H et al (2014) Luteolin nanoparticle in chemoprevention: in vitro and in vivo anticancer activity. Cancer Prev Res (Phila) 7(1):65–73

Martins S, Costa-Lima S, Carneiro T et al (2012) Solid lipid nanoparticles as intracellular drug transporters: an investigation of the uptake mechanism and pathway. Int J Pharm 430 (1–2):216–227

Mehrabi M, Esmaeilpour P, Akbarzadeh A et al (2016) Efficacy of pegylated liposomal etoposide nanoparticles on breast cancer cell lines. Turk J Med Sci 46(2):567–571

Mitra S, Gaur U, Ghosh PC et al (2001) Tumour targeted delivery of encapsulated dextran–doxorubicin conjugate using chitosan nanoparticles as carrier. J Control Release 74 (1–3):317–323

Mohanty SK, Swamy MK, Sinniah UR et al (2017) Leptadenia reticulata (Retz.) Wight & Arn. (Jivanti): botanical, agronomical, phytochemical, pharmacological, and biotechnological aspects. Molecules 22(6):1019

Mostafa DM, Ammar NM, Abd El-Alim SH et al (2014) Transdermal microemulsions of Glycyrrhiza glabra L.: characterization, stability and evaluation of antioxidant potential. Drug Deliv 21(2):130–139

Mostafa DM, Abd El-Alim SH, Asfour MH et al (2015a) Transdermal nanoemulsions of Foeniculum vulgare Mill. essential oil: preparation, characterization and evaluation of antidiabetic potential. J Drug Deliv Sci Technol 29:99–106

Mostafa DM, Ammar NM, Basha M et al (2015b) Transdermal microemulsions of Boswellia carterii bird: formulation, characterization and in vivo evaluation of anti-inflammatory activity. Drug Deliv 22(6):748–756

Mostafa DM, Kassem AA, Asfour MH et al (2015c) Transdermal cumin essential oil nanoemulsions with potent antioxidant and hepatoprotective activities: in-vitro and in-vivo evaluation. J Mol Liq 212:6–15

Mostafa DM, Abd El-Alim SH, Kassem AA (2017) Nanoemulsions: a new approach for enhancing phytonutrient efficacy. In: Nanotechnology applications in food. Elsevier, Saint Louis

Mukherjee PK (2015) Evidence-based validation of herbal medicine. Elsevier, Amsterdam

Nankar R, Prabhakar PK, Doble M (2017) Hybrid drug combination: combination of ferulic acid and metformin as anti-diabetic therapy. Phytomedicine 37(15):10–13

Natarajan V, Krithica N, Madhan B et al (2011) Formulation and evaluation of quercetin polycaprolactone microspheres for the treatment of rheumatoid arthritis. J Pharm Sci 100 (1):195–205

Ochi MM, Amoabediny G, Rezayat SM et al (2016) In vitro co-delivery evaluation of novel pegylated nano-liposomal herbal drugs of silibinin and glycyrrhizic acid (nano-phytosome) to hepatocellular carcinoma cells. Cell J 18(2):135–148

Odeh F, Ismail SI, Abu-Dahab R et al (2012) Thymoquinone in liposomes: a study of loading efficiency and biological activity towards breast cancer. Drug Deliv 19(8):371–377

Panahi Y, Farshbaf M, Mohammadhosseini M et al (2017) Recent advances on liposomal nanoparticles: synthesis, characterization and biomedical applications. Artif Cells Nanomed Biotechnol 45(4):788–799

Panda VS, Naik SR (2008) Cardioprotective activity of Ginkgo biloba Phytosomes in isoproterenol-induced myocardial necrosis in rats: a biochemical and histoarchitectural evaluation. Exp Toxicol Pathol 60(4):397–404

Pandita D, Neelam P, Sandeep K et al (2014) Solid lipid nanoparticles enhance oral bioavailability of resveratrol, a natural polyphenol. Food Res Int 62:1165–1174

Pardeike J, Hommoss A, Müller RH (2009) Lipid nanoparticles (SLN, NLC) in cosmetic and pharmaceutical dermal products. Int J Pharm 366(1–2):170–184

Paul S, Bhattacharyya SS, Boujedaini N et al (2011) Anticancer potentials of root extract of *Polygala senega* and its PLGA nanoparticles-encapsulated form. Evid Based Complement Alternat Med 2011:1–13 (517204)

Piazzini V, Cinci L, D'Ambrosio M et al (2019) Solid lipid nanoparticles and chitosan-coated solid lipid nanoparticles as promising tool for Silybin delivery: formulation, characterization, and in vitro evaluation. Curr Drug Deliv 16(2):142–152

Prabhu N, Raj DT, Yamuna GK et al (2010) Synthesis of silver phyto nanoparticles and their antibacterial efficacy. Dig J Nanomater Biostruct 5(1):185–189

Priprem A, Watanatorn J, Sutthiparinyanont S et al (2008) Anxiety and cognitive effects of quercetin liposomes in rats. Nanomedicine 4(1):70–78

Rattanapak T, Young K, Rades T et al (2012) Comparative study of liposomes, transfersomes, ethosomes and cubosomes for transcutaneous immunisation: characterisation and in vitro skin penetration. J Pharm Pharmacol 64(11):1560–1569

Rebolleda S, Sanz MT, Benito JM et al (2015) Formulation and characterisation of wheat bran oil-in-water nanoemulsions. Food Chem 167:16–23

Ruan JQ, Leong WI, Yan R et al (2010) Characterization of metabolism and in vitro permeability study of notoginsenoside R1 from Radix notoginseng. J Agric Food Chem 58(9):5770–5776

Sailor G, Seth AK, Parmar G et al (2015) Formulation and in vitro evaluation of berberine containing liposome optimized by 3^2 full factorial designs. J Appl Pharm Sci 5(7):023–028

Sato Y, Joumura T, Nashimoto S et al (2018) Enhancement of lymphatic transport of lutein by oral administration of a solid dispersion and a self-microemulsifying drug delivery system. Eur J Pharm Biopharm 127:171–176

Scognamiglio I, De Stefano D, Campani V et al (2013) Nanocarriers for topical administration of resveratrol: a comparative study. Int J Pharm 440(2):179–187

Setthacheewakul S, Mahattanadul S, Phadoongsombut N et al (2010) Development and evaluation of self-microemulsifying liquid and pellet formulations of curcumin, and absorption studies in rats. Eur J Pharm Biopharm 76(3):475–485

Shah R, Daniel E, Enzo P et al (2015) Lipid nanoparticles: production, characterization and stability. Springer International Publishing, New York

Shao B, Cui C, Ji H et al (2015) Enhanced oral bioavailability of piperine by self-emulsifying drug delivery systems: in vitro, in vivo and in situ intestinal permeability studies. Drug Deliv 22 (6):740–747

Shen LN, Zhang YT, Wang Q et al (2014) Enhanced in vitro and in vivo skin deposition of apigenin delivered using ethosomes. Int J Pharm 460(1):280–288

Shi J, Wang Y, Luo G (2012) Ligustrazine phosphate ethosomes for treatment of Alzheimer's disease, in vitro and in animal model studies. AAPS PharmSciTech 13(2):485–492

Shukla M, Jaiswal S, Sharma A et al (2017) A combination of complexation and self-nanoemulsifying drug delivery system for enhancing oral bioavailability and anticancer efficacy of curcumin. Drug Dev Ind Pharm 43(5):847–861

Siddiqui AA, Iram F, Siddiqui S et al (2014a) Role of natural products in drug discovery process. Int J Drug Dev Res 6(2):172–204

Siddiqui IA, Bharali DJ, Nihal M et al (2014b) Excellent anti-proliferative and pro-apoptotic effects of (−)-epigallocatechin-3-gallate encapsulated in chitosan nanoparticles on human melanoma cell growth both in vitro and in vivo. Nanomedicine 10(8):1619–1626

Silva AH, Filippin-Monteiro FB, Mattei B et al (2012a) In vitro biocompatibility of solid lipid nanoparticles. Sci Total Environ 432:382–388

Silva HD, Cerqueira MÂ, Vicente AA (2012b) Nanoemulsions for food applications: development and characterization. Food Bioprocess Technol 5(3):854–867

Singer JW, Baker B, de Vries P et al (2004) Poly-(l)-glutamic acid-paclitaxel (CT-2103) [XYOTAX™], a biodegradable polymeric drug conjugate. In: Polymer drugs in the clinical stage. Springer, Boston, MA

Singh SK, Vuddanda PR, Singh S, Srivastava AK (2013) A comparison between use of spray and freeze drying techniques for preparation of solid self-microemulsifying formulation of falsartan and in vitro and in vivo evaluation. Biomed Res Int 909045:1–13

Singh PK, Rawat P (2017) Evolving herbal formulations in management of dengue fever. J Ayurveda Integr Med 8(3):207–210

Sinico C, De Logu A, Lai F et al (2005) Liposomal incorporation of Artemisia arborescens L. essential oil and in vitro antiviral activity. Eur J Pharm Biopharm 59(1):161–168

Snima KS, Arunkumar P, Jayakumar R et al (2014) Silymarin encapsulated poly(D, L-lactic-co-glycolic acid) nanoparticles: a prospective candidate for prostate cancer therapy. J Biomed Nanotechnol 10(4):559–570

Song J, Shi F, Zhang Z et al (2011) Formulation and evaluation of celastrol-loaded liposomes. Molecules 16(9):7880–7892

Swamy MK, Sinniah UR (2016) Patchouli (Pogostemon cablin Benth.): botany, agrotechnology and biotechnological aspects. Ind Crop Prod 87:161–176

Tang L, Feng Q, Zhao J et al (2012) Involvement of UDP-glucuronosyltransferases and sulfotransferases in the liver and intestinal first-pass metabolism of seven flavones in C57 mice and humans in vitro. Food Chem Toxicol 50(5):1460–1467

Telange DR, Patil AT, Pethe AM et al (2017) Formulation and characterization of an apigenin-phospholipid phytosome (APLC) for improved solubility, in vivo bioavailability, and antioxidant potential. Eur J Pharm Sci 108:36–49

Thanki K, Gangwal RP, Sangamwar AT et al (2013) Oral delivery of anticancer drugs: challenges and opportunities. J Control Release 170(1):15–40

Tiwari SB, Amiji MM (2006) Nanoemulsion formulations for tumor-targeted delivery. In: Nanotechnology for cancer therapy. CRC Press, Boca Raton, FL

Trotta V, Pavan B, Ferraro L et al (2018) Brain targeting of resveratrol by nasal administration of chitosan-coated lipid microparticles. Eur J Pharm Biopharm 127:250–259

Venishetty VK, Chede R, Komuravelli R et al (2012) Design and evaluation of polymer coated carvedilol loaded solid lipid nanoparticles to improve the oral bioavailability: a novel strategy to avoid intraduodenal administration. Colloids Surf B Biointerfaces 95:1–9

Vigna GB, Costantini F, Aldini G et al (2003) Effect of a standardized grape seed extract on low-density lipoprotein susceptibility to oxidation in heavy smokers. Metabolism 52 (10):1250–1257

Wang B, Yu XC, Xu SF et al (2015) Paclitaxel and etoposide co-loaded polymeric nanoparticles for the effective combination therapy against human osteosarcoma. J Nanobiotechnol 13(1):22

Wang L, Yan W, Tian Y (2020) Self-microemulsifying drug delivery system of phillygenin: formulation development, characterization and pharmacokinetic evaluation. Pharmaceutics 12 (2):130

Wu H, Lu C, Zhou A et al (2009) Enhanced oral bioavailability of puerarin using microemulsion vehicle. Drug Dev Ind Pharm 35(2):138–144

Wu H, Zhou A, Lu C et al (2011) Examination of lymphatic transport of puerarin in unconscious lymph duct-cannulated rats after administration in microemulsion drug delivery systems. Eur J Pharm Sci 42(4):348–353

Xiao L, Zhang Y, Jianchen XU et al (1994) Preparation of floating rutin-alginate-chitosan microcapsule. Zhong Cao Yao 2

Xiao-Ying L, Luo JB, Yan ZH et al (2006) Preparation and in vitro and in vivo evaluations of topically applied capsaicin transfersomes. Yao Xue Xue Bao 41(5):461–466

Xu J, Zhao JH, Liu Y et al (2012) RGD-modified poly(D,L-lactic acid) nanoparticles enhance tumor targeting of oridonin. Int J Nanomedicine 7:211–219

Yallapu MM, Gupta BK, Jaggi M et al (2010) Fabrication of curcumin encapsulated PLGA nanoparticles for improved therapeutic effects in metastatic cancer cells. J Colloid Interface Sci 351(1):19–29

Yan F, Li L, Deng Z et al (2013) Paclitaxel-liposome–microbubble complexes as ultrasound-triggered therapeutic drug delivery carriers. J Control Release 166(3):246–255

Yanyu X, Yunmei S, Zhipeng C et al (2006) The preparation of silybin–phospholipid complex and the study on its pharmacokinetics in rats. Int J Pharm 307(1):77–82

Yen FL, Wu TH, Lin LT et al (2008) Nanoparticles formulation of Cuscuta chinensis prevents acetaminophen-induced hepatotoxicity in rats. Food Chem Toxicol 46(5):1771–1777

Yen FL, Wu TH, Lin LT et al (2009) Naringenin-loaded nanoparticles improve the physicochemical properties and the hepatoprotective effects of naringenin in orally-administered rats with CCl (4)-induced acute liver failure. Pharm Res 26(4):893–902

Yim ZH, Tiong CB, Mansa RF et al (2010) Release kinetics of encapsulated herbal antioxidants during gelation process. J Appl Sci 10(21):2668–2672

You J, Han X, Wang YS et al (2006) Study of the preparation of sustained-release microspheres containing zedoary turmeric oil by the emulsion–solvent-diffusion method and evaluation of the self-emulsification and bioavailability of the oil. Colloids Surf B Biointerfaces 48(1):35–41

Zhang Q, Ye M (2009) Chemical analysis of the Chinese herbal medicine Gan-Cao (licorice). J Chromatogr A 1216(11):1954–1969

Zhang JF, Hou SX, Liu HL (2007) Comparison of preparing two polylactide nanoparticles loaded lipophilic anti-cancer herb drug by nanoprecipitation method. Zhongguo Zhong Yao Za Zhi 32 (4):303–306

Zhang C, Peng F, Liu W et al (2014) Nanostructured lipid carriers as a novel oral delivery system for triptolide: induced changes in pharmacokinetics profile associated with reduced toxicity in male rats. Int J Nanomedicine 9(1):1049–1069

Zhao L, Zhang L, Meng L et al (2013) Design and evaluation of a self-microemulsifying drug delivery system for apigenin. Drug Dev Ind Pharm 39(5):662–669

Zheng Y, Hou SX, Chen T et al (2006) Preparation and characterization of transfersomes of three drugs in vitro. Zhongguo Zhong Yao Za Zhi 31(9):728–731

Zhu Y, Xu W, Zhang J et al (2019) Self-microemulsifying drug delivery system for improved oral delivery of limonene: preparation, characterization, in vitro and in vivo evaluation. AAPS PharmSciTech 20(4):153

Liposomes: As a Potential Drug Carrier 12

Piyushkumar K. Sadhu, Nirmal Shah, and Hiral M. Manani

Abstract

Since many decades, liposomes have been playing a tremendous role in therapeutic drug delivery system. Liposomes are closed bilayer vesicles which were first discovered in 1965. These possess many clinical properties and due to this, these have become more and more useful in the delivery of many therapeutic drugs. Liposomes are relatively stable and these encapsulate lipophilic, hydrophilic, and/or amphiphilic natured therapeutics in their bilayer lipid as well as aqueous context. Moreover, liposomes are novel drug delivery systems and among various talented new drug therapies, liposomes are characterized as advanced technology. Nowadays, research on liposome therapy has progressed from traditional techniques to new generation. This chapter includes comprehensive updated details on liposome technology with focus on current development in the liposome technology, methodology, characterization of liposomes, various new techniques for the manufacture of liposomes, current therapeutic applications, various marketed products, and patents.

Keywords

Liposomes · Stealth liposome · Novel drug delivery system · Patents · Nanotechnology · Marketed formulations

P. K. Sadhu (✉) · N. Shah
Department of Pharmacy, Sumandeep Vidyapeeth Deemed to be University, Piparia, Vadodara, Gujarat, India
e-mail: piyush.dop@sumandeepvidyapeethdu.edu.in; nirmalshah.dop@sumandeepvidyapeethdu.edu.in

H. M. Manani
Department of Pharmaceutical Sciences, Saurashtra University, Rajkot, Gujarat, India

N. Shah (ed.), *Nanocarriers: Drug Delivery System*,
https://doi.org/10.1007/978-981-33-4497-6_12

12.1 Introduction

The drugs which are available for commercial clinical practice against various diseases are administered in different forms according to the nature of the active medicament. There are certain properties of drugs such as poor solubility, lesser half-life, and low bioavailability. Such properties make them undesirable to conventional dosage forms. Thus, to overcome such limitations, the novel drug delivery systems came in pharmaceutical practice, particularly the vesicular systems like liposomes, ethosomes, niosomes, and nanoparticles.

Among all, the liposome vesicles are formed by self-assembling of two components, the phospholipids and the cholesterol, into bilayers with a nonaqueous core. The concept of liposomal drug delivery system was first given by *Alec D. Bangham, in 1965* (Bangham et al. 1965). He discovered that the phospholipid molecules in contact with water spontaneously form bilayer vesicles. Liposomes are available with a size range from micrometer to nanometer and have been identified as suitable drug carriers (Li et al. 2019) and the encapsulation of hydrophilic and lipophilic drugs in the aqueous and/or lipid phase or bilayer membrane phase has also been reported by using the affinity of different parts of vesicles, and so the liposome was introduced as a suitable drug carrier.

The components such as phosphatidylcholines and phosphatidylglycerols obtained from natural, semisynthetic, or fully synthetic sources, cholesterol, and PEG-ylated phosphatidylethanolamines are the materials commonly used in liposome formulation for drug delivery (Joseph et al. 2018). Choline, the polar group, is a hydrophilic head and glycerol and phosphate are the two hydrophobic tails with essential fatty acid chain. A self-alignment of the natural or synthetic phospholipids is needed to form the bilayer. The charged groups will be facing the aqueous core towards the exterior while the (Li et al. 2010) uncharged groups are within the bilayer interacting with each other. It can entrap both hydrophilic and hydrophobic drugs due to the amphiphilic nature of liposomes. The incorporation of cholesterol helps to increase the distance between the polar heads and thus reduce the interactions between the electrostatic force and hydrogen bonding. It also helps to prevent the leakage of drugs as it gives tight packing of the membrane (Li et al. 2019). Due to their small size, various chemotherapeutic agents like doxorubicin, vincristine, gemcitabine, or cisplatin have unfavorable pharmacokinetics and a suboptimal biodistribution, as exemplified by a short blood half-life and prominent off-target accumulation in multiple healthy organs. Furthermore, the drugs which are encapsulated into the liposome vesicles show reduced kidney excretion and so prolonged blood half-life.

There are several advantages of liposomes as a carrier, such as cell-like membrane structure, high biocompatibility, low immunogenicity, protection of the drugs or active groups, prolongation of drug half-life, reduced toxicity, and increased efficiency. For better liposome delivery systems, structural modification and surface modification by classical lipid molecules have been carried out to generate novel liposomes with specific biological effects, which greatly expand the application of liposomes in biomedicine. Since the first liposomes encapsulating drugs that entered

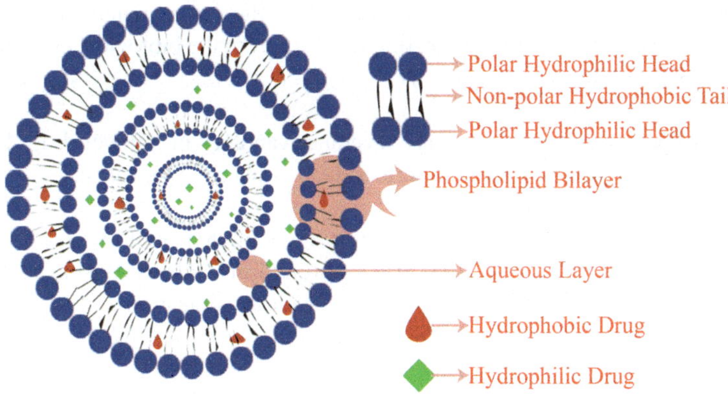

Fig. 12.1 Large multiple layer liposome

the clinical trial stage in 1985 (Li et al. 2019), more than 40 liposome-loaded drugs have been successfully marketed or are in indifferent clinical research stages. The basic structure of multiple layer liposomes is shown in Fig. 12.1.

12.2 Significance and Restraints of Liposomes

The various merits and demerits of liposomes are described as follows (Akbarzadeh et al. 2013; Bulbake et al. 2017; Daraee et al. 2016; Yadav et al. 2017):

There are several advantages of liposomal drug delivery system:

- Liposomal dispersion can be emulsified in a nonaqueous phase. This helps to regulate the delivery rate of various drugs and the administration of normal vesicles can be possible in the external nonaqueous phase.
- Liposome increases the efficacy and therapeutic index of drugs.
- The vesicle suspension is a water-based vehicle. This offers high patient compliance in comparison with oily dosage forms.
- They are osmotically active and stable, as well as increase the stability of the entrapped drug.
- They can be made to reach the site of action by oral, parenteral, as well as topical routes.
- The surfactants are biodegradable, biocompatible, and non-immunogenic.
- Liposomes are capable to incorporate hydrophilic, lipophilic, and amphiphilic moieties together resulting to accommodate the drug molecules with a vast range of solubilities.
- The characteristics of the vesicle formulation are variable and controllable. Altering vesicle composition, size, lamellarity, tapped volume, surface charge, and concentration can control the vesicle characteristics.
- The vesicles may act as a depot, releasing the drug in a controlled manner.

- The therapeutic tendency of the drug molecules increases by delaying the clearance from the circulation by protecting the drug from the biological environment, and restricting effects to target cells.

Though liposomal formulation has several advantages in a novel drug delivery system, there are a few factors as described below that limit its usage:

- The aqueous suspension of liposomes may have limited shelf life due to fusion, aggregation, leaking of entrapped drugs, and hydrolysis of encapsulated drugs.
- The preparations of multilamellar vesicles such as sonication and extrusion are time consuming and require special instruments.
- The formulation has less physical stability.
- Aggregation and fusion may occur in storage under longer and inappropriate conditions.
- It may give rise to chance of leaking of entrapped drug.
- The hydrolysis of encapsulated drug limits the shelf life of dispersion.
- The production cost is high.

12.3 Limitation of Conventional Dosage Forms

Any drug therapy for any disease is one which immediately attains the desired therapeutic concentration of drug in plasma, maintains its constant for the entire duration of treatment, and stands as an ideal dosage regimen (Shaheen et al. 2006). The administration of conventional dosage forms is supposed to do and attain the ideal dosage regimen through a particular dose and at a particular frequency. The frequency of administration or dose interval of any drug depends upon its half-life or mean residence time and its therapeutic index (Nisini et al. 2018). In most cases, dosing interval is much shorter than the half-life of the drug, resulting in some limitations associated with such a conventional dosage form which are as follows: (a) poor patient compliance: increased chances of missing the dose of a drug with a short half-life for which frequent administration is necessary; (b) a typical peak valley plasma concentration-time profile is obtained which makes attainment of steady-state condition difficult; (c) the unavoidable fluctuation in the concentration may lead to undermedication or overmedication as the steady-state concentration (C_{ss}) value falls or rises beyond the therapeutic range (Olusanya et al. 2018). The fluctuating drug level may lead to precipitation of adverse effects especially of a drug with a small therapeutic index whenever overmedication occurs.

12.4 Advantages Over Conventional Dosage Forms

To overcome the above-discussed limitations of conventional dosage forms, there is a need for the development of nonconventional dosage forms. There are two ways to overcome such situations, namely (1) development of new, better, and safer drugs with a long half-life and large therapeutic indices and (2) effective and safer use of existing drugs through concepts and techniques of sustained/controlled and targeted drug delivery systems (Daraee et al. 2016; Zylberberg and Matosevic 2016). Oral controlled/sustained-release dosage forms are being developed since the past three decades due to their advantages. The design of oral controlled/sustained-release drug delivery systems should primarily be aimed at achieving more predictable and increased bioavailability of drugs.

12.5 Classification/Types of Carrier System

On the basis of the liposome vesicle size (varying from very small 0.025 μm to large 2.5 μm) and number of bilayers, liposomes are classified into two categories: (1) multilamellar vesicles (MLV) and (2) unilamellar vesicles with a size range of 20–250 nm. Unilamellar vesicles are further classified into two categories: (1) large unilamellar vesicles (LUV) and (2) small unilamellar vesicles (SUV). Unilamellar liposomes have a single phospholipid bilayer sphere-enclosed aqueous solution. Multilamellar liposome vesicles have an onion-like structure which is characterized by two or more concentric lipid bilayers with a size range of 2–5 μm. Several unilamellar vesicles will form on the inside of the others, making a multilamellar structures of phospholipid separated by a layer of water that preferentially entraps lipid-soluble molecules (Akbarzadeh et al. 2013). Owing to high biocompatibility, biodegradability, low toxicity, and capability to encapsulate hydrophilic and hydrophobic compounds, liposomes constitute the most successful drug carrier system to date. The stability of conventional liposome formulation largely depends on the natural phospholipids and the many other types of phospholipids like phosphatidylcholine and sphingomyelin, and so various critical issues occur as mentioned previously. Due to this many new approaches have been developed to form improved liposomes for therapeutic delivery (Daraee et al. 2016)

Various types of liposomes are depicted in Fig. 12.2.

12.5.1 Stealth Liposomes

Conventional liposome, first-generation liposome, was observed with several limitations such as low stability, short half-life, and rapid clearance after administration. This is only due to the physical interactions of circulating protein in plasma with conventional liposomes and adsorption of protein on the liposome membrane, which then imparts their clearance. Therefore, to overcome such limitations, longer circulating liposome, also called as stealth liposome, was developed by modifying the composition. Further modification was included in size and charge of the regular liposome.

The liposome shell is coated with the inert biocompatible hydrophilic polymer such as polyethylene glycol (PEG), chitosan, and others called as stealth liposomes. These facilitate longer circulation time within the plasma by reducing the interaction

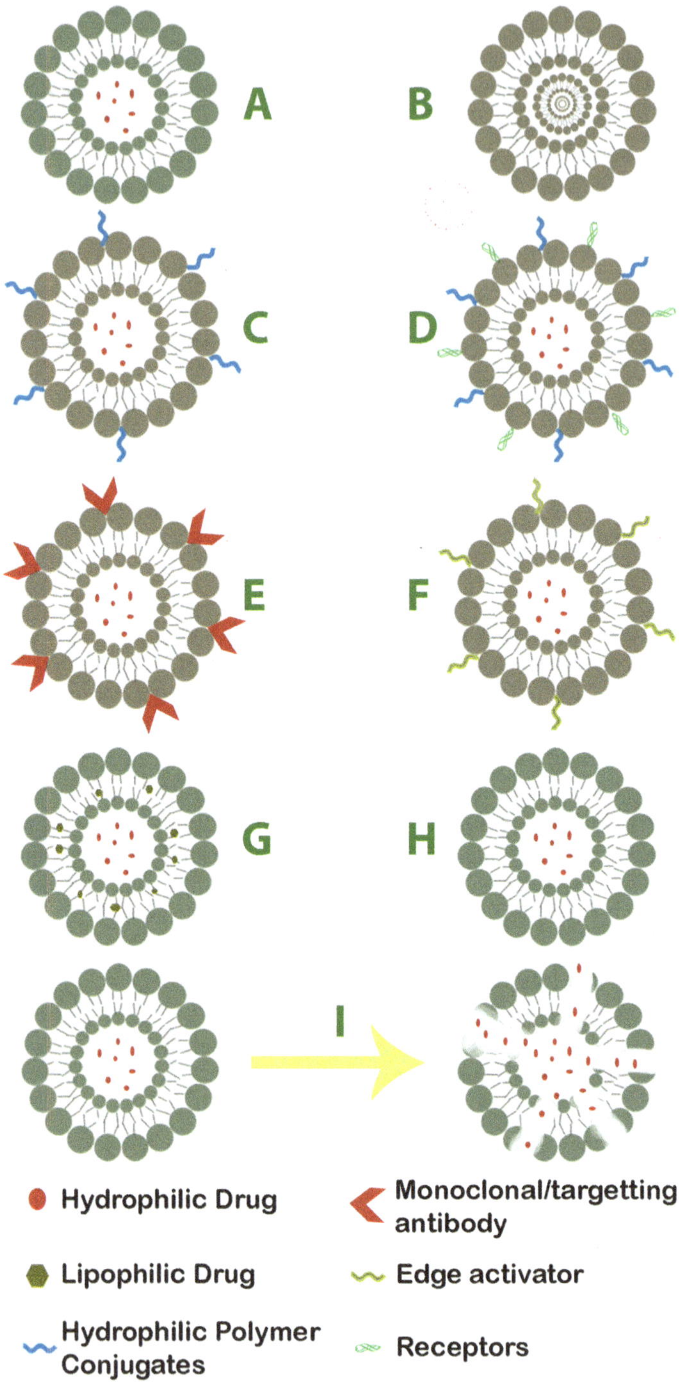

Fig. 12.2 Various types of liposomes as per the specific applications. (**a**) Traditional unilamellar liposome. (**b**) Traditional multilamellar liposome. (**c**) Stealth liposome. (**d**) Targeted liposome. (**e**) Immunoliposome. (**f**) Transfersome. (**g**) Ethosome. (**h**) Pharmacosome. (**i**) Stimulus-responsive liposome (entrapped drug is released on pH or temperature effect)

with the various blood proteins. This also reduces immunogenicity and macrophage uptake. This technology makes liposome capable to escape from phagocytosis, to reduce the toxicity because of the presence of charged lipids, and to increase the half-life within blood circulation. Stealth liposomes also demonstrate non-saturable, dose-independent, log-linear kinetic and increased bioavailability (Zylberberg and Matosevic 2016). Long-circulating liposomes can be prepared with polyethylene glycol chains covalently attached to various hydrocarbon chains. Improved properties of stealth liposomes are due to the properties of polyethylene glycol such as excellent biocompatibility, nonionic nature, and high solubility in aqueous as well as organic media. On the other hand, stealth liposomes have an important disadvantage, i.e., large biodistribution in tissues. This is because of the encapsulation of the liposome shell with bioactive compounds that cannot be selectively distributed to target cells (Nisini et al. 2018).

12.5.2 Targeted Liposomes

To overcome the limitation of stealth liposomes, targeted liposomes are designed. This targeted liposome counterbalances the large body distribution of stealth liposomes. These are characterized by the presence of membranes, functionalized with various glycoproteins, polysaccharides, and/or ligands for specific receptors. This determines the preferential accumulation of liposomes in selected tissues, so that the encapsulated drugs in liposome vesicles can be preferentially released in predetermined time and predetermined target cells or organs (Zylberberg and Matosevic 2016).

12.5.3 Immunoliposomes

The next liposome technology to deliver entrapped drug in the desired tissue or cell is immune liposomes. These immune liposomes are functionalized by antibodies or antibody fragments (Nisini et al. 2018).

12.5.4 Transfersomes

Over the past few years, the transdermal drug delivery system has proven an emerging technology and strategy for drug therapy across a strong barrier: the skin. Transfersomes signify the therapeutics need, which is prevented by the low permeation efficiency of therapeutic drugs across the skin. There are various advantages of transdermal drug therapy (Seth 2019) and to meet these all advantages, a new strategy with a modified structure of conventional liposomes was invented, called as transfersomes. They are ultra-deformable vesicles with high permeation capacity of active ingredients. They were developed to favor skin penetration. This strategy is met by the addition of membrane modifiers, also called edge activators. The edge activators are typically single-chain surfactants with a high radius of curvature. The edge activators allow liposomes to squeeze between the skin layers by destabilizing the lipid bilayers and also by increasing the deformability of membranes (Duangjit et al. 2011a, b).

12.5.5 Ethosomes

Touitou et al. first described ethosomes in 1997. The size of these ethosomes ranged from 10 nm to 10 μm. These were developed to improve the penetration of novel traditional liposomes by utilizing the ethanol's penetrating properties. They are the phospholipid-based elastic novel liposomes containing a high content of ethanol (approx. 20–45%) (Touitou et al. 2000).

12.5.6 Pharmacosomes

Pharmacosomes are amphiphilic phospholipid-drug complexes that bind through covalent, electrostatic, or hydrogen bonding (Semalty et al. 2009). This was first described in 1980. Pharmacosomes are as common as conventional liposomes. Most of the drugs bind through covalent, electrostatic, or hydrogen bonding with the amphiphilic phospholipid and make complexes. Pharmacosomes can exist as an ultrafine micelle or hexagonal aggregates due to their chemical structure (Shivhare et al. 2012).

12.5.7 Stimulus-Responsive Liposomes

A further strategy to deliver an entrapped drug in the desired tissue/cell is represented by stimulus-responsive liposomes. Examples of stimulus-responsive liposomes are the pH-sensitive liposomes (Lu et al. 2014), which undergo conformational and chemical changes in response to acid pH. Another approach in this strategy is temperature-sensitive liposomes (Li et al. 2010), which keep their cargo encapsulated at body temperature, but discharge it upon local heating.

12.6 Materials for the Preparation of Liposomes

Liposomes are the unique novel dosage form whose stability, release kinetics, and behavior depend on the excipient's quality and its composition (Anderson and Omri 2004). It was found that the entrapped drug loss can be minimized by the modification in lipid compositions (Omri and Ravaoarinoro 1998; Gregoriadis 1973).

The permeability and firmness of liposomes are enhanced by the rigidity of the lipid bilayer. The phase transition temperature of lipids has an important impact on

Table 12.1 List of lipids and phospholipids for fabrication of liposomes

Class of lipids	Variety of each class of lipids
Sterols	Cholesterol
Sphingolipids	Sphingomyelin
Charge inducing lipids	Dioctadecyldimethylammonium bromide/chloride (DODAB/C) Dioleoyl trimethyl ammonium propane (DOTAP)
Natural phospholipids	Phosphatidylcholine (lecithin) Phosphatidylethanolamine Phosphatidylglycerol Phosphatidylserine Phosphatidylinositol
Synthetic phospholipids	Distearoyl phosphatidylcholine (DSPC) Dipalmitoyl phosphatidylglycerol (DPPG) Dipalmitoyl phosphatidylcholine (DPPC) Dipalmitoyl phosphatidic acid (DPPA) Dimyristoyl phosphatidylcholine (DMPC) Dipalmitoyl phosphatidylserine (DPPS)
Unsaturated phospholipids	Dioleoyl phosphatidylcholine
Glycosphingolipids	Gangliosides

the selection of lipid as it depends on the acyl chain length of lipid. The lipid having long acyl chain length is mainly used due to high phase transition temperature. However such parameters cannot be achieved by utilizing only one lipid; instead a proper composition of mixed lipids is necessary (Yadav et al. 2011).

Usually, liposome is composed of natural and/or synthetic phospholipids. Liposome may also contain other important components such as cholesterol, hydrophilic polymer-conjugated lipids, and water (Laouini et al. 2012).

Table 12.1 shows an infinite number of lipids and other substances utilized for the preparation of stable liposomes.

Lecithin is probably the most common phospholipid. It is originated in egg yolks, wheat germ, and soybeans. Lecithin is also an important component in the lipid bilayers of cell membranes. Lecithin contains the ammonium salt of choline joined to the phosphate by an ester linkage.

Cholesterol has been mainly used to enhance bilayer properties of the liposomes. It is reported that cholesterol significantly enhances the membrane fluidity and bilayer stability and diminishes the permeability of hydrophilic molecules all the way through the membrane. An obvious benefit of liposomes is the reality that the lipid membrane is composed of physiological lipids which lessen the risk of acute and chronic toxicity.

12.7 Methods of Preparation

General method for preparation of liposomes

The preparation of liposomes mainly involves three basic stages (Sharma et al. 2010):

1. Drying of lipids from an organic solvent
2. Lipid dispersion in aqueous media
3. Separation and purification of liposomes

In this method lipid is dissolved in organic solvent in a proper ratio in a round-bottom flask. The solvent is evaporated and a thin film of lipids is formed on the wall of the round-bottom flask. Drug loading in liposomes can be attained by two different methods. In passive loading, drug is encapsulated during the liposome preparation or in active loading after liposome formation. Organic solvent can be utilized for solubilizing the hydrophobic drug and aqueous media or buffer can be utilized for hydrophilic drug molecule. Hydration of liposomes is attended by using different hydration media on the basis of release profile of the drug. The mixture is first agitated to produce large unilamellar vesicles followed by sonication or extrusion to get small unilamellar vesicles. The general method for liposome preparation is depicted in Fig. 12.3.

The various methods for the preparation of liposomes are described as follows (Kant et al. 2012):

1. *Passive loading techniques*
   (a) Mechanical dispersion
      - Lipid film hydration technique
        - Hand shaking method
        - Non-hand shaking method
      - Microemulsification
      - Sonication
      - French pressure cell
      - Membrane extrusion
      - Dried reconstituted vesicles
      - Freeze-thawed liposomes

   (b) Solvent dispersion
      - Ethanol injection
      - Ether injection
      - Double-emulsion method
      - Reverse-phase evaporation

   (c) Detergent removal technique
      - Detergent removal from mixed micelles by dialysis
2. *Active loading techniques*

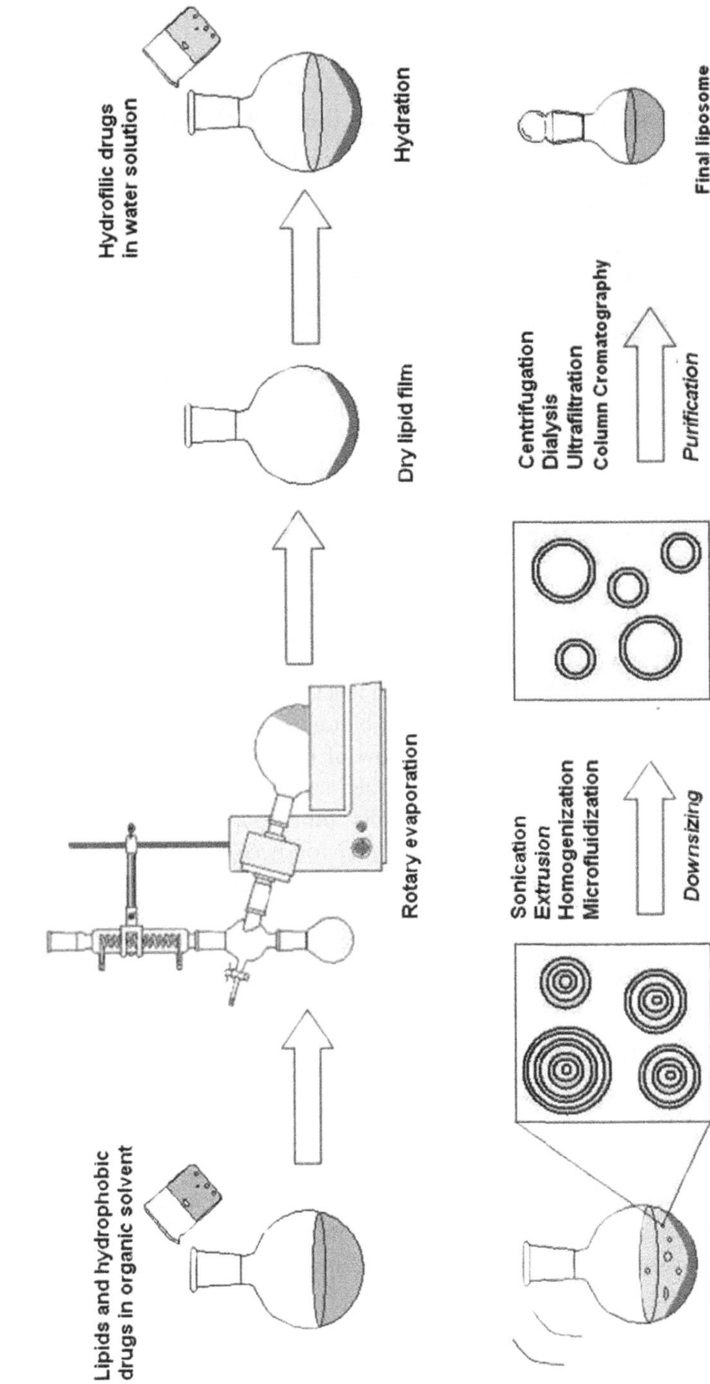

Fig. 12.3 Method of preparation by thin-film hydration (Lopes et al. 2013)

12.7.1 Passive Loading Techniques

It is classified on the basis of mode of dispersion as follows.

12.7.1.1 Mechanical Dispersion Methods (Kant et al. 2012)

12.7.1.1.1 Lipid Film Hydration Technique

Hand Shaking Method

Hand shaking method is the simplest and frequently used method. The mixture of lipids and drug is dissolved into the organic solvent such as chloroform, methanol, and ethanol in different ratios in a round-bottom flask. The flask is attached to a rotary evaporator connected with vacuum pump and rotated with specified rpm. The flask is rotated till the complete evaporation of solvent and the dry thin film is formed on the surface of the round-bottom flask. A specified volume of hydrating media is added to the flask for the hydration of lipid film. The flask is rotated for a specific period of time at fixed rpm till the lipid film has been removed from the wall of the flask. Finally milky white suspension is formed and allowed to stand for 2–3 h to complete the swelling process.

Non-hand Shaking Method

This method is same as the shaking method but a special care is taken during the swelling process. In this method, lipid is taken in a specified ratio of organic solvent as discussed above and spread on to the flat-bottom conical flask. The solution is evaporated at room temperature with the flow of nitrogen without disturbing the flask solution. On completion of drying, the water-saturated nitrogen is again passed through the flask until the opacity of the dry film disappears. Addition of the hydration media may result in swelling of lipid followed by formation of milky white suspension. This method is used for the preparation of multilamellar vesicles. To convert the MLVs to LUVs the following procedures are used.

12.7.1.1.2 Microemulsification of Liposomes

Microfluidizer is used to prepare small vesicles from the concentrated lipid suspension. Microfluidizer pumps the fluid at very high pressure of around 10,000–12,000 psi through a 5 μm orifice. It is forced along defined microchannels which direct two streams of fluid that collide together at the right angles at a very high velocity, thereby affecting an efficient transfer of energy. Lipids can be introduced into the microfluidizer as a slurry of a hydrated lipid in organic medium or multilamellar vesicle dispersion. The fluid collected can be recycled through the pump until vesicles of spherical dimensions are obtained. The diameter changes after a single cycle pass and the size of vesicles is reduced to 0.1–0.2 μm (Dwivedi and Verma 2013).

12.7.1.1.3 Sonication

In this method multilamellar vesicles are transformed to the small-unit lamellar vesicles. The ultrasonic irradiation is provided to the MLVs to get SLVs.

There are two methods: (1) bath sonication and (2) probe sonication.

1. Bath sonication
   In bath sonication, multilamellar vesicles are converted into the smaller lamellar vesicles. In this method ultrasonic irradiation is provided to the sample for reducing the size. The water level in the bath sonicator should be maintained properly for the experiment to be conducted. The bath sonicator shall be covered properly with the lid during the use of organic solvents.
2. Probe sonication
   This probe sonicator is used for dispersion that mainly requires high energy in small volume containing high concentration of lipids or viscous dispersion. Probe sonicator provides high energy to the dispersion that leads to heat production causing size reduction in liposomal formulation. The tip of the sonicator also releases titanium into the dispersion; it can be removed by centrifugation before use. Due to this reason, bolt connectors are widely used. MLV sonication is accomplished by putting dispersion into the bath sonicator or taking the dispersion in test tube and probe sonicating for 5–10 min. The resulting dispersion is centrifuged and vesicles are settled as per their sizes (Dwivedi and Verma 2013).

12.7.1.1.4 French Pressure Cell

French pressure cell method works on the mechanism of high pressure. This will give the unilamellar or oligo-lamellar liposomes of medium size of 30–80 nm. Liposomes produced from this method are more stable as compared to the liposomes produced from the sonication. This method is costly as compared to sonication method.

12.7.1.1.5 Membrane Extrusion Technique

This method is mainly used to convert MLVs to SLVs. In this method, size of liposomes is reduced by passing them through different sizes of membrane filters at a lower pressure of <100 psi. Liposomes prepared by this method are termed as LUVETs. This method is mainly used for the production of large unilamellar vesicles and small lamellar vesicles for the in vivo and in vitro studies (Vyas and Khar 2007).

12.7.1.1.6 Dried Reconstituted Vesicles (DRVs)

Liposomes prepared by freeze-drying technique are reconstituted for further use and converted into uni- or oligo-lamellar vesicles. Whenever freeze-drying technique is used the freeze-dried membrane is utilized for the purpose. This dried membrane is rehydrated with the use of aqueous media containing the drug to be entrapped. This leads to the formation of small lamellar vesicles.

12.7.1.1.7 Freeze-Thaw Sonication

This method is based on the freezing of unilamellar dispersion followed by thawing of dispersion at room temperature for 15 min and subjecting to a sonication which reduces the permeability of the liposome membrane. This technique has modified the vesicle size up to less than 10–50 μm to incorporate a dialysis step against hypo-osmolar buffer in the place of sonication. Freeze-thaw method is simple, rapid, and useful for the study of membrane transport phenomenon (Dwivedi and Verma 2013).

12.7.1.2 Solvent Dispersion Method

12.7.1.2.1 Ether and Ethanol Injection Method

Ethanol injection method was first described in 1973. In ethanol method, lipids are dissolved into the organic solvent followed by injection into an aqueous solvent. Liposome prepared by ethanol injection method is of less than 100 nm size.

Ether injection method is different as compared to ethanol injection method because ether is immiscible with aqueous solvent, so as the solvent is removed from the liposomal formulation by heating the liposomal dispersion on hot plate. In this method, ether-lipid solution is added into warmed aqueous phase above the boiling point of the ether. Ether vaporizes upon contacting with the warm aqueous phase and the dispersed lipid forms into unilamellar liposomes. This technique is rapid, simple, reproducible, and easy to get small lamellar vesicles. These techniques also require a large amount of organic solvents, which are harmful for both the environment and humans and also require residual solvent method for complete removal of solvent (Vyas and Khar 2007).

12.7.1.2.2 Double-Emulsion Method

This method is emulsification of an organic solution in water in which multicompartment vesicles are formed. The organic solution comprising water droplet is introduced into an excess aqueous phase by mechanical dispersion method. The obtained dispersion is called W/O/W system. Small proportion of water droplets are collapsed by continuous shaking using mechanical vortex mixer. The vesicle forms are unilamellar and have a diameter of 0.3–0.5 μm (Dwivedi and Verma 2013).

12.7.1.2.3 Reverse-Phase Evaporation Technique

This is one of the methods that provided a breakthrough in liposome formulation development. This method allows liposomes with more aqueous space-to-lipid ratio and allows entrapping a high percentage of the hydrophilic drug. This method is based on the micelle formation. The mixture consists of aqueous phase containing water-soluble molecule and an organic solvent in which lipids are solubilized with the help of sonication. Slow removal of organic solvent leads to transformation of micelles into a gel-like viscous dispersion. Due to excess of lipids, formation of a complete bilayer around the micelles takes place, which results in the formation of liposomes. This method is mainly used to entrap large, small, and macromolecules. This method has a drawback of denaturation of some proteins due to exposure to sonication for a longer period of time. Modified reverse-phase evaporation technique was presented by Handa et al. (2006), and found to have high encapsulation efficiency (Handa et al. 2006; Dwivedi and Verma 2013).

12.7.1.3 Detergent Removal Methods

12.7.1.3.1 Detergent Removal from Mixed Micelles by Dialysis

Phospholipid-containing mixtures can form large unilamellar vesicles upon removal of nonionic detergent using selected adsorbents.

Dialysis

The detergents at their critical micelle concentration have been used to solubilize the lipids. When the detergent is segregated, the micelles become better off in phospholipid and at last associated to form LUVs. The detergents are removed by the dialysis technique (Daemen et al. 1995). Dialysis is performed by using the dialysis bags in large amount containing buffer (Shaheen et al. 2006). The detergent is removed by shaking the micelle solution using organic polystyrene beads, XAD-2 beads, and bio-beads (Akbarzadeh et al. 2013).

Gel Permeation Chromatography

In this method, detergent is utilized on the basis of size of the chromatography. Sephadex G-50 or Sephadex G-100 can be used for the gel filtration chromatography. In this method the liposomes can percolate from the inter-bead spaces (Alpes et al. 1986).

12.8 Reported Scale-Up Techniques of Liposome Preparation

The following methods are successfully simplified for the large-scale production of liposomes (Huang et al. 2014):

1. Spray drying method
2. Supercritical anti-solvent technique (SAS)
3. Supercritical reverse-phase evaporation (SPER) method
4. High-pressure homogenizer

12.9 Characterization of Liposomes

12.9.1 Vesicle Shape and Lamellarity

Vesicle shape and bilayers present in liposomes can be found by P-31 nuclear mass spectroscopy, freeze electron microscopy, and inverted microscopy. In the NMR technique the lamellarity is recorded before and after with the addition of manganese ions that interact with bilayers (Kant et al. 2012).

12.9.2 Particle Size, Polydispersity Index, and Zeta Potential

Particle size and polydispersity are measured by zeta sizer which works on the light scattering principle, called as laser light scattering technique. This method is very simple and rapid as compared to other methods. The drawback of this method is that it measures the particle size only in bulk dispersion. Zeta sizer also measures the zeta potential which is of prime importance as far as the stability of liposomes is concerned (Sharma 1994).

12.9.3 Entrapment Efficiency

It describes the amount of drug ultimately entrapped during the formation of liposomes and is expressed as % entrapment. The entrapment efficiency is obtained by two different methods; one is minicolumn centrifugation method and the other is protamine aggregation method.

The entrapment efficiency is calculated by using the following formula:

$$\%\ \mathrm{EE} = \left(C_{\mathrm{entrapped}}\right)/\left(C_{\mathrm{total}}\right) \times 100 \tag{12.1}$$

where

$C_{\mathrm{entrapped}}$ = Concentration of drug in liposomes
C_{total} = Concentration of drug added into the liposomes

12.9.4 In Vitro Drug Release Study

The dialysis sac method is used to check the drug release from liposomal formulation. The dialysis sac undergoes pretreatment before the usage. It is soaked overnight in distilled water. The formulation dispersion is placed in an appropriate-sized sac which is tied from both ends with the help of a thread. The aliquots are taken at specific time interval and the drug content is measured using sophisticated analytical instrument.

12.10 Therapeutic Application of Liposomes

12.10.1 Parenteral Drug Delivery

For parenteral administration the liposomal formulation can be utilized for a variety of routes such as intramuscular, intravenous, and intraperitoneal. List of approved parenteral liposomal preparation is given in Table 12.2.

Table 12.2 List of approved parenteral formulation (Mahfoozur et al. 2017)

Liposomal products and year	Route of administration	Active ingredient	Lipid composition	Indication for use
Doxil (1995)	IV	Doxorubicin	HSPC: cholesterol:PEG 2000-DSPE (56:39:5 molar ratio)	Ovarian, breast cancer, Kaposi's sarcoma
DaunoXome (1996)	IV	Daunorubicin	DSPC and cholesterol (2:1 molar ratio)	AIDS-related Kaposi's sarcoma
Depocyt (1999)	Spinal	Cytarabine/ Ara-C	DOPC, DPPG, cholesterol, and triolein	Neoplastic meningitis
Myocet (2000)	IV	Doxorubicin	EPC:cholesterol (55:45 molar ratio)	Combination therapy with cyclophosphamide in metastatic breast cancer
Mepact (2004)	IV	Mifamurtide	DOPS:POPC (3:7 molar ratio)	High-grade, resectable, nonmetastatic osteosarcoma
Marqibo (2012)	IV	Vincristine	SM:cholesterol (60:40 molar ratio)	Acute lymphoblastic leukemia
Onivyde (2015)	IV	Irinotecan	DSPC:MPEG-2000:DSPE (3:2:0.015 molar ratio)	Combination therapy with fluorouracil and leucovorin in metastatic adenocarcinoma of the pancreas
Abelcet (1995)	IV	Amphotericin B	DMPC:DMPG (7:3 molar ratio)	Invasive severe fungal infections
Ambisome (1997)	IV	Amphotericin B	HSPC:DSPG: cholesterol: amphotericin B (2:0.8:1:0.4 molar ratio)	Presumed fungal infections
Amphotec (1996)	IV	Amphotericin B	Cholesteryl sulfate: amphotericin B (1:1 molar ratio)	Severe fungal infections
Visudyne (2000)	IV	Verteporfin	Verteporfin: DMPC and EPG (1:8 molar ratio)	Choroidal neovascularization
DepoDur (2004)	Epidural	Morphine sulfate	DOPC, DPPG, cholesterol, and triolein	Pain management

(continued)

Table 12.2 (continued)

Liposomal products and year	Route of administration	Active ingredient	Lipid composition	Indication for use
Exparel (2011)	IV	Bupivacaine	DEPC, DPPG, cholesterol, and tricaprylin	Pain management
Epaxal (1993)	IM	Inactivated hepatitis A virus (strain RGSB)	DOPC:DOPE (75:25 molar ratio)	Hepatitis A
DaunoXome (1996)	IV	Daunorubicin	DSPC and cholesterol (2:1 molar ratio)	AIDS-related Kaposi's sarcoma
Depocyt (1999)	Spinal	Cytarabine/ Ara-C	DOPC, DPPG, cholesterol, and triolein	Neoplastic meningitis
Myocet (2000)	IV	Doxorubicin	EPC:cholesterol (55:45 molar ratio)	Combination therapy with cyclophosphamide in metastatic breast cancer
Mepact (2004)	IV	Mifamurtide	DOPS:POPC (3:7 molar ratio)	High-grade, resectable, nonmetastatic osteosarcoma
Marqibo (2012)	IV	Vincristine	SM:cholesterol (60:40 molar ratio)	Acute lymphoblastic leukemia
Onivyde (2015)	IV	Irinotecan	DSPC:MPEG-2000:DSPE (3:2:0.015 molar ratio)	Combination therapy with fluorouracil and leucovorin in metastatic adenocarcinoma of the pancreas
Abelcet (1995)	IV	Amphotericin B	DMPC:DMPG (7:3 molar ratio)	Invasive severe fungal infections
Ambisome (1997)	IV	Amphotericin B	HSPC:DSPG: cholesterol: amphotericin B (2:0.8:1:0.4 molar ratio)	Presumed fungal infections
Amphotec (1996)	IV	Amphotericin B	Cholesteryl sulfate: amphotericin B (1:1 molar ratio)	Severe fungal infections
Visudyne (2000)	IV	Verteporfin	Verteporfin: DMPC and EPG (1:8 molar ratio)	Choroidal neovascularization

(continued)

Table 12.2 (continued)

Liposomal products and year	Route of administration	Active ingredient	Lipid composition	Indication for use
DepoDur (2004)	Epidural	Morphine sulfate	DOPC, DPPG, cholesterol, and triolein	Pain management
Exparel (2011)	IV	Bupivacaine	DEPC, DPPG, cholesterol, and tricaprylin	Pain management
Epaxal (1993)	IM	Inactivated hepatitis A virus (strain RGSB)	DOPC:DOPE (75:25 molar ratio)	Hepatitis A

12.10.2 Oral Drug Delivery

Oral route is convenient for patients as compared to other routes and it also increases patient compliance. Conventional dosage forms have limitation due to first-pass metabolism and thereby the bioavailability may get reduced. Many researches are available that provide evidences of liposomes for efficient oral administration. Liposomes increase the lymphatic circulation and due to that the bioavailability of drug entrapped in liposomes increases (Akbarzadeh et al. 2013).

12.10.3 Pulmonary Drug Delivery

Liposomes are nanocarriers so with the help of a nebulizer it can be used as the pulmonary drug delivery system. Liposomes have advantages for pulmonary drug delivery due to its high stability, small particle size, and high affinity to control the delivery of drug to the lung tissue with high absorption and local action in the respiratory tract. Pulmonary delivery of liposomal formulation can be achieved using the following inhalers (Kant et al. 2012):

(a) Dry-powder inhalers
(b) Pressurized metered-dose inhalers
(c) Nebulizers

12.10.4 Transdermal Drug Delivery

In liposomes, phospholipids play a vital role in efficient vesicular drug delivery. Lipids enhance the penetration power and diffusion of drug through the skin and avoid first-pass metabolism. It provides sustained-release dosage form of drug which has low half-life and poor solubility and improves bioavailability of drugs such as aceclofenac and nicotine (Dwivedi and Verma 2013).

Table 12.3 List of liposomal marketed products (Nekkanti and Kalepu 2015)

Name of active pharmaceutical ingredient	Brand name	Name of pharmaceutical industry	Application of product
Doxorubicin hydrochloride	Lipodox (PEGylated liposomes)	Sun Pharmaceutical Ind. Ltd., India	Metastatic cancer
Mitoxantrone	LEM ETU	Neo Pharm Inc., USA	Antineoplastics
Cisplatin	Lipoplatin™ (liposomal injection)	Regulon Inc., Greece	Pancreatic cancer
Doxorubicin hydrochloride	Myocet (liposomal powder)	Teva Pharma B. V., India	Metastatic cancer
Amphotericin B	Amphotec	Sequus Pharmaceuticals Inc., USA	Fungal infections with leishmaniasis
Doxorubicin hydrochloride	Doxil (STEALTH® liposomal injection)	Ben Venue Laboratories Inc., USA	Ovarian cancer, AIDS-related Kaposi's sarcoma, multiple myeloma
Cisplatin	Nanoplatin	Regulon Inc., Greece	Lung cancer
Daunorubicin citrate	DaunoXome™	NeXstar Pharmaceuticals Inc., USA	Kaposi's sarcoma in AIDS
Amikacin	Mikasome®	NeXstar Pharmaceuticals Inc., USA	Bacterial infection
Prostaglandin-E1	Ventus	The Liposome Company, USA	Systemic inflammatory disease
Estradiol	Estrasorb	Novavax, USA	Menopausal therapy
Daunorubicin citrate	Daunosome®	Galen Ltd., UK	Kaposi's sarcoma in AIDS

12.10.5 Mucosal Drug Delivery

Lipids used in liposomes have nontoxic and nonirritant nature with high affinity for the mucosal membrane. Liposomes provide prolong-release drug delivery for vaginal and nasal routes and increase the drug retention at the site of action.

12.11 Marketed Formulations of Liposomes

Owing to certain tremendous advantages of liposomes, it is used frequently in the commercial market as an efficient nanocarrier for the treatment of a variety of diseased conditions as shown in Table 12.3.

Table 12.4 List of recent patents of liposomes

Patent/publication number	Title of patent	Name of inventor/s	Year of publication	Reference
JPWO2018181963A1	Liposome composition and pharmaceutical composition	Kasagi et al.	2020	Kasagi et al. (2020)
WO2018127016A1	Light-responsive liposome, preparation method, and application thereof	Zhang et al.	2018	Zhang et al. (2018)
WO2018045094A1	Ladderane lipid compounds and liposomes and methods of preparing and using the same	Burns et al.	2018	Burns et al. (2018)
WO2018177140A1	Use of liposome for treatment of chronic viral hepatitis B	Wu	2018	Wu (2018)
WO2018136002A1	Hyperstabilized liposomes increase targeting of mitotic cells	Ng and Cheong	2018	Ng and Cheong (2018)
WO2018033118A1	All-trans-retinoic acid liposome preparation, and preparation and application thereof	Xu et al.	2018	Xu et al. (2018)
WO2018011705A1	Liposome-based eye drops and use thereof for in vivo evaluation of the drug efficacy of medical and surgical antiglaucoma therapy	Ambrosone et al.	2018	Ambrosone et al. (2018)
WO2018193458A1	Liposome compositions and uses thereof	Katz et al.	2018	Katz et al. (2018)
WO2018150429A1	Liposomal drug delivery vehicles	Klein et al.	2018	Klein et al. (2018)
WO2017034418A1	Acoustic driven drug delivery systems	Reynolds et al.	2017	Reynolds et al. (2017)
WO2017115381A1	Degradable or transformable gold-coated liposomal nano-construct and a process for its preparation	Srivastava et al.	2017	Srivastava et al. (2017)
WO2017115358A1	Liposomes for treatment of an autoimmune disease	Baru	2017	Baru (2017)
EP1746976B1	Liposomes useful for drug delivery	Hong et al.	2017	Hong et al. (2017)
WO2016013031A1	Liposome composition and method of preparing the liposome	Bhowmick et al.	2016	Bhowmick et al. (2016)

(continued)

Table 12.4 (continued)

Patent/publication number	Title of patent	Name of inventor/s	Year of publication	Reference
WO2015090572A1	Survivin-directed cancer vaccine therapy	Geissler et al.	2015	Geissler et al. (2015)
EP2680820A1	Advanced active liposomal loading of poorly water-soluble substances	Gaillard et al.	2014	Gaillard et al. (2014)
EP2682106A1	Method of solubilizing biologically active compounds	Leigh et al.	2014	Leigh et al. (2014)

12.12 Patents Related to Liposomes

Various patents related to liposomes are shown in Table 12.4.

References

Akbarzadeh A, Rezaei-Sadabady R, Davaran S et al (2013) Liposome: classification, preparation, and applications. Nanoscale Res Lett 8(1):102

Alpes H, Allmann K, Plattner H et al (1986) Formation of large unilamellar vesicles using alkyl maltoside detergents. Biochim Biophys Acta 862:294

Ambrosone L, Costagliola C, Zeppa L et al (2018) Liposome-based eye drops and use thereof for in vivo evaluation of the drug efficacy of medical and surgical anti-glaucoma therapy. WO2018011705, 18 Jan 2018

Anderson M, Omri A (2004) The effect of different lipid components on the in vitro stability and release kinetics of liposome formulations. Drug Deliv 11:33–39

Bangham AD, Standish MM, Watkins JC (1965) Diffusion of univalent ions across the lamellae of swollen phospholipids. J Mol Biol 13(1):238–252

Baru M (2017) Liposomes for treatment of an autoimmune disease. WO2017115358, 6 Jul 2017

Bhowmick S, Umrethia MK, Maiti K et al (2016) Liposome composition and method of preparing the liposome. WO2016013031, 28 Jan 2016

Bulbake U, Doppalapudi S, Kommineni N et al (2017) Liposomal formulations in clinical use: an updated review. Pharmaceutics 9(2):12

Burns N, Shuken SR, Mercer JAM et al (2018) Ladderane lipid compounds and liposomes and methods of preparing and using the same. WO2018045094, 8 Mar 2018

Daemen T, Hofstede G, Ten Kate MT et al (1995) Liposomal doxorubicin induced toxicity: depletion and impairment of phagocytic activity of liver macrophages. Int Cancer 61:761–721

Daraee H, Etemadi A, Kouhi M et al (2016) Application of liposomes in medicine and drug delivery. Artif Cells Nanomed Biotechnol 44(1):381–391

Duangjit S, Opanasopit P, Rojanarata T et al (2011a) Characterization and in vitro skin permeation of meloxicam-loaded liposomes versus transfersomes. J Drug Deliv 2011:418316

Duangjit S, Opanasopit P, Rojanarata T et al (2011b) Effect of edge activator on characteristic and in vitro skin permeation of meloxicam loaded in elastic liposomes. Adv Mater Res 194–196:537–540

Dwivedi C, Verma S (2013) Review on preparation and characterization of liposomes with application. J Sci Innov Res 2(2):486–508

Gaillard PJ, Appeldoorn CCM, Rip J (2014) Advanced active liposomal loading of poorly water-soluble substances. European Patent 2680820, 8 Jan 2014
Geissler S, Boniforte P, Plaschke J et al (2015) Survivin-directed cancer vaccine therapy. WO2015090572A1, 25 Jun 2015
Gregoriadis G (1973) Drug entrapment in liposomes. FEBS Lett 36(3):292–296
Handa T, Naito S, Hiramatsu M et al (2006) Thermal SiO and $H^{13}CO^{+}$ line observations of the dense molecular cloud G0.11-0.11 in the galactic center region. Astrophys J 636:261–266
Hong K, Drummond DC, Kirpotin DB (2017) Liposomes useful for drug delivery. European Patent 1746976B1, 11 Jan 2017
Huang Z, Li H, Zhang T et al (2014) Progress involving new techniques for liposome preparation. Asian J Pharm Sci 9:176–182
Joseph J, Vedha Hari BN, Ramya Devi D (2018) Experimental optimization of Lornoxicam liposomes for sustained topical delivery. Eur J Pharm Sci 112:38–51
Kant S, Kumar S, Prashar B (2012) A complete review on: liposomes. Int Res J Pharm 3(7):10–16
Kasagi N, Yamada N, Mori M et al (2020) Liposome composition and pharmaceutical composition. Japanese Patent WO2018181963A1, 16 Jan 2020
Katz Y, Gavish M, Allon N et al (2018) Liposome compositions and uses thereof. WO2018193458A1, 25 Oct 2018
Klein J, Goldberg R, Lin W (2018) Liposomic drug-delivery vehicles. WO2018150429A1, 23 Aug 2018
Laouini A, Jaafar-Maalej C, Limayem-Blouza I et al (2012) Preparation, characterization and applications of liposomes: state of the art. J Coll Sci Biotechnol 1(2):147–168
Leigh S, Leigh MLS, Schaich M (2014) Method of solubilizing biologically active compounds. European Patent 2682106A1, 8 Jan 2014
Li L, ten Hagen TLM, Schipper D et al (2010) Triggered content release from optimized stealth thermosensitive liposomes using mild hyperthermia. J Control Release 143(2):274–279
Li M, Du C, Guo N et al (2019) Composition design and medical application of liposomes. Eur J Med Chem 164:640–653
Lopes SCA, Giuberti CS, Rocha TGR et al (2013) Liposomes as carriers of anticancer drugs. In: Rangel L (ed) Cancer treatment—conventional and innovative approaches. IntechOpen, Rijeka, pp 86–124
Lu Y, Sun W, Gu Z (2014) Stimuli-responsive nanomaterials for therapeutic protein delivery. J Control Release 194:1–19
Mahfoozur R, Sarwar B, Vaerma A et al (2017) Chapter 4 -liposomal-based therapeutic carriers for vaccine and gene delivery. 151–166
Nekkanti V, Kalepu S (2015) Recent advances in liposomal drug delivery: a review. Pharm Nanotechnol 3(1):35–55
Ng CZA, Cheong SYI (2018) Hyperstabilized liposomes increase targeting of mitotic cells. WO2018136002, 26 Jul 2018
Nisini R, Poerio N, Mariotti S et al (2018) The multirole of liposomes in therapy and prevention of infectious diseases. Front Immunol 9:155
Olusanya TOB, Haj Ahmad RR, Ibegbu DM et al (2018) Liposomal drug delivery systems and anticancer drugs. Molecules 23(4):907
Omri A, Ravaoarinoro M (1998) Les liposomes: Interets et limites en pharmacologie dans la therapeutique des infections bacteriennes. Can J Clin Pharmacol 5:231–234
Reynolds JNJ, Tan EW, Hyland BI et al (2017) Acoustic driven drug delivery systems. WO2017034418A1, 2 Mar 2017
Semalty A, Semalty M, Rawat BS et al (2009) Pharmacosomes: the lipid-based new drug delivery system. Expert Opin Drug Deliv 6(6):599–612
Seth AK (2019) Transdermal drug therapy: emerging techniques and improved patient compliance. In: Misra A, Shahiwala A (eds) Novel drug delivery technologies. Springer, Singapore, pp 261–289

Shaheen SM, Shakil Ahmed FR, Hossen MN et al (2006) Liposome as a carrier for advanced drug delivery. Pak J Biol Sci 9(6):1181–1191

Sharma A (1994) Novel taxol formulations: preparation and characterization of taxol containing liposomes. Pharm Res 11:889–896

Sharma VK, Mishra DN, Sharma AK (2010) Liposomes: present perspective and future challenges. Int J Curr Pharm Rev Res 1(2):6–15

Shivhare R, Pathak A, Shrivastava N et al (2012) An update review on novel advanced ocular drug delivery system. World J Pharm Pharm Sci 1(2):545–568

Srivastava R, Banerjee R, Rengan AK et al (2017) Degradable or transformable gold coated liposomal nano-construct and a process for its preparation. WO2017115381A1, 6 Jul 2017

Touitou E, Dayan N, Bergelson L et al (2000) Ethosomes-novel vesicular carriers for enhanced delivery: characterization and skin penetration properties. J Control Release 65(3):403–418

Vyas SP, Khar RK (2007) Targeted controlled drug delivery novel carrier systems. CBS Publication, New Delhi, pp 173–206

Wu Y (2018) Use of liposome for treatment of chronic viral hepatitis B. WO2018177140A1, 4 Oct 2018

Xu Y, Zheng A, Chen X (2018) All-trans retinoic acid liposome preparation, and preparation and application thereof. WO2018033118, 22 Feb 2018

Yadav A, Murthy MS, Shete AS et al (2011) Stability aspects of liposomes. Indian J Pharm Educ Res 45(4):402–413

Yadav D, Sandeep K, Pandey D et al (2017) Liposomes for drug delivery. J Biotechnol Biomater 7 (4):276

Zhang G, Ge X, Gao W (2018) Light-responsive liposome, preparation method and application thereof. WO2018127016A1, 12 Jul 2018

Zylberberg C, Matosevic S (2016) Pharmaceutical liposomal drug delivery: a review of new delivery systems and a look at the regulatory landscape. Drug Deliv 23(9):3319–3329

Part IV

Nanocarriers in Biotechnology

Potential Applications of Cationic Lipids in Nucleic Acid-Based Therapeutic Delivery System

13

Sunil Kardani and Devendra Vaishnav

Abstract

Gene therapy can prevent or cure diverse pathological conditions associated with defects in gene expression. Three main delivery systems used to deliver genes to target cells include mechanical, biological, and chemical method of DNA transfection. However viral vectors were the most studied and reported; associated side effects and limitations of viral vectors (high risk of mutagenicity, immunogenicity, low production yield, limited gene-loading capacity, and poor host range) have led to the development of nonviral vectors. On the light of the above background, cationic lipids may be alternatively used as promising carriers for nucleic acid delivery. With certain advantages over viral vectors, such as low immunogenicity, high loading capacity, broad range of host cell, being cheap, and easy reproduction, cationic lipid will be the choice for future gene delivery system. This chapter provides an overview of recent developments employed for in vitro and in vivo delivery of therapeutically important nucleic acids using different types of cationic lipids.

Keywords

Cationic lipid · Nucleic acid · Plasmid · Transfection · Vector

S. Kardani (✉)
Department of Pharmacy, Sumandeep Vidyapeeth Deemed to be University, Vadodara, Gujarat, India
e-mail: sunilk.dop@sumandeepvidyapeethdu.edu.in

D. Vaishnav
Department of Pharmaceutical Sciences, Saurashtra University, Rajkot, Gujarat, India

N. Shah (ed.), *Nanocarriers: Drug Delivery System*,
https://doi.org/10.1007/978-981-33-4497-6_13

13.1 Introduction

The sophistication of instrumentation and other techniques uplifts the knowledge of biological science. The discoveries and innovation in the biological field with advanced technologies unrevealed the secret of life encoded in cell in the form of DNA. A successful decoding of human genome identified approximately 21,000 genes responsible for complex biology of humans (Kumar 2020). Defective genes are associated with a wide variety of disorders; gene therapy is a promising treatment or prevention for these defective expressions. In 1989, the first human gene therapy trial was reported by Edelstein and co-workers (2004) using retrovirus as a carrier to insert the gene coding for resistance to neomycin into human tumor-infiltrating lymphocytes before infusing them into five patients with advanced melanoma (Rosenberg et al. 1986). This research opens a new door for the treatment of various genetic disorders. It works by inserting a therapeutic gene encoding a functional protein into effective cells, followed by expression and production of corrected proteins in targeted cells. Gene delivery can be done by three methods: (a) physical method, (b) chemical method, and (c) biological method (Fig. 13.1) (Jinturkar et al. 2011; Guo and Huang 2012).

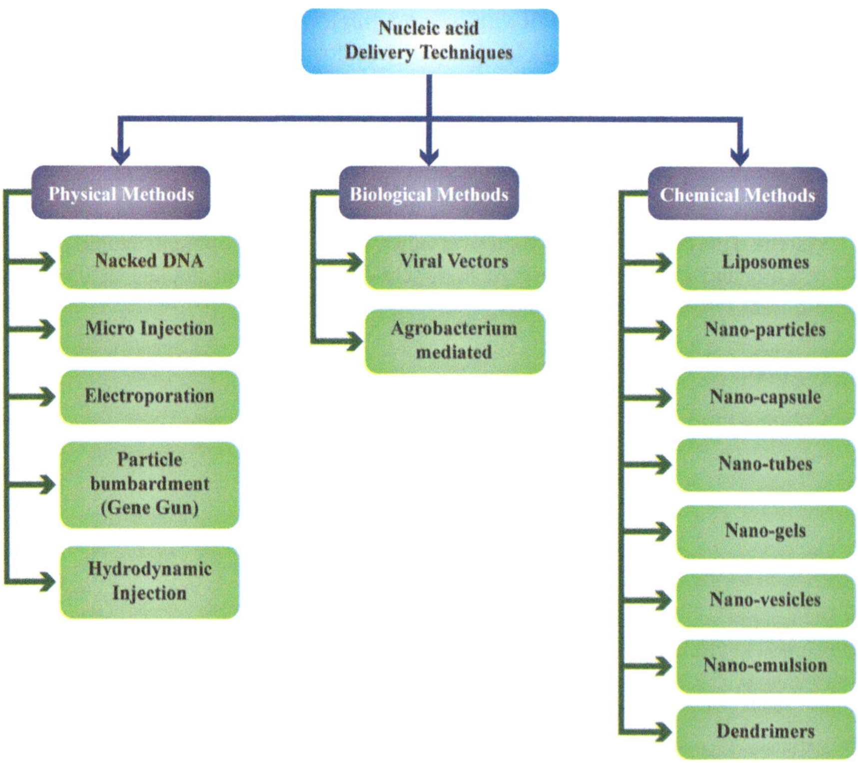

Fig. 13.1 Different techniques of nucleic acid delivery systems

Even though viral vectors have excellent delivery capacity, various disadvantages of viral vectors such as low nucleic acid-loading capacity, immunogenicity, and toxicity promote the development of nonviral vectors (Edelstein et al. 2004; Guo and Huang 2012). In this chapter, we discuss various nanocarrier techniques used for gene delivery and approaches to improve this promising technique. Various nanocarrier techniques include polymeric or solid lipid nanoparticles, lipid or albumin nanocapsules, nanotubes, nanogels, nanovesicles, nanoemulsions, liposomes, micelles, and dendrimers.

13.2 Chemical Methods for Nucleic Acid Delivery System

Gene expression occurs as beneficial genes are transferred to a cell nucleus requiring a safe and potential transporter system. As of 2017, over 2600 clinical trials have been performed or are being carried out mostly using viral vectors (Ginn et al. 2018). The death of an 18-year-old and fairly healthy gene trial patient has questioned the safety of viral vectors (Marshall 1999). Over viral and physical nucleic delivery approaches, chemical methods provide the benefit of consistency, ease of processing, and low toxicity in the transmission of viral genes. The term nonviral vector is widely used for a nonviral vector with a chemical mediation. Chemical methods are economic, simple to formulate and scale up, nonimmunogenic, and relatively safe for human administration. Nanocarrier system for nucleic delivery can be prepared using either cationic lipid, cationic polymers, or a combination of both.

13.3 Liposomes

Liposomes are vesicular structures that can form via the accumulation of lipids interacting with one another in an energetically favorable manner. Cationic lipids possess a positive charge through one or more amines present in the polar head group. These positively charged amines enable binding with anions such as those found on nucleic acids. The liposome thus formed is a result of energetic contributions by van der Waals forces and electrostatic binding to the nucleic acid which partially dictates liposome shapes. In aqueous solution spontaneous leaflets which bound together eventually form bilayer membrane and subsequently liposomes (Fig. 13.2) (Margineanu 1987).

Compared to viral vectors, liposomes are an attractive alternative because of high loading capacity of nucleic acids of any size; they are safe, biodegradable, versatile, cheap, and easy to manufacture in large scale (Balazs and Godbey 2011).

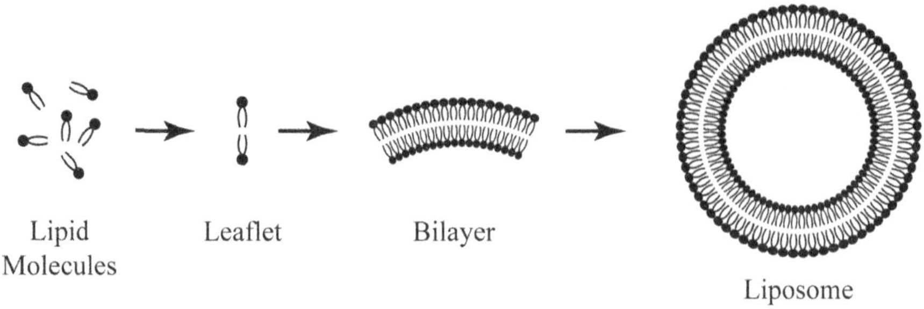

Fig. 13.2 Common mechanism of liposome formation

13.4 Cationic Lipid-Based Nanocarrier System

The first use of cationic lipid for the delivery of DNA was demonstrated by Felgner et al. (1987). Subsequently, Malone et al. (1989) reported the use of cationic lipid to deliver *Photinus pyralis* luciferase mRNA. Cationic lipids are also the primary carriers of gene transmission and by altering each of their constituent domains they can be quickly synthesized and widely facilitated (Lalani and Misra 2011). Chemically, cationic lipids are amphiphilic molecules possessing hydrophilic head and hydrophobic tail groups connected by either stable or degradable linkages (Pun and Hoffman 2013).

Since the introduction of *N*-[1-(2,3-dioleyloxy) propyl]-*N*,*N*,*N*-trimethylammonium chloride (DOTMA) for gene delivery, a new era of nucleic acid delivery has been opened and many cationic lipids have been introduced for the successful in vitro and in vivo delivery of nucleic acid. The composition of DOTMA is made up of two unsaturated oleoyl chains (C18:al 9), connected by an ether bond to a glycerol's three-carbon backbone, with a quaternary amine as the cationic head component.

Cationic lipids are amphiphilic molecules; common cationic lipid structure can be classified into three parts: (1) hydrophobic domain, (2) linker bond, and (3) hydrophobic domain (Zhi et al. 2013) (Fig. 13.3).

Significant progress has been made in the design and functionalization of several cationic lipids. Commonly used cationic lipids for nucleic acid delivery include 2,3-dioleyloxy-*N*-[2-spermine carboxamide] ethyl-*N*,*N*-dimethyl-1-propanammonium trifluoroacetate (DOSPA, Lipofectamine); 1,2-dioleoyl-3-trimethylammonium-propane (DOTAP); *N*-[1-(2,3-dimyristyloxy) propyl]-*N*,*N*-dimethyl-*N*-(2-hydroxyethyl) ammonium bromide (DMRIE); 3-β-[*N*-(*N*,*N*-′-dimethylaminoethane) carbamoyl] cholesterol (DC-Chol); dioctadecylamidoglycerylspermine (DOGS, Transfectam); dimethyldioctadecylammonium bromide (DDAB); pyridinium lipid, 4-((9Z,28Z)-heptatriaconta-9,28-dien-19-yl)-1-methylpyridin-1-ium (SAINT 2); bis-guanidium-spermidine-cholesterol (BGSC); dioleyldimethylammonium bromide (DODAB);

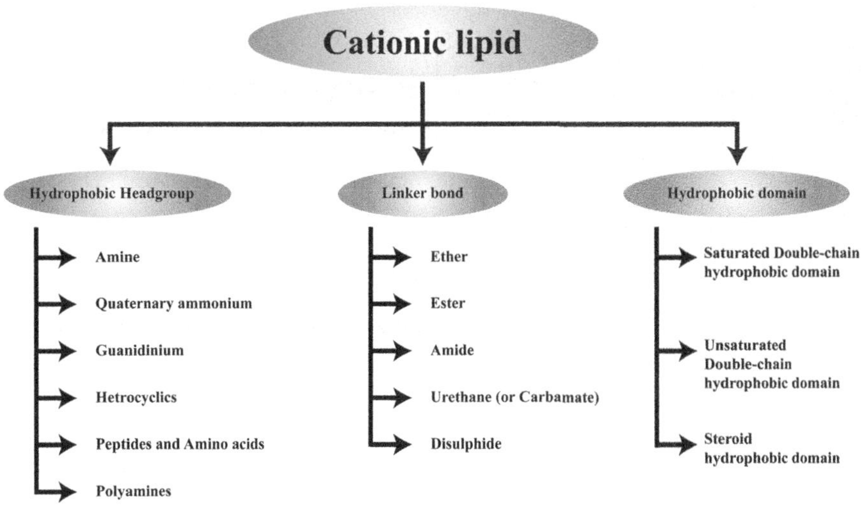

Fig. 13.3 Classification of cationic lipids based on hydrophobic head, linker bond, and hydrophobic domain

1,3-di-oleoyloxy-2-(6-carboxy-spermyl) propylamid (DOSPER); 1-[2-(9(Z)-octadecenoyloxy)-ethyl]-2-(8(Z)-heptadecenyl)-3-(2-hydroxyethyl) imidazolinium chloride (DOTIM); and dipalmytoylphosphatidylethanolamylspermine (DPPES).

Despite the fact that all the reported cationic lipids are structurally distinct, most comprise three basic components, the cationic head group, the hydrophobic domain, and the hydrophobic linker connecting the head group to the hydrophobic group. The efficient nucleic acid delivery and cytotoxicity of corresponding complex formed between cationic lipid and nucleic acid are influenced by the three parts of selected lipid. The detailed knowledge of these structural parameters enhances the effectiveness of cationic lipid-based transfection reagents.

Scientific studies revealed that the transfection efficiency of a cationic lipid greatly depends on (1) the extent of DNA or RNA condensation; (2) enhanced cellular uptake resulting from ionic and/or hydrophobic interaction with biological surfaces; and (3) membrane fusion by means of transient membrane destabilization to achieve delivery into cytoplasm while avoiding degradation in the lysosomal compartment (Zhu and Mahato 2010).

13.4.1 Quaternary Ammonium Salt Lipids

13.4.1.1 DOTMA

DOTMA is the first cationic lipid bearing quaternary ammonium head group and two oleoyl chains linked by a glycerol spacer reported for the successful delivery of nucleic acid belonging to the class of quaternary ammonium salt (Felgner et al. 1987). The quaternary ammonium head groups of DOTMA containing one or more

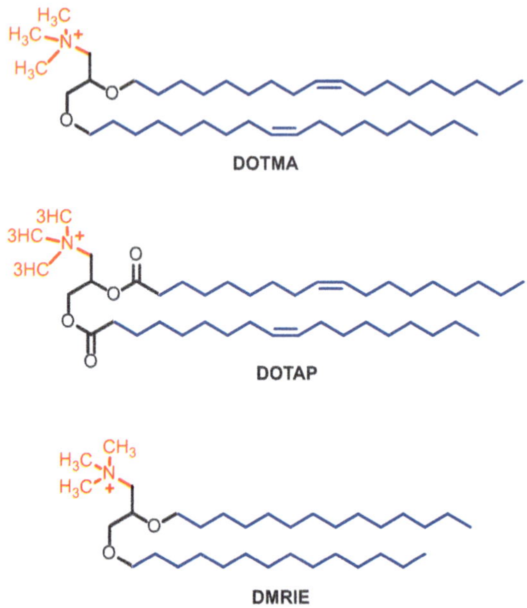

Fig. 13.4 Chemical structure of quaternary ammonium lipids. Red color: Hydrophilic head, black color: linker group, blue color: hydrophobic domain

hydroxyl moieties are responsible for higher transfection efficiency (Fig. 13.4). Duzgunes et al. (1989) studied the behavior of unilamellar liposomes composed of DOTMA and phosphatidylcholine or phosphatidylethanolamine. The outcome of the study suggests that the interaction of the negatively charged phosphatidyl serine polar group with the positively charged trimethyl ammonium of DOTMA is sufficient to mediate fusion between the two membranes containing these lipids and that the nature of the zwitterionic phospholipid component of these vesicles is an additional determinant of membrane fusion. The interaction study of DOTMA liposome-cell revealed endocytosis of the liposome-DNA complexes in coated pits. Membranes which form after exposure to DOTMA-containing liposomes were 10 nm in thickness as compared to the approx. 8 nm thickness of endogenous cellular membranes. Findings prove that the DOTMA liposomes entered into cell by endocytosis (Friend et al. 1996).

The first cationic lipid-based gene delivery was reported by Felgner et al. (1987). They studied transfection efficacy of DOTMA-DNA liposomes using mouse L cells. Malone et al. (1989) developed a high-efficiency RNA transfection system using DOTMA-containing liposomes. The local delivery of antisense oligodeoxynucleotide phosphorothioate ISIS 3466 in mice using DOTMA was studied (Saijo et al. 1994). DOTMA significantly increased the oligonucleotide cellular uptake (4–10 times) in LOX ascites tumors in an IP/IP model. These liposomes are 100–1000-fold more effective in RNA delivery compared to the DEAE-dextran method. Ren et al. (2000) studied the structural aspects of DOTMA for its high intravenous DNA transfection efficacy in mouse. Findings showcase how structural characteristics of DOTMA such as hydrocarbon chains to the cationic head group, two ether bonds, and paired oleyl chains as the hydrophobic anchor set the basis for its higher transfection activity. However, changes in any of the above do not affect the in vitro transfection activity; they do in fact decrease intravenous transfection activity. Kawakami et al. (2004) studied the transfection efficacy of DOTMA in rabbits. They found that pDNA complexed with DOTMA/DOPE liposomes could not be prepared with pDNA greater than 60 μg. Among these experiments, pDNA (85 μg) complexed with DOTMA/Chol liposomes (pDNA:cationic liposome charge ratio (−:+) = 1.0:2.0) showed the highest transfection efficiency in the ocular tissue and its transfection-mediated luciferase activity peaked at 3 days. Among the ocular tissues, the highest gene expression was observed in the aqueous humor.

A hybrid vector prepared out with DOTMA and polyethylenimine (PEI) showed high transfection and gene expression in various organs of mice such as liver, spleen, and lung (Matsumoto et al. 2008).

Liu et al. (1997) have studied different factors affecting the transfection efficacy of cationic lipid carrier systems following intravenous administration. DOTMA-Tween 80 complex as a carrier system and cDNA of luciferase or β-galactosidase gene as a reporter were used and the importance of DOTMA-to-DNA ratio and DOTMA-to-Tween 80 ratio in the lipid formulation was investigated in determining the site and level of transgene expression following intravenous administration. Randomized, single-blind, placebo-controlled, dose-rising designed phase I clinical trials of hIL-2 plasmid at four dose levels formulated in DOTMA/Chol liposomes were studied in head and neck squamous cell cancer by Wollenberg et al. (1999).

Sakurai et al. (2001) have studied three cationic lipid vectors, DNA–DOTMA/Chol liposome complexes, DNA–DOTMA liposome complexes, and DNA–DOTMA/DOPE liposome complexes. The finding shows that DOTMA/Chol and DOTMA complexes with a higher in vivo transfection activity do not induce fusion between erythrocytes, whereas DOTMA/DOPE complexes, a less efficient vector in vivo, induce fusion between the erythrocytes after a short incubation period. Mashal et al. (2017) have studied the characterization of a novel nonviral formulation based on DOTMA cationic lipid and polysorbate 60 nonionic surfactant. They found that lycopene incorporation in the above liposomes increases the transfection efficacy of DNA in both in vivo and in vitro (ARPE-19 cells) conditions. This formulation gives new hope for the treatment of many inherited diseases by safe nonviral formulations. DOTMA was found to enhance by at least 1000 times the potency of an antisense oligonucleotide (ISIS 1570) that hybridizes to the AUG translation initiation codon of human intercellular adhesion molecule-1 (Bennett et al. 1992).

13.4.1.2 DOTAP

When diether bonds of DOTMA were replaced by diester bonds, they formed a biodegradable agent known as DOTAP. It is widely known as transfection lipid, made of a monocationic trimethylammonium head group and two unsaturated hydrocarbon chains, derived from oleic acid (Fig. 13.4). Interesting findings regarding the stability of DNA in various delivery systems were reported by Moret et al.

(2001). The necked DNA was degraded rapidly in mouse plasma followed by rat and human plasma. In contrast to that, DOTAP liposomes protect DNA in all species of serum compounds. DOTAP complex also protects DNA in the presence of DNase I.

Templeton et al. (1997) have studied the systemic delivery of a chloramphenicol acetyltransferase (CAT) DNA reporter plasmid and in vivo gene expression in mice. Outcome of the study revealed 50-fold higher CAT expression in mice lungs in DOTAP:Chol-DNA liposome treatment group compared to other liposome complexes. In vitro experiment revealed the stability of DOTAP:Chol-formed colloidal complexes over a wide range of DNA:liposome ratios. Further in vivo study showed excellent expression of CAT in 5 mM DOTAP:Chol liposomes complexed to 150 μg of DNA in a 200 μL final volume. Porteous et al. (1997) reported the comparable efficacy of DOTAP-mediated DNA delivery with that of viral vectors and absence of any adverse effects.

Regelin et al. (2000) studied the transfection efficacy of DOTAP and its analogues. Crook et al. (1998) studied the transfection efficacy of DOTAP complex and DOTAP/cholesterol complex. Both complexes showed resistance to DNase degradation. Even-Chen and Barenholz (2000) have studied the electrostatic interaction of DNA with monocationic DOTAP, and polycationic DOSPA liposomes. The study focused on how DNA in two different forms, supercoiled and nicked-relaxed (open circular), bound the cationic liposomes of plasmid DNA in the nicked-relaxed over the supercoiled form. Interestingly they found that DOTAP formulation preferably binds to the relaxed DNA plasmid which suggests that the binding of supercoiled DNA is weaker and easier to dissociate from the complex. Remaut et al. (2006) have studied the effect of topology of plasmid DNA on the transfection properties of DOTAP/DOPE lipoplexes. Previous reports suggest that higher transfection efficiencies are achieved with smaller pDNA molecules (Kreiss et al. 1999; Darquet et al. 1999), which was attributed to a better diffusion through the cytoplasm of the cells or an enhanced uptake through the nuclear pore complexes. Findings suggest that supercoiled (SC) pDNA is likely more efficient in passing the nuclear pore complexes than open circular (OC) or linearized pDNA.

Ott et al. (2002) prepared cationic submicron emulsion using MF59/DOTAP for effective DNA vaccine delivery. HIV p55 gag DNA-immunized mice and rabbits were treated with prepared formulation and serum IgG response was recorded. MF59/DOTAP emulsion showed significant level of IgG production in immunized mice compared to mice treated with naked DNA. Schäfer et al. (2010) have reported the clathrin-mediated endocytosis for the uptake of poly-L-lysine-labeled DOTAP/DNA lipoplexes and the simultaneous internalization of PEI polyplexes by clathrin-dependent as well as clathrin-independent mechanisms. Hybrid system consists of PLGA nanoparticles and DOTAP potentiates the gene silencing efficacy of siRNA (Jensen et al. 2012). In addition, spray-dried siRNA is used in the delivery vehicle with mannitol under formulation and processing conditions that preserve the integrity of the siRNA and gene silencing.

13.4.1.3 DMRIE

DMRIE has a quaternary nitrogen adjacent to a primary alcohol, thus imparting a pH-independent positive charge. DMRIE has a polar hydroxyethyl substituent on the quaternary ammonium group (Fig. 13.4) which makes it a more effective transfection efficacy in many cases compared to conventional cytofectines. Wheeler (2013) has received patent for delivering anionic molecules into cell by DMRIE.

Galanis et al. (1999) constructed a plasmid DNA using cytomegalovirus (CMV)-IE promoter which drives the expression of the IL-2 cDNA combined with DMRIE/DOPE to perform phase I and II clinical trials on patients having metastatic solid tumors or lymphomas. Intratumoral administration of the IL-2 gene using DMRIE/DOPE cationic system seems to be safe and well tolerated in the study population. Further immunohistochemistry of post-treatment tumor samples showed that tumor cells, rather than the surrounding lymphocytes, were predominantly IL-2 positive, which suggests transgene expression of IL-2.

The USFDA-approved treatment of metastatic renal cell carcinoma is IL-2. Recombinant IL-2 treatment shows beneficiary in only 10–20% of patients; occasionally it showed life-threatening toxicities in patients, which limits its usability. Gene delivery of IL-2 may be beneficiary in the treatment of metastatic renal cell carcinoma. Hoffman and Figlin (2000) have studied the efficacy of DMRIE/DOPE lipid complexed with plasmid DNA expression vector (VCL1102, 30) encoding human IL-2 (Leuvectin) in phase I and II clinical trials. A total of 52 patients were enrolled in the study and were evaluated for toxicity; none showed grade 4 toxicity related to the study drug.

13.4.2 Lipoamines

13.4.2.1 DOGS

Behr (1986) proposed DNA-binding vesicles containing lipopolyamines. DOGS is the first example of lipoamine head consisting of cationic lipid, commercially known as Transfectam™ (Fig. 13.5). DOGS showed high transfection activity over quaternary ammonium salts.

Functional expression of pdoc-i in HCPC-1 cells was studied using Transfectam system (Todd et al. 1995). Plasmid (pGL2 (containing nucleotides 2100–2300 of the bovine GLUT1 3′-UTR inserted at the Pfl MI site within the luciferase 3′-UTR))-Transfectam complex (1:5 v/v) was incubated with C6 cells plated at 60% confluency and luciferase activity was measured for 10 s in duplicates of 20 pl of lysate with 100 pl of luciferase reagents (Promega), using Luminometer Monolight 2010 (Tsukamoto et al. 1997). Outcome of the Tsukamoto and co-workers' experiment provides an evidence that nucleotides 2181–2190 of the bovine GLUT1 mRNA 3′-UTR form a complex with brain tumor cytosolic proteins that serves to increase GLUT1 gene expression at the posttranscriptional level (Tsukamoto et al. 1997).

In et al. (1997) transfected p5LO-CAT plasmid into HeLa cells infected with recombinant vaccinia virus containing human Sp1 cDNA when they reached

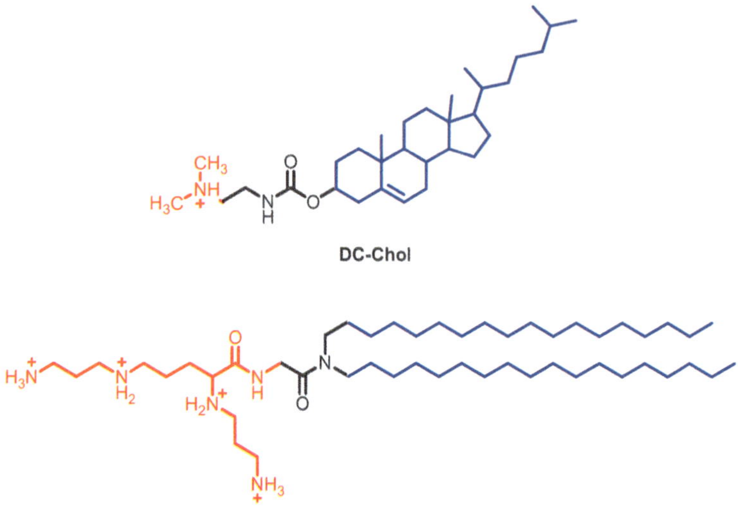

Fig. 13.5 Chemical structure of lipoamines

50–70% confluence using Transfectam. After 48-h incubation, cells were harvested and lysed and lysates were analyzed for CAT activity which provides a potential way to link a given patient's clinical response to treatment modifying the 5-LO pathway and their genotype at the 5-LO locus.

13.4.2.2 DC-Chol

The lipoamine DC-Chol was prepared by tertiary amine linked through a spacer to a cholesteryloxy-carbonyl lipid (Fig. 13.5). DC-Chol was used as a coformulate with other cationic lipids to obtain good transfection of genetic material as it does not confer enough compactness to DNA.

13.4.3 Quaternary Ammonium Salt and Lipoamines

13.4.3.1 DOSPA

DOSPA is the first cationic lipid containing both quaternary ammonium salt and polyamines in one lipid, commercially known as Lipofectamine™ (Fig. 13.6).

Hofland et al. (1996) have studied the successful transfection efficacy of DOSPA on different cell types, HT1080 cells, HepG2 cells, mouse B16 melanoma cells, mouse renal cell carcinoma cells, mouse colorectal cancer cells, primary human fibroblasts, and primary human melanoma cells. Plasmid SSV9-MD-2 consisting of cytomegalovirus (CMV) immediate early gene promoter/enhancer 13-globin intervening sequence 2 (IVS2) and a human 13-globin polyadenylation signal was selected for the study. The human placental alkaline phosphatase (mAP) as a reporter

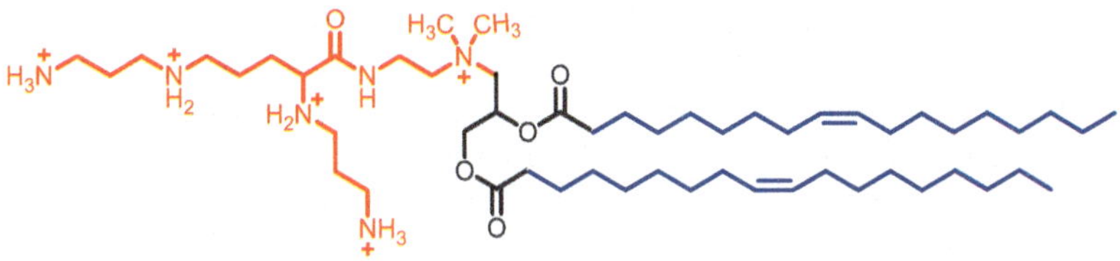

DOSPA (Lipofectamine)

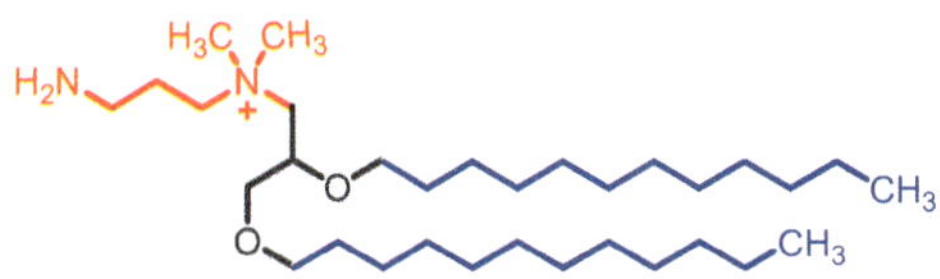

GAP-DLRIE

Fig. 13.6 Chemical structure of DOSPA (Lipofectamine) and GAP-DLRIE

gene was inserted into the plasmid and used to check transfection efficacy of DOSPA in vitro using NIH3T3 cells and in vivo using female Balb/C mice (Hofland et al. 1997). Although DOSPA has excellent transfection capability, high toxicity limits its usability.

13.4.3.2 GAP-DLRIE

GAP-DLRIE is the second example of cationic lipoid containing both quaternary ammonium salt and lipoamines which displayed a high level of transfection with low cytotoxicity compared to DOSPA. Chemically it has two hydrophobic chains appended to a quaternary ammonium moiety via a polar dioxy-propyl group in a manner affording a central glycerol-like structure (Fig. 13.6).

Wheeler et al. (1996) have studied in vitro transfection efficacy of GAP-DLRIE by conjugating it with plasmid vector pCMVP3 (Clontech) in OptiMEM using CFT1 cells. They also checked transfection efficacy in vivo in female BALB/c mice. The in vitro studies showed maximum transfection efficacy at a 1:3 molar ratio of GAP-DLRIE/DOPE whereas 1:1 molar ratio was the most effective in vivo. Stephan et al. (1996) have studied in vivo arterial gene transfer in male and female Yorkshire pigs. They constructed eukaryotic expression vector plasmid encoding a chloramphenicol acetyl transferase (CAT) gene under the control of a cytomegalovirus promoter and enhancer (pCMV-CAT). They divided animals into three groups and arterial gene transfer was performed in the right and left iliofemoral arteries of each pig. No toxicity was observed in test animals and significant transfection capacity was observed compared to other nonviral vectors and almost equal efficacy was observed with adenoviral vectors.

Norman et al. (1997) have studied in vivo transfection-promoting activity of GAP-DLRIE:DOPE and DC-cholesterol:DOPE. The lipids were complexed with plasmid DNA; the formulation was administered intranasally to anesthetized BALBlc mice. After 2 days, the lungs were harvested and assayed for CAT activity. GAP-DLRIE:DOPE showed 450-fold higher transfection capacity than naked DNA and 150-fold greater activity than DC-cholesterol:DOPE.

Souza et al. (2002) have synthesized tuberculosis DNA vaccine using GAP-DLRIE:DOPE for intranasal instillations. Hong et al. (2002) studied overexpression of interleukin 10 (IL-10) in the donor heart which prolongs allograft survival in animals using different cationic lipids and plasmid pSV IL-10 containing human recombinant IL-10 cDNA coupled to the simian virus 40 (SV40) early promoter. They found that GAP:DLRIE is the best cationic liposome for ex vivo gene transfection in hypothermic conditions.

13.4.4 Pyridinium Amphiphiles

13.4.4.1 SAINT

SAINTS, also abbreviated and known as pyridinium amphiphiles, have efficient transfection capability (Fig. 13.7). Šmisterová et al. (2001) have synthesized several analogues of SAINTS and studied their transfection efficacy using pCMVβ-gal on COS-7 cells. They found that SAINT 2 showed excellent transfection efficacy against all analogues.

13.4.5 Bis-Guanidinium Cholesterol Derivatives

13.4.5.1 BGSC

BGSC is a cholesterol derivative with guanidinium polar head groups (Fig. 13.8). Lehn et al. (2000) received patent for therapeutic gene transfection using BGSC and

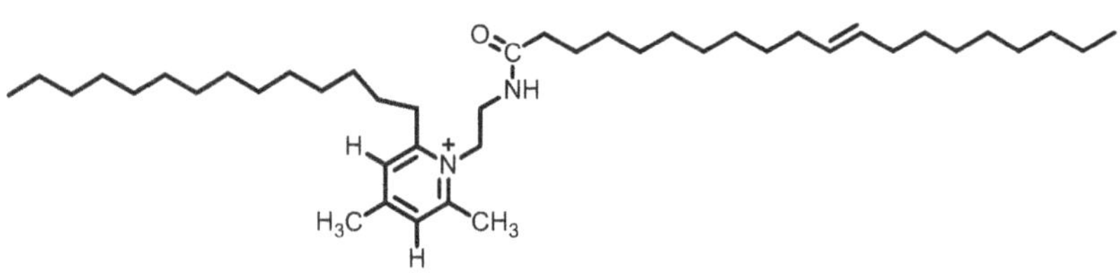

Pyridinium lipid

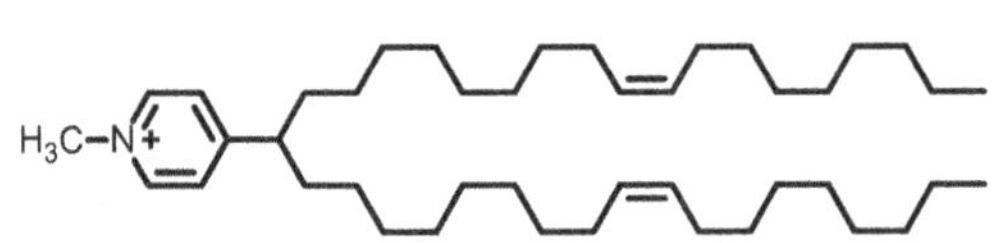

4-((9Z,28Z)-heptatriaconta-9,28-dien-19-yl)-1-methylpyridin-1-ium (SAINT 2)

Fig. 13.7 Chemical structure of pyridinium amphiphiles

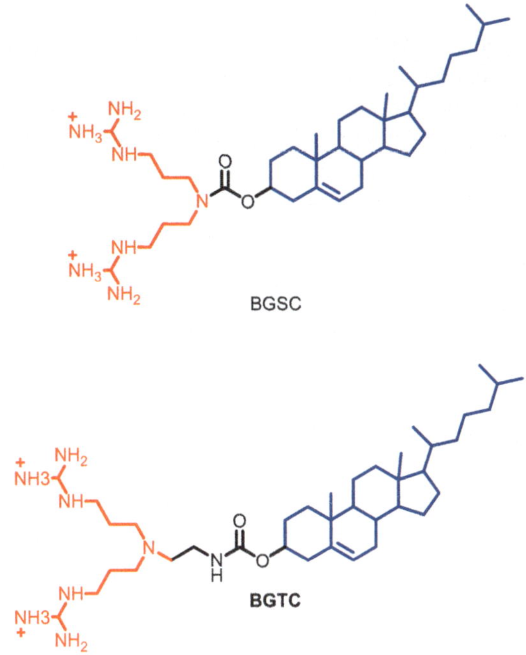

Fig. 13.8 Chemical structure of bis-guanidinium cholesterol derivatives

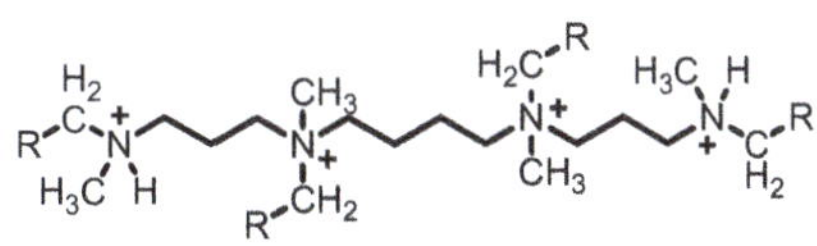

Tetramethyltetralaurylspermine (TMTLS) R = $(CH_2)_{10}CH_3$.
Tetramethyltetramyristylspermine (TMTMS) R = $(CH_2)_{12}CH_3$.
Tetramethyltetrapalmitylspermine (TMTPS) R = $(CH_2)_{14}CH_3$.
Tetramethyltetraoleoylspermine (TMTOS) R = $(CH_2)_7CH = CH(CH_2)_7CH_3$

Fig. 13.9 Chemical structure of tetramethyltetraalkylspermine analogues.

BGTC. BGSC has good gene transfection efficacy but very few studies reported the transfection efficacy of BGSC compared to BGTC.

13.4.5.2 BGTC

First synthesis of BGTC was proposed by Almeida et al. (1988). BGTC is a cholesterol derivative with guanidinium polar head groups (Fig. 13.9). Vigneron et al. (1996) tested the in vitro gene transfection efficacy of BGTC on HeLa cells, 3T3 cells, and NB 2A cell line using pRSV-Luc plasmid. Research outcomes suggest that the transfection efficacy of BGTC is relatively insensitive to variations of pH during the in vitro formation of the DNA/reagent aggregates and the trafficking in the cell toward the nucleus protects the DNA against degradation. BGTC can be used as an aqueous solution for the transfection of DNA. Pitard et al. (1999) studied the structural characteristics of supramolecular assemblies formed by BGTC for gene transfection. The multivalent nature of BGTC forms multilamellar DNA lipoplexes which may escape from intracytoplasmic vesicles by means of membrane disruption, as these complexes might have highly original fusion/destabilization properties. In utero gene transfection efficacy of BGTC was tested in fetal sheep airway (Luton et al. 2004). BGTC/DOPE complexed with plasmids expressing the *Escherichia coli* chloramphenicol acetyltransferase (CAT) reporter gene was introduced in utero to fetal sheep at 70 days of gestation via surgical replacement of the airway fluid by the transfection mixture followed by tracheal occlusion. The experiment successfully demonstrated the in utero gene transfection efficacy of BGTC. Literature revealed that BGTC was an efficient and versatile reagent for gene transfection in vitro, in vivo, and in utero.

13.4.6 Tetramethyl Tetraalkylspermine (TMTAS) Analogues

McCluskie et al. (1998) have formulated various derivatives of TMTAS by alkylating spermine with desired acyl chloride followed by lithium aluminum hydride reduction to obtain tetraalkylspermine further alkylated with methyl iodide to obtain tetramethyltetraalkylspermine. Different derivatives, namely TMTLS, tetramethyltetralaurylspermine; TMTMS, tetramethyltetramyristylspermine; TMTOS, tetramethyltetraoleoylspermine; and TMTPS, tetramethyltetrapalmitylspermine, were synthesized (Fig. 13.9) and tested for transfection efficacy with one of DOPE, DPAPE, DMPE, and DLPE.

13.4.7 DOSPER

DOSPER is a polycationic liposomal compound (Fig. 13.10) used for nucleic acid delivery in vivo and in vitro. Lampela et al. (2002) have studied the transfection efficacy of DOSPER alone and in conjugation with polyethylenimines (PEIs) using TKBPVlacZ expression plasmid on subconfluent CV1-P (monkey fibroblastoma cell line) or SMC (rabbit aortic smooth muscle cell line) cells. The study concluded that the combination of PEI and DOSPER induced severalfold increase in transfection efficiency compared with the effect of either component alone with the minimum toxicity in vitro. In a subsequent study Lampela et al. (2003) found that the synergism of PEI–DNA-DOSPER complexes in gene transfection is not due to

Fig. 13.10 Chemical structure of DOSPER

differences in the cellular uptake, DNA condensation, or complex size. The study revealed that PEI2K improves the intracellular kinetics of DNA without affecting cellular uptake, whereas DOSPER is responsible for the cellular uptake of PEI-DNA complex.

13.4.8 DODAB

Kunitake et al. (1977) have synthesized dimethyldioctadecylammonium bromide (DODAB), the first cationic amphiphile (Fig. 13.11), with two long hydrocarbon-saturated chains.

The in vivo gene transfection efficacy of DODAB was evaluated by Das et al. (2016). The data revealed that compared to naked DNA injection, the GFP fluorescence was around 3.5-fold higher for CeO_2/DODAB; however, the in vivo transfection efficiency of CeO_2/DODAB was only 1.23-fold less than the commercial in vivo-jeiPEI reagents. Gene delivery system did not show any cytotoxicity after intravenous injection. Based on its high biocompatibility for gene transfection efficacy Das et al. (2016) recommended CeO_2/DODAB as a new class of nonviral vectors for therapeutic gene delivery applications.

13.4.9 DDAB

Mizuarai et al. (2001) have studied dimethyldioctadecylammonium bromide (DDAB) (Fig. 13.11) for direct DNA transfection into cultured cell. Transfection efficacy of plasmid DNA–DDAB complex was studied on COS-7, NIH3T3, and CH0-K1 cells using recombinant plasmids pCMVp and pCMV-GFP for transient expression and recombinant plasmid, pmiwZ, which encodes the 3-galactosidase gene under the control of chicken 3-actin and Rous sarcoma virus (RSV) hybrid promoter, was used for gene transfer of quail embryos. Outcome of the study suffices the importance of protamine treatment on different liposomes. The transfection efficacy of all liposomes drastically decreases in the presence of 10% serum. Protamine not only protects liposomes in the presence of serum but also maintains the transfection efficacy of each vesicle system (Kamihira et al. 2004; Ohama et al. 2005). Liposomes containing DDAB are an inexpensive, highly efficient, and

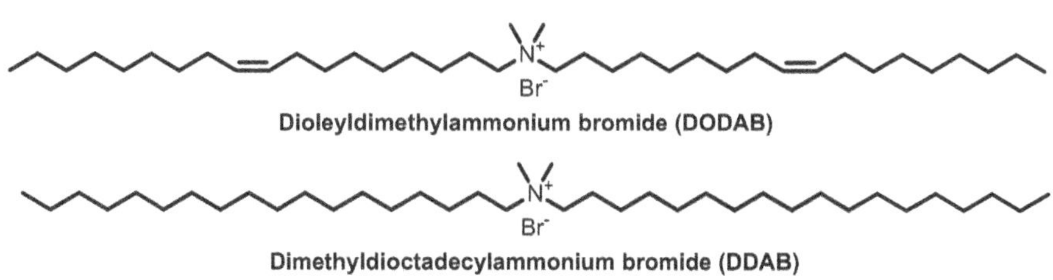

Fig. 13.11 Chemical structure of dimethyldioctadecylammonium bromide (DDAB)

reproducible alternative for the transfection of animal cells and are well suited for use with the vaccinia virus/T7 expression system (Rose et al. 1991).

13.5 Conclusion

Nonviral vectors possess great potential for nucleic acid delivery in a diverse host range. Reproducibility, cheapness, high efficiency, and low side effects attract many scientists to develop diverse formulations for delivery of different types of vectors in different host cells. Cationic lipids are more advantageous as compared to viral vectors because of their low immune response, high capacity of nucleic acid load, and broad range of host cells. Further studies are required to establish a strong foundation for the use of cationic lipid as an alternate of viral vector in the DNA vaccine and treatment of genetic disorders, cancer, hypertension, neurological disorders, etc.

References

Almeida MLS, Grehn L, Ragnarsson U (1988) Selective protection of polyamines: synthesis of model compounds and spermidine derivatives. J Chem Soc Perkin Trans 1(7):1905–1911

Balazs DA, Godbey WT (2011) Liposomes for use in gene delivery. J Drug Deliv 2011:1–12

Behr JP (1986) DNA strongly binds to micelles and vesicles containing lipopolyamines or lipointercalants. Tetrahedron Lett 27(48):5861–5864

Bennett CF, Chiang MY, Chan H et al (1992) Cationic lipids enhance cellular uptake and activity of phosphorothioate antisense oligonucleotides. Mol Pharmacol 41(6):1023

Crook K, Stevenson BJ, Dubouchet M et al (1998) Inclusion of cholesterol in DOTAP transfection complexes increases the delivery of DNA to cells *in vitro* in the presence of serum. Gene Ther 5 (1):137–143

Darquet AM, Rangara R, Kreiss P et al (1999) Minicircle: an improved DNA molecule for *in vitro* and *in vivo* gene transfer. Gene Ther 6(2):209–218

Das J, Han JW, Choi YJ et al (2016) Cationic lipid-nanoceria hybrids, a novel nonviral vector-mediated gene delivery into mammalian cells: investigation of the cellular uptake mechanism. Sci Rep 6(1):29197

Duzgunes N, Goldstein JA, Friend DS et al (1989) Fusion of liposomes containing a novel cationic lipid, N-[2,3-(dioleyloxy)propyl]-N,N,N-trimethylammonium: induction by multivalent anions and asymmetric fusion with acidic phospholipid vesicles. Biochemistry 28(23):9179–9184

Edelstein ML, Abedi MR, Wixon J (2004) Gene therapy clinical trials worldwide 1989-2004—an overview. J Gene Med 6(6):597–602

Even-Chen S, Barenholz Y (2000) DOTAP cationic liposomes prefer relaxed over supercoiled plasmids. Biochim Biophys Acta 1509(1):176–188

Felgner PL, Gadek TR, Holm M et al (1987) Lipofection: a highly efficient, lipid-mediated DNA-transfection procedure. Proc Natl Acad Sci U S A 84(21):7413–7417

Friend DS, Papahadjopoulos D, Debs RJ (1996) Endocytosis and intracellular processing accompanying transfection mediated by cationic liposomes. Biochim Biophys Acta 1278 (1):41–50

Galanis E, Hersh EM, Stopeck AT et al (1999) Immunotherapy of advanced malignancy by direct gene transfer of an interleukin-2 DNA/DMRIE/DOPE lipid complex: phase I/II experience. J Clin Oncol 17(10):3313–3323

Ginn SL, Amaya AK, Alexander IE et al (2018) Gene therapy clinical trials worldwide to 2017: an update. J Gene Med 20(5):e3015

Guo X, Huang L (2012) Recent advances in nonviral vectors for gene delivery. Acc Chem Res 45 (7):971–979

Hoffman DMJ, Figlin RA (2000) Intratumoral interleukin 2 for renal-cell carcinoma by direct gene transfer of a plasmid DNA/DMRIE/DOPE lipid complex. World J Urol 18(2):152–156

Hofland HE, Shephard L, Sullivan SM (1996) Formation of stable cationic lipid/DNA complexes for gene transfer. Proc Natl Acad Sci 93(14):7305

Hofland HEJ, Nagy D, Liu JJ et al (1997) *In vivo* gene transfer by intravenous administration of stable cationic lipid/DNA complex. Pharm Res 14(6):742–749

Hong YS, Laks H, Cui G et al (2002) Localized immunosuppression in the cardiac allograft induced by a new liposome-mediated IL-10 gene therapy. J Heart Lung Transplant 21(11):1188–1200

In KH, Asano K, Beier D et al (1997) Naturally occurring mutations in the human 5-lipoxygenase gene promoter that modify transcription factor binding and reporter gene transcription. J Clin Invest 99(5):1130–1137

Jensen DK, Jensen LB, Koocheki S et al (2012) Design of an inhalable dry powder formulation of DOTAP-modified PLGA nanoparticles loaded with siRNA. J Control Release 157(1):141–148

Jinturkar KA, Rathi MN, Misra A (2011) Gene delivery using physical methods. In: Misra A (ed) Challenges in delivery of therapeutic genomics and proteomics. Elsevier, London, pp 83–126

Kamihira M, Nishijima KI, Iijima S (2004) Transgenic birds for the production of recombinant proteins. In: Kobayashi T (ed) Recent progress of biochemical and biomedical engineering in Japan II. Springer, Berlin, pp 171–189

Kawakami S, Harada A, Sakanaka K et al (2004) *In vivo* gene transfection via intravitreal injection of cationic liposome/plasmid DNA complexes in rabbits. Int J Pharm 278(2):255–262

Kreiss P, Mailhe P, Scherman D et al (1999) Plasmid DNA size does not affect the physicochemical properties of lipoplexes but modulates gene transfer efficiency. Nucleic Acids Res 27 (19):3792–3798

Kumar D (2020) The human genome and molecular medicine. In: Kumar D (ed) Clinical molecular medicine. Academic Press, London, pp 3–16

Kunitake T, Okahata Y, Tamaki K, Kumamaru F, Takayanagi M (1977) Formation of the bilayer membrane from a series of quaternary ammonium salts. Chem Lett 6(4):387–390

Lalani J, Misra A (2011) Gene delivery using chemical methods. In: Misra A (ed) Challenges in delivery of therapeutic genomics and proteomics. Elsevier, London, pp 127–206

Lampela P, Räisänen J, Männistö PT et al (2002) The use of low-molecular-weight PEIs as gene carriers in the monkey fibroblastoma and rabbit smooth muscle cell cultures. J Gene Med 4 (2):205–214

Lampela P, Elomaa M, Ruponen M et al (2003) Different synergistic roles of small polyethylenimine and Dosper in gene delivery. J Control Release 88(1):173–183

Lehn JM, Lehn P, Vigneron JP (2000) Compounds related to the amidinium family, pharmaceutical compositions containing same, and uses thereof. US Patent 6143729, 7 Nov 2011

Liu F, Qi H, Huang L et al (1997) Factors controlling the efficiency of cationic lipid-mediated transfection *in vivo* via intravenous administration. Gene Ther 4(6):517–523

Luton D, Oudrhiri N, de Lagausie P et al (2004) Gene transfection into fetal sheep airways in utero using guanidinium-cholesterol cationic lipids. J Gene Med 6(3):328–336

Malone RW, Felgner PL, Verma IM (1989) Cationic liposome-mediated RNA transfection. Proc Natl Acad Sci 86(16):6077

Margineanu DG (1987) Equilibrium and non-equilibrium approaches in biomembrane thermodynamics. Arch Physiol Biochem 95(4):381–422

Marshall E (1999) Gene therapy death prompts review of adenovirus vector. Science 286 (5448):2244–2245

Mashal M, Attia N, Puras G et al (2017) Retinal gene delivery enhancement by lycopene incorporation into cationic niosomes based on DOTMA and polysorbate 60. J Control Release 254:55–64
Matsumoto M, Kishikawa R, Kurosaki T et al (2008) Hybrid vector including polyethylenimine and cationic lipid, DOTMA, for gene delivery. Int J Pharm 363(1):58–65
McCluskie MJ, Chu Y, Xia JL et al (1998) Direct gene transfer to the respiratory tract of mice with pure plasmid and lipid-formulated DNA. Antisense Nucleic Acid Drug Dev 8(5):401–414
Merlin JL, Dolivet G, Dubessy C et al (2001) Improvement of nonviral p53 gene transfer in human carcinoma cells using glucosylated polyethylenimine derivatives. Cancer Gene Ther 8 (3):203–210
Mizuarai S, Ono KI, You J et al (2001) Protamine-modified DDAB lipid vesicles promote gene transfer in the presence of Serum1. J Biochem 129(1):125–132
Moret I, Esteban Peris J, Guillem VM et al (2001) Stability of PEI–DNA and DOTAP–DNA complexes: effect of alkaline pH, heparin and serum. J Control Release 76(1):169–181
Norman JA, Hobart P, Manthorpe M et al (1997) Development of improved vectors for DNA-based immunization and other gene therapy applications. Vaccine 15(8):801–803
Ohama Y, Heike Y, Sugahara T et al (2005) Gene transfection into HeLa cells by vesicles containing cationic peptide lipid. Biosci Biotechnol Biochem 69(8):1453–1458
Ott G, Singh M, Kazzaz J et al (2002) A cationic sub-micron emulsion (MF59/DOTAP) is an effective delivery system for DNA vaccines. J Control Release 79(1–3):1–5
Pitard B, Oudrhiri N, Vigneron JP et al (1999) Structural characteristics of supramolecular assemblies formed by guanidinium-cholesterol reagents for gene transfection. Proc Natl Acad Sci 96(6):2621
Porteous DJ, Dorin JR, McLachlan G et al (1997) Evidence for safety and efficacy of DOTAP cationic liposome mediated CFTR gene transfer to the nasal epithelium of patients with cystic fibrosis. Gene Ther 4(3):210–218
Pun SH, Hoffman AS (2013) B.8 - Nucleic acid delivery. In: Ratner BD, Hoffman AS, Schoen FJ et al (eds) Biomaterials science, 3rd edn. Academic Press, Waltham, MA, pp 1047–1054
Regelin AE, Fankhaenel S, Gürtesch L et al (2000) Biophysical and lipofection studies of DOTAP analogs. Biochim Biophys Acta 1464(1):151–164
Remaut K, Sanders NN, Fayazpour F et al (2006) Influence of plasmid DNA topology on the transfection properties of DOTAP/DOPE lipoplexes. J Control Release 115(3):335–343
Ren T, Song YK, Zhang G et al (2000) Structural basis of DOTMA for its high intravenous transfection activity in mouse. Gene Ther 7(9):764–768
Rose JK, Buonocore L, Whitt MA (1991) A new cationic liposome reagent mediating nearly quantitative transfection of animal cells. BioTechniques 10(4):520–525
Rosenberg SA, Spiess P, Lafreniere RJS (1986) A new approach to the adoptive immunotherapy of cancer with tumor-infiltrating lymphocytes. Science 233(4770):1318–1321
Saijo Y, Perlaky L, Wang H et al (1994) Pharmacokinetics, tissue distribution, and stability of antisense oligodeoxynucleotide phosphorothioate ISIS 3466 in mice. Oncol Res 6(6):243–249
Sakurai F, Nishioka T, Saito H et al (2001) Interaction between DNA–cationic liposome complexes and erythrocytes is an important factor in systemic gene transfer via the intravenous route in mice: the role of the neutral helper lipid. Gene Ther 8(9):677–686
Schäfer J, Höbel S, Bakowsky U et al (2010) Liposome–polyethylenimine complexes for enhanced DNA and siRNA delivery. Biomaterials 31(26):6892–6900
Šmisterová J, Wagenaar A, Stuart MC et al (2001) Molecular shape of the cationic lipid controls the structure of cationic lipid/dioleylphosphatidylethanolamine-DNA complexes and the efficiency of gene delivery. J Biol Chem 276(50):47615–47622
Souza S, Rosseels V, Denis O et al (2002) Improved tuberculosis DNA vaccines by formulation in cationic lipids. Infect Immun 70(7):3681–3688
Stephan DJ, Yang ZY, San H et al (1996) A new cationic liposome DNA complex enhances the efficiency of arterial gene transfer *in vivo*. Hum Gene Ther 7(15):1803–1812

Templeton NS, Lasic DD, Frederik PM et al (1997) Improved DNA: liposome complexes for increased systemic delivery and gene expression. Nat Biotechnol 15(7):647–652

Todd R, McBride J, Tsujl T et al (1995) Deleted in oral cancer-1 (doc-l), a novel oral tumor suppressor gene. FASEB J 9(13):1362–1370

Tsukamoto H, Boado RJ, Pardridge WM (1997) Site-directed deletion of a 10-nucleotide domain of the 3′-untranslated region of the GLUT1 glucose transporter mRNA eliminates cytosolic protein binding in human brain tumors and induction of reporter gene expression. J Neurochem 68 (6):2587–2592

Vigneron JP, Oudrhiri N, Fauquet M et al (1996) Guanidinium-cholesterol cationic lipids: efficient vectors for the transfection of eukaryotic cells. Proc Natl Acad Sci 93(18):9682

Wheeler CJ (2013) Complex cationic lipids having quaternary nitrogens therein. US Patent 8,541,628, 24 Sept 2013

Wheeler CJ, Felgner PL, Tsai YJ et al (1996) A novel cationic lipid greatly enhances plasmid DNA delivery and expression in mouse lung. Proc Natl Acad Sci 93(21):11454

Wollenberg B, Kastenbauer D, Mundl H et al (1999) Gene therapy—phase I trial for primary untreated head and neck squamous cell cancer (HNSCC) UICC stage II-IV with a single intratumoral injection of hIL-2 plasmids formulated in DOTMA/Chol. Hum Gene Ther 10 (1):141–147

Zhi D, Zhang S, Cui S et al (2013) The headgroup evolution of cationic lipids for gene delivery. Bioconjug Chem 24(4):487–519

Zhu L, Mahato RI (2010) Lipid and polymeric carrier-mediated nucleic acid delivery. Expert Opin Drug Deliv 7(10):1209–1226

Nanocarriers in Protein and Peptide Drug Delivery

14

Dipti Gohil and Theivasanthi Thirugnanasambandan

Abstract

Protein and peptide are biomolecules that determine and control biological functions, thereby playing a significant role in treating various pathological conditions. Current research focuses on the development of dosage form for efficient delivery of proteins and peptides. Characteristics of biomolecules make them challenging therapeutic agents to formulate and to deliver. Stability of protein and peptide molecules is a preliminary requirement, ranging from development of dosage form to availability of drug at the site of action. The most convenient oral route is not suitable to deliver protein and peptide molecules due to poor stability of drug in harsh gastric environment. Parenteral route has the potential application to deliver biomolecules, and circumvent drawbacks associated with oral delivery of proteins and peptides. Parenteral route can elicit pharmacological action even with minimal dose of drug, preserving the integrity of molecules. To overcome the drawback of short half-life and patient compliance, associated with intravenous route, an alternate route has also been developed to deliver drugs. Extensive studies have been conducted to deliver protein and peptide molecules effectively using various delivery systems. Among a variety of systems, nanocarriers have emerged as a compelling tool due to very fine particle size, flexibility with a variety of dosage forms and routes of administration, and long circulatory half-life due to avoidance by reticular endothelial system. Further, nanocarriers can also deal with targeted and site-specific drug

D. Gohil (✉)
Department of Pharmacy, Sumandeep Vidyapeeth Deemed to be University, Vadodara, Gujarat, India
e-mail: diptigohil.dop@sumandeepvidyapeethdu.edu.in

T. Thirugnanasambandan
International Research Centre, Kalasalingam Academy of Research and Education (Deemed University), Krishnankoil, Tamil Nadu, India

N. Shah (ed.), *Nanocarriers: Drug Delivery System*,
https://doi.org/10.1007/978-981-33-4497-6_14

delivery. Physicochemical properties of biomolecules play a vital role in the selection of polymer, method of preparation, kind of carrier molecules, and drug delivery system to be used.

Keywords

Nanocarriers · Protein and peptide · Approaches · Method of preparation

14.1 Introduction

Peptides consist of 2–50 amino acids that are the building blocks of proteins. Peptides are easy to absorb by the body than proteins. They are broken down into smaller units easily than proteins. However, the oral bioavailability of most peptides and proteins is less than 1%. Hence, it is difficult for them to protect the drug from the acidity and pepsin activities existing in the stomach. Insulin, gamma globulin, and protein-containing vaccines are some examples of protein-based drugs. The advantage of proteins as therapeutic agents is their compatibility with living systems.

Mucoadhesive nanoparticles are suitable nanocarriers for protein and peptide drugs, due to their increased retention time in the GI tract. Also, it promotes absorption and easily attaches with the mucus layer that increases the concentration gradient of the drug. Cao et al. have reported that protein and peptide drugs possess high bioactivity, specificity, strong solubility, and low toxicity. However, chemical modification of nanocarriers is done to improve the bioavailability of proteins and peptides. Figure 14.1 exhibits the barriers and transport mechanisms in drug delivery (Cao et al. 2019).

The limitations for the transportation of protein drugs in the body are (1) high molecular weight that prevents them to cross tissue barriers, (2) short lifetime due to immunoresponse, and (3) enzymatic degradation. Due to the limitations mentioned above, protein drugs are delivered primarily using parenteral route. Further, parenteral route suffers from some drawbacks like low patient compliance, pain at the site of injection, less stability, and quite expensive dosage form. Figure 14.2 exhibits the structure of a unilamellar liposome utilized in protein-based drug delivery (Solaro et al. 2010).

The structure possesses the protein drug that has been covered by phospholipid layer with an outermost polymer (PEG) coating. The targeting moiety has been interlinked with phospholipid layer by polymer (Solaro et al. 2010).

Peptides are used as drugs for diabetes, oncology, and cardiovascular and infectious diseases. Polymer–peptide conjugation is performed by physical encapsulation techniques, which are divided into surfactant-based techniques and polymer carriers. In surfactant-based techniques the carriers for peptides are liposomes, microemulsion, and solid-lipid nanoparticles. The other carriers for peptides include polymer-decorated liposomes, solid microspheres, polyelectrolyte complex, emulsions, hydrogels, and injectable polymers (Du and Stenzel 2014). Proteins

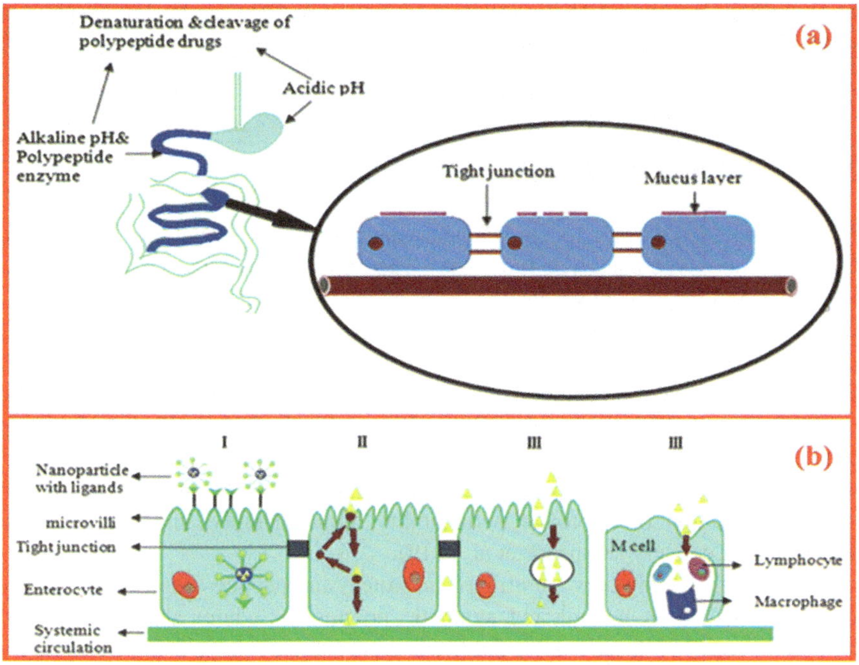

Fig. 14.1 Schematic diagram shows the drug delivery. (**a**) Main barriers in the oral delivery of peptide- and protein-based drugs, (**b**) transport mechanisms: (I) Receptor-mediated transport. (II) Carrier-mediated transport. (III) Paracellular transport. (IV) Phagocytosis by M cells (Cao et al. 2019)

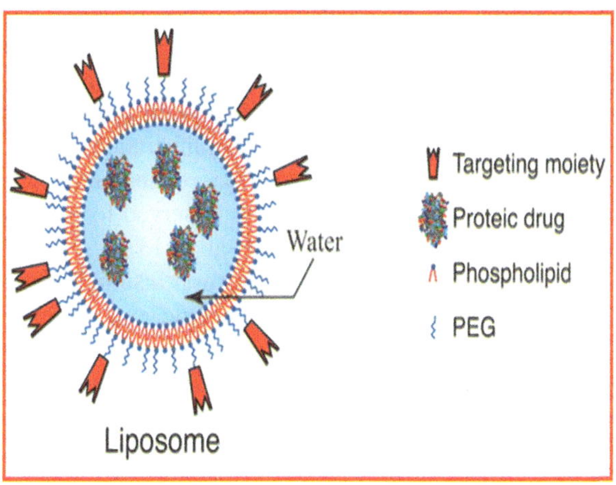

Fig. 14.2 Schematic diagram shows the structure of a unilamellar liposome utilized in drug delivery

and peptides offer more specific mode of action and so they can be administered at relatively low doses for therapeutic effects (Tan et al. 2010).

14.2 Material Used to Prepare Nanocarriers

Major population is affected by diabetes and insulin is a vital therapy for the management of type I and type II diabetes. The only efficient route to deliver insulin is parenteral route. To deliver insulin using an alternate route through nanotechnology is an emerging field. Combination of polymers and nanocarriers plays a significant role in the delivery of insulin. Natural polymers like chitosan, alginate, hyaluronic acid, dextran, gelatin, and synthetic polymers like polylactic-co-glycolic acid (PLGA), polycaprolactone (PCL), and polyvinyl alcohol (PVA) which are biodegradable in nature are widely used to prepare nanocarriers to deliver protein- and peptide-based drugs. Nanotechnology-based drug delivery system is developed with the intention of improved permeability, bioavailability, and modified and targeted drug delivery (Mansoor et al. 2019).

Biodegradable polymers are utilized in protein and peptide drug delivery. They can be classified as natural and synthetic polymers. Sodium alginate, dextran, hyaluronic acid, chitosan, and gelatin are some natural polymers. PLGA—poly (lactic-co-glycolic acid), PCL—poly(e-caprolactone), polyaminoacid, pluronic P-123, and polyvinyl alcohol are some synthetic polymers. Chemical structure of these natural and synthetic polymers is shown in Fig. 14.3.

Insulin- and glucose-specific enzyme-loaded dextran nanoparticles were prepared using electrostatic interaction. Oppositely charged nanoparticles interact forming a porous structure designed to release insulin in hyperglycemic condition through glucose-specific enzyme which can convert glucose to glucuronic acid. Acid degradation of polymer is responsible for the release of insulin from nanoparticles. Release of drug continued until there was a change in concentration of glucose. Developed nanoparticles upon subcutaneous administration in mice with type I diabetes showed sufficient reduction in glucose concentration level. Consistent blood glucose concentration (<200 mg/dL) was observed for 10 days upon single administration of developed dosage form (Gu et al. 2013).

Chitosan is a biopolymer that can be used for oral insulin delivery because it is more biocompatible, biodegradable, non-immunogenic, and nontoxic. For example, modified chitosan, namely succinyl chitosan, possesses good pH-sensitive swelling due to the presence of carboxyl and provides effective protection to the insulin in the highly acidic environment of the stomach (Mukhopadhyay et al. 2012).

PLGA, a biodegradable polymer, is used to prepare nanoparticles and allowed to get adsorbed on the surface of porous microsphere. Leuprolide acetate was incorporated by immersion method on the surface of previously prepared blank nanoparticles. Higher encapsulation efficiency and thereby sustained drug release were observed. Drug release kinetic study confirmed that Higuchi model showed Fickian drug release. Developed process provided potential alternative to deliver

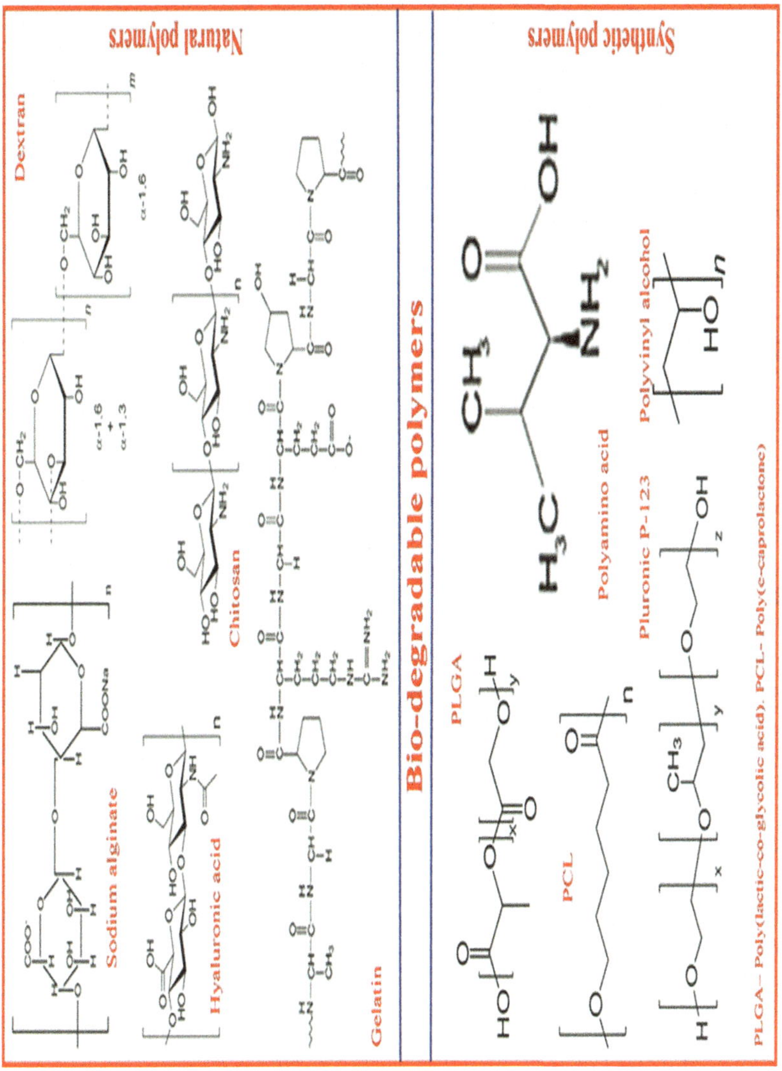

Fig. 14.3 Chemical structure of various biodegradable polymers utilized in the nano-insulin formulation delivery

protein- or peptide-based drug using depot preparation, which can be delivered by intramuscular or subcutaneous route (Alcalá-Alcalá et al. 2013).

The polyelectronic nanocomplexes made of CS-g-polyethylene glycol monomethyl ether (mPEG) copolymers with different mPEG graft ratios are synthesized by self-assembly method for oral insulin delivery. Polyelectronic complexes loaded with insulin were prepared by using electrostatic interaction between oppositely charged polymer and drug. Their physicochemical properties, surface hydrophilicity, and interaction with mucus are analyzed. The enhanced absorption of the drug is achieved for mPEG graft ratio of 10% (Liu et al. 2019).

Exenatide was delivered orally using nanoparticles for type 2 diabetes to improve oral absorption. Nanoparticles were designed to enhance intestinal absorption by using synthesized block copolymer, CSKSSDYQC-dextran-poly(lactic-co-glycolic acid) (CSK-DEX-PLGA). Modified nanoparticles prepared using synthesized block copolymer showed targeted drug delivery to intestinal epithelial cell. Further, in vivo study confirmed effective drug delivery with optimal hypoglycemic activity (Song et al. 2019).

Polyelectrolyte complexes (PEC) have the potential for effective oral delivery of insulin with additional advantage of controlled drug delivery. Chitosan-coated nanoparticles and alginate-coated nanoparticles were prepared separately. Double-emulsion method was used to self-assemble two oppositely charged nanoparticles by electrostatic interaction. Developed pH-sensitive nanoparticles containing insulin showed sufficient hypoglycemic activity for sustained period of time upon oral administration to diabetic rats (Chen et al. 2019).

Modification of chitosan is done by ionic cross-linking with hydroxypropyl methylcellulose phthalate (HPMCP) as a pH-sensitive polymer and applied for the oral delivery of insulin. The resulting nanocarrier shows a superior acid stability in in vitro studies. It also shows a significant control over insulin release and degradation in simulated acidic conditions with or without pepsin. Also, fluorescently labeled CS/HPMCP NPs showed a two- to fourfold increase in the intestinal mucoadhesion and penetration compared to chitosan/tripolyphosphate NPs (Makhlof et al. 2011).

To improve the stability of insulin, chitosan- and mucin-grafted microparticles were prepared by double-emulsion method. Prepared nanoparticles were subjected to various characterization parameters. Result exhibited developed nanoparticles with sufficient loading and encapsulation efficiency. In vitro drug release behavior was assessed in gastric and intestinal environment. In vitro and in vivo drug release showed satisfactory result, up to 12 h for sustained drug delivery (Mumuni et al. 2019).

14.3 Approaches Used to Deliver Proteins and Peptides

14.3.1 Chemical Modification

Chemical modification is addition or removal or conjugation of some functional groups to modify the structure of the drug. Modification in the chemical structure of drug leads to alteration in physical, chemical, and biological property like solubility, melting point, susceptibility to enzymatic hydrolysis, permeability, and therapeutic effect. Modification in the chemical structure of protein and peptide on amino acid part leads to change in immunogenicity and pharmacological activity (Mahato et al. 2003).

The most commonly used method is to change the isomeric form of drug, which alters solubility, permeability, and pharmacological activity also in some cases. It can be explained by the common example of peptide desmopressin which is a hormone analogue of vasopressin. To synthesize desmopressin, N-terminal amino acid of vasopressin was deaminated, and the C-terminal L-arginine residue was substituted with a D-arginine residue. Arginine vasopressin is a synthetic analogue of naturally occurring hormone vasopressin, which is mainly responsible for antidiuretic activity. Common symptoms of diabetic patients are polyuria, polyphagia, and polydipsia. Synthetic arginine vasopressin can be utilized to treat symptomatic condition but due to physicochemical characteristics, the only available formulation is parenteral injection. Due to great hydrophilicity, it shows very short biological half-life and hence requires frequent administration. In contrast, upon changing isomeric form, use of D-arginine vasopressin known as desmopressin exhibits less hydrophilic character and improved bioavailability due to less proteolytic degradation upon oral administration (Shaji and Patole 2008). Another method utilizes introduction of hydrophobic moiety in the structure of protein and peptide. As a hydrophobic moiety, fatty acid and fatty acid ester can be incorporated in the structure of peptide which increases lipophilicity of drug and thereby also influences permeability, circulatory time, and stability (Deb et al. 2014).

14.3.2 PEGylation

Polyethylene glycol (PEG) is available in different physical forms ranging among liquid, semisolid, to solid depending on their molecular weight; chemical structure can be linear or branched form. PEG is categorized as a semipolar solvent due to its solubility in both aqueous and organic solvents depending on the number of ethylene glycol unit present in its structure. PEGylation is the process of attachment of one or more polyethylene glycol (PEG) moiety to drug by covalent bonding. PEGylation of protein and peptide molecules can alter pharmacokinetic profile by providing long circulatory time in systemic circulation upon parenteral administration. This characteristic of PEGylation makes it a suitable approach for delivery of protein and peptide for oral administration. PEGylation of biomolecules enhances chemical and physical stability by reducing susceptibility to proteolysis and also preventing agglomeration (Roberts et al. 2002; Veronese 2001). PEG increases the size of biomolecules and thereby is not recognized by reticuloendothelial system (RES) which prevents it from clearance by excretory organs. PEG molecule is biodegradable and non-immunogenic in nature. Process of PEGylation covers the therapeutic agent with large structure of polyethylene glycol and protects the biomolecules from recognition by immune system. Further, PEG keeps the structure of biomolecules intact, thereby preserving the biological activity of molecules. Approved marketed formulation of PEGylated insulin shows efficient hypoglycemic effect (Hamman et al. 2005).

14.3.3 Enzyme Inhibitor

Upon oral administration of protein and peptide, enzymes present in gastrointestinal tract cause degradation and denaturation of biomolecules. Protease enzyme is responsible for hydrolysis of peptide bond present in the structure of proteins and peptides.

This leads to reduction in the stability of biomolecules. Further, very less amount of drug would be available for absorption and thereby less amount of bioavailable drug may not exhibit effective pharmacological activity. The approach used for successful delivery of protein and peptide molecules by oral administration requires incorporation of enzyme inhibitor along with biomolecules. Sufficient higher concentration of enzyme inhibitor can potentially prevent degradation of biomolecules (Park et al. 2011).

Major drawback to the use of enzyme inhibitor is nonspecificity of enzyme to degrade protein. Enzyme inhibitors cause inhibition of a variety of enzymes present in gastrointestinal tract which leads to unobvious absorption of protein without metabolism. Protein absorption without degradation causes metabolic changes to occur in GIT and may lead to toxic effects. Enzyme is active at specific pH conditions. This principle can be used to modify the functioning of enzymes. A strategy can be used to alter the pH of biological fluid in specific parts of gastrointestinal tract, compared to use of very large amount of enzyme inhibitors (Renukuntla et al. 2013).

14.3.4 Absorption Enhancer

Absorption enhancers are substances that enhance the absorption of therapeutic proteins and peptides when administered with biomolecules. Absorption enhancers mainly act by disrupting the physical barrier present on intestine wall and thereby facilitate absorption of biomolecules, though disruption of intestinal wall is reversible in nature.

This provides a path for absorption of biomolecules after oral administration (Williams and Barry 2004). Absorption enhancers use various mechanisms for functioning. One of the mechanisms is the temporary opening in the epithelial cell of intestinal wall which allows the absorption of proteins and peptides. Structural modification in the membrane of epithelial cell allows improved absorption of protein and peptide through the cell, i.e., intracellular pathway. Another mechanism involves absorption of biomolecules through tight junctions of epithelial cells. This strategy involves movement of drug through intercellular space of epithelial cell of intestine, i.e., intercellular pathway (Hamman and Steenekamp 2011).

Apart from the above-discussed mechanism, based on other mechanisms, there are various types of absorption enhancers used. Some other possible mechanisms are used as chelating agents, such as ethylenediaminetetraacetic acid (EDTA). Chelating agent forms a complex with metal ion present on cell membrane, which causes opening of membrane-gated channels that allows the movement of biomolecules through tight junction of epithelial cells. Another possible mechanism is the use of surfactant like sodium lauryl sulfate, bile salts, fatty acid, polysorbate, and tweens. Surfactant damages cell membrane and thereby improves penetration of protein and peptide.

Absorption enhancers work by various mechanisms but more or less involve alteration of cell structure or damaging or rupturing of cell membrane which ultimately causes damage to intestinal cell wall. Hence, it is having limited application for enhancing the bioavailability of proteins and peptides (Renukuntla et al. 2013).

14.4 Nanocarriers Used for Delivery of Proteins and Peptides

14.4.1 Polymeric Nanoparticles

Polymeric nanoparticles are nanoparticulate systems composed of small colloidal particles synthesized using polymers. There are various types of polymers which can be used to prepare nanoparticles. Polymers can be classified as natural, semisynthetic, and synthetic polymers. Another class has also emerged as biodegradable polymer which has potential application in preparing nanoparticles and to be incorporated in other dosage forms. Biodegradable polymers are widely used due to its characteristics like compatibility with biological medium, flexibility with a variety of dosage forms, non-mutagenicity, and sustained and controlled drug release. Polymeric nanoparticles are also used in the preparation of novel dosage forms like depot injection, implant, and biological products like vaccine. There are various methods to prepare polymeric nanoparticles. Some of them are discussed below.

The small intestine can transport the L-forms of amino acids against a concentration gradient and so L-valine is used as a ligand for the transport of insulin-loaded PLGA nanoparticles. L-Valine-conjugated PLGA nanoparticles are synthesized by double-emulsion solvent evaporation method. Polymer is conjugated with amino acid using carbodiimide coupling process, and involves addition of activated polymer solution in amino acid solution under magnetic stirring for a specific period of time. Addition of diethyl ether, previously stored at freezing temperature, precipitates out PLGA-valine conjugate in the form of nanoparticles. Obtained nanoparticles are washed with deionized water and subjected to subsequent centrifugation and lyophilization (Vishwakarma et al. 2018).

Chitosan (CS) and poly(γ-glutamic acid) (γ-PGA) were used to prepare nanoparticles using simple ionic gelation method. Aqueous solution of poly (γ-glutamic acid) (γ-PGA) was added dropwise into the aqueous solution of chitosan under magnetic stirring at room temperature. Due to surface charge present on polymer, nanoparticles were formed by interaction between oppositely charged particles and ionic bond. Obtained nanoparticles were collected by ultracentrifugation and washed with deionized water (Lin et al. 2007).

Primary emulsion was prepared by mixing aqueous phase of insulin and oil phase of polymer (PLGA/HP55 in methylene chloride:acetone at a ratio of 3:2) under sonication for a specific period of time. Primary emulsion is incorporated in external aqueous phase containing 1% polyvinyl alcohol to form w/o/w emulsion. Multiple emulsions are stirred to evaporate oil phase and thereby precipitated nanoparticles are collected by centrifugation (Wu et al. 2012).

Nanoparticles of chitosan and poly(γ-glutamic acid) (γ-PGA) are synthesized by ionic gelation method for oral insulin delivery. Fourier transform infrared (FT-IR) spectra reveal that CS and γ-PGA are ionized at pH 2.5–6.6. The XRD pattern shows the disruption of crystal structure of chitosan after it is combined with γ-PGA. The in vivo results show that insulin-loaded NPs reduce blood glucose level in a diabetic rat model (Lin et al. 2007).

Dextran sulfate and chitosan, two oppositely charged polymers, were used to form block copolymer. Block polymer was formed by ionic gelation at a specified condition of temperature and pH. Insulin was preciously dissolved in dextran sulfate solution before dropwise addition in chitosan solution under magnetic stirring. Both polymeric solutions were maintained at different pH conditions. Prepared nanoparticles were evaluated for hypoglycemic activity in diabetic rats (Sarmento et al. 2007).

14.4.2 Anionic Nanoparticles

Anionic nanoparticles are negatively charged colloidal particles, usually of very small size. Even with less particle size, they are capable of entrapping therapeutic agents with large molecular size and large molecular weight like protein and peptide molecules. Due to negative surface charge, permeability of nanoparticles is enhanced through tight junctions of epithelial cell of intestinal wall upon oral administration of biomolecules.

Anionic nanoparticles allow oral delivery of proteins by the process of tight junction relaxation and increasing intestinal permeability. This permeation-enhancing effect depends on the nanoparticle size, charge, and more negative particles such as silica. The oral delivery of insulin and exenatide is tested in mice with silica nanoparticles. In healthy, hyperglycemic, and diabetic mice, the oral delivery of 10 U/kg insulin shows bioactivity of, respectively, 35%, 29%, and 23%

that of the subcutaneous injection of 1 U/kg insulin. The permeation-enhancing effect of the nanoparticles is attributed to the binding to integrin on the surface of epithelial cells (Lamson et al. 2020).

14.4.3 Microemulsion

Microemulsion is a thermodynamically stable system consisting of oil phase and aqueous phase with greater proportion of surfactant and co-surfactant. It is widely used to deliver biomolecules which are prone to degradation. This approach has potential application in delivering protein and peptide molecules due to some characteristics like ease of preparation, less viscosity, high viscosity, sustained release, protection of drug in internal phase, and prevention from exposure to external condition which can degrade drug.

Water-in-oil type of emulsion is widely used to deliver peptide-based drugs to improve oral absorption through GI tract. Microemulsion used light liquid paraffin as an oil phase; Tween 80 and snail mucin powder were used as surfactant and co-surfactant, respectively. Insulin was incorporated in internal aqueous phase containing surfactant. Aqueous phase was dispersed in oil phase using homogenizer for a specific period of time (Momoh et al. 2020).

14.4.4 Liposomes

Liposomes are spherical vesicles mainly composed of phospholipid bilayer and cholesterol. Liposome has emerged as a potential drug delivery system due to small size and its capacity to deliver hydrophilic and hydrophobic drugs efficiently. Liposomes have different applications depending on particle size, surface charge, and method of preparation used. Based upon selected composition of excipients for preparation of liposomes, vesicles show rigidity or fluidity. Liposomes can be widely used to deliver a variety of drugs for site-specific as well as targeted drug delivery.

Liposomes are used to deliver insulin specifically to hepatocytes because they selectively target insulin to the liver by increasing oral absorption of insulin (Spangler 1990). The surface of liposomes is coated with poly (ethylene glycol) or sugar chain of mucin to enhance the oral absorption of insulin. Liposome is prepared by thin-film hydration evaporation method. Phospholipid, cholesterol, and surface-coating material (cetyl-mucin or DSPE-PEG) are dissolved in chloroform and subjected to rotary evaporator under vacuum. Thin film is formed due to evaporation of solvent which is finally hydrated using phosphate buffer solution containing insulin. Prepared liposomes are extracted from suspension using centrifugation and characterized for various parameters. PEG-Lip interacts strongly with the intestinal mucous layer, thus rendering slow transit in the intestine (Iwanaga et al. 1999).

Liposomes coated with silica nanoparticles have more stability and encapsulation efficiency and can be used as protein delivery vehicles. Thin-film hydration is performed with insulin and fluidic phosphatidylcholine lipid vesicles at pH 2.5. A layer of silica is formed above lipid bilayer by acid catalysis. The presence of silica coating and insulin is confirmed by confocal electron microscopy and confocal micro-Raman spectroscopy. Silica coating is helpful to increase the stability of liposomes and the formulations reduce the glucose level effectively which is confirmed through in vivo studies (Dwivedi et al. 2010).

14.5 Route of Administration

Various routes of administration are used to deliver proteins and peptides using nanocarriers:

(a) Oral route
(b) Nasal and pulmonary route
(c) Parenteral route
(d) Ocular route
(e) Transdermal route

14.5.1 Oral Route

Oral route is the most convenient noninvasive route for administration of drug with added advantage of greater patient compliance (Kaintura et al. 2015; Gulfam et al. 2012). But protein and peptide show poor bioavailability due to physicochemical barriers like enzymatic degradation, short plasma half-life, ion permeability, immunogenicity, denaturation, aggregation, and poor GI absorption due to large molecular size and high molecular weight (Mirshahi et al. 2002). Most of the proteins and peptides degrade in harsh environment of gastrointestinal tract in the presence of gastric acid and protease enzyme which are responsible for degradation and denaturation. Due to these constraints, it is very difficult to attain sufficient oral bioavailability. Numerous researches have been carried out for successful delivery of protein and peptide molecules by oral route (Morishita and Peppas 2006).

14.5.2 Nasal and Pulmonary Route

Application of nasal and pulmonary route involves administration of drug using novel approaches for local and systemic drug delivery showing promising result. Nasal route is noninvasive for effective drug delivery with potential advantage of large self-administration, larger surface area, high vascularization, bypass of first-pass hepatic metabolism, and low enzymatic activity compared to GIT, etc. Nasal route is proven to be having comparable bioavailability to that of systemic administration of drug; very small dose of drug is also effective to elicit therapeutic effect. Development of nasal and pulmonary administration of drug requires focusing on physiological conditions, physicochemical properties of drugs, dosage form design, and type and design of device used to deliver the drug. Drug administered through this route has also proven to cross blood-brain barrier (BBB) through olfactory region, i.e., nose-to-brain drug delivery (Wu et al. 2008).

14.5.3 Parenteral Route

Parenteral route is the most effective route for administration of proteins and peptides as physicochemical barriers of oral route like exposure of drug to highly acidic media, presence of protein-degrading enzyme, first-pass hepatic metabolism, and less absorption due to large molecular weight and size can be overcome. Intramuscular and subcutaneous routes are two most common routes of administration. Intravenous route offers the advantages of rapid onset of action, nearly 100% bioavailability with low chances of degradation, and pharmacologic effectiveness at even low dose of drug. For administration by intramuscular and subcutaneous routes, drugs can be formulated as solution or suspension. At the site of injection, formulation acts as a reservoir from which the drug is slowly released into systemic circulation for a sustained period of time. This approach is used for controlled/ sustained drug delivery. It has been observed that solubility and stability are major challenges to formulate a dosage form. Solution can be easily formulated by selecting appropriate excipients. To increase the stability of dosage form, freeze-drying can be a suitable alternate (Jain et al. 2019; Agrawal et al. 2011).

14.5.4 Ocular Route

Some of the peptides used to treat ophthalmological condition can be delivered through this route. As eye is a site of action, drug can be easily administered for local action. Patient compliance is observed to be very low as eye is a very sensitive organ to the presence of any particulate matter, foreign particles, pH, osmolarity of dosage form, and size of suspended particles. Approaches to improve ocular absorption can be use of nanocarriers like nanoparticles, liposomes, ocular insert, and mucoadhesive system (Mandal et al. 2018).

14.5.5 Transdermal Route

Transdermal route is an alternate route to oral route to deliver drug through the skin. Transdermal route offers the advantage of ease of administration of drug, delivery of a variety of drugs, and also bypass of the drawbacks of oral administration. Though a number of drugs are available as marketed formulation to be delivered via topical route, it is very difficult to permeate through stratum corneum due to physicochemical barriers present in the skin. Transdermal drug delivery involves dermatological application with the objective to deliver drug systematically by transiting through dermal layers of skin. Physicochemical characteristics of drug like molecular weight, molecular size, partition coefficient, and physiological condition of the skin are the factors that affect permeation of drug to the skin. Some of the approaches have been used to deliver protein and peptide through transdermal route like iontophoresis, sonophoresis, microneedle array, use of prodrug, or permeation enhancer (Florence and Attwood 2011).

14.6 Marketed Formulation

A number of protein- and peptide-based dosage forms has been commercialized. Few examples of marketed formulation of peptides are discussed in Table 14.1.

Table 14.1 Examples of marketed formulation of protein and peptide drugs

Route of administration	Drug	Dosage form	Trade name	Company name
Oral	Cyclosporine	Capsules	Neoral® Soft Gelatin Capsules	Novartis Pharmaceuticals UK Ltd.
	Cyclosporine	Oral solution	Neoral® Oral Solution	Novartis Pharmaceuticals UK Ltd.
	Desmopressin acetate	Tablet	Minirin	Ferring Pharmaceuticals Pvt. Ltd.
Nasal	Desmopressin acetate	Nasal spray	Minirin	Ferring Pharmaceuticals Pvt. Ltd.
	Calcitonin (salmon)	Nasal spray, MDI	Miacalcin	Physicians Total Care, Inc.
	Nafarelin acetate	Nasal spray, MDI	Synarel	Pfizer Canada Ulc
	Oxytocin	Nasal spray	Syntocinon	Novartis
Parenteral	Calcitonin (salmon)	Intramuscular; subcutaneous	Calcimar	Sanofi Aventis
	Oxytocin	Intramuscular, intravenous	Pitocin	Baxter Laboratories
	Leuprolide acetate	Subcutaneous	Lupron®	Sun Pharmaceutical Industries, Inc.
Ocular	Cyclosporine	Solution/drops	Cequa	Sun Pharmaceutical Industries, Inc.

14.7 Commercial Challenges of Protein Drug Delivery

Nowadays, proteins and peptides are mainly used for diagnostic purpose as well as treatment of many diseases like diabetes and cancer which require long-term treatment. Hence, dosage form development requires focusing on delivery of drug for longer period of time in sustained/controlled manner. In 1922, insulin was first discovered as a therapeutic protein to be delivered to patients. With time, many of the protein and peptide candidates have been investigated to be commercialized. To resemble the natural structure of protein and peptide, biotechnology-based recombinant DNA technology is widely used to develop and commercialize protein and peptide candidate production at large scale with high purity. Primary route to deliver protein and peptide is parenteral route. Most of the protein and peptide candidates are commercially developed to be delivered by parenteral route. It is found to be the most suitable, very successful, and effective route for delivery of drug though invasive in nature. Some noninvasive route was also investigated for delivery of protein therapeutics. Nasal, ocular, transdermal, and even oral routes were found to be effective to deliver few protein and peptide molecules. Few molecules have also been used for sustained/controlled drug delivery for a longer period of time using polymeric nanoparticles. Some of the approaches like PEGylation, glycosylation, conjugation with amino acid, and chemical modification are also used to improve bioavailability of protein and peptide.

Scale-up of protein and peptide therapeutics to be delivered via nanocarriers will remain limited due to hurdles like solubility, permeability, stability, chemical diversity, and pharmacodynamic and pharmacokinetic pattern of therapeutic molecules (Usmani et al. 2017).

References

Agrawal H, Thacker N, Misra A (2011) Parenteral delivery of peptides and proteins. In: Challenges in delivery of therapeutic genomics and proteomics. Elsevier, Amsterdam, pp 531–622

Alcalá-Alcalá S, Urbán-Morlán Z, Aguilar-Rosas I et al (2013) A biodegradable polymeric system for peptide–protein delivery assembled with porous microspheres and nanoparticles, using an adsorption/infiltration process. Int J Nanomedicine 8:2141

Cao SJ, Xu S, Wang HM et al (2019) Nanoparticles: oral delivery for protein and peptide drugs. AAPS PharmSciTech 20(5):190

Chen T, Li S, Zhu W et al (2019) Self-assembly pH-sensitive chitosan/alginate coated polyelectrolyte complexes for oral delivery of insulin. J Microencapsul 36(1):96–107

Deb PK, Ahmad J, Dina ER et al (2014) Molecular docking studies and comparative binding mode analysis of FDA approved HIV protease inhibitors. Asian J Chem 26:6227–6232

Du AW, Stenzel MH (2014) Drug carriers for the delivery of therapeutic peptides. Biomacromolecules 15(4):1097–1114

Dwivedi N, Arunagirinathan MA, Sharma S et al (2010) Silica-coated liposomes for insulin delivery. J Nanomater 2010:652048

Florence AT, Attwood D (2011) Peptides, proteins and other biopharmaceuticals. In: Physicochemical principles of pharmacy. Pharmaceutical Press Pub, London, pp 451–476

Gu Z, Aimetti AA, Wang Q et al (2013) Injectable nano-network for glucose-mediated insulin delivery. ACS Nano 7(5):4194–4201

Gulfam M, Kim JE, Lee JM et al (2012) Anticancer drug-loaded gliadin nanoparticles induce apoptosis in breast cancer cells. Langmuir 28(21):8216–8223

Hamman JH, Steenekamp JH (2011) Oral peptide drug delivery: strategies to overcome challenges. In: Peptide drug discovery and development: translational research in academia and industry. Wiley, Weinheim, pp 71–90

Hamman JH, Enslin GM, Kotzé AF et al (2005) Oral delivery of peptide drugs. Bio Drugs 19 (3):165–177

Iwanaga K, Ono S, Narioka K et al (1999) Application of surface-coated liposomes for oral delivery of peptide: effects of coating the liposome's surface on the GI transit of insulin. J Pharm Sci 88 (2):248–252

Jain D, Mahammad SS, Singh PP et al (2019) A review on parenteral delivery of peptides and proteins. Drug Dev Ind Pharm 45(9):1403–1420

Kaintura R, Sharma P, Singh S et al (2015) Gelatin nanoparticles as a delivery system for proteins. J Nanomed Res 2(1):1–3

Lamson NG, Berger A, Fein KC et al (2020) Anionic nanoparticles enable the oral delivery of proteins by enhancing intestinal permeability. Nat Biomed Eng 4(1):84–96

Lin YH, Mi FL, Chen CT et al (2007) Preparation and characterization of nanoparticles shelled with chitosan for oral insulin delivery. Biomacromolecules 8(1):146–152

Liu C, Kou Y, Zhang X et al (2019) Enhanced oral insulin delivery via surface hydrophilic modification of chitosan copolymer based self-assembly polyelectrolyte nanocomplex. Int J Pharm 554:36–47

Mahato RI, Narang AS, Thoma L et al (2003) Emerging trends in oral delivery of peptide and protein drugs. Crit Rev Ther Drug Carrier Syst 20(2–3):153–214

Makhlof A, Tozuka Y, Takeuchi H (2011) Design and evaluation of novel pH-sensitive chitosan nanoparticles for oral insulin delivery. Eur J Pharm Sci 42(5):445–451

Mandal A, Pal D, Agrahari V et al (2018) Ocular delivery of proteins and peptides: challenges and novel formulation approaches. Adv Drug Deliv Rev 126:67–95

Mansoor S, Kondiah PP, Choonara YE et al (2019) Polymer-based nanoparticle strategies for insulin delivery. Polymers 11(9):1380

Mirshahi T, Irache JM, Nicolas C et al (2002) Adaptive immune responses of legumin nanoparticles. J Drug Target 10(8):625–631

Momoh MA, Franklin KC, Agbo CP et al (2020) Microemulsion-based approach for oral delivery of insulin: formulation design and characterization. Heliyon 6(3):e03650

Morishita M, Peppas NA (2006) Is the oral route possible for peptide and protein drug delivery? Drug Discov Today 11(19–20):905–910

Mukhopadhyay P, Mishra R, Rana D et al (2012) Strategies for effective oral insulin delivery with modified chitosan nanoparticles: a review. Prog Polym Sci 37(11):1457–1475

Mumuni MA, Kenechukwu FC, Ernest OC et al (2019) Surface-modified mucoadhesive microparticles as a controlled release system for oral delivery of insulin. Heliyon 5(9):e02366

Park K, Kwon IC, Park K et al (2011) Oral protein delivery: current status and future prospect. React Funct Polym 71(3):280–287

Renukuntla J, Vadlapudi AD, Patel A et al (2013) Approaches for enhancing oral bioavailability of peptides and proteins. Int J Pharm 447(1–2):75–93

Roberts MJ, Bentley MD, Harris JM et al (2002) Chemistry for peptide and protein PEGylation. Adv Drug Deliv Rev 54(4):459–476

Sarmento B, Ribeiro A, Veiga F et al (2007) Oral bioavailability of insulin contained in polysaccharide nanoparticles. Biomacromolecules 8(10):3054–3060

Shaji J, Patole V (2008) Protein and peptide drug delivery: oral approaches. Indian J Pharm Sci 70 (3):269–277

Solaro R, Chiellini F, Battisti A (2010) Targeted delivery of protein drugs by nanocarriers. Materials 3(3):1928–1980

Song Y, Shi Y, Zhang L et al (2019) Synthesis of CSK-DEX-PLGA nanoparticles for the oral delivery of exenatide to improve its mucus penetration and intestinal absorption. Mol Pharm 16 (2):518–532
Spangler RS (1990) Insulin administration via liposomes. Diabetes Care 13(9):911–922
Tan ML, Choong PF, Dass CR (2010) Recent developments in liposomes, microparticles and nanoparticles for protein and peptide drug delivery. Peptides 31(1):184–193
Usmani SS, Bedi G, Samuel JS et al (2017) THPdb: database of FDA-approved peptide and protein therapeutics. PLoS One 12(7):e0181748
Veronese FM (2001) Peptide and protein PEGylation: a review of problems and solutions. Biomaterials 22(5):405–417
Vishwakarma P, Mishra S, Kulkarni S et al (2018) L-valine combined PLGA nanoparticles for oral insulin delivery. J Drug Deliv Ther 8(6-A):93–101
Williams AC, Barry BW (2004) Penetration enhancers. Adv Drug Deliv Rev 56(5):603–618
Wu Y, MacKay JA, McDaniel JR et al (2008) Fabrication of elastin-like polypeptide nanoparticles for drug delivery by electrospraying. Biomacromolecules 10(1):19–24
Wu ZM, Ling L, Zhou LY et al (2012) Novel preparation of PLGA/HP55 nanoparticles for oral insulin delivery. Nanoscale Res Lett 7(1):1–8

Part V

Nanocarriers in Oral Cancer

Nanotechnology in Oral Cancer Treatment 15

Chandramani B. More, Rahi M. Brahmbhatt, and Naman R. Rao

Abstract

Oral cancer (OC) is one of the leading causes of increasing global mortality. The nature of its local invasion following metastasis is a growing concern for the entire medical fraternity. Despite multiple approaches towards improvement in conventional diagnostics and treatment modalities, global mortality has not shown a significant downfall. Hence, this research focuses on developing novel diagnostic techniques and drug delivery systems to improve the morbidity and mortality numbers. The utilization of nanotechnology for such innovative diagnostic and therapeutic methods has become increasingly valuable for current researchers. This chapter provides in-depth scientific literature regarding the use of nanotechnology in OC treatment.

Keywords

Head and neck cancer · Nanotechnology · Oral cancer · Oropharyngeal cancer · Treatment modalities · Target delivery

C. B. More
Department of Oral Medicine and Maxillofacial Radiology, K. M. Shah Dental College and Hospital, Sumandeep Vidyapeeth Deemed to be University, Vadodara, Gujarat, India
e-mail: drchandramanimore@sumandeepvidyapeethdu.edu.in

R. M. Brahmbhatt
Department of Neuroscience, University of Illinois at Chicago, Chicago, IL, USA
e-mail: rbrahm2@uic.edu

N. R. Rao (✉)
Harvard Medical School, Harvard University, Boston, MA, USA

Henry M. Goldman School of Dental Medicine, Boston, MA, USA
e-mail: namanrao@bu.edu; naman.rao0@gmail.com

N. Shah (ed.), *Nanocarriers: Drug Delivery System*,
https://doi.org/10.1007/978-981-33-4497-6_15

15.1 Introduction

Oral cancer (OC) is a deleterious condition that forms in the tissues of the oral cavity or the oropharynx. The main subsites include the lip, oral cavity, nasopharynx, and pharynx. Per the World Health Organization (WHO), there are an estimated 657,000 new cases every year, and more than 330,000 deaths globally (WHO). Several risk factors for OC include tobacco use, heavy alcohol use, human papillomavirus (HPV) infection, etc. (Ram et al. 2011). The prevalence of OC goes in hand with these factors. Per the National Institute of Dental and Craniofacial Research (NIDCR), about 10.5 adults per 100,000 will develop OC with higher predilection in males. OC rates also increase with age with prevalence increasing after the age of 50 and peaking between ages 60 and 70 (NIDCR). Due to the advanced ages, the early symptoms of OC are often confused for other problems such as the common cold or tooth pain. In turn, these symptoms can be persistent for weeks or months before a diagnosis of OC is considered. The symptoms may range from persistent mouth pain in the teeth or jaw to nonhealing mouth sores or a lump/thickening in the cheek or neck (Rivera 2015). Tremendous research has been conducted to further our knowledge of this deleterious condition.

In addition to the variable symptoms of OC, there are many divergent conventional methods of treatment. As common with other cancers, surgery and radiation therapy are standard forms of treatment (Omura 2014). Targeted therapy can be combined with radiation as initial treatment in some cases. For advanced cases, chemotherapy and immunotherapy are considered (Ow and Myers 2011). With so many discontinuities in the presentation of diagnosing early OC modalities and OC treatment, it has become essential to design innovative approaches to this enigmatic disease. One such brazen design is the branch of nanotechnology. Nanotechnology is the science and tech that focuses primarily on the control of matter on the atomic and molecular scale. The growing research on the utilization of nanotechnology for the treatment of OC is at present improving the overall outcomes. This chapter presents the latest literature on the utilization of nanotechnology for therapeutics and management of OC.

15.2 Risk Factors and Pathogenesis

15.2.1 Risk Factors

The etiology of OC in humans is unspecified. However, chronic irritation, which is observed in most of the OC patients, is considered as a chief contributing risk factor (Karaca and Ozturk 2019). The other contributing associated risk factors are consumption of substance (tobacco, snuff, and areca nut) and alcohol, poor nutrition status, ionizing radiation at therapeutic, chronic exposure to actinic radiation, action of specific infectious viral traits of human papillomavirus (HPV), genetic factors, and prolonged defective immune response (Rao et al. 2017).

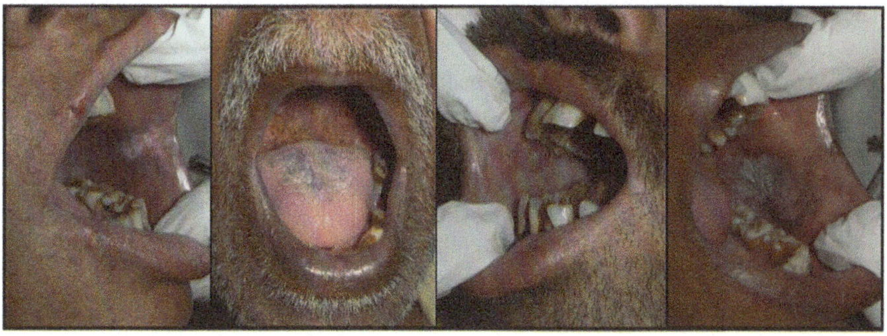

Fig. 15.1 Clinical presentations of oral potentially malignant disorders

Chronic irritation and trauma: Chronic irritation and trauma from sharp teeth, dentures, faulty restorations, and constant trauma from other sharp objects are considered as major risk factors for the development of OC (Piemonte et al. 2010).

Consumption of substance and alcohol: Consumption of large quantities of smokeless and smoking tobacco, areca nut, and other related products and alcohol increases the risk of developing oral potentially malignant disorders like homogeneous leukoplakia, nonhomogeneous leukoplakia, erythroplakia, verrucous leukoplakia, lichen planus, oral submucous fibrosis, and chronic traumatic ulcers (Fig. 15.1) (More et al. 2020). All these diseases are associated with malignant transformation and subsequently its metastasis.

Poor nutritional status: Poor nutritional status, especially in low-income countries, due to lack of consumption of a balanced diet and access to vegetables and fruits, is an associated risk factor for OC (McLaughlin et al. 1988). However, in high-income countries, high consumption of processed food with a lack of natural fibers, proteins, and vitamins may account as a risk factor for OC. Studies also show a reduction of OC to 40–50% on high consumption of fruits and vegetables (Lucenteforte et al. 2009).

Viral infection: Recent studies show that HPV infection may play as an independent risk factor for oral and oropharyngeal cancers. HPV may modulate the carcinogenesis process in the substance-consuming population and potentially act as a primary oncogenic agent amongst the non-substance consumers (McKaig et al. 1998).

Genetic factor: Any inherited gene with its particular form especially deteriorates the cytochrome P450 system in the liver (where the majority of carcinogens are metabolized) and increases the risk for developing OC (Lu et al. 2011).

Defective immune response: A defective immune response as observed in the human immunodeficiency virus (HIV)-infected and acquired immunodeficiency syndrome (AIDS) individual may predispose to cancer (Scully et al. 1991). The most common oral malignancy observed in the HIV-infected patients is Kaposi's sarcoma. OC of the lips is commonly observed in transplant recipients due to the nature of immunosuppressive therapy prior to the transplant.

15.2.2 Pathogenesis

OC is a progressive disease that arises because of multimolecular events that develop from the collective effect of genetic predisposition and environmental exposures to carcinogens (Califano et al. 1996). Cellularly, it initiates from the normal epithelium and passes through several stages starting from hyperplasia to finally transform into its invasive types and their metastasis (Fig. 15.2). Growing genetic and proteomic research has unfolded the molecular pathogenesis of OC. Numerous studies have reported the significance of heredity in oral carcinogenesis. However, the relative risk shows a 1.1–3.8 odds ratio in the first-degree relatives of the patient (Garavello et al. 2007).

There are several genes like glutathione S-transferase mu 1 (GSTM1), cytochrome P450 1A1 (CYPIA 1), and alcohol dehydrogenase 3 genotype that are associated with the genetic predisposition of OC and oropharyngeal cancers (Markopoulos 2012). These genetic damages have the potential for amplification of oncogenes (Table 15.1), which promotes survival and proliferation of the cell. It can also inactivate tumor-suppressor genes that are implicated in cell proliferation inhibition. Largely, these events develop cell dysregulation to an extent where the

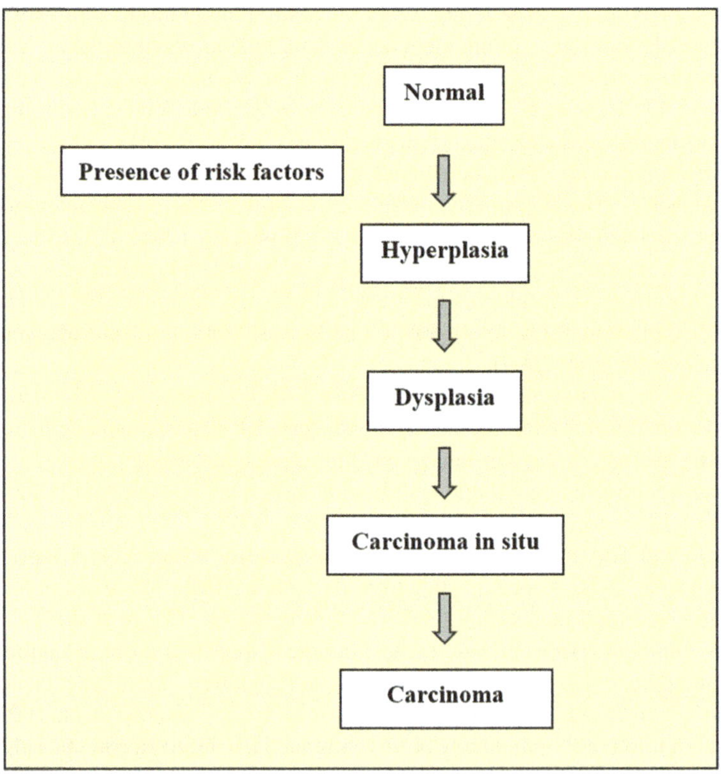

Fig. 15.2 Cellular pathogenesis of OC

Table 15.1 Oncogenes and their functional categories

Oncogenes	Functional category
hst-1, int-2, EGFR/erbB, c-erbB-2/Her-2, sis	Growth factors or growth factor receptors
myc, fos, jun, c-myc	Transcription factors
ras, raf, stat-3	Intracellular signal transducers
bcl-2, bax	Inhibitory factors of apoptosis
Cyclin D1	Cell cycle regulators

growth becomes autonomous and further develops invasive nature. As OC grows and invades, angiogenesis begins (i.e., an essential part of the formation of a tumor).

15.3 Clinical Presentation and Diagnosis

Diagnosis of OC is based on clinical, radiographical, and histological evaluation. Clinically, the frequent sites detected with OC are tongue, the floor of the mouth, and the buccal region (Bagan et al. 2010). The most common symptom of OC is painful or painless nonhealing ulcerated hemorrhagic lesions in addition to cervical lymphadenopathy, tooth mobility, halitosis, breathing problems, dysphagia, speaking difficulties, trismus, paresthesia, and ear pain. Severe bleeding, anemia, cachexia, and fistula are also some symptoms observed in advanced stages. In general, OC is characterized as endophytic, exophytic, or mixed. The endophytic lesions commonly appear as grayish or red ulcers with healed edges, which have a high tendency to bleed easily and high incidence of deep infiltration. However, the exophytic type is warty in appearance and slower growing and has less potential for infiltration.

Conducting biopsy followed by its histopathological examination is considered as a standard confirmatory step for the diagnosis of OC. Squamous cell carcinoma (SCC) is the most detected histopathological type of OC (Schmidt 2006). In addition to the confirmation, it is empirical to identify the staging of OC for designing treatment plans for the patient. Tumor Node and Metastasis (TNM) classification is one such commonly utilized staging instrument for OC (Silva et al. 2011). This staging chiefly depends on the physical examination of the lesion, cervical lymph nodes, and radiological evaluation. The other diagnostic methods include vital staining with toluidine blue, cytology, molecular and salivary biomarkers, salivary biomarkers, and optical techniques (Scully and Bagan 2009).

15.4 Treatment Approaches

The treatment of OC is based on its stage of diagnosis. In the primary stage (stages I and II), surgery and/or radiotherapy is the treatment of choice, whereas a combination of surgery, radiotherapy, or chemotherapy is preferred for stages III and IV (Markopoulos 2012). Due to the exposure to chemotherapeutic agents and radiations, complications like mucositis, xerostomia, and osteonecrosis are very commonly observed. Accordingly, a continuous observation of the oral cavity should be considered as of extreme importance (Wong 2014).

The new-generation therapeutics are constantly working on altered fractionated radiotherapy or concomitant chemoradiotherapy (CT-RT) to increase the efficacy of the radiotherapy, especially for the locally advanced cases (Mazeron et al. 2009). From chemotherapy standpoint, cisplatin-based chemotherapy remains the standard of care for locoregionally advanced head and neck OC (Specenier and Vermorken 2009). Surgically, constant efforts are made to develop techniques for selective neck dissection to reduce the morbidities of radical neck dissection. Recently, there has been tremendous work going on for targeted molecular therapy (gene therapy and monoclonal antibodies) due to its less reported side effects in comparison to the other approaches. This therapy is chiefly focused on four molecules (i.e., cyclooxygenase-2 (COX-2), epidermal growth factor receptor (EGFR), progesterone receptor, and peroxisome proliferator-activated receptor γ (PPARγ)) that are closely associated with differentiation and proliferation of OC cells (Hamakawa et al. 2008).

15.5 Nanotechnology in OC Treatment

Per the National Nanotechnologies Initiative definition, nanoscience and nanotechnology are the study and application of extremely small things and can be used across all the other science fields, such as chemistry, biology, physics, materials science, and engineering. Nanomaterials and nanoparticles are the main terminologies that define nanoscale structures. The growing research has revealed that these nanoscale structures may be utilizable in several ways for the management of cancer.

15.5.1 Nanotherapeutics

- *Nanocarriers*: There are very limited, yet vital, clinical trials performed using nanocarriers. Damascelli and colleagues performed such a trial that introduced intra-arterial paclitaxel combined with human albumin nanocarriers in 23 cases of advanced-stage oral cancer. Of all, four patients showed a complete response with microscopic residual tumor after surgical removal. The study reported very few side effects, which indicates it as a potential treatment for OC (Damascelli et al. 2003).
- Many in vivo and in vitro studies have been performed over a period, amongst which the drug delivery system using a monomeric self-assembled nucleoside nanoparticle and 5-fluorouracil introduced by Zhao et al. (2015) has shown significantly improved antitumor activity and high stability of blood. The study reported minimal side effects (Zhao et al. 2015).

15.5.2 Nanomaterials for Drug Delivery System and Targeting

Nanomaterials have been useful in drug delivery methods due to their unique ability for individual optimization, based on the needs of the pharmacological profile of the drug. Additionally, their nature to improve bioavailability and distribution of the drug from early degradation permits active, passive, and selective targeting that reduces harmful side effects.

Drug targeting through nanomaterials can be done by three approaches: (1) active targeting/ligand-conjugated targeting, (2) passive targeting, and (3) stimulus response targeting (Allen 2004). *Active targeting/ligand-conjugated targeting*: This targeting method initially termed as "magic bullet" was introduced by Paul Ehrlich. This method was introduced to improve the efficacy of the anticancer agents/drugs without harming the normal surrounding tissues. *Passive targeting*: This is a chief targeting method for first-generation nanomaterials based on therapeutic and chemotherapeutic agents. As it is well understood that solid tumors have a disorganized microenvironment (i.e., cancerous and noncancerous cells), it becomes difficult for active and stimulus-responsive targeting. To overcome these challenges, passive targeting has been considered a suitable method. *Stimulus-responsive targeting*: Although active and passive targeting methods can achieve the desired action on cancerous tissues, they lack functional predictability. To overcome this, stimulus-responsive targeting method was introduced. This stimulus can be induced either chemically or physically. These nanomaterials can be chiefly activated by external stimuli and internal stimuli. External stimuli include enzyme activity, change in pH, and reactive oxygen radicals and internal stimuli include light, ultrasound waves, temperature, and irradiation. Study by Cheng and colleagues recommended potential benefits of multi-stimulus-responsive polymeric nanomaterials with incorporation of previous stimuli for programmed drug delivery methods (Cheng et al. 2013).

15.5.3 Nanomaterials as an Adjunct Therapy with Conventional Management

Nanomaterials have also been observed to improve the efficacy of conventional management, especially with radiotherapy and photothermal therapy (Eskiizmir et al. 2017).

Radiotherapy: It is the most commonly used method for treating oral cancer (Grégoire et al. 2015). The major shortcoming of this method is its posttreatment complications. These complications can range from localized toxicities to resistance, and ultimately treatment failure. Accordingly, to reduce these complications, nanomaterials were tested in combination with radiotherapy. The studies with combination showed increased efficacy and improved outcomes with reduced complications. Au, silver, and bismuth are some of those regarded and dense inorganic nanomaterials used as an adjunct to radiotherapy (Retif et al.

2015). Amongst all, Au is highly regarded due to its biocompatibility, high dose enhancement factor, and delivering of high-Z material (Hainfeld et al. 2008).

Photothermal therapy: This therapy is usually used for local treatments causing minimal toxicity. The therapy typically follows the principle of hyperthermia. Resultantly, it may cause cellular lysis, protein denaturation, cytosol evaporation, and cell death. To prevent these harmful effects, nanomaterial-based laser-activated photothermal therapy was introduced. However, the studies are limited to the bench-side research (Pitsillides et al. 2003).

15.5.4 Nanoparticles for Oral Cancer Management

Nanoparticles like gold nanoparticles, magnetic nanoparticles, carbon nanoparticles, and graphene nanoparticles have been a focus for researchers due to its benefits in therapeutic efficacy for cancer management.

Gold nanoparticles: These are usually preferred for the drug delivery systems, radiotherapy, and hyperthermia treatment. The chief benefit of using gold nanoparticles is its high biocompatibility, less technique-sensitive functionality, and large surface/volume ratio (Jain et al. 2012).

Carbon nanotubes: They are preferred as a promising agent for photothermal therapy. The chief benefit of using carbon nanotubes is its superior light absorption and chemical, mechanical, and physical properties, which enable its beneficial functionality in biomedicine. Carbon nanotubes are available in two forms, namely single-walled and multi-walled carbon nanotubes. Both forms have been studied at various levels and have proved to be beneficial at animal model studies (Eatemadi et al. 2014).

Graphene nanoparticles: They are preferred for biomarker sensors, drug delivery systems, and hyperthermia treatment. The chief benefit of using graphene nanoparticles is its high planner surface area, higher mechanical strength, and superior thermal conductivity, which enables them to be suitable for the application in molecular sensing, catalysis, and electronics. A study performed by Zhang et al. also showcased minimal toxicity caused by graphene nanoparticles (Zhang et al. 2010).

Moreover, due to the complexity and involvement of multiple head and neck regions, an interdisciplinary approach should be followed (Rao et al. 2020). This team should involve oral oncologists, head and neck surgeons, facial reconstructive surgeons, physical therapists, speech pathologists, and maxillofacial prosthodontists for satisfactory outcomes (i.e., removal of pathology, functional stability, and cosmetic outcome).

15.6 Prevention

Primary, secondary, and tertiary preventions of OC are extremely important. The primary preventions include cessation of known carcinogens (areca nut and its related products, smoking and smokeless tobacco, and alcohol) (More and Rao 2019). The secondary preventions include public health awareness of symptoms for potentially malignant disorder symptoms and early diagnosis of OC. The tertiary preventions include regular after-treatment follow-up. This level of prevention aims to early detect the locoregional recurrence, distant metastasis, secondary primary cancer, and complications from the treatment.

15.7 Conclusion

Global morbidity and mortality due to OC remain alarming. The only way to prevent and manage this disease is to have (1) a vigorous public health promotion at early education centers, (2) training dentist with the ability to identify OPMDs and OCs for early intervention, (3) improving diagnostic technologies with high specificity and sensitivity incorporating nanotechnology, (4) translational bench-side studies to bedside applications, and (5) interprofessional approach for overall management. Per the recent evidence, the application of nanotechnology in the management of OC, especially at preclinical studies, has shown promising outcomes. Expectantly, they will be translated to bedside applications in upcoming management modalities.

References

Allen TM (2004) Drug delivery systems: entering the mainstream. Science 303:1818–1822

Bagan J, Sarrion G, Jimenez Y (2010) Oral cancer: clinical features. Oral Oncol 46:414–417

Califano J, van der Riet P, Westra W et al (1996) Genetic progression model for head and neck cancer: implications for field cancerization. Cancer Res 56:2488–2492

Cheng R, Meng F, Deng C et al (2013) Dual and multi-stimuli responsive polymeric nanoparticles for programmed site-specific drug delivery. Biomaterials 34:3647–3657

da Silva SD, Ferlito A, Takes RP et al (2011) Advances and applications of oral cancer basic research. Oral Oncol 47:783–791

Damascelli B, Patelli GL, Lanocita R et al (2003) A novel intraarterial chemotherapy using paclitaxel in albumin nanoparticles to treat advanced squamous cell carcinoma of the tongue: preliminary findings. Am J Roentgenol 181:253–260

Eatemadi A, Daraee H, Karimkhanloo H et al (2014) Carbon nanotubes: properties, synthesis, purification, and medical applications. Nanoscale Res Lett 9:393

Eskiizmir G, Ermertcan AT, Yapici K (2017) Nanomaterials: promising structures for the management of oral cancer. In: Andronescu E, Grumezescu AM (eds) Nanostructures for oral medicine. Elsevier, Amsterdam, pp 511–544

Garavello W, Foschi R, Talamini R et al (2007) Family history and the risk of oral and pharyngeal cancer. Int J Cancer 122:1827–1831

Grégoire V, Langendijk JA, Nuyts S (2015) Advances in radiotherapy for head and neck cancer. J Clin Oncol 33:3277–3284

Hainfeld JF, Dilmanian FA, Slatkin DN et al (2008) Radiotherapy enhancement with gold nanoparticles. J Pharm Pharmacol 60:977–985

Hamakawa H, Nakashiro K, Sumida T et al (2008) Basic evidence of molecular targeted therapy for oral cancer and salivary gland cancer. Head Neck 30:800–809

Jain S, Hirst DG, O'Sullivan JM (2012) Gold nanoparticles as novel agents for cancer therapy. Br J Radiol 85:101–113

Karaca IR, Ozturk DN (2019) Oral cancer: etiology and risk factors. J Cancer Res Ther 15:739

Lu D, Yu X, Du Y (2011) Meta-analyses of the effect of cytochrome P450 2E1 gene polymorphism on the risk of head and neck cancer. Mol Biol Rep 38:2409–2416

Lucenteforte E, Garavello W, Bosetti C et al (2009) Dietary factors and oral and pharyngeal cancer risk. Oral Oncol 45:461–467

Markopoulos AK (2012) Current aspects on oral squamous cell carcinoma. Open Dent J 6:126–130

Mazeron R, Tao Y, Lusinchi A et al (2009) Current concepts of management in radiotherapy for head and neck squamous-cell cancer. Oral Oncol 45:402–408

McKaig RG, Baric RS, Olshan AF (1998) Human papillomavirus and head and neck cancer: epidemiology and molecular biology. Head Neck 20:250–265

McLaughlin JK, Gridley G, Block G et al (1988) Dietary factors in oral and pharyngeal cancer. J Natl Cancer Inst 80:1237–1243

More CB, Rao NR (2019) Proposed clinical definition for oral submucous fibrosis. J Oral Biol Craniofacial Res 9:311–314

More C, Rao NR, More S et al (2020) Reasons for initiation of areca nut and related products in patients with oral submucous fibrosis within an endemic area in Gujarat, India. Subst Use Misuse 55(9):1413–1421

Omura K (2014) Current status of oral cancer treatment strategies: surgical treatments for oral squamous cell carcinoma. Int J Clin Oncol 19:423–430

Ow TJ, Myers JN (2011) Current management of advanced resectable oral cavity squamous cell carcinoma. Clin Exp Otorhinolaryngol 4:1–10

Piemonte ED, Lazos JP, Brunotto M (2010) Relationship between chronic trauma of the oral mucosa, oral potentially malignant disorders and oral cancer. J Oral Pathol Med 39:513–517

Pitsillides CM, Joe EK, Wei X et al (2003) Selective cell targeting with light-absorbing microparticles and nanoparticles. Biophys J 84:4023–4032

Ram H, Sarkar J, Kumar H et al (2011) Oral cancer: risk factors and molecular pathogenesis. J Maxillofac Oral Surg 10:132–137

Rao NR, Parikh A, Patel A, Hirenkumar N, Dave PM (2017) Oral cancer—a review of recent non-invasive diagnostic methods. Int J Curr Res 9:59973–59976

Rao NR, Villa A, More CB et al (2020) Oral submucous fibrosis: a contemporary narrative review with a proposed inter-professional approach for an early diagnosis and clinical management. J Otolaryngol Head Neck Surg 49:3

Retif P, Pinel S, Toussaint M et al (2015) Nanoparticles for radiation therapy enhancement: the key parameters. Theranostics 5:1030–1044

Rivera C (2015) Essentials of oral cancer. Int J Clin Exp Pathol 8:11884–11894

Schmidt BL (2006) Molecular biology and clinical behavior of oral cancer. Oral Maxillofac Surg Clin North Am 18:483–491

Scully C, Bagan J (2009) Oral squamous cell carcinoma overview. Oral Oncol 45:301–308

Scully C, Laskaris G, Pindborg J et al (1991) Oral manifestations of HIV infection and their management. I. More common lesions. Oral Surg Oral Med Oral Pathol 71:158–166

Specenier PM, Vermorken JB (2009) Current concepts for the management of head and neck cancer: chemotherapy. Oral Oncol 45:409–415

Wong HM (2014) Oral complications and management strategies for patients undergoing cancer therapy. Sci World J 2014:1–14

Zhang Y, Ali SF, Dervishi E et al (2010) Cytotoxicity effects of graphene and single-wall carbon nanotubes in neural phaeochromocytoma-derived PC12 cells. ACS Nano 4:3181–3186

Zhao H, Feng H, Liu D et al (2015) Self-assembling monomeric nucleoside molecular nanoparticles loaded with 5-FU enhancing therapeutic efficacy against oral cancer. ACS Nano 9:9638–9651

Part VI

Nanocarriers for Site Specific Delivery

Nanocarriers in Transdermal Drug Delivery 16

Jagruti L. Desai, Tosha Pandya, and Ashwini Patel

Abstract

The transdermal route of drug administration is the most widely used route for systemic administration of drugs due to the convenience and noninvasiveness it provides. Skin is one of the largest organs offering multiple sites for drug delivery and acting as "reservoir" for sustained delivery of drugs. However, the penetration of drugs through skin is hindered due to stratum corneum, which acts as an efficient barrier. In order to overcome this barrier, nanocarriers (liposomes, nanoparticles, ethosomes, dendrimers, etc.) have emerged as an efficient tool for delivering a range of drugs through stratum corneum. In this chapter, various nanocarriers for transdermal delivery, their applications, commercial formulations, safety, and toxicity are described.

Keywords

Nanocarriers · Transdermal delivery · Drug penetration · Permeation enhancement

J. L. Desai (✉)
Ramanbhai Patel College of Pharmacy, Charotar University of Science and Technology, Changa, Gujarat, India
e-mail: jagrutidesai.ph@charusat.ac.in

T. Pandya
L.J. Institute of Pharmacy, Ahmedabad, Gujarat, India
e-mail: tosha.pandya_ljip@ljinstitutes.edu.in

A. Patel
Babaria Institute of Pharmacy, Babaria Institute of Technology, Vadodara, Gujarat, India
e-mail: ashwinipatel.bip@bitseducampus.ac.in

N. Shah (ed.), *Nanocarriers: Drug Delivery System*,
https://doi.org/10.1007/978-981-33-4497-6_16

16.1 Introduction

To enhance the drug efficacy in patients, it must deliver the drug in the right amount at the right time to the right targeted site. Researchers gain interest in the formulation of various novel dosage forms for available existing molecules because the discovery of novel drugs is a very time-consuming and expensive exercise. Conventional methods of dosage forms have certain limitations that they are multidose systems and not delivering an optimum amount of drugs. Redesigning of this dosage form is in less demand and to facilitate drug molecules into the body is a more tedious task. To overcome these limitations, various approaches like novel drug delivery system, controlled drug delivery system, and systemic drug delivery systems have been investigated which release drug into systemic circulation at predetermined time and interval at unit concentration. This evolution not only improves the safety and efficacy of drugs but also improves patient compliance and welfare.

It is important to deliver a drug in a superlative and adopted manner to the patient, which is a challenge to the researchers. This is because the concentration of the drug at the target site plays an important role in producing pharmacological responses. If the drug is released slowly, it may not be absorbed or if the release is too fast the patient may suffer unwanted side effects and ultimately desired effect of the drug may not last as long as required. Due to that, the frequency of dosing is increased which may lead to unwanted side effects. Variation in bioavailability and absorption of the drug through controlled-release oral dosage form is observed, as they have to follow first-pass metabolism of the liver. Transdermal drug delivery system is one of the solutions developed which deliver drugs into systemic circulation via the skin at required rate and maintain minimum effective concentration for a prolonged period. Administration through this route alleviates hazards and discomfort associated with the parenteral route. It improves patient compliance, as it is easy to apply and remove a patch when the need arises.

Transdermal drug delivery system (TDDS) provides several advantages over conventional methods of drug administration. It enhances efficacy, safety, patient compliance, and convenience. It also shows a steady flow of drugs into the bloodstream and reduces retention and metabolism of the drug through the skin. This drug delivery route provides a continuous flow of drugs in a controlled manner and eliminates pulse entry into the blood. Hence, the transdermal delivery route is recognized as one of the potential routes for local and systemic administration of the drug. This chapter reviews the skin anatomy, percutaneous absorption and its routes, kinetics of transdermal permeation, various nanocarriers for TDDS, applications, commercial formulations in TDDS, safety, and toxicity.

16.2 Skin Anatomy

Anatomically and physiologically, the skin is a multilayered complex organ. It receives about one-third of total blood circulation. It is a readily available organ with thickness about a few millimeters. It serves some of the important functions in

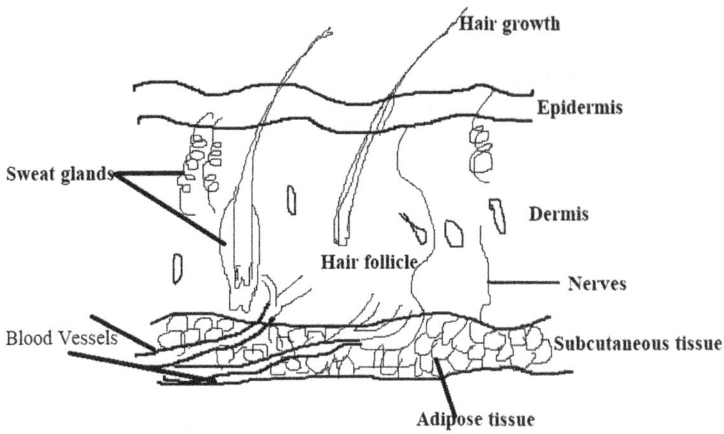

Fig. 16.1 Histology of human skin comprises mainly three layers: (1) the stratified vascular epidermis, (2) underlying dermis of connective tissue, and (3) hypodermis

the body, namely it serves as a barrier against physical, chemical, and microbial attack; separates blood network from the outer environment; regulates blood pressure and body temperature; and gives protection against UV rays.

The stratification is the beetling morphological feature of the skin. Macroscopically, there are two distinct parts of the skin, namely upper epidermis and lower epidermis. Microscopically, skin is a multilayered organ composed of various layers. The histological arrangement of skin is illustrated in Fig. 16.1.

16.2.1 Epidermis

It is multilayered and varies in thickness in the overall body depending on size and number of cells. It is comprised of stratum corneum and stratum germinativum. The stratum corneum is a thick and dry layer also known as the horny layer since it consists of keratin, cytoplasmic protein matrices logged in extracellular lipid. Keratinized cells called corneocytes are flexible but impermeable. Lipids in corneocytes are arranged in multiple bilayers like brisk in a mortar and have sufficient amphiphilic lipid materials. Thus, stratum corneum forms a major permeability barrier for the external environment. The cells of this corneum are formed and continuously replenished by upward migration of cells produced by basal cell layers in stratum germinativum. Most of the human corneum lipids consist of ceramides and neutral lipids like free fatty acids and triglycerides. The remainder is made up of phospholipids, glycosphingolipids, and cholesterol sulfate. All this contributes to the tightness and impermeability characteristics of intact skin. It is also split by hair follicles, sebum glands, and sweat glands which provide less resistance to the diffusion pathway (Yousef and Sharma 2019; Geerligs 2010).

Following the stratum corneum, there are several layers, which are collectively known as viable epidermis (includes stratum lucidum, stratum granulosum, stratum spinosum, and stratum basale). Among these, stratum granulosum is physiologically as important as stratum corneum because removal of these upper three layers is responsible for water loss and an enhancement of transdermal permeability (Tanwar and Sachdeva 2016).

16.2.2 Dermis

It is the second outer layer just beneath the epidermis. It is about 3–5 mm thick and is composed of collagen and elastin fibers inserted into mucopolysaccharide matrices. It contains blood vessels, lymph vessels, and nerve endings. It provides nutrients and oxygen to the skin and removes toxins and waste materials. It also maintains blood pressure by continuously supplying blood to the cutaneous blood vessels. This layer is not considered as a barrier for drug penetration in vivo because blood supply keeps the dermal concentration of a permeant very low and the resulting concentration difference across the epidermis provides the essential concentration gradient for transdermal permeation (Tanwar and Sachdeva 2016).

16.2.3 Hypodermis

The hypodermis is present just beneath the dermis. It is also known as the subcutaneous fat layer made up of areolar tissue (superficial fascia) and supports the dermis and epidermis. It serves as fat storage, which carries the main blood vessels and nerves to the skin and may contain the main sensory pressure (Ledger 1992).

16.3 Percutaneous Absorption

For transdermal drug delivery, a drug must penetrate through the stratum corneum of epidermis that is called percutaneous absorption. It is defined as the penetration of substances into various layers of skin and permeation across the skin into the systemic circulation (Jain and Umamaheshwari 2006). However, it was recognized that the absorption of drugs is nonidentical at different skin layers. This percutaneous absorption plays a vital role in the development of the transdermal delivery system because the drug has to absorb at an adequate rate and extent, and maintain a steady systemic therapeutic level throughout the period of course for the potential use of the drug. Once the drug penetrates into epidermis bypassing stratum corneum, it can easily enter into blood circulation through papillary and capillary layers of the dermis and deep blood vessels into subdermal tissues (Mehta 2004). The diagrammatic representation of this pathway is demonstrated in Fig. 16.2.

Accordingly, series of steps are involved in percutaneous absorption as follows (Ramteke et al. 2012):

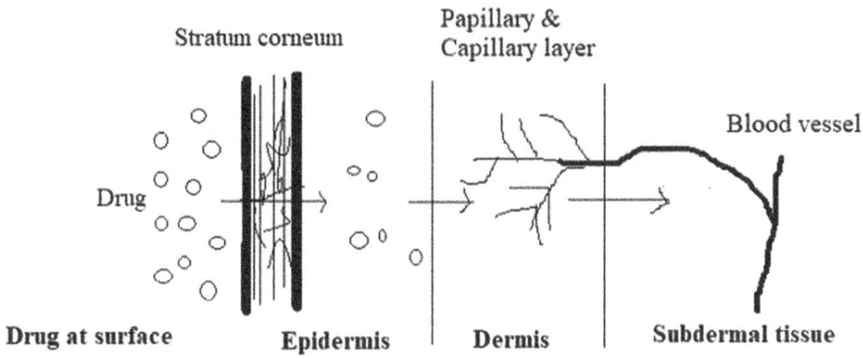

Fig. 16.2 Diagrammatic representation of percutaneous absorption

1. Sorption of drug molecule on stratum corneum for penetration
2. Slow diffusion towards nonviable dermis to viable dermis
3. Drug penetration into papillary and capillary systems because of concentration gradient
4. Entry of drug into systemic circulation from dermis

16.4 Routes of Drug Percutaneous Penetration

Percutaneous absorption through skin involves passive diffusion of drug molecules that does not require any energy for the penetration. Different layers and varieties of cells are present in the epidermis like hair follicles, sweat glands, and sebaceous glands; the penetration mainly involves two pathways: transepidermal route and transfollicular route.

16.4.1 Transepidermal Route

This involves permeation through an intact horny layer which follows micro-routes, i.e., transcellular and intracellular pathways. Various pathways are involved in permeation through stratum corneum which includes permeation between two cells of stratum corneum, and penetration through hair follicles, through sweat and sebaceous glands, and through cells of stratum corneum.

Different mechanisms are involved in the diffusion of polar and nonpolar molecules through the skin. Polar molecules diffuse through bound water available within the cells of stratum corneum, i.e., intracellular route, while nonpolar molecules first dissolve into lipid material present in the cells of stratum corneum, i.e., intercellular route. Thus, the permeation of materials depends on the partition coefficient (log k). However, the intercellular route is considered as the principal pathway and a major barrier for permeation of most drugs (Jain 1997).

16.4.2 Transfollicular Route

This route consists of transport via diffusion of molecules through the hair shaft openings, which are presumed to be filled with sebum. This route is the shortest route as they occupy approximately about 0.1% area of the total skin. This route seems to be most important for ions and large polar molecules that find difficulty in permeation through stratum corneum (Jain 1997).

The rate of permeation across the skin (*dQ*/*dt*) is determined by the following equation:

$$\left(\frac{dQ}{dt}\right) = P_s C_s \tag{16.1}$$

where P_s is the overall permeability constant of the skin tissue and C_s is the concentration of molecules at stratum corneum. Moreover, the membrane limited flux (J) under steady-state conditions is described by the following equation:

$$J = \frac{DCK}{h} \tag{16.2}$$

where "*J*" is the amount of drug diffused through stratum corneum, "*D*" is the diffusion coefficient, "*C*" is the concentration of drug at the surface, "*K*" is the partition coefficient of the drug molecule, and "*h*" is the thickness of the skin (Vyas and Khar 2011).

16.5 Factors Affecting Permeation Through Skin

Drug molecules that permeate through the skin via diffusion maintain steady-state concentration at stratum corneum. Various factors that affect the percutaneous absorption are the physicochemical nature of the drug molecules and its formulations and the physiology and pathological condition of the skin.

16.5.1 Physicochemical Nature of the Drug

Drug molecules have to penetrate through each layer of skin. The layers of the skin are characterized by diffusion resistance (R) to the layer, thickness of the layer (h), diffusion coefficient within the layer (D_S), and partition coefficient between two adjacent layers (K_S). Hence, diffusion resistance plays an important role in penetration and is described as follows:

$$R = \frac{h_S}{D_S K_S} \tag{16.3}$$

Also, permeability of molecules is related to membrane flux (J) which is described as

$$J = AP_s(C_p - C_r) \tag{16.4}$$

A is the area of applicant
$C_p - C_r$ is the concentration difference of permeant at two different layers
P_s is the partition coefficient
R is the diffusion resistance

But

$$P_s = \frac{D_s K_s}{h_s}$$

Hence,

$$\frac{J}{A} = \frac{D_s K_s (C_p - C_r)}{h_s} \tag{16.5}$$

From the above equation, it can be concluded that major variables that account for different penetration rates of drugs through the skin are diffusivity of drug, partition coefficient between stratum corneum and vehicle, and concentration gradient. The partition coefficient of the drug indicates the ability of the drug to be dissolved in the aqueous and organic phases. The water/lipid partition coefficient of about one or more is identical for the formulation of transdermal dosage form. The movement of the drug through the different layers of the skin mainly depends on the diffusion coefficient. It depends, to a greater extent, on the degree of interaction between diffusant and the surrounding medium which is directly proportional to the volume of diffusant.

The dissociation of an ionic drug from the formulation and its skin permeation depends on the pH of the formulation as well as the skin. For example, scopolamine permeation is increased with increasing pH up to 1.2 higher than their pK_a values. But a further increase in pH has no additional effect.

16.5.2 Physicochemical Nature of the Formulation

The release of drug from the formulation and its composition can largely affect the drug permeation through the skin. The presence of organic solvents and surfactants alters the skin permeability and percutaneous absorption.

If the release of active medicament from its formulation is high, greater will be the skin permeation. This release mainly depends on the presence of a vehicle in the formulation which implies that the molecule in the vehicle should remain in dissolved or suspended form. The interfacial partition coefficient between drug and formulation also plays a vital role in drug release. The pH of the vehicle can also influence the drug release. More acidic or basic drugs have an affinity towards vehicles while poorer would be released rapidly due to less affinity.

The composition of drug molecules can also alter the drug release and its penetration through the skin. Changing the skin hydration, presence of natural lipids, and permeation enhancer alter the permeation. For example, methyl salicylate is lipophilic and when it is administered with a fatty acid, a higher percutaneous absorption is observed compared to propylene glycol preparations. Permeability of fluocinolone acetamide and its ester increases with an increase in the concentration of propylene glycol in the vehicle. But, up to a certain level, the permeability is increased if further increase in concentration does not affect the permeability. This indicates the effect of thermodynamic activity and partition behavior of the drug in the formulations.

Modification of skin barrier by chemical and physical means alters the permeability of the skin. The physical means include techniques such as iontophoresis and sonophoresis while chemical means include various chemical penetration enhancers like solvents or surfactants which solubilize the surface lipid and enhance the porosity of skin. Solvents like lower alcohol like methanol and ethanol enhance the permeability due to their lipid-extracting ability, and hydrophobic cosolvents like *n*-hexane also enhance the permeation. Other solvents like propylene glycol, acetone, and dimethyl sulfoxide markedly enhance the permeation. Surfactants due to their surface tension-lowering ability are presumed for the enhancement of skin penetration. Some surfactants enhance the wetting of skin which provides better contact with skin and some anionic surfactants can modify the stratum germinativum.

16.5.3 Biological Properties of Skin

Physiological factors deal with the properties of the barrier itself like the location on the body and age which is not controllable. However, several factors can affect the rate and extent of percutaneous absorption. The most significant factor affecting the permeation of drug through the skin is hydration of stratum corneum. As the presence of water increases, the salvation of the polar region of lipid occurs and opens up cell density of closely packed cells that enhances the porosity and ultimately enhances permeation.

The other factor is the site of delivery since skin permeation is not uniform to the entire body. This is due to the difference in thickness, the number of cell layers and its stacking, and presence of relatively different amounts of intracellular lipid molecule. Skin permeability may be affected by the temperature and pathological injury of the skin. With the increase in temperature, skin permeability is increased because the temperature may enhance the diffusivity and solubility of drug molecules in the skin and also increase the blood flow. Pathological injury on skin disrupts the continuity of stratum corneum and enhances the permeability because injury increases the vasodilation at the site due to removal of barrier level.

Nature of the skin layer also plays an important role in permeation. A horny layer of skin acts as a reservoir for many drugs. For example: for scopolamine, the reservoir may lead to irreversible binding of the drug to skin and permeation is less. But the presence of surfactant like SLS will improve the permeability by decreasing the reservoir effect. Apart from this, the surfactant modifies the structure of the stratum germinativum layer by denaturing the present protein. The viable epidermis is the most metabolically active layer of skin which has the ability to modify the structure of drug molecules. Oxidation, reduction, hydrolysis, and conjugation of all these types of metabolic reactions are catalyzed by the enzymes present in the skin. This may be a disadvantage of this route for some drug molecules as the latter may be deactivated by the metabolic process. However, it may be beneficial for the development of a transdermal system of prodrug. Moreover, other factors like age, sex, and race may alter the rate and extent of permeation of drug molecules. All these factors may not have a significant effect on permeation as there is no such evidence that may prove that male and female skin anatomy is different in terms of skin permeability.

16.6 Advantages of Transdermal Drug Delivery

The positive features for the delivery of drug molecules through the skin are summarized as follows:

1. Enables the avoidance of presystemic metabolism, i.e., degradation in the gastro-intestinal tract by enzymes and first-pass metabolism in the liver. This, in turn, reduces the dosing frequency.
2. The lack of peak plasma concentration, i.e., maintained systemic level within the therapeutic window for a prolonged period.
3. Less fluctuation in plasma level is observed which reduces the risk of side effects of drugs. A consistent plasma level is an ideal candidate for these dosage forms.
4. The duration of the drug action with a single administration can be enhanced.
5. Less inter- and intrapatient variability, particularly in terms of drug release from the dosage form and diffusion of drug to the stratum corneum.
6. Termination of therapy is easy and convenient by removing the patch.
7. Improved patient compliance and acceptability.

16.7 Disadvantages of Transdermal Drug Delivery

1. The limitation of this dosage form is mainly dependent on its barrier skin which actually absorbs a certain amount of drug from its dosage form during the dosing period.

2. The physicochemical properties of drug molecules affect drug penetration. Hence, molecular weight and solubility of drug molecules in both lipophilic and aqueous environments determine diffusion through the skin.
3. The drug must not be locally irritating or sensitizing to the skin membrane because the drug reaction beneath the skin will prevent its regulatory approval.
4. This delivery is limited only for potent drug molecules which have dosing requirement of about 10 mg or less and hence effective plasma concentration may be achieved.

16.8 Nanocarriers for Transdermal Delivery of Drugs

Currently, in an attempt to favor the transport of drugs through the skin, facultative drug retention, increased half-life and bioavailability, increased stability, and in some cases sustained release, nanosized drug carrier systems have been propounded. Nanocarriers have the ability to circumvent the immune system and to deliver the drug in the targeted organ owing to their small size, thus lowering the drug dose and related side effects (Escobar-Chavez et al. 2012a). The cellular uptake of nanocarriers is highly influenced by their size, rigidity, shape, and change in the surface properties.

Nanocarriers are being investigated for many transdermal drug delivery applications, due to greater patient compliance compared to the parenteral route (Kumar et al. 2010). Thus, for the TDDS of vaccinations, antihypertensive drugs, anti-parkinsonian drugs, and chemotherapeutics, nanocarriers have been explored (Palmer and DeLouise 2016). Nanoformulations for TDDS can be classified as vesicular transport systems, nanoparticles, nanoemulsions, and dendrimers. Vesicular transport systems include carriers such as liposomes, transferosomes, niosomes, and transfersosmes. Nanoparticulate formulations include solid lipid nanoparticles, nanostructured lipid carriers, and polymeric nanoparticles. The most widely explored formulations for TDDS in the pharmaceutical field are shown in Fig. 16.3 and enlisted in Table 16.1.

In the last three decades, research and patents have been going on in this technology for diagnosis and prevention of health issues using noninvasive methods (Brower 2006). This involves manipulation in the size of drug and other excipients can change the basic properties and bioavailability of active molecules. Increased surface area, solubility, targeted drug release, and controlled release characteristics are the gifts of nanotechnology and manipulate drug delivery systems. In health science, new devices and chips are used for diagnosis, and new materials for substituting body structure and vesicles for transporting active drug material to the body or any other targeted part of the body (Escobar-Chavez et al. 2012b). These nanostructures are called nanocarriers. These nanocarriers are available in a size range of 10–1000 nm. They allow sustained release and targeted drug delivery systems along with protection of the labile group, reduced toxicity, and great adhesivity to skin. The different physicochemical properties like rigidity, hydrophobicity, size, and charge play an important role in skin permeability

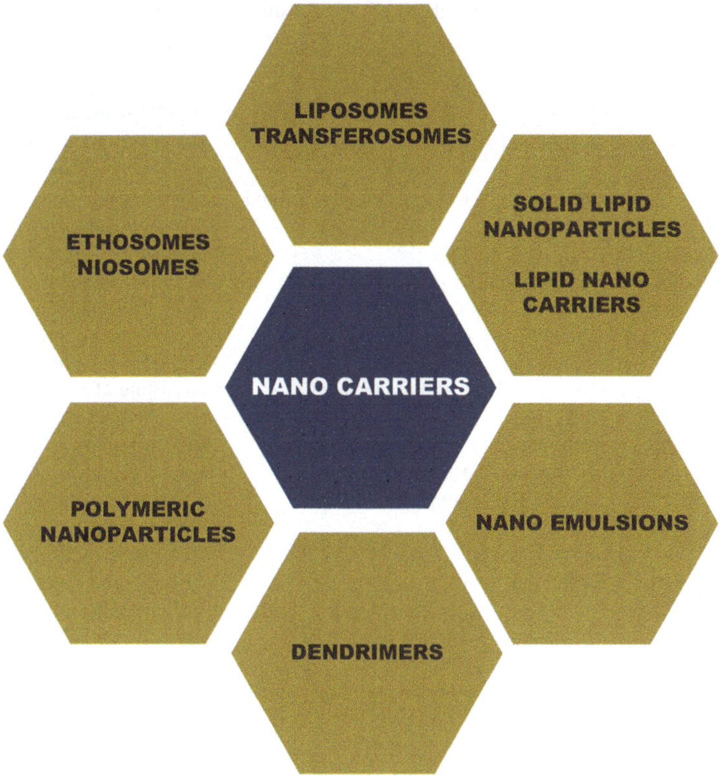

Fig. 16.3 Nanocarriers used in transdermal drug delivery

(Borm et al. 2006). These nanostructured carriers are a boon in transdermal drug delivery by increasing the specificity of drugs and thus reducing the side effects and decreasing the dose of administered drugs (Lam et al. 2004). Many different types of nanocarriers are widely used in the effective management of such diseases where conventional treatment is not enough for curing and prevention of diseases. These systems are designed for mainly two reasons: temporal delivery and spatial location.

16.8.1 Vesicular Transport Systems

Vesicles are highly ordered water-filled colloidal particles composed of multiple concentric bilayers created due to the self-assembling of amphiphilic molecules. Vesicle-based delivery systems have the ability to localize at the site of action, thus providing targeted delivery of drugs by lowering its concentration at other sites in the body. With the aim of increasing the penetration function of the components, vesicles can accommodate both water-soluble and lipid-soluble drugs. With respect to TDDS, the absorption rate can be regulated through the multilayer structure of the

Table 16.1 Characteristics of most widely used transdermal formulation

Nanocarriers	Size range	Preparation methods	Characteristics
Liposomes	25 nm–100 μm	Sonication extrusion, Mozafari method	Vesicles carrying one or more concentric lipid bilayers
Niosomes	10–1000 nm	Self-assembly of nonionic surfactant	Bilayered systems composed of nonionic surfactant vesicles
Ethosomes	<400 nm	Cold method, hot method	Noninvasive delivery carriers that enable drugs to reach the deep skin layers and/or the systemic circulation
Transfersomes	<400 nm	Thin-film hydration method, modified hand shaking, lipid-film hydration technique	Recent novel drug delivery system and are special types of liposome (flexible liposomes), consisting of phosphatidylcholine and an edge activator
Solid lipid nanoparticles	50–1000 nm	High-pressure homogenization	Solid or hollow particles which have entrapped, bound, or encapsulated drugs, composed of solid lipids
Nanostructured lipid carriers	10–500 nm	High-pressure homogenization Solvent emulsification evaporation Spray drying Phase inversion Lyophilization Solvent evaporation	Composed of solid lipid and a spatially different liquid lipid as hybrid carrier
Nanoemulsions	20–200 nm	High-pressure homogenization, microfluidization, phase inversion temperature	Submicron emulsions o/w or w/o
Dendrimers	3–10 nm	Polymerization	Macromolecular high-branched structures

vesicles (Rizvi and Saleh 2018). Due to the presence of different components; the vesicular system can be divided into the following several types.

16.8.1.1 Liposomes

Liposomes are spherical vesicles comprising one or more lipid bilayer structures enclosing an aqueous core. They are majorly composed of phospholipids and cholesterol. The phospholipid molecules are arranged in layers or sheets and the molecules aligned side by side, in which the hydrophilic heads of phospholipid and their hydrophobic tails face downward. They can be surface-charged as neutral, negative, or positive, depending on the functional groups and pH of the medium. The formation of liposomes occurs spontaneously upon the reconstitution of dry lipid films in an aqueous solution. This unique structure allows liposomes to encapsulate both water-soluble and lipid-soluble substances in a stable manner (Sharma et al. 2012).

There are small unilamellar vesicles (25–100 nm), medium-sized unilamellar vesicles (100 and 500 nm), large unilamellar vesicles, giant unilamellar vesicles, oligolamellar vesicles, large multilamellar vesicles, and multivesicular vesicles (500 nm to μm). These shapes and sizes depend on the preparation technique, lipids used, and process variables. Liposomes are structurally related to cell membrane "phospholipid bilayer" which is a great advantage for targeting lipophilic drugs. The degree of transdermal drug penetration is affected by the lamellarity, lipid composition, surface charge, mode of application, and total lipid concentrations. Liposomes have distinct advantages for TDDS, such as improvement of drug release, drug accumulation, and skin penetration. Thus, liposomes are one of the best alternatives for drug delivery as they are nontoxic and remain inside the bloodstream for a long time.

Local anesthetic agents (Foldvari et al. 1993), antileprotic agents, antifungal agents, antibiotics, antineoplastic agents (Glavas-Dodov et al. 2003), vitamins, and peptide/proteins are among the substances whose liposomal application holds promise when applied topically for localized drug delivery.

Mechanism of action: The possible mechanisms of liposomes promoting transdermal absorption of drugs are discussed. The phospholipids contained in liposomes have high compatibility with the lipids of the SC that allows lipophilic active molecules "to flow" more easily; this is called the fusion mechanism (Artmann et al. 1990). At the surface of SC, the structure of liposomes breaks up, thereby enabling phospholipids to penetrate within the SC. This allows them to form a flat granular structure in a way that the drugs encapsulated in liposomes can easily enter the skin through the lipid granular space while promoting skin permeation of the effective payload. Secondly, the hydration mechanism, wherein the liposomes can increase the moisture of the cuticle, promotes hydration of the skin, changes the orderly structure of the lipid layer between the keratinocytes, reduces the density, and enhances the permeability of drugs (Paliwal et al. 2015). Thirdly, in the penetrating mechanism, some researchers believe that the liposomes can penetrate through keratinocytes directly into the deep layer of skin (Weiner 1998). In addition, they can also enter the subcutaneous layer directly through the channels in the skin appendages to achieve the transdermal effect.

16.8.1.2 Transfersomes

The primitive form of modified liposomes is transfersomes, having ideal characteristics for TDDS like flexibility, deformability, and enormous elasticity, making them suitable for transdermal drug delivery. Transfersomes, unlike other formulations, are able to squeeze through small preexisting channels in the SC due to their deformability and adaptability (Cevc and Blume 2004). The deformability is imparted by different surfactants like sodium cholate, sodium deoxycholate, different spans and Tweens, and dipotassium glycyrrhizinate embedded in its structure, which acts as an "edge activator" making them capable of destabilizing the lipid bilayer. The transfersomes do not get impaired during the transportation of the loaded drug to the target tissues during TDDS. The movement of transfersome towards the highly hydrated environment of deeper SC is driven by osmotic gradient across the skin due to the dehydration of carriers (Eldhose et al. 2016). The use of flexible liposomes (transformable liposomes) is an indispensable strategy to reach the aim of TDDS, as they are proficient to penetrate intact skin in vivo, thus delivering therapeutic concentrations of drugs and macromolecules comparable to subcutaneously administered molecules. Macromolecules such as insulin, corticosteroids, ketoprofen, and anticancer drugs can be delivered locally and systematically by transfersomes. It is also exhibited that transfersomes are able to penetrate SC and supply the nutrients locally to maintain its functions resulting in nurturing of skin.

Mechanism of action: The mechanism by which transfersomes efficiently penetrate the SC can be summed up in three aspects. Firstly, surfactants perforate into and interact with the SC causing swelling, after which the keratin turns into swollen uncoiled fibers, extracting lipids from the SC and disturbing its organization. The permeability of the SC to hydrophilic substances is increased as a result of the fluidity of lipids due to crystalline and gel states of the long-chain molecules that produce or expand the hydrophilic voids (Malakar et al. 2012). Secondly, the transport is driven by the water gradient created between SC (water content: 15%) and other layers of the epidermis (water content: 75%). The skin hydration gradient is naturally formed, under nonocclusive conditions, during the transdermal action of transfersomes. Due to the high deformability, transfersomes can squeeze through the gaps in SC. Lastly, it has been reported that the hair follicle pathway is the main route for the passage of transfersomes.

16.8.1.3 Niosomes

Niosomes are synthetic microscopic vesicles prepared from self-assembly of hydrated nonionic surfactants, composed of an aqueous core enclosed in a bilayer made of cholesterol and nonionic surfactants. The incorporation of nonionic surfactants in their structure imparts more stability as compared to liposomes, and makes the structure less rigid, thus imparting greater penetrability. Initially, niosomes found use in the cosmetics industry, but now they are used in various areas due to their versatility. They offer versatility in the route of drug delivery like oral, parenteral, ocular, vaginal, and transdermal (Muzzalupo et al. 2011). A wide variety of drugs with different solubility can be incorporated. In transdermal drug delivery, niosomes are of prime importance due to their ability to deliver antiaging agents and antifungal molecules. Biodegradability and minimal toxicity are among the other advantages of the niosomes. The process of scale-up is simplified and the economies of their production and storage are low. Niosome-entrapped enoxacin was delivered at a higher rate from either liposomes or simple drug solutions (Fang et al. 2001). The niosomal formulations were also reported superior in the permeation of capsaicin over 12 h as compared to microemulsions. These results reveal the superiority and favorability of niosomes as drug delivery systems.

Mechanism of action: The proximity between the niosomes and the SC of the skin and their adsorption and fusion on the surface of the skin generate an increase in the thermodynamic activity gradient of the drug at the interface, leading to increased permeation of lipophilic drugs. Moreover, niosomes impart smoothness through replenishment of lost skin lipids by reducing the transepidermal water loss which increases SC hydration and loosens its closely packed cellular structure. Thirdly, the lipid bimolecular layer of noisome acts as a rate-limiting membrane and provides a slow-release effect locally (Abdelkader et al. 2014). The circulation time of the drug encapsulated in niosomes is prolonged, thus enhancing the penetration of the drug to the target organ and increasing the curative effect.

16.8.1.4 Ethosomes

Ethosomes, as the name suggests, are ethanolic liposomes that contain alcohol in the lipid bilayer imparting greater flexibility and ability to deform under pressure. They consist mainly of phospholipids, alcohol, and water. The main characteristics of these nanocarrier systems are malleability and softness. Ethosomes possess small particle size, stable structure, and high entrapment efficiency, the characteristics that can delay the drug release (Touitou et al. 2000). The high concentration of alcohol in ethosomes increases their permeability as compared to liposomes, thus allowing extension of drugs to deep skin layers. Ethosomes have higher stability and longer retention periods compared to transfersomes (Samala and Pannala 2011). Ethosomal formulation includes cholesterol (vesicle membrane stabilization), carbopol, pluronic F 127, and poloxamer 407 (gel formers used to produce vesicular gels in order to increase residence time) as the key excipients. A diverse range of drugs, more importantly highly water-insoluble drugs like testosterone, minoxidil, and cationic molecules such as propranolol and trihexyphenidyl, can be delivered by ethosomes.

Mechanism of action: The high concentration of alcohol in their structure enhances lipid fluidity and flexibility of the membranes, thus imparting enhanced permeability. The ethosomes are distorted during transmission, due to decreased density of the intercellular lipid domains. They create their own paths through altered intercellular lipid lamella in order to reach deep skin layers. Secondly, the solubility of the drug is increased, promoting permeation for a long time due to a high concentration of alcohol (Paolino et al. 2005). Moreover, the percutaneous absorption of the drug is promoted owing to the enhanced fusion of ethosomes with the lipids of SC. So, the higher the alcohol content, the greater the permeation flux.

16.8.1.5 Nanoemulsions

Nanoemulsions (NE) or ultra-emulsions are isotropic dispersed systems of two immiscible phases stabilized by an interfacial film of surfactant or co-surfactant molecules of extremely small droplet size (100 nm). They are low-viscosity mixtures consisting of transparent or translucent oil globules dispersed in the aqueous phase (o/w nanoemulsion) or an aqueous system dispersed in an oily phase (w/o nanoemulsion). The transparency of nanoemulsion is due to the droplet size, which is less than 25% of the visible light wavelength. Three main methods, viz. high-pressure homogenization, micro-fluidization, and phase inversion temperature, are employed for NE. High energy is required for the generation of nanoemulsions, and thus they are thermodynamically unstable in contrast to microemulsions (Shakeel and Ramadan 2010). Despite this, the use of adequate surfactants can impart stability for long periods due to their extremely small size. The low surface tension of NE gives them good wettability to retain them in close contact with the skin. Compared to commonly used topical skin preparations, the transdermal time of NE is shorter and the effect of percutaneous absorption is better.

They have higher solubilization capacity and can entrap both hydrophilic and hydrophobic drugs. It has been demonstrated that nanoemulsions coated with alginate or chitosan increased the transdermal delivery of polypeptides and proteins (Li et al. 2013). Several drugs like gamma tocopherol, caffeine, plasmid DNA, aspirin, methyl salicylate, insulin, and nimesulide are described to be delivered transdermally using nanoemulsions (Phatak and Chaudhari 2012).

Mechanism of action: The components of NE like the surface-active agent, co-surfactant, and oil phase disturb the arrangement of lipids in the SC, thus promoting permeability. Nanoemulsions, due to their unique structure, can accumulate both liposoluble and water-soluble drugs. This leads to an increased permeable concentration gradient inside and outside the skin, which in turn enhances the penetration of drugs. Thirdly, the constant driving force of drug diffusion is maintained due to the storage function of the inner phase, which prolongs drug absorption. Fourthly, NE expands the connections between SC by using the powerful combination with water to produce a faster percutaneous passage through the keratinocytes (Sonneville-Aubrun et al. 2004). They also release the collagen fibers in the dermis to a certain extent. Lastly, good wettability and increased contact with the skin are imparted by the small particle size, large specific surface area, and low surface tension of NE.

16.8.2 Nanoparticulate Drug Delivery System

Nanoparticles are nanosized carriers with particle size ranging from 1 to 1000 nm having great targeting ability. They can be categorized as nanospheres or nanocapsules based on their structure. Nanospheres have drugs embedded in solid polymer matrix whereas nanocapsules have drugs loaded in a hollow shell within and in inner space. The drugs can be encapsulated, adsorbed, or coupled on the surface of nanoparticles (Baroli et al. 2007). Depending on their physicochemical properties, methods such as emulsification–solvent diffusion, emulsification–polymerization, in situ polymerization, gelation, nano-precipitation, solvent evaporation/extraction, inverse salting out, and dispersion polymerization are employed for formulation of nanoparticles. Apart from providing targeted and controlled release, nanoparticles modulate the in vivo drug dynamics and increase the benefit-to-risk ratio of drugs (Neagu et al. 2018). Solid lipid nanoparticles (SLN) and nanostructure lipid carriers (NLC) are the two main alternatives for TDDS. Polymeric nanoparticles are also a very good option for TDDS, as they can be tailor-made in different sizes and their surface properties can be modified to enhance skin penetration. Nanoparticles are also a promising tool for immunomodulation as they can travel from the skin to lymph nodes.

16.8.2.1 Lipid Nanocarriers

Solid lipid nanoparticles (SLN) and nanostructured lipid carriers (NLC) are the complex forms of colloidal drug delivery systems having lipid as a carrier. SLN, the first-generation lipidic nanocarriers are submicron (50–1000 nm) in size and composed mainly of lipids like triglycerides, partial glycerides, fatty acids, steroids, and wax, having good physiological compatibility. Other ingredients include various emulsifiers (stabilizers), co-emulsifiers, and water (Schafer-Korting et al. 2007). The permeation of drugs is enhanced due to the interaction of phases at the interfaces and better stability. The SLNs can be easily upgraded to production scale as compared with liposomes due to the introduction of high-pressure homogenization technology. Recently Salve et al. described SLN-loaded formulation for the transdermal delivery of rivastigmine tartrate, as an alternative to oral and parenteral routes, to achieve brain drug delivery for the treatment of Alzheimer's disease (Salve et al. 2016).

Nanostructured lipid carriers (NLC) are the second-generation lipid nanoparticles that prevail over SLN, having high drug-loading capacity (Wissing et al. 2004). The inner structures of SLN and NLC are different. NLCs are usually composed of up to 30% of oil apart from solid lipids. NLCs promote the access of drugs to deeper layers by widening the inter-keratinocyte gaps. They are suitable candidates for long-term skin administration due to the chemical similarity to skin lipids, high specific surface area, and biocompatibility they possess. NLCs are particularly useful for targeting water-soluble drugs, cosmetics, foods, substances for agricultural use, anti-inflammatory drugs, etc. (Beloqui et al. 2016).

Mechanism of action: NLCs form a mono-layered lipid film due to good skin adhesion, which increases contact between drugs and SC. Lipid nanoparticles provide "occlusion effect" by forming hydrophobic film on the surface of the skin, which enhances deep skin penetration of drugs. Moreover, the small particle size enhances the "occlusive effect" by providing a higher specific surface area, thus promoting transdermal penetration. Thirdly, the thermodynamic stability of SLN and NLC is enhanced by the nonionic surfactants present in their composition which aid in permeability (Bhaskar et al. 2009). The properties of carrier, type, concentration, etc. of the lipid are some of the other factors affecting the mechanism. Fourthly, they release the active substance in a biphasic regimen—an initial burst release from the surface, followed by reservoir effect. According to some reports, the drug in the NLC remains in the liquid lipid surrounded by solid lipid. This organization offers some degree of mobility to the drug and stability even when the solid lipid undergoes polymorphic changes. In short, the skin adhesion and the "occlusion effect" of the lipid nanoparticles are the main mechanisms enhancing permeability (Uner et al. 2014).

16.8.2.2 Polymeric Nanoparticles

Polymeric nanoparticles are solid colloidal carriers which may be prepared from a polymeric matrix composed of either synthetic polyesters (polylactic acid, poly

(DL-lactide-co-glycolide), polycaprolactones (PCL), etc.), natural polyesters (chitosan, gelatin, alginate, etc.), or other polymeric carriers. Based on their method of preparation and architecture, the polymeric NP can be classified into nanospheres, nanocapsules, and polymeric micelles. Polymeric NPs offer unique advantages over other lipid systems like protection to unstable drugs, continuous release, and increase in concentration gradient to promote percutaneous permeation of drugs. The polymer membrane structure is highly entangled imparting them high mechanical strength and also non-deformability. They retain drugs for a longer period of time and diffuse them into deep skin layers. Both hydrophilic and lipophilic drugs can be encapsulated in NPs. They are successfully used in the treatment of diseases such as cancer and diabetes (Shah et al. 2012).

Mechanism of action: The mechanism of drug penetration by polymeric nanoparticles is as follows: The polymeric nanoparticles assemble on the surface of the skin or hair follicle and are then directed to a drug reservoir for the gradual release of drugs. This leads to the higher local concentration of the drug contained in the polymeric NP, as the drug diffuses into the active layer of the skin at the concentration gradient. Secondly, if the skin integrity can be modulated, the percutaneous penetration of drugs in the nanoparticles can be facilitated (Jung et al. 2015). Thirdly, the highly cationic components on the surface of polymeric nanoparticles promote the interaction with the negatively charged skin surface, leading to an increase in drug release. Lastly, the permeation is further enhanced through the hair follicle.

16.8.2.3 Dendrimers

The term dendrimer is from a Greek word "dendra" means tree and "meros" means part. Dendrimers are nanosized micellar monodisperse structures, typically composed of a symmetric central core and branched structures. Due to the branched structure, conjugation with numerous molecules such as drugs and solubilizing groups is possible in a defined manner based on desired application. Dendrimers are classified based on the number of generations. The stepwise synthesis after the core generation is called first generation, after which every consecutive addition of monomers creates the next generation (Svenson and Tomalia 2012).

Dendrimers have proved to be good carriers for hydrophobic and labile molecules. The final architecture and features of dendrimers depend on the kind of polymers chosen. The most widely used material for dendrimer fabrication is poly (amidoamine). Dendrimers are potent skin permeation enhancers. Dendrimers have been investigated for drug delivery, gene therapy, and delivery of contrast agents (Choksi et al. 2013). They are found to improve the aqueous solubility of many lipophilic drugs like sulfamethoxazole, furosemide, ketoprofen, ibuprofen, albendazole, and ketoconazole.

Mechanism of action: The physicochemical characteristics of dendrimers like particle size, molecular weight, surface charge, and composition determine their penetration through the skin. The 3D and fractal architecture, as well as the peripheral functional groups, provides dendrimers with important physical and chemical properties. In comparison with linear polymers, dendritic structures have "dendritic voids" which impart many useful features (Bosnjakovic et al. 2011). These spaces inside dendrimers perform analogous molecular recognition performed by natural proteins. The ionizable groups impart high surface charge density to dendrimers that help them attach drugs by electrostatic forces. This dendrimer-drug association provides an increase in solubility, transport through biological membranes, and stability. Dendrimers are versatile in their complexing ability with different types of drugs, owing to the amphiphilic nature. In general, tiny size, high surface energy, composition, good penetrability, architecture, and attached molecules are the advantages of using nanocarriers for transdermal drug delivery (Gupta et al. 2006). Table 16.2 shows the advantages and disadvantages of these carrier systems.

16.9 Applications of Nanocarriers for Transdermal Drug Delivery

Topical applications of active constituents are the delivery of the drug directly to the site of action for producing higher local concentrations of the drug at pathological sites. TDDS provides systemic action of topically applied drugs due to penetration through the skin. This route is more beneficial in the delivery of agents which cause gastrointestinal tract (GIT) irritancy and susceptible candidate for hepatic first-pass metabolism. Additionally, transdermal delivery provides a sustained and controlled release for medication, easy termination of drug release in the case of sensitivity, certain acute toxic effect of medication, etc. Various carrier systems, for enabling drug retention and providing controlled release, have been proposed to favor the transport of drugs through the skin. Table 16.3 summarizes the applications of nanocarrier systems for transdermal drug delivery.

Liposomes are drug carriers that reduce toxicity and improve efficacy. They are used successfully in cancer therapy and in skin melanoma as they remain in the bloodstream for a longer period of time. Multilamellar liposomes improved drug release and skin penetration of quercetin compared to normal drug solutions (Seong et al. 2018), suggesting the possibility of developing transdermal delivery for effectively delivering poorly soluble drugs.

Allantoin, a drug used for skin ulcers, has low skin penetration due to low log *P*. Incorporation of allantoin in liposomes prolonged the residence time and doubled the local accumulation (-3.4 g/cm^2, $p < 0.05$) compared to commercial Sameplast gel (Manca et al. 2016). Liposomal formulation of betamethasone dipropionate used in the treatment of atopic eczema increased the concentration and release of the drug. Liposomes are also used in the delivery of many analgesic and anesthetic agents. Liposomal formulation of epidural morphine (DepoDur) prolonged analgesia by a minimum of 2 days. Liposomal formulations were also found to be effective in the delivery of lidocaine, estradiol, tretinoin, cyclosporine, etc. Recently, modified liposomes have found wide applications due to many advantages. Transfersomes, which are flexible liposomes, have major benefits in TDDS compared to conventional liposomes. Increased level of cyclosporin A was delivered by transfersomes compared to conventional vesicles.

Table 16.2 Advantages and disadvantages of nanocarriers used in transdermal delivery

Nanocarriers	Advantages	Disadvantages
Liposomes	• Easy manufacture • High biocompatibility • High drug loading • Controlled release • Nontoxic • Long circulation times • High stability due to protein carriers	• Susceptibility to physical instability • Polymorphic issues due to lipid crystallization • High-pressure homogenization leads to decreased stability of high-weight molecules • Variable kinetics of distribution process
Transfersomes, niosomes, ethosomes	• Biodegradability, low toxicity • Easy to prepare • Softness, malleability • Both hydrophilic and lipophilic • Drugs can be encapsulated • Targeted drug delivery • Extremely high flexibility of their membrane	• Predisposition to oxidative degradation • Purity of natural phospholipids • Formulations are expensive
Nanoparticle lipid nanocarriers (NLC, SLN), polymeric nanoparticles	• Wide range of biodegradable materials for manufacture • Different methods of preparation • Can offer targeted delivery by surface attachment of antibodies • Both hydrophilic and hydrophobic drugs can be loaded • Occlusivity • Avoidance of immune clearance	• Lack of enough toxicological assessment • It is difficult to develop an analytical method for drug delivery • Some processes are difficult to scale up • Sometimes, the size they reach is not enough to avoid the immune system
Nanoemulsions	• They can be formulated as foams, liquids, creams, and sprays • They are nontoxic and nonirritant • Easily applied to skin and mucous membranes	• They are susceptible to Ostwald ripening • Surface charge has a marked effect on stability • Variable kinetics of distribution processes and clearance
Dendrimers	• High stability of therapeutic agents • Easy to prepare • They increase bioavailability of drugs • They covalently associate drugs • Dendrimers also act like solubility enhancers, increasing the permeation of lipophilic drugs	• Cellular toxicity • Elimination and metabolism could be a problem • Depending on the generation and materials • High synthesis costs • Hemolytic effects can be found • They are not good carriers for hydrophilic drugs

Table 16.3 Applications of nanocarriers for transdermal delivery

Nanocarriers	Applications
Liposomes	Examples of drugs delivered throughout the skin by using liposomes are melatonin, indinavir, methotrexate, amphotericin B, ketoprofen, estradiol, clindamycin hydrochloride, and lignocaine (Sharma et al. 1994). Currently, positively charged liposomes have been used for DNA delivery in gene therapy. Also, liposomes are being used for many antifungal and anticancer applications.
Transfersomes, niosomes, ethosomes	• Examples of transdermal drug delivery using transformable liposomes (transfersomes) are diclofenac (Boinpally et al. 2003), insulin, tetanus toxoid, corticosteroids, superoxide dismutase, DNA, triamcinolone acetonide, ketoprofen, interleukin-2, and ketotifen fumarate (Trotta et al. 2004). • Ethosomes could be used in the treatment of atopic dermatitis. Furthermore, ethosomes can be used for Parkinsonian syndrome and for dystonia therapy. Examples of transdermal drug delivery using ethosomes are tacrolimus, clotrimazole (Maheshwari et al. 2012), trihexyphenidyl HCl, ketoprofen, and testosterone. • Niosomal formulations have greater potential for drug cutaneous targeting and could be used as a feasible cargo carrier for the topical delivery of minoxidil in skin diseases such as hair loss. Examples of transdermal drug delivery using niosomes are minoxidil and ellagic acid (Junyaprasert et al. 2009).
Nanoparticles, lipid nanocarriers (NLCs, SLNs), polymeric nanoparticles	Nanoparticles have been used successfully in the treatment of diseases such as cancer and diabetes. Furthermore, polymeric nanoparticles are used to deliver therapeutic agents for various types of tumors, bone healing, and vaccination (Santander-Ortega et al. 2010). Examples of drugs delivered throughout the skin by using nanoparticles are minoxidil, DNA, triamcinolone acetonide acetate, dexamethasone phosphate, cyclosporin A, flufenamic acid, testosterone, caffeine, 5-fluorouracil, artemether (Aditya et al. 2010), chlorhexidine, econazole nitrate, insulin (Huang et al. 2009), celecoxib, coenzyme Q10, and triclosan.
Nanoemulsions	Some examples of drugs using nanoemulsions for transdermal drug delivery are gamma tocopherol, caffeine, plasmid DNA, aspirin, methyl salicylate, insulin, and nimesulide (Li et al. 2013).
Dendrimers	Dendrimers have been used in numerous applications such as gene therapy, delivery of contrast agents, controlled drug delivery, light-harvesting agents, catalysts, chemical sensors, and cross-linking agents. Additionally, dendrimers can be used in antiviral and anticancer pharmaceutical therapies, including vaccines. Examples of drugs delivered throughout the skin by using dendrimers are tamsulosin, indomethacin, ketoprofen, diflunisal, 5-fluorouracil, and peptides (Gupta et al. 2006).

Sildenafil citrate-loaded nanotransfersomal transdermal films enhanced and controlled the permeation of drug (Badr-Eldin and Ahmed 2016). The degradation of drug was also reduced as compared to oral drug delivery, thus improving drug bioavailability. The permeation of itraconazole was 2.2 times greater than regular liposomes. Niosomal formulation showed sustained release and is a better alternative in terms of efficacy, bioavailability, and permeation. Improved systemic delivery of lopinavir in the treatment of HIV infection was found using niosomes.

Simvastatin-loaded niosomal gels exhibited improved hypolipidemic efficacy (Zidan et al. 2016). The bioavailability of simvastatin was threefold augmented with transdermal niosomal formulation compared to oral suspension. This was found promising in the treatment of hyperlipidemic pediatric patients.

Ethosomes can deliver a wide range of highly lipophilic drugs. The transdermal penetration of testosterone from the ethosomal patch was increased compared to the commercial patch. The antidiabetic effect of repaglinide was prolonged by ethosomal formulation (Bodade et al. 2013). There was an increase in the permeation of administered dose (64–97%) than free drug and alcoholic solutions. Thus there was a reduction in dosing frequency in type 2 diabetes mellitus.

Lipid nanoparticles enhance the chemical stability of compounds sensitive to light, oxidation, and hydrolysis. Many cosmetics such as coenzyme Q10, ascorbyl palmitate, vitamin E, and vitamin A are delivered by lipid nanoparticles. Aconitine-loaded solid lipid nanoparticles were found to have increased safety, permeability, and skin deposition in vitro. Higher concentration was found compared to tincture and increased anti-inflammatory and analgesic effect was demonstrated in in vivo mouse model of pain (Ma et al. 2017). Increased efficacy of clobetasol propionate loaded in SLNs was found compared to conventional cream in the treatment of eczema. Higher recovery from skin carcinoma was found with paclitaxel-loaded SLN in carbopol gel. Increased skin deposition and localization were found using lidocaine-loaded nanolipid carriers (NLC) (entrapment efficiency 69.80%). Transdermal delivery of pioglitazone-loaded NLCs lowered blood glucose and provided sustained and prolonged drug release compared to tablet formulation (Alam et al. 2016).

Polymeric nanoparticles deliver therapeutic agents for tumors, diabetes, bone healing, vaccination, etc. Shetty et al. developed novel sunscreen creams containing polymeric NP of morin (an important plant flavonoid possessing both antioxidant and UVR protection properties) to optimize the delivery of the material into the skin. The amount of morin permeated from the nanoparticulate suspension at the end of 12 h was much higher than that observed with the morin plain suspension (Shetty et al. 2015). This may be due to the increased ability of polymeric nanoparticles to penetrate the skin. PLGA nanospheres loaded with vitamin derivatives including vitamin C, vitamin E, and vitamin A are used for skin whitening and antiaging applications.

Nanoemulsions have great potential for enabling new advances in pharma science especially with regard to transdermal applications. NE increases the solubility and transdermal delivery of poorly soluble drug carvedilol. Drug penetration follows the order NE > NLC > SLN > gel. Nanogel-entrapped aceclofenac reduces ulcers and gastric bleeding. Nanoemulsion-loaded gel containing clobetasol propionate and calcipotriol showed higher anti-psoriatic activity compared to marketed formulation (Kaur et al. 2017). However, due to stability issues, nanoemulsion is not much in use as nanoparticles or liposomes.

Dendrimers are nanocarriers which improve the aqueous solubility of many drugs. Promising results have been found in the delivery of drugs such as tamsulosin, indomethacin, ketoprofen, and diflunisal. They are mainly used to transport photosensitizers for photochemical therapy and antifungal molecules. Increased transdermal delivery of ketoprofen was found using polyamidoamine dendrimer complex. The bioavailability of indomethacin was increased by its dendrimer loading (Gillies and Frechet 2005). Overall, nanocarriers are promising systems for transdermal drug delivery of different molecules, suitable for different purposes and for several disease conditions.

16.10 Current Market and Clinical Trial Scenario of Nanocarriers for Transdermal Delivery

The transdermal products of only those drug categories, which show acceptable physicochemical and pharmacokinetic characteristics meeting with prerequisites for effective delivery through the skin, are able to undergo successful transformation to market product. Based on this requirement, only a limited number of drugs (approximately <20) are being able to deliver systematically using the transdermal route. About half of the clinical studies for transdermal delivery have one of the four molecules such as nicotine, estradiol, fentanyl, and testosterone (https://clinicaltrials.gov). Only a limited number of studies are able to reach phase 1V, while many are terminated. Representative successful stories include TransdermScop® which was introduced by Alza Corp. for the delivery of scopolamine, in 1981 (Tanner and Marks 2008). Formulation development of transdermal products required the cautious approach to manage skin permeation variability between patients and also different sites on the same patient which suffer from variation in absorption profiles. Although the market and clinical trial status of the transdermal product shows improvement towards nanocarrier-mediated drug delivery with an attempt to counterbalance the problems persisting with the existing conventional transdermal delivery system these limitations would be running simultaneously. Table 16.4 summarizes various transdermal formulations available in the market.

Table 16.4 Novel transdermal formulations in the market

Brand name	Nanocarrier system	Disease	Company
Sileryst®	Nanocrystalline silver	Atopic dermatitis	Nucryst Pharmaceuticals
VivaGel®	SPL7013, or astodrimer sodium	Antimicrobial agent	Starpharma
Diractin	Transfersome® gel (KTG)	Pain related to eccentric muscle contractions	Idea AG
Evamist	Estradiol transdermal spray	Moderate-to-severe vasomotor symptoms associated with menopause	Perrigo Pharma International
Aczone	Dapsone topical gel 5%	Treatment of acne vulgaris	Allergan, Inc.
Daytrana®	Methylphenidate transdermal patch system	CNS stimulant indicated for the treatment of attention-deficit hyperactivity disorder	Noven Therapeutics, LLC

16.11 Toxicity

New technology always has a wonderful and great impact on its positive side but still they have some dark side, which would require some attention for their proper usage. These new nanocarriers or vehicles are unintentionally responsible for the generation of new allergens, irritants, cross-reactants, haptens, different particle-particle interactions, and ultimately new suffering or diseases. They are also very harmful because of their extremely small size. The small size increases surface area and hence they are not only permitted to applied area and show effect, but also many times they cannot be logged out and lead to nanopollution. Along with this its particle size is responsible for its reactivity and toxicity (Nighswonger 1999; Oberdorster 2006). For example enhanced oxidative stress by the reaction between nanoparticles and cell leads to endocytosis and modification of protein, rendering them as an antigen, thus promoting an autoimmune response (Klimuk et al. 2000). The research finding focused that nanoparticles might be transferred to the mother's fetus and show cytotoxic effect towards mammalian germline stem cells and many more as they can be present in the body for longer periods of time and their catalytic activity disturbs the host enzyme cascade.

16.12 Conclusion

Altogether, the transdermal drug delivery offers convincing opportunities to overcome the low bioavailability of many oral drugs, the pain and inconvenience of injections, and the limited controlled release options of both oral and parenteral routes. Nanocarriers in medicine are promising as they provide diverse opportunities for variations and advancements in order to improve the quality of life. The administration of drugs via transdermal route is constricted because of skin acting as a barrier. Nanocarrier delivery via this route is so designed that it overcomes the permeability complications for drugs to cross the skin and effectively reach the targeted site. The diversity options available for nanocarriers including varying compositions, size, surface charge, and surface modification using different ligands provide great opportunities for potential targeting of molecules via transdermal route. However, one has to circumvent certain limitations of these deliveries like irritation and hypersensitivity. Furthermore, continuous research in the field of materials and production methodology of transdermal nanocarriers will ensure their application in clinic in the coming time for effective treatment of various diseases via noninvasive transdermal route.

References

Abdelkader H, Alani AW, Alany RG (2014) Recent advances in non-ionic surfactant vesicles (niosomes): self-assembly, fabrication, characterization, drug delivery applications and limitations. Drug Deliv 21:87–100

Aditya NP, Patankar S, Madhusudhan B et al (2010) Arthemeter-loaded lipid nanoparticles produced by modified thin-film hydration: pharmacokinetics, toxicological and in vivo antimalarial activity. Eur J Pharm Sci 40:448–455

Alam S, Aslam M, Khan A et al (2016) Nanostructured lipid carriers of pioglitazone for transdermal application: from experimental design to bioactivity detail. Drug Deliv 23:601–609

Artmann C, Roding J, Ghyczy M et al (1990) Liposomes from soya phospholipids as percutaneous drug carriers. 1st communication: qualitative in vivo investigations with antibody-loaded liposomes. Arzneimittelforschung 40:1363–1365

Badr-eldin S, Ahmed OA (2016) Optimized nano-transfersomal films for enhanced sildenafil citrate transdermal delivery: ex vivo and in vivo evaluation. Drug Des Devel Ther 10:1323–1333

Baroli B, Ennas MG, Loffredo F et al (2007) Penetration of metallic nanoparticles in human full-thickness skin. J Invest Dermatol 127:1701–1712

Beloqui A, Solinis MA, Rodriguez-gascon A et al (2016) Nanostructured lipid carriers: promising drug delivery systems for future clinics. Nanomedicine 12:143–161

Bhaskar K, Anbu J, Ravichandiran V et al (2009) Lipid nanoparticles for transdermal delivery of flurbiprofen: formulation, in vitro, ex vivo and in vivo studies. Lipids Health Dis 8:6

Boinpally RR, Zhou SL, Poondru S et al (2003) Lecithin vesicles for topical delivery of diclofenac. Eur J Pharm Biopharm 56:389–392

Bodade SS, Shaikh KS, Kamble MS et al (2013) A study on ethosomes as mode for transdermal delivery of an antidiabetic drug. Drug Deliv 20:40–46

Borm P, Klassig FC, Landry TD et al (2006) Research strategies for safety evaluation of nanomaterials, part V: role of dissolution in biological fate and effects of nanoscale particles. Toxicol Sci 90:23–32

Bosnjakovic A, Mishra MK, Ren W et al (2011) Poly (amidoamine) dendrimer-erythromycin conjugates for drug delivery to macrophages involved in periprosthetic inflammation. Nanomedicine 7:284–294

Brower V (2006) Is nanotechnology ready for primetime? J Natl Cancer Inst 98:9–11

Cevc G, Blume G (2004) Hydrocortisone and dexamethasone in very deformable drug carriers have increased biological potency, prolonged effect, and reduced therapeutic dosage. Biochim Biophys Acta 1663:61–73

Choksi A, Sarojini K, Vadnal P et al (2013) Comparative anti-inflammatory activity of poly (amidoamine) (PAMAM) dendrimer–dexamethasone conjugates with dexamethasone-liposomes. Int J Pharm 449:28–36

Eldhose MP, Mathew F, Mathew NJ (2016) Transfersomes—a review. Int J Pharm Pharm Res 6:436–452

Escobar-Chavez JJ, Rodriguez-cruz IM, Dominguez-delgado CL et al (2012a) Nanocarrier systems for transdermal drug delivery. In: Recent advances in novel drug carrier systems. InTech Publisher, London, pp 201–240

Escobar-Chavez JJ, Rodriguez Cruz IM, Dominguez-delgado CL (2012b) Chemical and physical enhancers for transdermal drug delivery. In: Pharmacology. InTech Publisher, London, pp 397–434

Fang JY, Hong CT, Chiu WT et al (2001) Effect of liposomes and niosomes on skin permeation of enoxacin. Int J Pharm 219:61–72

Foldvari M, Jarvis B, Oguejiofor CJ (1993) Topical dosage form of liposomal tetracaine: effect of additives on the in vitro release and in vivo efficacy. J Control Release 27:193–205

Geerligs M (2010) Skin layer mechanics. PhD Thesis, TU Eindhoven, Eindhoven. ISBN: 978-90-74445-92-4

Gillies ER, Frechet JM (2005) Dendrimers and dendritic polymers in drug delivery. Drug Discov Today 10:35–43

Glavas-Dodov M, Fredro-kumbaradzi E, Goracinova K et al (2003) 5-Fluorouracil in topical liposome gels for anticancer treatment-formulation and evaluation. Acta Pharma 53:241–250

Gupta U, Agashe HB, Asthana A et al (2006) Dendrimers: novel polymeric nano architectures for solubility enhancement. Biomacromolecules 7:649–658

Huang X, Du Y, Yuan H et al (2009) Preparation and pharmacodynamics of low-molecular-weight chitosan nanoparticles containing insulin. Carbohydr Polym 76:368–373

Jain NK (1997) Controlled and novel drug delivery. CBS Publishers & Distributors Pvt. Ltd., New Delhi

Jain NK, Umamaheshwari RB (2006) Control and novel drug delivery systems. In: Pharmaceutical product development, vol 21. CBS Publishers, New Delhi, pp 419–455

Jung SM, Yoon GH, Lee HC et al (2015) Thermodynamic insights and conceptual design of skin-sensitive chitosan coated ceramide/PLGA nanodrug for regeneration of stratum corneum on atopic dermatitis. Sci Rep 5:180–189

Junyaprasert VB, Teeranachaideekul V, Souto EB et al (2009) Q10-loaded NLC versus nanoemulsions: stability, rheology and in vitro skin permeation. Int J Pharm 377:207–214

Kaur A, Katiyar SS, Kushwah V et al (2017) Nanoemulsion loaded gel for topical co-delivery of clobetasol propionate and calcipotriol in psoriasis. Nanomedicine 13:1473–1482

Klimuk SK, Semple SC, Nahirney PN et al (2000) Enhanced anti-inflammatory activity of a liposomal intercellular adhesion molecule-1 antisense oligodeoxynucleotide in an acute model of contact hypersensitivity. J Pharmacol Exp Ther 292:480–488

Kumar JA, Pullakandam N, Prabu SL (2010) Transdermal drug delivery system: an overview. Int J Pharm Sci Rev Res 3:49–54

Lam CW, James JT, Mccluskey R et al (2004) Pulmonary toxicity of single-wall carbon nanotubes in mice 7 and 90 days after intratracheal instillation. Toxicol Sci 77:126–134

Ledger PW (1992) Skin biological issues in electrically enhanced transdermal delivery. Adv Drug Deliv Rev 9:289–307

Li X, Qi J, Xie Y et al (2013) Nanoemulsions coated with alginate/chitosan as oral insulin delivery systems: preparation, characterization, and hypoglycemic effect in rats. Int J Nanomedicine 8:23–32

Ma M, Di HJ, Zhang H et al (2017) Development of phospholipid vesicle-based permeation assay models capable of evaluating percutaneous penetration enhancing effect. Drug Dev Ind Pharm 43:2055–2063

Maheshwari RGS, Tekade RK, Sharma PA et al (2012) Ethosomes and ultradeformable liposomes for transdermal delivery of clotrimazole: a comparative assessment. Saudi Pharm J 20:161–170

Malakar J, Sen SO, Nayak AK et al (2012) Formulation, optimization and evaluation of transfersomal gel for transdermal insulin delivery. Saudi Pharm J 20:355–363

Manca ML, Matricardi P, Cencetti C et al (2016) Combination of argan oil and phospholipids for the development of an effective liposome-like formulation able to improve skin hydration and allantoin dermal delivery. Int J Pharm 505:204–211

Mehta R (2004) Topical and transdermal drug delivery: what a pharmacist needs to know. INET Continuing Education, pp 1–10

Muzzalupo R, Tavano L, Cassano R et al (2011) A new approach for the evaluation of niosomes as effective transdermal drug delivery systems. Eur J Pharm Biopharm 79:28–35

Neagu S, Preda S, Zaharescu M et al (2018) The effect of titanate nanotubes towards moderately halophilic bacteria. Rom Biotechnol Lett 23:13814–13822

Nighswonger G (1999) Medical device link MD & DI column: new polymers and nanotubes add muscle to prosthetic limbs. http://www.devicelink.com/mddi/archive/99/08/004.html

Oberdorster G (2006) Toxicology of air born environment and occupational particles. Part Fibre Toxicol 5:83–91

Paliwal SR, Paliwal R, Vyas SP (2015) A review of mechanistic insight and application of pH-sensitive liposomes in drug delivery. Drug Deliv 22:231–242

Palmer BC, Delouise LA (2016) Nanoparticle-enabled transdermal drug delivery systems for enhanced dose control and tissue targeting. Molecules 21:1719

Paolino D, Lucania G, Mardente D et al (2005) Ethosomes for skin delivery of ammonium glycyrrhizinate: in vitro percutaneous permeation through human skin and in vivo anti-inflammatory activity on human volunteers. J Control Release 106:99–110

Phatak AA, Chaudhari PD (2012) Development and evaluation of nanogel as a carrier for transdermal delivery of aceclofenac. Asian J Pharm Technol 2:125–132

Ramteke K, Dhole S, Patil S (2012) Transdermal drug delivery system: a review. J Adv Sci Res 3:22–35

Rizvi SA, Saleh AM (2018) Applications of nanoparticle systems in drug delivery technology. Saudi Pharm J 26:64–70

Salve P, Pise S, Bali N (2016) Formulation and evaluation of solid lipid nanoparticle based transdermal drug delivery system for Alzheimer's disease. Res J Pharm Dosage Forms Technol 8:73–80

Samala U, Pannala S (2011) Ethosomes, a novel transdermal drug delivery system: a review. Drug Invent Today 4:4628–4630

Santander-Ortega MJ, Stauner T, Loretz B et al (2010) Nanoparticles made from novel starch derivatives for transdermal drug delivery. J Control Release 141:85–92

Schafer-korting M, Mehnert W, Kortinh HC (2007) Lipid nanoparticles for improved topical application of drugs for skin diseases. Adv Drug Deliv Rev 59:427–443

Seong JS, Yun ME, Park SN (2018) Surfactant-stable and pH-sensitive liposomes coated with N-succinyl-chitosan and chitooligosaccharide for delivery of quercetin. Carbohydr Polym 181:659–667

Shah PP, Desai PR, Patel AR et al (2012) Skin permeating nanogel for the cutaneous co-delivery of two anti-inflammatory drugs. Biomaterials 33:1607–1617

Shakeel F, Ramadan W (2010) Transdermal delivery of anticancer drug caffeine from water-in-oil nanoemulsions. Colloids Surf B Biointerfaces 75:356–362

Sharma BB, Jain SK, Vyas SP (1994) Topical liposome system bearing local anaesthetic lignocaine: preparation and evaluation. J Microencapsul 11:279–286

Sharma A, Kumar S, Mahadevan N (2012) Nanotechnology: a promising approach for cosmetics. Int J Recent Adv Pharm Res 2:54–61

Shetty PK, Venuvanka V, Jagani HV et al (2015) Development and evaluation of sunscreen creams containing morin-encapsulated nanoparticles for enhanced UV radiation protection and antioxidant activity. Int J Nanomedicine 10:6477

Sonneville-Aubrun O, Simonnet JT, Lalloret F (2004) Nanoemulsions: a new vehicle for skincare products. Adv Colloid Interf Sci 108:145–149

Svenson S, Tomalia DA (2012) Dendrimers in biomedical applications-reflections on the field. Adv Drug Deliv Rev 64:102–115

Tanner T, Marks R (2008) Delivering drugs by the transdermal route: review and comment. Skin Res Technol 14:249–260

Tanwar H, Sachdeva R (2016) Transdermal drug delivery system: a review. Int J Pharm Sci Res 7:2274–2290

Touitou E, Dayan N, Bergelson L et al (2000) Ethosomes—novel vesicular carriers for enhanced delivery: characterization and skin penetration properties. J Control Release 65:403–418

Trotta M, Peira E, Carlotti ME et al (2004) Deformable liposomes for dermal administration of methotrexate. Int J Pharm 270:119–125

Uner M, Karaman EF, Aydogmus Z (2014) Solid lipid nanoparticles and nanostructured lipid carriers of loratadine for topical application: physicochemical stability and drug penetration through rat skin. Trop J Pharm Res 13:653–660

Vyas S, Khar RK (2011) Controlled drug delivery: concepts and advances, 1st edn. Vallabh Prakashan, New Delhi, pp 38–50

Weiner N (1998) Targeted follicular delivery of macromolecules via liposomes. Int J Pharm 162:29–38

Wissing S, Kayser O, Muller R (2004) Solid lipid nanoparticles for parenteral drug delivery. Adv Drug Deliv Rev 56:1257–1272

Yousef H, Sharma S (2019) Anatomy, skin (integument), epidermis. StatPearls Publishing, Treasure Island, FL. https://www.ncbi.nlm.nih.gov/books/NBK470464/

Zidan AS, Hosny KM, Ahmed OA et al (2016) Assessment of simvastatin niosomes for pediatric transdermal drug delivery. Drug Deliv 23:1536–1549

Leveraging Nanotechnology in Cosmeceuticals: Formulation, Characterisation, Regulatory Status and Toxicity

17

Pranav Shah, Jaimini Gandhi, Yashwini Kansara, Kalyani Barve, and Shweta Gupta

Abstract

The Federal Food Drug, and Cosmetic Act 2018 clearly defines the terms "drugs" and "cosmetics". Surprisingly, "cosmeceuticals" do not find any mention under this Act. The cosmetic industry commonly uses this word to refer cosmetic products that have medicinal or drug-like benefits. Cosmeceuticals are the fastest growing segment of the personal care industry. A new generation of cosmeceuticals containing more efficacious and stable active ingredients incorporated into versatile nanocarriers like liposomes, nanoparticles, buckyballs, nanoemulsions, dendrimers, fullerenes, microgels, nanogels, nanocrystals, nanogold and nanosilver have come into existence. They have been termed as "nanocosmeceuticals". A wide range of nanocosmeceuticals are presently available as antiaging products, skin cleansers, moisturisers and haircare products as colour cosmetics. There are more than 20 nanocosmeceutical based products commercially available and hundreds of patents pertaining to nanocosmeceuticals. Nanocosmeceuticals exhibit improved activity because of better entrapment efficiency, enhanced skin penetration, and retention leading to prolonged release of active ingredients. In spite of their fascinating and innumerable advantages, they have lot of safety concerns which should never be

P. Shah (✉) · J. Gandhi · Y. Kansara
Maliba Pharmacy College, Uka Tarsadia University, Surat, Gujarat, India
e-mail: pranav.shah@utu.ac.in

K. Barve
Shobhaben Pratapbhai Patel School of Pharmacy & Technology Management SVKMs NMIMS, Mumbai, Maharashtra, India
e-mail: kalyani.barve@nmims.edu

S. Gupta
Ideal College of Pharmacy and Research, University of Mumbai, Mumbai, Maharashtra, India
e-mail: shwetagupta@idealcollege.in

N. Shah (ed.), *Nanocarriers: Drug Delivery System*,
https://doi.org/10.1007/978-981-33-4497-6_17

overlooked. There are reports suggesting damage to body tissues, denaturation of proteins, alteration in genetic material, abnormal immune system reactions and increased oxidative stress. Threats to the marine life and soil have also been reported due to their casual disposals to streams, oceans and seas. However if used cautiously this advanced miniaturization technology has the potential to bring significant advances in the cosmetic industry. It is foreseen that, in the coming years, the market will be captured exclusively by nanocosmeceuticals.

Keywords

Nanocosmeceuticals · Formulation · Characterisation · Regulatory status · Toxicity

17.1 Introduction

The term cosmetics is defined in Federal Food, Drug, and Cosmetic Act 2018 as "articles intended to be rubbed, poured, sprinkled, or sprayed on, introduced into, or otherwise applied to the human body for cleansing, beautifying, promoting attractiveness, or altering the appearance" (USFDA 2018a).

The term "cosmeceutical", although widely used in the cosmetics industry, is not clearly defined under the Act. However, the cosmetics industry often refers the term to the products which have medicinal benefits like drugs but are also used as cosmetics. The applications of nanotechnology and nanomaterials in many cosmetic products have led to the emergence of "nanocosmeceuticals" (Kaul et al. 2018). Nanoparticles as UV filters and nanocarriers for site-specific delivery of active ingredients have been the two principal uses of nanotechnology in cosmetics (Duarah et al. 2016; Santos et al. 2019).

Newly developed nanocarrier-based cosmetics comprising microemulsions, nanoemulsions, liposomes, niosomes, solid lipid nanoparticles, nanostructured lipid carrier and nanospheres are gradually replacing conventional cosmetic products (Verma and Pathak 2010; Garg et al. 2015).

These novel nanocarriers possess the advantage of higher entrapment efficiency, improved skin penetration, and prolonged and controlled release of active ingredients with enhanced stability (Verma and Pathak 2010; Santos et al. 2019).

However, nanotoxicological researches have indicated concerns regarding the possibilities of health hazards with the excessive use of these nanocosmeceuticals due to higher rate of absorption of nanomaterials in the body. In order to better understand the benefits and health hazards of different nanocosmeceuticals, we must understand the formulation of nanocosmeceuticals and anatomy as well as physiology of body parts like skin, hair, lips and nails generally intended for application of cosmetic products.

17.2 Anatomy and Physiology of Body Parts Intended for Cosmetic Application

17.2.1 Anatomy and Physiology of Skin

The largest organ covering the body is skin. It comprises two layers: epidermis, outermost avascular layer, and dermis, deeper thick vascular layer. Below the dermis, we find the hypodermis/subcutaneous layer made up of areolar connective tissue and adipose tissue. This layer serves to store fats and also supplies blood to the skin.

The epidermis is made up of four types of cells: keratinocytes which are present in four to five layers producing keratin, melanocytes producing melanin, macrophages also known as Langerhans cells which destroy the invading microbes by mounting an immune response and epithelial cells present in the deepest layer connecting to the sensory nerve which help in detecting touch sensation. Hence these epithelial cells are also known as tactile epithelial cells.

The keratinocytes of the epidermis are arranged as follows starting from the underlying deepest layer to the outermost layer:

Stratum basale: Keratinocytes in this layer are cuboidal or columnar, some of which have the capacity to divide and form new keratinocytes. This layer also contains melanocytes and tactile epithelial cells.

Stratum spinosum: Here, the keratinocytes become flatter and are arranged in 8–10 rows. This layer provides strength and flexibility to the skin due to specific arrangement of the keratinocytes. Langerhans cells and melanocytes are a part of this layer.

Stratum granulosum: The cells of this layer are more flattened and are arranged in 3–5 layers. They contain a protein called as keratohyalin and also release a lipid-rich secretion which imparts water repellence to the skin. The cells of this layer continuously undergo apoptosis.

Stratum corneum: It consists of 25–30 multiple layers of flattened dead keratinocytes, which protects the underlying tissue from injury and microbial invasion. The cells in this layer are continuously replaced by cells of the underlying layer.

The dermis has mainly collagen and elastic fibres which impart the required extensibility and firmness to the skin and are attached to the underlying subcutaneous layer. The dermis which is essential for the existence of the epidermis can be divided into papillary region and reticular region. The papillary region which forms the dermal ridges consists of the areolar connective tissue, free nerve endings and corpuscles of touch. The reticular region consists of dense irregular connective tissue, adipocytes, nerves, hair follicles and glands. The dermis also embeds fibroblasts, macrophages and blood vessels.

Functions of skin are as follows:

1. Regulation of body temperature
2. Protection of underlying tissue from external environment
3. Excretion and absorption of substances
4. Synthesis of vitamin D
5. Detection of cutaneous sensations

Nanotechnology is used not only to enhance the chemical and physical stability of the drug but also to modify its penetration and permeation through the skin. The *stratum corneum* acts as an important barrier for the penetration of substances through the skin; however a drug molecule can permeate by either of the following routes: intercellular (through solubilisation in the lipids present in the membrane bilayer), transcellular (through the corneocytes) and appendageal (through hair follicles or sweat glands). The first two are important pathways for a drug to cross the *stratum corneum* (Schaefer et al. 2011).

17.2.2 Anatomy and Physiology of Hair

Hair is present on all body surfaces in varying degree, more on the scalp, eyebrows, armpit and around the external genitalia whereas completely absent on the palms and feet. The pattern of hair distribution and thickness depends on the genes and hormones. Hair is nothing but dead keratinised cells and comprises (1) shaft, the portion that projects above the skin (Buzea et al. 2007), and (2) root, the portion of hair below the skin, penetrating the dermis/subcutaneous layer. Both these portions are made up of three concentric layers: the innermost medulla surrounded by cortex and the outermost cuticle. The medulla is made up of 2–3 layers of cells and contains the pigment melanin. The cortex forms a major part of the shaft and comprises elongated cells whereas thin, flat, heavily keratinised cells form the cuticle.

The hair follicle surrounds the hair root. This follicle comprises epithelial and dermal root sheath, and the latter arises from the dermis. The epithelial root sheath has two parts: the external sheath which is the extension of the epidermis and internal sheath which arises from the matrix. The base of each hair follicle is called hair bulb. It is made up of areolar connective tissue, blood vessels to nourish the hair and matrix cells, which arise from the *stratum basale* and are responsible for the growth of existing hair and formation of new hair. The hair follicles are surrounded by dendrites of the neurons forming the hair root plexus. Sebaceous glands and arrector pili muscle are also associated with hairs.

17.2.2.1 Functions of Hair

Hair on the scalp protects the scalp from the dangerous UV rays and other types of injuries and decreases heat loss. Hair present on the eyebrow, eyelashes, nostrils and external ear canal protects the respective organs from injury by any foreign object. They also serve the function of light touch sensation.

17.2.3 Anatomy and Physiology of Lips

Lips are composed of surface epidermis made up of stratified squamous epithelial cells, connective tissue and orbicularis oris muscle that extends inwards in the cheek region. It forms the margin of the mouth of most of the vertebrates. The reddish skin depicts the transition from the hairy skin to the inner mucosa. It is also referred to as vermilion border and has abundant nerve endings. Lips have abundant blood vessels and nerve endings.

17.2.3.1 Functions of Lips

1. Help in the mastication
2. Help in speech
3. Provide sensory information about food and functions as a tactile organ
4. Have a cosmetic appeal
5. Help in expression of emotions

17.2.4 Anatomy and Physiology of Nails

Nails are made up of dead keratinised epidermal cells. The visible part of the nail similar to the *stratum corneum* of the skin is called nail body, below which lies an epithelium and the dermis of the skin. The free edge is that part which extends beyond the digits and is white in colour due to no blood circulation. The nail gets attached to the fingertip by the hyponychium which lies beneath the free edge. The nail root is buried deep into the skin, and proximal to the nail root lies the nail matrix. The matrix cells divide for the formation of new nails or growth of the existing nails.

17.2.4.1 Functions of Nails

1. Protect the fingertips
2. Help to grasp small objects
3. Help to scratch
4. Enhance touch perception

17.3 History of Cosmeceuticals

Raymond Reed, who was a cosmetic chemist, was a part of the founding member of US society who composed the word "cosmetics" during the year 1961. Cosmetic products are well defined as the products which upon application improve the texture of the dermis, escalate the cleaning and also glamorise the dermis layer of skin (Gautam and Singh 2011).

It was recorded that the use of cosmetics was first started by the Egyptians around 4000 BC and later they were accompanied by the Greeks, the Romans, the Chinese, the Japanese and the Americans. The early nineteenth century was marked by the use of cosmetics by the housekeepers secretly in the Western countries and the usage of cosmetics was widened in the twentieth century. By the twenty-first century cosmetic products had a huge use and modern cosmetic formulations were developed through advances in technology by incorporating newer technologies (Dureja et al. 2005; Sharma 2012).

Throughout the period of 4000 BC, utilisation of nanotechnology had been registered by the Egyptians and the Greeks, along with the Romans, with the perception of hair dye formulation with the application of nanotechnology (Bangale et al. 2012).

Nanocosmeceuticals was a considerate part of the science and technology which was useful to advance and evolve the particle units in the range of 1–100 nm size (Maynard 2006, Logothetidis 2012) which comprises biologically active elements possessing therapeutic advantages when applied on the surface (Mukta and Adam 2010).

They have determinate therapeutic effectiveness on the dermis of skin, as drugs and preparations have widened their range from the dermis layer of skin to body and to hair and they can be made use in the treatment of a variety of cases like wrinkles, photoaging, skin dehydration, dark spots, uneven complexion, hyperpigmentation and hair damage (Srinivas 2016).

Nanocosmeceuticals offer a number of advantages, namely:

1. It regulates the supply of active material through the control, in combination with a physical or chemical interface, of drug, polymer and additive composition, proportion, and preparatory process, of pharmacy releases of carriers. They are used in hair products for hair loss treatment, and for preventing hair from turning grey like Origem shampoo, and Nirvel hair loss shampoo.
2. Nanocosmeceuticals avail the odours last long, for example Allure Perfume and Allure Eau Perfume body spray by Chanel. These products avail the dermis layer care preparations extra efficacious and upsurge the usefulness of sunscreens by ameliorating UV shield inside these. Implementation of minor magnitude of the particle units makes the surface range amplified which in turn permits the active transportation of the active elements into the dermis layer. Obstruction contributes to the enrichment of the diffusion process and dermis layer water penetration is improved.
3. High levels of trapping effectiveness and advantageous sensory property are present in nanocosmeceuticals and are more consistent than standard cosmetics.
4. Nanoparticles are mostly suited for both lipophilic and hydrophilic drug conveyance. Nanomaterials are extensively utilised in the composition of moisturising lotion, skin-whitening cream, anti-wrinkle cream, hair-repairing shampoos, hair serums and conditioners (Nohynek et al. 2007; Mu and Sprando 2010; Antonio et al. 2014).

Nanocosmetics are expected to be the fastest growing portion of personal care manufacturing in the market. Regardless of the huge advantages of nanoparticles, there is still much to be found on the impacts on the atmosphere and on humans on short- and long-term health. As a result of reported toxicity and potential dangers of nanomaterials, care and security issues have been brought up.

17.4 Regulations on Nanocosmeceuticals

Cosmetics are not exposed to severe analytical conditions imposed by the FDA for their consent, unlike drugs/pharmaceutical products. The Federal Food, Drug, and Cosmetics Act and the FDA are unfamiliar with the concept of "cosmeceuticals" and therefore the appealing and practical advantages are enjoyed by articles that do not transform into counter-drugs K (Abbott et al. 2006).

More than few cosmeceutical products change the physiological procedures in the skin, but productions abstain from grasping clinical trials and creating the exact claims to escape exposing their articles to pricey and prolonged approval procedure by FDA (Sandoval 2009).

Some jurisdiction creates additional class for harbour cosmeceuticals or borderline products. Every country has its necessities for these types of articles: In Japan: the articles lying amid cosmetics and drugs are called "quasi-drugs". Elements must be allowed before they are included in the quasi-drugs and must be pre-approved before they are traded on the marketplace (Khaiat 2014). Korea classifies cosmeceuticals as "functional cosmetics" through Korea Food and Drug Administration (Marchant and Sylvester 2006).

Thailand: Conferring to the elements utilised in cosmeceuticals, they are categorised as "controlled cosmetics". Before being marketed in Thailand, control cosmetics need precise components that involve the notification from FDA for the practice of the products.

New Zealand classifies cosmeceuticals by the term "related products".

Australia: Goods in Australia are characterised on the basis of claims regarding the product and product composition; the borderline articles are known as "therapeutic goods". Only permitted components are made use of for the fabrication of these articles. Register of Therapeutic Goods in Australia records the therapeutic goods (Newgreen 2005).

Canada: Pharmaceutical cosmetics in Canada are named as "dermo-cosmetics". Pharmaceutical cosmetics here cannot be grouped as an independent cosmetic category; Canadian healthiness system has recognised Category V for the products falling in the class of both drugs and cosmetics. Lesser regulatory necessities are vital for regulation of these articles.

United States: In the United States, they have three classes like the cosmetics, drugs and OTC drugs in the market, where there is no legal description of cosmeceuticals as per USFDA (Raj and Chandrul 2016).

European Union: The EU does not categorise any special group to be referred to as cosmeceuticals, but it does have strict laws requiring proof of any company claim. Cosmeceutical regulation coined that any article containing nanoparticles as its constituents should be stated visibly and that the word "nano" should be inserted in brackets after listing of ingredients (Chappell 2012; Dhull et al. 2015).

China: Cosmeceutical products are known as special-use cosmetics. Special-purpose cosmetics undergo tests such as microbiological test, toxicological test, cancer testing and State Food and Drug Administration (USFDA) authorisation, according to their federal guidelines prior to selling them on a market. It is necessary to obtain a marketing certificate for cosmetic hygiene by Health Management Division of the National Council-SFDA (Hammes 1997).

FDA may ban the selling and making of the product or take other decisions such as prohibition on ingredients, seizure of unsafe products, mandatory warning labels and even worldwide prohibition of the product if it finds any safety concern with regard to the use of any cosmetics or of components, including nanoparticles. The United States Environmental Protection Agency (EPA) developed a new study strategy to investigate the effects on the environment and health owing to the use of nanoparticles in cosmetics, sunscreens, lacquers, etc. (Millikan 2001).

EPA is mainly concerened on research on nanomaterials including titanium dioxide, silver nanoparticles, nanotubes, cerium oxide, fullerenes and zero-valents present in cosmeceuitcals which may severly impact the environment and human health. Consumer Products Scientific Committee (SCCP) has also raised concerns about the consumption of insoluble nanoparticles used in cosmetics topically applied for poisonous purposes. The oldest science organisation in royal society in the world also asked about nanoparticles whether they can enter, absorb cells and communicate its effect in the bloodstream.

At the same time, it also expresses the additional exploration to report the chronic effects that might be caused by people around the globe via its long-term use (Liang and Hartman 1999).

In India, cosmetics are directed under Schedule "S" of Drugs and Cosmetics Act 1940 and Rules of 1945 but there is no distinct provision for safety or quality assessment of nanocosmetics. In the United States, corporations that intend to market have a legal responsibility to establish that their products and ingredients are safe and properly labelled, including nanoscale materials.

According to the USFDA and FD&C Act, only colour additives and preservatives that would be used in cosmetics need pre-approval (FDA 2018). In June 2014, FDA issued three final guidances related to nanotechnology applications in FDA-regulated products, together with cosmetics and food products (FDA 2018).

The characterisation of these particles has been underlined by the FDA. The guiding principles suggest that characterisation should involve measurements of particle dimensions; dispersal, accumulation and agglomeration features; chemicals in surfaces (Zeta potential, surface load, surface coating); catalytic action; morphology; solubility; density; and porosity (shape, surface area, surface topology and crystallinity).

17.5 Classification of Cosmeceutical Products

The cosmeceuticals are divided into eight categories.

17.5.1 Retinoids

Retinoids, premier evidence-based cosmeceuticals, are amongst the most prevalent products available in the market. They are either natural or synthetic derivatives of vitamin A and function through surface cell receptor interaction to produce a clinically defined effect. Other retinoids such as niacinamide and panthenol (pro-B vitamins) function in different manners by enhancing barrier properties of the uppermost layer of the skin (stratum corneum) through physical means.

17.5.2 Sunscreens

Sunscreens are recommended by dermatologists as the single most important dosage form that should be applied on a daily basis. UV-A and UV-B radiations cause disruption of the extracellular matrix, a hallmark of photoaging. Broad-spectrum sunscreens that provide protection against UV-A and UV-B are the cornerstones of photoaging therapy. Dioxybenzone, oxybenzone and sulisobenzone (benzophenones) provide protection in the UV-A and UV-B range of 320–340 nm.

17.5.3 Moisturisers

Moisturisers are one of the most useful categories of cosmetic products being employed for the management of various skin disorders like psoriasis, pruritus, atopic dermatitis and ageing skin. Products included in this category include humectants, emollients and occlusive. Majority of moisturisers act by augmentation of skin barrier function. Moisturisers are useful for providing hydrated, softer, smooth, less wrinkled, more radiant and firmer skin. The ingredients commonly employed in the manufacturing of moisturisers include petrolatum, silicon, mineral oil and glycerine.

17.5.4 Antioxidants

This category of cosmeceutical products upon topical application enhances the skin's natural antioxidant protection. Antioxidants cause reduction in free radical damage by blocking the oxidative processes in skin cells. Thus, this category of cosmetics provides protection to the skin from cancer, photodamage and photoaging. They also inhibit inflammatory responses that cause depletion of collagen. Vitamins A, C and E; alpha-lipoic acid and lactobionic acid; ubiquinone (coenzyme

Q-10); idebenone; and polyphenols (e.g. catechins, flavonoids) have been the commonly employed antioxidants.

17.5.5 Hydroxy Acids

α-Hydroxy acids (glycolic acid, lactic acid) and β-hydroxy acids (salicylic acid) have been employed since age-old times in dermatological and cosmetic formulations. Their exact mechanism of action is unknown and remains controversial. Some of the experts opine that α-hydroxy acids enhance the synthesis of glycosaminoglycans (GAGs), which in turn improves the quality of elastic fibres, and increases the density of collagen. On the other hand, β-hydroxy acids possess dermolytic properties and help in various ichthyotic and xerotic disorders.

17.5.6 Depigmentation Agents

Hydroquinone, a widely employed depigmentation agent, is employed for treatment of melasma post-inflammatory hyperpigmentation. The mechanism of action involves inhibition of conversion from tyrosine to melanin. Aloesin, arbutin, azelaic acid, glycolic acid and kojic acids are amongst other depigmentation agents employed.

17.5.7 Proteins and Peptides

Peptides belonging to the category of signal peptides, carrier peptides and neurotransmitter-inhibiting peptides have been employed in cosmetics due to their ability to improve the appearance of ageing skin. These peptides are known to trigger the wound-healing mechanism which in turn activates fibroblasts in response to fragmented chains of elastin and collagen. Peptides are also known to cause increment in the production of collagen to improvise the appearance of skin, resulting in smoother skin.

17.5.8 Growth Factors

Epidermal growth factor (EGF) is known to cause stimulation of epidermal growth. It is employed in the treatment of excision wounds and burns, where it accelerates the process of re-epithelialisation. Transforming growth factor (TGF) is known to stimulate normal skin growth and is responsible for cellular growth and repair. TGF is known to exert positive regulatory effects on accumulating body's extracellular matrix proteins. TGF mediates fibrosis (repair tissue formation) and angiogenesis (development of new blood cells) as well as promotes healing of wounds.

17.6 Different Nanosystems for Cosmeceuticals

Different nanotechnological products and novel nanodrug delivery systems like nanoemulsions, microemulsions, liposomes, niosomes, solid lipid nanoparticles (SLNs), nanostructured lipid carrier and nanospheres are increasingly utilised in cosmetic industries these days to design different cosmetic formulations which are replacing the usage of conventional forms. These newer nanosystems have greater potential to fulfil the increasing demands of consumers to improve their appearance by owing to enhanced penetration through skin and having controlled-release profile of different active cosmetic ingredients (ACI) (Kaul et al. 2018).

However, greater permeability of excipients along with ACI may also enhance the toxic effects to the body. Hence, the toxic effects of different nanocosmeceutical products need to be rationally analysed along with the benefits of these products. There are several available nanotechnological based marketed formulations used in the cosmetic industry.

The nanosystems used in cosmetics can be categorised broadly into four different groups: polymeric nanosystems, lipidic nanosystems, metallic nanosystems and miscellaneous other nanosystems (Santos et al. 2019).

17.6.1 Polymeric Based Nanosystems/Nanoparticles

Polymeric nanoparticles (PNP) are colloidal particles (<1000 nm) consisting of active ingredients either entrapped or encapsulated within a macromolecular matrix (usually a biocompatible and biodegradable polymer of natural or synthetic origin). The polymeric nanoparticles may exist as nanospheres (matrix system) or nanocapsules (reservoir system). These nanosystems are mostly utilised to enhance the absorption of active ingredients at target sites owing to their small size and for controlled release of these ingredients as per need (Nagavarma et al. 2012).

Controlled release of active ingredients reduces the frequency of application of cosmetic products, may it be sunscreens, anti-ageing products, perfumes or any other product. These delivery systems can further be incorporated in powder, gel, cream, lotions, etc., for enhancing its efficiency, stability, convenience, cosmetic appeal and user's compliance (Garg et al. 2015; Santos et al. 2019).

A film is formed on the skin surface due to the rigidity and bioadhesiveness of nanoparticles leading to enhanced retention and efficient release of actives (Santos et al. 2019). Various cosmetic products such as moisturisers, anti-ageing and other skincare products also find applications of these PNPs. The investigations with PLGA nanoparticles loaded with a natural polyphenol having photoprotection and antioxidant properties (morin) indicated a significant effect of polymeric nanoparticles in sunscreen creams (Afaq and Katiyar 2011; Watkins et al. 2015).

Chitosan is another commonly used natural biopolymer obtained by deacetylation of chitin because of its biocompatibility, biodegradability, and antimicrobial and wound-healing properties. It is a cationic polysaccharide which binds to proteins, lipids, nucleic acids and metal ions and expresses a wide spectrum of antimicrobial activities against microorganisms. Chitosan being a weak base can bind to carboxylic groups and this results in salt formation. Hence, it has the capacity to bind to hyaluronic acid which has anti-wrinkle properties and is commonly used in the cosmetic industry and thus chitosanic nanoparticles also hold a good potential in the development of novel cosmetic products (Watkins et al. 2015; Aranaz et al. 2018). Polymeric nanoparticle-based hair products are also under investigations. Nanoparticles of chitosan encapsulating the drug minoxidil have showed promising results for the treatment of alopecia in some of the investigations by researchers. High follicular uptake of nanoparticles was observed with PLGA and chitosan-coated nanoparticles loaded with ovalbumin (Matos et al. 2015).

17.6.2 Lipid-Based Nanosystems

Lipidic nanosystems have widespread uses as a carrier system for dermal applications because of their characteristics like biocompatibility, improved penetrability, improved bioavailability, stability and potential to be incorporated in different conventional and novel dosage forms (Gupta et al. 2013).

17.6.2.1 Solid Lipid Nanoparticles

Solid lipid nanoparticles are particles having an average diameter of about 10–500 nm possessing a core matrix of solid lipid which can solubilise lipophilic active ingredients. The lipid core is stabilised by surface active agents or emulsifiers. These lipids may include physiological and biocompatible lipids like triglycerides, diglycerides, monoglycerides, fatty acids, steroids and waxes. Different emulsifiers or combination of emulsifiers are utilised to prevent particle agglomeration and to stabilise lipid dispersion (Gupta et al. 2017).

The SLNs were developed as a substitute to other carriers such as liposomes, emulsions and polymeric nanomaterials. SLN offers various advantages like controlled release properties, targeted drug delivery, enhanced absorption owing to small size and low toxicity along with considerably good physical stability (Mehnert and Mäder 2012; Nikam et al. 2014; Gupta et al. 2017). SLNs also have great potential in the field of cosmeceuticals because of their submicron size which provides adhesiveness and greater penetration through skin, pearl-like appearance, and occlusive, hydrating and protective properties. The occlusive properties provided by SLN bring increased skin hydration. SLNs have UV-resistant properties; hence they may act as ultraviolet ray blockers and can even combine with organic sunscreens to improve the UV shield in skincare products (Duarah et al. 2016; Santos et al. 2019). Solid lipid nanoparticles are reported to be used in sunscreens developed using 3,4,5-trimethoxybenzoylchitin and vitamin E and have been found to enhance UV protection (Patidar et al. 2010). For anti-ageing products, SLNs have shown its impact with the launch of Nano Repair Q10 cream and Nano Repair Q10 Serum by Dr. Kurt Richter Laboratories, Germany, in 2005 (Mei et al. 2005; Attama et al. 2012). SLNs also provide advantages in perfume formulations by delaying the release of perfume over longer period of time.

17.6.2.2 Nanostructured Lipid Carrier (NLC)

Nanostructured lipid carriers (NLCs) may be considered as the next generation of lipid nanoparticles. They are prepared by entrapping the active ingredients in the combination of both solid and liquid lipids. These were developed to overcome the drawback of crystalline structure rearrangements which was observed in SLN (Kumar and Randhawa 2013). These rearrangements happen for achieving more ordered and stable form leading to reduced matrix imperfections and space to accommodate the active ingredient. In NLC, matrix of lipids is lesser organised than SLN. The distorted structure of NLC creates more space and contributes to its higher loading capacity compared to SLN. NLC also prevents the expulsion of drug during storage as observed with SLN (Kovacevic et al. 2011). NLC offers long-term stability and this also makes them better than SLNs in various cosmeceuticals. SLN and NLC are most frequently incorporated in moisturising creams, sunscreens and other skincare products.

17.6.2.3 Nanoemulsions

Nanoemulsions are clear biphasic metastable dispersions formulated spontaneously without requirement of energy where droplets of one liquid are placed inside another liquid having single tangible and consistency properties. The arrangement can be manipulated as per the technique of formation to offer articles with discrete features, e.g. water-like fluids or gels. Nanoemulsions have discrete rewards over conventional emulsions as they are smaller in size, have higher stability and have greater suitability to load active elements due to larger surface area. With nanoemulsions, the proportion of adsorbed film depth to droplet area is higher; therefore, the steric repulsion is sturdy enough which prevents any flocculation in the structure.

Nanoemulsions grew momentum as a hopeful transporter for accurate, harmless, effective and rational transfer of actives to the dermal composition as it incorporates the GRAS listed ingredients. Ceramide 3 and palmitic acid are such lipids that are naturally present in the dermis layer. It was detected that addition of these cholesterol and lipids advances water retention and bounciness of skin to an enormous level and further augments diffusion. In a view of formulation aspects non-ionic emulsifiers (ethoxylated) preferably for topical preparation are mostly used for preparation of nanoemulsions. A common cosmetic item, Korres Red Vine Hair Sunscreen, enables utilisation of nanoemulsions which requires very less concentration of surfactant. Nanoemulsions encompass superior sensation properties (swift penetration, merger textures) and comprise bioactive effects (specifically hydration power) making them innovative as compared to all other transporters. Nanoemulsions for haircare products have a size range from 50 to 200 nm. Formulation containing viscosity enhancers like ceteareth-20, ceteareth-12, glyceryl stearate, cetyl alcohol and cetyl palmitate and low quantity of non-ionic surfactants provides increased absorption through the topical route, smoothness to the tissues and nourishments (Jean et al.).

The major challenge with these types of formulations is its production, as it requires extensively costly instruments and consumption of energy inputs too with high-pressure homogenisation. While formulating and characterising nanoemulsions, Kabri et al. (2011) determined that homogenisation is the utmost favoured technique to attain the chosen dimensions of nanoemulsion globule. Alain et al. formulated a nanoemulsion that can be utilised in cosmetics and dermo pharmaceuticals. Formulation containing an amphiphilic lipid part is based on at least one non-ionic lipid amphiphilic in nature. The lipid is liquid at temperature not exceeding 45 °C. It has better stability than those containing an amphiphilic lipid phase composed of phosphoglycerides, water and oil. The resulting nanoemulsions are used as impregnating lotions for wet wipes. So far two commercial products are on hand in the market, TEGO® Wipe DE and TEGO® Wipe DE PF. "Phase inversion concentration" concept is relatively novel and cost effective through the use of very less surfactant at the same instance being comparatively useful for the production of small emulsion droplets.

Nanoemulsions are being enrolled in therapeutic cosmetics as surface medications to deal and cure dermatological diseases such as psoriasis, keratosis, actinic keratosis, basal cell cancer, squamous cell cancer, neurodermatitis, Bowen's disease, vulvar intraepithelial neoplasia and nodular or subcutaneous carcinoma.

17.6.3 Metal-Based Nanosystems

Along with the polymeric and lipidic nanoparticles, inorganic, metallic nanoparticles like nanoparticles of silver, gold, titanium oxide, zinc oxide and others are also emerging in cosmetic products. These NPs can be present in different formulations like in sunscreens, hair cosmetics and other herbal product-based formulations (Lohani et al. 2018). The toxicological aspects of these nanoparticles must be assessed before their widespread use in common public. Silver is reported to have antimicrobial and antifungal activity and hence silver nanoparticles are commonly found in various products used in cosmetics like creams, lotions, deodorants, soaps or toothpastes (Raj et al. 2012; Gupta 2018). Gold nanoparticles are also finding a wide range of applications. In cosmetic industries, these are of great interest to researchers owing to their antioxidant and antimicrobial properties, high drug-loading capacity and ability to penetrate through skin. Moreover, they are reported to improve skin firmness and elasticity, and are used in creams, lotions and deodorants.

Titanium oxide and zinc oxide nanoparticles are used in cosmetic products as UV filters. Zinc oxide is reported to reflect UVA radiation while titanium dioxide has the ability to reflect UVB radiation (Duarah et al. 2016). They are also claimed to have good spreadability and texture, and are non-irritant to skin which is usually observed with many chemical UV filters. Zinc oxide also has astringent and antimicrobial activity. Incorporating silver or nanoparticles of metal oxide in nail paints for the treatment of toenails due to fungal infections is also a new strategy being adopted. Such nanosystem-based nail paints have various other advantages like ease of application due to elasticity, fast dryness, improved toughness, chip resistance and durability (Lohani et al. 2018). Silica nanoparticles also find their application in cosmetic products, skincare, hairstyling, deodorants and toothpastes because of low-cost production, large-scale synthesis and sustained-release potential (Nafisi and Maibach 2015).

17.6.4 Vesicular and Miscellaneous Other Nanosystems

Novel nanocarriers like microemulsion, nanoemulsion, liposomes, niosomes, ethosomes, nanosomes, microgels, nanospheres, nanocrystals and nanofibres have the potential to play an important role in the cosmetic industry. These may be used in restoring gloss and texture, refurbishing damaged cuticles, making hair non-greasy, shiny and less brittle and many more.

17.6.4.1 Liposomes

The structure of liposomes encompasses spherical vesicular structures wherein the aqueous core is enclosed with a phospholipid bilayer. The outer lipidic bilayer supports hydrophobic or lipophilic active ingredient transport while the hydrophilic active ingredients may be incorporated in their aqueous core. According to the number of bilayers, liposomes are categorised as unilamellar vesicles (ULV) or multilamellar vesicles (MLV). These nanosystems also offer various advantages like high loading capacity of lipophilic as well as hydrophilic active ingredients, biocompatibility, protection of ingredients from degradation, improved penetration and controlled release (Allen and Cullis 2013). They find application in beauty products including sunscreens, moisturising creams, anti-ageing products and hair loss prevention products and in distributing fragrances in body sprays, deodorants, antiperspirants and lipsticks. Lecithin and phospholipids are of natural origin and the optimal choice for natural cosmetics (Schaffer 2007).

Endonuclease from *Micrococcus luteus* is found to be having activity toward ultraviolet-irradiated deoxyribonucleic acid. Hence, ultrasomes which are liposomes encapsulating these endonucleases were developed (Carrier et al. 1970; Santos et al. 2019; Afaq and Katiyar 2011). Similarly, photosomes are a type of liposomes which find application in sunscreen products, where photolytic enzymes extracted from—*Anacystis nidulans*—marine plant are encapsulated and released. Liposomal amphotericin B (L-AmB) is reported to be efficacious and safe for treating visceral leishmaniasis topical infection (Guery et al. 2017). A few major drawbacks associated with liposomal delivery systems are low reproducibility and stability issues.

17.6.4.2 Niosomes

Niosomes are similar to liposomes in their structure having self-assembled vesicular nanostructure, the difference being the bilayer which is formed by non-ionic surfactants instead of phospholipids in case of niosomes. Cholesterol and polyethylene glycol are commonly incorporated in niosomes to provide them more rigidity and improved stability and mechanical properties. The presence of surfactants increases the permeability of niosomes through the skin (Seleci et al. 2016). Niosomes are used in various beauty and haircare products. These nanosystems are highly used in chemotherapy treatments as well (Arul and Shanmuganathan 2015).

17.6.4.3 Ethosomes

Ethosomes are also vesicular systems like liposomes and niosomes generally consisting of phospholipids, ethanol and water and have the characteristics of a deformable structure which make them more penetrable through skin. Classical ethosomes consist of high amount of ethanol, binary ethosomes consist of propylene glycol or any other alcohol and transethosomes are the most flexible vesicular systems (ethosomes) consisting of surfactants as well. The ethosomes may prove to be of high potential in dermatological products (Zhang et al. 2017).

17.6.4.4 Nanofibers

As the name suggests, they are fibres with diameters ranging in nanometres which may be formed from different polymers and depending on the type of material used, they have different physical properties and thus different applications. The polymers used in their structure can be obtained from natural source (such as chitosan, collagen, silk) or from synthetic origin like polylactic-co-glycolic acid (PLGA), polyvinyl alcohol (PVA) or polyvinyl pyrrolidone (PVP) (Meireles et al. 2018). Nanofibres form one of the widest class of nanomaterials owing to their exclusive characteristics. High surface area-to-volume ratios, low diameters, highly porous structure, high strength, high loading capacities, high permeability of oxygen and water vapour, high liquid absorption properties and small pore sizes make them good carriers for numerous applications including cosmetic products. Nanofibres are used in skin wound dressing, facial masks, cleansing skin and nanocarrier for delivery of active ingredients for skin therapy (Yilmaz et al. 2016).

17.6.4.5 Nanocrystals

Nanocrystals are particles of pure drug which is stabilised by surfactants without any surrounding polymer or lipid matrix and are composed of atoms in either a single or polycrystalline arrangement. The main advantage of nanocrystals over other delivery systems is enhanced solubility. Different techniques are utilised for processing poorly soluble compounds as nanocrystals. Homogenisation, solvent evaporation, media milling and microfluidisation are some of these techniques. Increased solubility increases penetrability of active ingredients into various layers of skin. Flavonoids (antioxidants), beta-carotene, coenzyme Q10 and many other active ingredients are reported to be successfully produced as nanocrystals and observed to show improved drug and cosmetic characteristics (Duarah et al. 2016).

17.6.4.6 Dendrimers

Dendrimers have a well-defined extremely ramified symmetry of approximately 20 nm in size, with a homogenous construction and a large concentration of efficient terminal sets on their edge. They have many external organisations suitable for multifunctional purposes (Tournilhac and Simon 2001; Michael et al. 2000). They contain a well-defined symmetric structure, which is frequently branched and provides different benefits for cosmetic preparations when thin films such as nail enamel and mascara are required. The preference of external groups of function (amine, carboxyl and hydroxyl) depends on the molecular size and weight of the dendrimers, the most significant development and application parameters of different cosmetics. L'Oréal has patented a design of extensively highly branched polymers and dendrimers forming a fine coat on the substrate (Magnan and Genard 2002). A severe issue is the fast growth of the film after implementation. This is why film-forming agents for various applications are part of the new preparations. Most of their use is seen in cosmetics for artificial tanning, hair, skin and clots (Raj et al. 2012).

17.6.4.7 Microgels/Nanogels

They are micronetwork-forming structures prepared using inverse microemulsion polymerisation technique. They have very confined dispersal and establish balanced suspensions in water. The denser the network, the lesser the grade of microgels. They have sizable domain as they possess exposed network arrangement and act as transporters to carry drugs. Coupling lipophilic category to their structure hopefully alters their action; electronegative group is fused in their main chain.

17.6.4.8 Nanocapsules

Nanocapsules are vesicular structures consisting of a polymeric membrane where an inner core of liquids may be aqueous or oily. They are capsulated at the nanoscale division (10–1000 nm) (Fontana et al. 2009) and are built on encapsulation technology which can transfer medicine payloads for locally acting or specific-site drug delivery. The nature of the substance which has to be inculcated depends solely on the various types of nanocapsules to be developed. Companies like Exlica Ltd. and MiCapt are looking at many different materials to be used, such as polymer microbeads, silicon nano-shells or microbial cell walls, as nanocapsule shells. The formulation of an emulsion gel comprising octyl methoxycinnamate-loaded nanocapsules was based on butyl glucoside and polyacrylamide/C13–14 isoparaffin/laureth-7 (Sepigel® 305) (Vettor et al. 2010).

The sunscreen safety has been improved by this preparation and contact with skin layers has been reduced without diminishing its efficiency. The sunscreen capsules of benzophenone-3 were inculcated into hydrophile gels and the impact of the sunscreen's nano-encapsulation together with the efficiency of sunscreen was assessed. The use of nanocapsules reduces the penetration of UV octyl methoxycinnamate filters in pig skin compared to conventional emulsions (Hwang and Kim 2008).

Nanocapsules laden with coenzyme Q10 changed the rheological drift pattern of Carbopol® 940 gels from the yield-pseudoplastic to the pseudoplastic structure (Terroso et al. 2009). There has also been reports of inculcation into shampoos of hinokitiol-loaded nanocapsules and of hair tonic for hair growth therapy (Hwang and Kim 2008).

The impact of hair development on both mice preparations was similar to the beneficial control of the 3% minoxidil solution used. As per the authors, electrostatic communication amongst the nanocapsules (positively charged) and dermis layer influences the hair growth.

17.6.4.9 Buckyballs

Interpretation of fullerenes can be done as "Compounds composed solely of an even number of carbon atoms, which form a cage-like fused-ring polycyclic system with twelve five-membered rings and the rest six-membered rings. The archetypal example is C60 fullerene, where the atoms and bonds delineate a truncated icosahedron" (Rania et al. 2007).

Fullerenes are immensely lipophilic so are insoluble in water-based systems that originally restricts its significance but the use of surface active agents or surface modifications has expanded its dissolving property in aqueous medium and increased their potential pharmaceutical utilisation (Usenko et al. 2007). Buckyballs have various applications in cosmeceuticals due to their oxidation inhibitor characteristics. They exhibit strong removing action against reactive radicals and they are applied in the composition of skin-revitalising cosmetics (Dhawan et al. 2006).

17.6.4.10 Nanopigments

Nanopigments of gold and silver were formulated to produce new coloured pigments for lipstick and have proved harmless, effective and dispersive. For instance, when the dimensions of gold particles become several hundred nanometres of sphere shape, different shades of red are presented. In silver, nanoparticles have the same colour of yellow and not the colour of grey silver. As gold and silver are harmless, unlike a standard pigment, and have a high degree of disinfection and stability, the cosmetics and personal care sector has a vast and varied future. Zinc oxide and titanium dioxide are the two metal oxides unchanged in sunscreen pigments for the physical environment. Decreasing the particle dimensions in nano range advances spreadability and also offers transparent texture to the article.

17.6.4.11 Carbon-Based Nanomaterials

Carbon nanoparticles together with fullerenes represent carbon-based nanomaterials (Baroli 2010). Carbon nanotubes are hollow pipes with multiple walls or single walls, whereas sphere-shaped fullerenes have very small mean diameters (<100 nm) (Crosera et al. 2009).

These structures contain carbon rings which are oddly numbered (like pentagonal shape, heptagonal shape), granting a three-dimension sphere-like shape (Aillon et al. 2009). Thus, these systemic arrangements are called fullerenes or "buckyballs". Fullerenes have a very hydrophobic nature; thus, they are not soluble in hydrophilic preparations. This contributes to its disadvantage, but by using surfactants the capacity of fullerenes to solubilise in water is improved (Xiao et al. 2005; Kaur et al. 2007). Studies of the antioxidant impact were conducted on the use of keratinocyte (human skin) culture (HaCaT) and demonstrated their capacity in the presence of UV-B radiation to decrease reactive oxygen levels (Xiao et al. 2005). When the skin was previously flexed, particles entered the dermis, when no penetration in the dermis of inflexed skin was reported (Sung 2013). The degree of diffusion of fullerenes seems to be liable on the solvent type (Xia et al. 2010). Inui et al. in 2011 described radical scavenging potential of fullerene. Fullerene gel suppressed acne vulgaris by reducing the infiltration of neutrophils and sebum manufacturing while increasing the skin water content (Inui et al. 2011). Derivatives and fullerene-based molecules protect skin against free radicals, melanin synthesis and cell death (Xiao et al. 2005). Bio Fullerene™, called Radical Sponge®, which was made by the Japanese business Vitamin C60 Bio Research Corporation, is progressively used as anti-ageing agents by virtue of its ROS-scavenging capacity. The rolled graph with SP2 hybridisation can be labelled as carbon nanotubes (CNTs). These are ongoing cylindrical hollow fibres, consisting of graphic walls, which roll at particular and discreet chiral corners as a hexagonal carbon grid. Discrete carbon nanotubes naturally line up into pi-stacked "ropes". The diameter varies between 0.7 and 50 nm with length measuring in the 10 μm range (Ibrahim 2013; Kaushik and Majumder 2015).

The weight of carbon nanotubes is very light. Several patents have been submitted for carbon nanoparticles in cosmeceuticals such as hair dyes and cosmetic formulations, which include carbon nanoparticles and carbon peptide nanotubes, and their use in hair colours and cosmeceutical products (Huang et al. 2005, 2006).

Nano-diamond units have unbelievable absorption properties, can augment mechanical properties, offer improved bonding of certain biological materials and shield a subject from ultraviolet radiation. Sung in 2013 patented a sunscreen composition which comprises functionalised nano-diamond for protection against some forms of skin cancer. The particles are in the range of 0.5 nm to about 50 nm. It comprises an optional dispersant to prevent aggregation and flocculation of particles due to the electrical charge existent on the surface (Sung 2013).

17.7 Popular Categories of Nanocosmetics

These nanostructured particles are found in a large sum of cosmeceutical articles that enhance the diffusion of rejuvenating or skin-brightening active chemical ingredients and promote continuous action of make-up; they act as transporters of bactericidal ACIs in body sprays and are also a constituent of UV shield articles.

17.7.1 Beauty Care

A tyrosinase inhibitor consisting of depigmentation molecule is 6-methyl-3-phenethyl-3,4-dihydro-1H-quinazoline-2-thione (JSH18). In this regard, to assess the reduction in melanin synthesis and its application in surface formulations, this molecule was inculcated inside SLNs. After a week of exposure to UV fallout and implementation of the mentioned preparation during 4 days, the dermis was assessed using reflectance spectrophotometry, demonstrating a complete retrieval of pigmentation caused by the sun (So et al. 2010). Jiménez et al. had researched the impacts of

Panax ginseng extract nanoparticles that are known for their antioxidant characteristics and have been integrated into cosmetic whitening products (Jimenez et al. 2018). The whitening characteristics were then explored and the activity of tyrosinase and melanin production by melanoma cells of skin showed significant decrease in levels, suggesting gold nanoparticles for cosmetic applications (Ayumi et al. 2019). By inhibiting extreme melanin manufacturing, α- and β-arbutin are renowned for their skin-brightening characteristics. Chitosan nanoparticles are reported to be capable in topical formulations to arbutin encapsulation structures with high stability capacities. Ethosomes and transfersomes (Celia et al. 2012) have also been the topic of an inquiry into skin-whitening formulations involving linoleic acid. Both NPs' lipid structure enabled a healthy permeation of the skin through corneocyte channels with its implementation in the improvement of skin hyperpigmentation illnesses. Nanoparticles have become an appealing approach for practical application in make-up articles (Dingley et al. 2018). Avon introduced a gel preparation comprising TiO_2 nanoparticles and colour pigmentation that form a space-filling film with optical properties when applied to the skin. It shifts the refractive pattern and the dissemination of light on the dermis layer, thus fetching a tool for concealing wrinkly skin and dark spots, hence starring a natural look. Hiroyuki and Takesuke (2010) outlined a way to decrease the output of skin oil to upsurge the impacts of make-up. A biocompatible polymer NP comprising an inhibitor of sebum manufacturing was created to allow this ACI to reach the pores of the skin. In this sense, in hydrophobic silicone-coated chitosan NPs, glycyrrhizin acid was encapsulated. The ACI is continually released as the biocompatible polymer is degraded and the inhibition of sebum manufacturing is prolonged.

Integrated into skin foundations, this structure can stop the collapse of make-up.

Hyun et al. (2009) revealed a lip care hurdle solution linked to toxic pigments that were initially used in lipstick manufacturing. Cosmetic pigments that are used in lipsticks, including Au and AgNPs, demonstrate a wide range of colours and are harmless to the skin. Reported by the writers of this job, AgNPs displayed yellow colour and AuNPs displayed red colour. The Au and AgNPs are therefore blended together and a varied range of colours can be produced as per the visible light band, depending on the ratio of the individual, thus enabling production of diverse colour lipsticks. By integrating metal NPs, Lau et al. (2017) identified laser-generated NPs as a manner to eliminate traditional nail colour. This gives optical effects, colour durability, hardness and damage or even antibacterial impact resistance.

17.7.2 Dental Care

Malarkodi et al. (2014) demonstrated zinc sulphide (ZnS) and cadmium sulphide (CdS) NP synthesis together with antimicrobial action accountable for dental illnesses such as *Candida albicans*, *Streptococcus* sp., *Lactobacillus* sp. and *Staphylococcus* sp. (Malarkodi et al. 2014). The uncommon operation showed excellent limits, particularly due to gutta-percha reinfection, the primary filler material used up to now. Nano-diamonds were found as a great substitute to solve the matter stated directly above.

Lee et al. explained the growth for infection of gutta-percha entrenched in amoxicillin-comprising nano-diamonds. A liquid antiseptic comprising AgNPs with distinct external charges was created and chlorohexidine (CHX) and sodium hypochlorite (NaOCl) were compared to their antibacterial properties (Abbaszadegan et al. 2015). The research indicated that the smallest MIC was positively charged and offered low deadliness to fibroblasts, making it a deserving means to be inculcated in tooth-shielding liquids for future use. The fluid formulation of ferrous oxide (FeO) nanoparticles in the aversion and therapy of biofilms was defined by Hyun et al. (2009). FeONPs can be introduced into elixirs to avert the development of biofilm, to hinder bacterial development and to circumvent the demineralisation of teeth. In oral disorders such as dental caries, FeONPs can therefore be chosen to be coupled with hydrogen peroxide and enzymes. Dental treatments can also be accompanied by NPs with whitening impact. Toothpaste was synthesised as a whitening agent to include activated carbon NPs.

The inclusion of these NPs overcomes peroxide and elevated friction and related disadvantages. Traditional whitening methods which were utilised resulted in poor performance and enamel damage.

17.7.3 Haircare

Studies were created to take into account the use of polymer NPs as hair follicular ACI-loaded nanosystems. In the assessment of penetration as the hair-restoring agent with dyes, Glowka et al. (2014) outlined the growth via use of roxithromycin polymeric NPs. AuNPs are scattered into an oil that forms a plate in this invention. These NPs encourage the flow of the blood, and stimulate the scalp, and also the separation of hormones and cells. The fresh hair becomes stronger by encouraging hair development, and therefore hair loss can be avoided owing to its density. Substances such as silicone have significant hair lubrication and conservation attributes. However, owing to their hydrophobicity, some problems have been discovered with their absorption. These problems can be solved by nanotechnology. The survey of silicone oil imprisoned in stable nanoemulsion O/W, developed as surfactants in Tween 80 and Span 80, was recorded by Hu et al. (2012). The reduction of particle size enables more interaction between nanoemulsion particles and improves the disposal of silicone oil. Furthermore, the temperature and storage time that demonstrates the nanoemulsion stability have not influenced this disposal.

17.7.4 Skincare

The supply of vitamin E, resveratrol and epigallocatechin gallate capsulated in SLNs and NLCs to the stratum corneum was researched by Chen et al. (2017). Increased rates of encapsulation were provided and the ACIs were protected from degradation.

The lipid NPs loaded with antioxidants showed great stability and consistency. The release data of research showed that lipid-based NPs received 70%, after 24 h of controlled release for resveratrol, indicating that these lipid-based NPs are appropriate carrying agents for skincare. The use of coenzyme Q10 (CoQ10) by nanoemulsion in order to check its impact as an anti-ageing product was suggested by El-Leithy et al. (2018). An O/W nanoemulsion package comprising CoQ10 was generated and formulation diffusion by the stratum corneum was examined. Studies have shown that these topical formulations improve skin permeability and solubility in anti-ageing treatments. As carriers with hyaluronic acid and epigallocatechin gallate, transfersomes were also an option in the formulation of anti-ageing skin cream (Avadhani et al. 2017).

17.7.5 Deodorants

Earlier carbon-based nanoparticles were used for removal of smell, for example chemical, physical, biochemical and sensory removals. Another invention is a deodorant with excellent dispersibility, heat resistance and stability consisting of Ag-TiO_2 NPs of antibacterial and body spray action (Soo 2006).

Hosseinkhani et al. (2015) have researched a technique for releasing fragrances in cosmeceutical articles, such as body sprays. Fragrance molecules are normally poorly aqueous and are stable, limiting their use. The perfume molecules were therefore entangled in the polymer NPs and were measured and found to be satisfactory for their application in the axillary microbiome.

17.7.6 Sunscreens

Sunscreens are part of a vast cosmetics sector where nanotechnological development has seen an enormous increase. The morin polymeric nanoparticles in a sunscreen were defined by Shetty et al. (2015). Morin is an antioxidant and UV-protection natural flavonoid. In this respect, PLGA NPs displayed large in vitro outcomes in comparison to morin's antioxidant action, skin deposition and penetration. The preparation cream has a non-cytotoxic antioxidant property (Borase et al. 2014). The AuNPs were integrated in their initial form and their use was assessed in a sunscreen without metal nanoparticles. There has been a dramatic increase in the sun protection factor (Deng et al. 2015).

17.8 Excipients Used for Nanocosmetic Products and Their Safety Profile

Excipients used for a vast variety of cosmetic products come from even more vast category of sources ranging amongst different plants, animals, insects, minerals and many more natural and synthetic origin. The use of different substances to improve personal appearance started in very early stages of civilisation without considering much for their toxicological aspects and safety profile. Earlier people may have started with simple ingredients from a few oil-based perfumes, emollients like castor oil and different herbs to provide attractive colours to their skin, but the range of excipients used in cosmetic products in the present scenario has increased tremendously and needs proper consideration for their safety issues. The most common excipient required for cosmetic products which led to exploration of various categories of ingredients is colourants (dyes and pigments). The colourants may include carmine extracted from crushed bodies of cochineal insect, metallic rust (iron oxide), erythrosine, tartrazine, metallic lakes or other synthetic organic colours, many of which may also consist of ingredients which can harm our body, so effective regulations are needed to control their usage (Allam and Kumar 2011).

There are wide range of colours used in food, personal care products, cosmetics, household products and fabric dyeing which are listed as numbers in the Colour Index International. In the United States and Canada colours are listed as FD&C colours. The synthetic colours are listed as a code and may contain trace amounts of lead or arsenic (Aldayel et al. 2018).

Even trace amounts of these ingredients may produce harmful effects when these are absorbed through the skin or swallowed along with drinking or eaten when used for lipstick colour. The permissible levels of these are mentioned in regulations and need to be monitored stringently. Even in many cases, the chemicals produced on long-term storage of these products may be less efficient or harmful. For example, some mascaras contain an ingredient that produces formaldehyde on chemical decomposition which prevents the growth of bacteria. However, if stored for a long time, it may not produce formaldehyde at later stages and allow bacteria to proliferate and cause infection. For this reason, it is often recommended to replace a tube of mascara every few months (Ben-noun 2016). The pearlescent effect in facial and nail cosmetics is mainly achieved through mineral mica covered by a thin layer of titanium dioxide. Substances like stearates and other waxy substances or shimmery substance obtained from fish scales are also used for this effect (Lohani et al. 2018).

Castor oil, animal fat and waxy base are other common excipients for facial cosmetics. Different herbs and their extracts and other fragrance materials are frequently used as perfumes in these cosmetic products. Aluminium-based salts are commonly used in antiperspirants but are proved to have detrimental effects in many species (Garg et al. 2015). In addition to these, the tremendous number and variety of excipients used in cosmetic products these days and the newly developed various nanocosmeceuticals are even more controversial until the safety of all these ingredients is well proven and regulated. However, some of the ingredients used have beneficial effects as well. Phosphatidylcholine is one of the key components of liposomes which are used in various skincare formulations like moisturising creams and haircare products like shampoos and conditioners due to its softening and conditioning properties. Vegetable phospholipids contain high amount of esterified essential fatty acids and are widely used for topical applications in cosmetics and dermatology. Flexible liposomes are proven to help in wrinkle reduction and enhancement of skin smoothness (Schaffer 2007).

It is observed that the health hazards and safety profile of various excipients used in earlier versions of cosmetic products, conventional forms and recently developed various nanocosmeceuticals are usually less considered as compared to those of food and other pharmaceutical products. The use of excipients in cosmetics should be done in such a way so as to fulfil its intended use rather than complicating the situation by long-term undesired effects of some of its ingredients. However, this has been realised and different guidelines and regulations are developed for cosmetic products as well by different countries.

17.9 Formulation of Nanocosmeceuticals

Generally speaking, nanosystem preparation methods are categorised into two groups, downward and upward. The latter process involves the production of nanostructures from higher proportioned particles and scaling them back into nanosize with the desired optimised characteristics. The technology therefore needs a great deal of energy to exceed inner forces. Top-down techniques could include physical lithography, such as photolithography, interfering lithography, lithography of beams of electro-depression, nano-stencil lithography, lithography of nanosphere, homogenisation and milling at high pressures, or (Buzea et al. 2007) chemical input which includes chemical reaction input power (Chan and Kwok 2011; Biswas et al. 2012).

On the conflicting note, bottom-up methods comprise the manufacture, without waste, of nanosized assembled particles, of microscopic or molecular substances. Examples include the deposition of nuclear layers, sol-gel nanofabrication, precipitating evaporation, molecule auto-assembly, precipitation of sonorous elements and drying by spray. Other techniques are hydration in thin layer, inverse-phase evaporating, pH gradient transmembrane and sonication in rotational evaporation (Jain et al. 2015). These approaches have their rewards and drawbacks. For example some methods may improve the regulation of functionality as required, such as morphology or dimensions, in contrast with low-income methods, but top-down techniques such as moulding are simpler to scale up. However, the enormous temperature and the pressure needed for other up-down techniques, such as high-pressurising homogenisation, can lead to working problems. The chosen technique can affect the final formulation and probably cause difficulties, such as the degradation of active ingredients (Abbasi et al. 2014). The most advantageous technique for producing NPs, namely in the fields of SLNs and NLCs, has been defined. In the pharmaceutical industry, these techniques are commonly used and therefore have no regulatory problems since it can avoid organic carriers and it is simple to gage up (Pentek et al. 2017).

17.10 Efficacy Evaluation of Nanocosmeceuticals

Both the Organisation for Economic Co-operation and Development (OECD) and the Cosmetic, Toiletry and Fragrance Association (CTFA) have set the rules for noxious tests based on the excipients' toxicology outline and their pathways. Dermis allergy, skin-sensitive test, ocular allergy test, phototoxicity test, mutagenicity and genotoxicity test are required to be done for nanocosmeceuticals. Traditional testing shall also be conducted frequently, such as acute toxicity, repeated-dose toxicity (21 and 28 days) and subchronic toxicity (90 days).

In vivo toxicity tests are useful when conducted to obtain information of the organ translocation, biodistribution, accumulation and clearance. The trial elements should be applied to the dermis layer and test drugs should be provided for the oral route either by gavage or in the diet to evaluate the dermal route of administration. The dose metrics for in vivo research (mass, quantity and amount of particles) are the key criteria. Animals have mainly been used to check the harmfulness of testing of new chemical substances, but since it is prohibited from using animals to that end, certain of the already authenticated non-animal substitute testing techniques for particular nanomaterials can be optimised. Some alternatives include re-built human skin such as EPISkint and EPIDermt, 3T3 NRPT3T3 fibroblast neutral phototoxicity (for testing for phototoxicity and ultraviolet-removing substances) and provision of re-fabricated human skin. The use of human/pig skin in dermal absorption trials is suggested whereas the potential for eye irritation is evaluated by means of bovine cornea opacity/permeability test and isolated eye of chicken (ICE).

17.10.1 Nanomaterial Characterisation

New testing strategies for identifying the processes will be required for the particular characteristics of nanomaterials. The principal parameters assessed for nanomaterial safety are as follows:

Physical/chemical characteristics: Physical characteristics such as the dimensions of a structure, structure and chemical characteristics such as structural formula and structured form, specific surface area, agglomeration status, size distribution, size morphology, solubility, nanomaterial composition, phase identity, surface chemistry, hydrophilicity and lipophilicity must be analysed. Nanoparticles are smaller in size: their tiny sizes are the primary feature of nanoparticles. Compared to their bigger parts, this can change physicochemical characteristics and give the possibility to improve the use and interaction with the biological tissues. Poison concerns in particular the creation of reactive oxygen species such as free radicals that lead to oxidative strain, swelling and resultant harm to proteins, membranes and DNA. As of their size, these nanoparticles can readily access bloodstream through dermis layer or breath and are transferred to the different organs from there. The nanoparticles' elevated dose and lengthy residence time in the essential organs can lead to their dysfunction. While 500 nm of titanium dioxide elements are only to some extent capable of initiating DNA strand rupture, 20 nm of titanium dioxide particles are proficient of causing full demolition of supercoiled DNA, even at low concentration and without UV exposure. It was discovered in another research that mice subacutely subjected to 2–5 nm TiO_2 nanoparticles showed a substantial but mild swelling reaction.

Shape of nanoparticles: Nanoparticles are formed in various shapes, like spheres, pipes and blades, and are a main source of being unwell. A research showed that mice's stomach cavity exposure to lengthy carbon nanotubes is associated with stomach wall swelling.

Surface area of nanoparticles: As the particle size reduces, its surface area rises and its reactivity rises. Due to their elevated surface area-to-mass proportion, nanomaterials are also extremely reactive, offering more area by weight for chemical reactions. Trainings have shown that some nanoscale elements may be possibly explosive and/or photoactive due to this enhanced reactivity. For example, if finely dispersed in the air, some nanomaterials, such as nanoscale titanium dioxide and silicon dioxide, may explode and come into interaction with a sufficiently powerful source of explosion.

17.10.1.1 Microscopic Techniques

Microscopic examination of skin post-treatment can provide more helpful data from in vitro research. While complete quantification may not be feasible, conception of the tissue to which an active was applied may provide useful insight. The techniques used for microscopic assessment are confocal microscopy laser scanning, electron microscopy of high-resolution transmission, X-ray emission-causing article, positron emitter radio labelling.

17.10.1.2 Mathematical Modelling

These predictive models vary from easy empirical algorithms to complicated mathematical equations that occasionally necessitate experimentally inaccessible parameters to be known and estimated. However, because data related to macromolecular compounds or structures of particles are not included in any of these models, they cannot be cast off with trust to forecast what may occur if such entities touch the skin.

17.10.1.3 In Vitro Techniques

Although a number of alternative processes and techniques are available for the analysis of molecular processes engaged in the biological activity of compounds, for cosmetic products only validated techniques are allowed. These validated techniques shall be used for the safety evaluation of cosmetic ingredients when testing is needed.

17.10.1.4 Diffusion of Nanoparticles via Skin

Research has shown that nanoparticles can enter the skin, particularly when the skin is stretched. Wrecked skin is a direct path even in a size of 7000 nm for the penetration of particles. The existence of acne, eczema and injuries can improve the absorption and further complications of nanoparticles in the bloodstream. An initial research discovered that the penetration of nanoparticles into the skin impacted by psoriasis was deeper than in the skin without effect. The basic carriers have recently been altered to improve the strength of the skin by integrating certain physical and chemical penetration stimulants as well as by formulating new vesicular devices with enhanced skin permeability, such as ethosomes and transfersomes. The skin infiltration of nanoparticle can be improved by flexing and massaging. One study has discovered that intact skin can even absorb particles up to 1000 nm in magnitude to reach cells when the skin is flexed. Cellular poison of zinc oxide and titanium dioxide nanoparticles: It has been discovered that nanoparticles of zinc oxide (ZnO) used in sunblock can harm or destroy stem cells in the brain of mice in a research released by Minghong Wu and co-workers of Shanghai University. Wu et al. have produced neural stem (NSC) mouse culture and treated it with zinc oxide nanoparticles ranging from 10 to 200 nm of size in order to inspect its potential neurotoxicity of ZnO nanoparticles. After 24 h, the cell feasibility analysis specified that ZnO nanoparticles manifested dose-reliant, but not dimension-dependent, harmful effects on NSCs. Many of the NSCs demonstrated vibrant indications of apoptosis through evaluation using confocal microscopy, electron transfer microscopy and fluid cytometry. The impact of the dissolved zinc ions in the culture medium or in cells was discovered to have this zinc oxide nanoparticle toxicity. Another study by Arnaud Magrez at the NN Research Group discovered that titanium dioxide-based nanofilament was cytotoxic and was also affected by the appearance of chemical-treated defects on the nanofilament's surface. The internalisation of nanofilament and changes in cell morphology have been noted. In a summary, there are methods listed below that can be illustrative representation for the characterisation of nanomaterials used for cosmetics:

- Scanning electron microscopy (SEM) to explore shape and morphology
- SEM coupled with energy-dispersive X-ray diffraction (EDAX) allows elemental information to be gained for individual particles
- Surface area measurements and powder X-ray diffraction (XRD) for crystallinity/morphology and size
- Particle size determination (PSD) technology such as laser diffraction methods or disc centrifugation
- Cryo-SEM and cryo-preparation to preserve nano-liposome microstructure and allow imaging
- Dynamic light scattering and other PSD methods to measure the size of the dispersed liposomes
- Wide range of high-performance liquid chromatography (HPLC)/ultra-performance liquid chromatography (UPLC) detection systems to quantify the levels of active present
- Sensitive inductively coupled plasma-optical emission spectroscopy (ICP-OES) or inductively coupled plasma-mass spectrometry (ICP-MS) to determine trace metals and quantify nanoactives such as nanosilver

17.11 Fate of Nanocomponents

Commonly used materials for nanoparticles are metals such as silver, gold and titanium; carbon or silicon; and polymers or proteins; however one is free to use any material which is free of health hazard in nanocosmeceutical. Three important characteristics that determine the interaction of nanocosmeceuticals with the biological surroundings are particle size, shape and chemical properties of the surface.

Nanoparticles (NPs) may be classified as follows:

1. Those which disintegrate on application to skin such as liposomes, microemulsions and nanoemulsion
2. Those which do not disintegrate or remain as insoluble particles such as fullerenes, TiO_2 and quantum dots

Nanocosmetics follow the same route as that of nanomedicine once in the biological system, namely absorption, distribution in the body, cellular uptake and finally elimination from the tissues. The nanocomponents in cosmetics can penetrate through the skin via hair follicles (750 nm–6 μm) and also through flexed (up to 1 μm) and broken skin (500 nm–7 μm) as happens in case of certain skin diseases like acne, eczema and wounds. They can permeate through transdermal route and interact with the fat and proteins present in the dermal layer of the skin. These absorbed components then get access to the lymphatic and blood circulation.

Being nano in size they are easily taken up by the cells. The major difference in the above-mentioned classes of NPs is their elimination mechanism. The second category NPs are cleared either by kidney or liver. Clearance from the tissue is decided by the particle size of the components; those having a particle size less than 10 nm are eliminated by kidney whereas those with a size more than 10 nm are taken care by liver and mononuclear phagocyte system (Rolfe et al. 2014; Fox et al. 2009).

Health concerns mainly arise for the insoluble particles which get deposited in the internal organ on repeated application of the cosmetic product. These particles are also responsible for the ecotoxic effect due to its insolubility. The data regarding pharmacokinetics of such NPs must be made available by the manufacturer.

Nanocomponents in cosmetics can enter human body through various routes and their toxicity depends largely on the route of exposure; for example nano zinc oxide, a component of lotions and sunscreens, has been reported to cause severe gastrointestinal upset resulting in vomiting and diarrhoea after ingestion whereas inhaled zinc oxide causes reduced lung function. However, the effects through skin absorption are unknown (Sass 2007).

Inhalation of nanoparticles while using deodorants, perfumes and powders may cause damage to the pulmonary tract or cause lung toxicity. These nanocomponents may get entry to the brain and also in the systemic circulation and cause serious health hazards. NPs can also get ingested intentionally or unintentionally via cosmetics applied on the lips such as lipstick, lip balm and lip gloss. Zinc oxide-, copper- and silver-based nanoparticles are commonly present in many cosmetics, all of which cause severe toxicity to the organs on chronic exposure. Fullerene-based peptides, quantum dots and metallic nanoparticles can very easily penetrate the dermis and not only have toxic effects on the keratinocytes and fibroblasts but can also enter the lymphatics and systemic circulation and cause damage to the immune system and other organs.

17.12 Safety Assessment and Toxicity Evaluation of Nanobased Cosmetics

17.12.1 Safety Assessment

The safety of a nanocosmeceutical must be evaluated by analysing physicochemical properties, appropriate toxicological endpoints, expected exposure levels and intended use of finished product. Important parameters that need to be evaluated for assessing the safety of NPs are (Raj et al. 2012):

1. Physical properties: particle size and distribution, shape, surface characteristics such as morphology, area and charge, crystallinity, amorphicity, topology, tendency to undergo aggregation or agglomeration, porosity, density and rheology
2. Chemical properties: composition, solubility, stability, dissolution kinetics, stoichiometry, impurities, surface characteristics—chemistry, catalytic activity, coating and adsorbents, hydrophilicity and lipophilicity

Microscopic evaluation of the skin after application of nanocosmeceuticals can provide valuable insight into direct effects of nanocomponents on the skin tissue. The most commonly used microscopic techniques are laser scanning confocal microscopy (LSCM), high-resolution transmission electron microscopy, particle-induced X-ray emission and radio labelling with post-iron emitter (Marty and Engelen 2007; Ansell et al. 2010).

In vitro methods for assessing the safety of cosmetic ingredients include (Marty and Engelen 2007):

(a) Skin erosion testing via transcutaneous electrical resistance (TER)
(b) Skin irritation testing via Episkin
(c) Genotoxicity/mutagenicity testing
(d) In vitro mammalian cell gene mutation test
(e) Embryo toxicity testing
(f) In vitro micronucleus test or in vitro chromosome aberration test
(g) Dermal absorption measurement
(h) Phototoxicity testing

All the above methods are routinely employed for safety assessment of cosmetics; however their utility for nanoparticles is yet to be validated. There are some non-validated methods too for safety assessment of cosmetic ingredients but are not applicable to nanoparticles. Thus, there is a need to develop validated in vitro methods specifically for nanomaterials used as cosmetic ingredients.

17.12.2 Toxicity Study

The toxicity evaluation of nanocosmetics must be individualised depending on the nanocomponents used. This is because every ingredient is different with respect to its chemical structure, composition, physicochemical properties, intended use and degree of exposure. Both acute and chronic toxicity studies need to be performed since preliminary information regarding many nanocomponents is unavailable. Also compatibility with other ingredients and the packaging material need to be evaluated. Cosmetic ingredients can be evaluated for their toxicity by following the guidelines laid down by the Cosmetic, Toiletry and Fragrance Association (CTFA), as well as the Organisation for Economic Co-operation and Development (OECD). Acute, sub-acute and chronic toxicity studies need to be done in the usual manner. In vivo toxicity testing will provide information regarding the basic pharmacokinetic processes of the nano-ingredients; however care should be taken while selecting the route of exposure, which will depend on the site of application (Nanda et al. 2016).

17.12.2.1 In Vitro Methods

17.12.2.1.1 Permeation/Penetration Studies

Membranes obtained from either animal, human, vegetable, synthetic or reconstructed tissues—Episkin®, Epiderm® and Skinethic®—are used to study skin permeation. However, since pig skin resembles human skin except for the follicular properties, it is most widely used to study dermal absorption of nanoparticles using a Franz diffusion cell. It is a simple assembly comprising two separate compartments that can be joined with a membrane or skin fitted in the centre. The compartment facing the epidermis serves as the donor compartment and the one facing the dermis serves as the receiver. Either the static or flow-through cells can be used for permeation studies. The drug concentration in the receiver compartment as well as in several layers of the skin can be assayed using suitable techniques such as tape stripping, and application of heat, pressure and force. Alternatively, the Saarbruecken diffusion model can also be made use of. It works on similar principle as that of the Franz diffusion cell. Instead of using two compartments of glass, two Teflon blocks are used and the product and skin are sandwiched in the two blocks. The skin is removed at timely intervals, after tape stripping, and is sectioned using a microtome (Duran et al. 2011). Each section is evaluated for drug content in order to understand penetration at each layer. However, these models need to be validated for nanomaterials used in cosmetics (Marty and Engelen 2007). As discussed above since NPs in cosmetics may also enter the lungs or GIT, penetration studies across the lungs and intestinal epithelium also need to be performed. For this, isolated primary bronchial and alveolar cell lines and $CaCo_2$ cell lines are used, respectively.

17.12.2.2 Cytotoxicity Studies

To study the cytotoxicity various cell lines are used depending on the target organ toxicity to be studied. The damage to cell membrane, intracellular metabolic changes and apoptosis are the parameters which are quantified. Membrane damage is based on active or passive dye uptake and its spectrophotometric determination or quantification of enzymes released due to cell damage. Measurement of the phase I and phase II enzyme activity, ATP content or MTT assay can help in quantifying metabolic changes. Apoptosis can be measured using the TUNEL assay or quantifying enzymes involved in apoptosis-like caspase or measuring the expression of pro-apoptotic Bcl2 proteins Bax and Bid or tumour suppressor p53.

17.12.2.3 Genotoxicity Studies

Bone marrow micronucleus test and liver UDS (unscheduled DNA synthesis) test are used for in vivo genotoxicity studies. These tests determine the systemic genotoxic effect and also toxicity to the bone marrow and liver. An example of cosmetic nano-ingredient is TiO_2 which on exposure to UV light induces DNA damage due to free radical formation; however the damage depends on the particle size and form; for example, 20 nm size can induce chromosomal damage whereas with 200 nm no genotoxicity is reported (Gurr et al. 2005).

Silver nanoparticles are reported to cause genetic aberrations (Singh et al. 2009). There is one report that states the photo-genotoxic potential of ZnO nanoparticles; however it lacks information on physic-chemical nature of the particles (Dufour et al. 2006).

17.12.2.4 Oxidative Stress

Since NPs are known to cause free radical-induced cell injury, several in vitro and in vivo methods are used to determine this aspect of NPs.

In vitro: Cell lines of macrophages, lung cells, bronchial epithelial cells or brain microglia have been used to study the toxic effect of manufactured NPs. Reactive oxygen species (ROS) can be quantified by measuring the levels of catalase and superoxide dismutase (SOD), reduced glutathione (GSH) and its oxidised form (GSSG) and lipid peroxidation.

Gold nanoparticles induce oxidative stress and can thereby cause indirect damage to the DNA. The type and extent of damage depend on the cell type or the particle size of the nanoparticles; for example, 3–8 nm particles display antioxidant effect (Shukla et al. 2005), whereas 20 nm particles display oxidative stress damage on embryonic lung fibroblasts (Li et al. 2008). Fullerenes are commonly used NPs in anti-ageing creams, which are also reported to induce DNA damage due to oxidative stress mechanism, but the response again depends on the exposure time, size and form of the NPs (Singh et al. 2009).

17.12.2.5 Inflammatory Studies

NPs having ROS-generating capacity can also increase the expression of pro-inflammatory mediators such as nuclear factor kappa B (NFkB), activator protein 1 and other cytokines and chemokines (Nel et al. 2006; Donaldson et al.

2004). Thus, the release of these mediators in the cell lines needs to be evaluated in order to understand the inflammatory potential of NPs on the cells. Cellular targets for inflammation could be macrophages, neutrophils or cells of target organs like kidney, brain, lungs and liver.

17.12.2.6 In Vivo Studies

The cell-based assays and toxicity studies cannot perfectly mimic the in vivo conditions and sometimes it becomes difficult to correlate in vivo effects with the in vitro data. At the same time, toxicity evaluation of cosmetic ingredients on animals is banned, unless it is used in some other product. Also the use of mammals for toxicity testing may not be economically feasible; hence an alternative to mammals is zebra fish.

17.12.2.6.1 Zebra Fish Model

This fish resembles many aspects of human physiology and hence is a good model to perform studies related to genetics, embryology, cell development and toxicity. Zebra fish are also susceptible to alterations in their environment and are once again a good model to study the environmental toxicity of nanocosmetics (Ferreira et al. 2011). It has been used to study the effects of nanocomponents used in cosmetics as shown in Table 17.1.

There are various ways to administer these NPs to the zebra fish. The substance can be either directly added into the medium of zebra fish embryo, or administered orally (Duran et al. 2011) after the larval stage or injected into the yolk sac or sinus venosus. After the required time of treatment, phenotypic changes in the cardiovascular system, central nervous system, neural crest and ear are examined (Peterson et al. 2000; Heiden et al. 2007).

Table 17.1 Effects of nanocomponents used in cosmetics

Nanocomponent	Effect observed	Reference
1-Phenyl-2-thiourea, arbutin, kojic acid, 2-mercaptobenzothiazole, haginin and YT16i (melanogenic inhibitors)	Inhibited the pigmentation in zebra fish. YT16i showed malformation and cardiac dysfunction	Choi et al. (2007a, b)
ZnO (sunscreen and antibacterial)	Delayed the hatching rate and development of zebra fish. Also caused embryo malformation	Zhu et al. (2008)
TiO_2 (sunscreen)	Affected the genes involved in ribosomal functioning	Griffitt et al. (2009)
Fullerene (antioxidant)	Necrosis and apoptosis of embryos	Usenko et al. (2008)
Retinyl palmitate (anti-ageing)	Oedema, blood accumulation and altered axial curvature	Usenko et al. (2008)

The National Institute of Public Health and Environment, Netherlands, has come up with a computer program that can be used for risk assessment of nanocosmetics; however it will require data such as physicochemical characterisation of the nanomaterial, possible toxicity and exposure to the consumer and environment. There are many challenges in developing this tool such as lack of animal toxicity data and lack of toxicity prediction based on physicochemical properties of the nanomaterials. This tool has been successful in predicting the toxicity of TiO_2-enabled sunscreen and some other non-cosmetic nanomaterials like nanosilver-enabled T shirts and test case coating boat (de Jong et al. 2016). A NP is considered to be toxic if it penetrates any physiological barrier, causes cytotoxicity, induces cellular stress or causes mutagenicity or genotoxicity (FDA 2013). The most important point in the toxicity testing is the availability of the reference material. The world's first reference repository for nanomaterials has been launched by the European Commission's Joint Research Centre (JRC) for safety assessment testing by national and international standardisation bodies and has currently 25 types of nanomaterials such as carbon nanotubes, silver NPs and TiO_2. This can be used as a reference material to develop, calibrate and validate instruments, new protocols and experiments (Nanda et al. 2016). Well-designed experiments with a broad approach encompassing a detailed physicochemical characterisation of NPs before and after experimentation, study of nanomaterial dynamics during the assay, inclusion of appropriate controls during biological experiments and a battery of genotoxicity assay are important for future predictions and extrapolation of ecotoxicity of NPs, adverse health effects in humans and other biological systems like invertebrates, fungi, plant, bacteria, reptiles and amphibians (Singh et al. 2009). Given the fact that water resources are more likely to be contaminated by industrial manufacturing units, it is necessary to test the effect of NPs on freshwater and seawater aquatic life.

17.13 Toxicity to Ecosystem

Nanomedicine, although promising for its therapeutic efficacy, has a negative impact not only on human health but also on the environment associated to its production, use and disposal. Complete information with respect to toxicity of nanocomponents is yet to be explored and recorded. There are several ways by which NPs used in cosmetics enter the ecosystem. They can leach from the formulation and disturb the ecosystem; for example, a report states that the carbon nanospheres get mixed in the soil and are taken up by earthworms and in this manner they can get access to higher order animals and humans (Brumfiel 2003) or accumulate in the soil over time and get an entry into the plant system. This not only causes a direct phytotoxic effect to the plant but also leads to biomagnification in food chains, although no conclusive findings are reported.

Many NPs end up in water treatment plants through processes of either bathing, swimming, showering or washing and can cause unanticipated long-term adverse effects on health or toxic effects on the ecosystem (Boxall et al. 2007; Mueller and Nowack 2008) enter the human system via drinking water and cause an indirect risk to the health of an individual due to release of metallic nanoparticles. This could lead to kidney damage, hypertension, gastrointestinal inflammation, neurological damage and even cancer. In fact, nanomaterials affect all levels of aquatic species right from algae to vertebrates. Toxicity of nanocomponents depends on various factors such as extent of exposure to the biological system which is determined by retention and clearance from the body and chemical structure and composition of nanoparticles. It also depends on the particle size, shape, chemical composition, surface characteristics, solubility and agglomeration as stated above.

Some examples of ecotoxicity of nanocomponents used in cosmetics are given below:

17.13.1 Titanium Oxide Nanoparticles

Titanium oxide nanoparticles are used in sunscreen and when mixed in the river waters or other natural source of water promote the formation of hydrogen peroxide which is harmful for the survival of planktons (Sanchez-Quiles and Tovar-Sanchez 2014). TiO_2 NPs end up in waste waters and due to their antibacterial effect they can possibly eliminate microorganisms and in turn disrupt the ecosystem and waste water treatment (Matthew 2009).

17.13.2 Silver Nanoparticles

Nanoscale silver ions are used in nanocosmeceuticals as an antibacterial. These nanoscale ions when mixed in water will have a negative impact on the ecosystem by destroying the beneficial microorganisms in the ecosystem and further disrupting some of the food chains. Silver NPs can equally be toxic to the keratinocytes and fibroblasts in concentrations used as antimicrobials (Buzea et al. 2007).

17.13.3 Fullerenes

Fullerenes have been found to be toxic to water fleas and have bactericidal effect and can have ecotoxicological effects in a manner similar to TiO_2 or AgNPs (Ernie 2004; Lena et al. 2009). An efficient biological model needs to be developed to understand the toxicity of nanomaterials on the environment. Aquatic species belonging to the genus *Daphnia* has been used by various researchers to predict the toxicity of nanomaterials. Immunological testing procedures and development of genomic biomarkers in aquatic vertebrae are some of the future methodologies for examining the toxicity of nanomaterials (Klaper and Divino 2014).

17.14 Regulatory Guidelines/Framework for Nanocosmetics

Regulations vary from location to location; however, at every place it is attempted to ensure safe use of cosmetics. Several countries have their formal regulations which regulate the use of certain ingredients in cosmetic products. It is noteworthy that while the FDA has imposed strict scrutiny requirements for the approval of drugs, it does not mandate approval for *cosmetic* products and ingredients before they go to the market with the exception of colour additives. Note that the cosmetics are regulated under the Federal Food, Drug and Cosmetic Act (FD&C Act). According to this law, cosmetics must be safe for consumers under labelled conditions of use and must not be adulterated or misbranded. Any colour additives they contain must be approved for the intended use. Packaging and labelling must not be deceptive. The safety and labelling of cosmetic products are the legal requirements from the manufacturer and seller (FDA 2012). As evident from the above, we can conclude that cosmetics do not require approvals presently but are regulated by FDA (USFDA 2018b). Any ingredient may be used in cosmetic, as long as it does not cause the product to be adulterated in any way (with the exception of colour additives and ingredients that are prohibited or restricted by FDA regulations) (USFDA 2009).

There are many cosmeceutical products which alter the physiological processes in the skin, and if such be the case, the manufacturers of such products should ensure safety by conducting proper clinical trials seeking necessary approvals from the authorities; however, some manufacturers tend to avoid this to circumvent the expensive and lengthy process in approval by concerned authorities.

There are different regulatory agencies for regulating cosmetics across the world. US Food and Drug Administration (USFDA), European Commission (Commission), Health Canada, Central Drugs Standard Control Organisation (CDSCO), Department of Health and Family Welfare in India, Ministry of Health in Japan, Department of Health in Australia, China FDA, Cofepris—Mexico, CTFA—South Africa, ASEAN Cosmetic Directive—Southeast Asian Nations, ANMAT—Argentina, and INMETRO—Brazil are some of the examples of such agencies in different parts of the world (Raj and Chandrul 2016).

17.14.1 USFDA Regulations

Cosmetic products and their ingredients except for colour additives are subject to FDA regulation only after their release into the market. The formulator may use any ingredient (other than colour additives and such ingredients which are prohibited from use in cosmetics through regulation), so that the final product and its ingredients are safe and properly labelled, and are not adulterated or misbranded under the FD&C Act and the Fair Packaging and Labelling Act (FPLA). The regulations related with colour additives, colour additive petitions, list of colour additives exempt from certification, list of colour additives subject to certification (the permitted uses, specifications and restrictions that apply to different colour additives), colour additives certification procedures, list of certified provisionally listed colours and specifications are mentioned in Title 21 of the Code of Federal Regulations Parts 70, 71, 73, 74, 80, 81 and 82, respectively.

The FDA may take strict actions against manufacturers if the product is found to be in violations of Federal Food, Drug and Cosmetics Act (FD&C Act) and Fair Packaging and Labelling Act (FPLA) through various means ranging from issuing warning letters to mandating a product recall to removing the product from shelves to detaining the production or import of unsafe products by a manufacturer/trader. FDA pursues its action through the US Department of Justice through the US Federal Court system. FDA also has the authority to inspect manufacturing facilities of the cosmetic products under scrutiny in their pursuit of establishing and enforcing product safety in public interest through the powers vested in them under FD&C Act or FPLA.

17.14.2 European Commission Regulations

Regulation No. 1223/2009 of the European Parliament and of the Council on cosmetic products which came into effect in July 2013 is the major regulatory outline for finished cosmetic products positioned on the EU market. The Regulation has been amended from time to time (for instance—2013, 2014, 2015, 2016) and it is the first legislative document to include provisions relating specifically to the use of nanomaterials in any product. This Regulation establishes rules which must be complied by all members wanting to launch any cosmetic product in the market, with the purpose of ensuring smooth functioning of products in the market and ensuring the well-being and protection of the human health and wellness. The Regulation replaces Directive 76/768/EC, which was adopted in 1976 and had been substantially revised many number of times. Amongst the significant changes introduced by this Regulation are enhanced safety requirements for cosmetics, introduction of the concept of "responsible person", introducing and adapting a unified EU's product notification portal (CPNP) for all cosmetic products placed on the EU market, introduction of SUE (serious undesirable effects) and revamped rules for the use of nanomaterials in cosmetic production.

Compared to legislative requirements and regulations of other countries, these are much more detailed one and are followed by all member states. This regulatory document is a comprehensive reference on the scope, terminologies, safety protocols, responsibility metrices, movement of products, safety assessment, product information file, notifications, applicable limitations and restrictions for certain substances, animal testing, information to consumers, market surveillance, non-compliance norms, safeguard clauses, administrative cooperation and implementation road map including final provisions. The document also includes detailed cosmetic product safety report contents; list of substances prohibited in cosmetic products; list of substances which cosmetic products must not contain; list of colourants, preservatives and UV filters allowed in cosmetic products; symbols used on packaging/container; list of alternative methods to animal testing which are validated; repealed directive with its successive amendments; and list of time limits for transposition into national law and application included as annexures to the document (Commission 2019). The Regulation defines "nanomaterial" as "an insoluble or bio-persistent and intentionally manufactured material with one or more external dimensions, or an internal structure, on the scale from 1 to 100 nm" (Commission 2019). For every cosmetic product containing nanomaterials, protection of human health shall be ensured with utmost importance. However, the provisions, here, would not apply to nanomaterials used as UV filters, colourants or preservatives regulated under Article 14 (Restrictions for substances listed in the Annexes), unless expressly specified. The responsible person is required to submit the information by electronic means to the Commission in the presence of substances in the form of nanomaterials 6 months prior to placing the cosmetic product on the market. A responsible person is usually the manufacturer or importer of finished cosmetic products based in EU. The information notified to the Commission shall contain at least "the identification of the nanomaterial including its chemical name (IUPAC) and other descriptors, the specification of the nanomaterial including size of particles, physical and chemical properties, an estimate of the quantity of nanomaterial contained in cosmetic products intended to be placed on the market per year, the toxicological profile of the nanomaterial, the safety data of the nanomaterial relating to the category of cosmetic product, as used in such products, the reasonably foreseeable exposure conditions". Another legal or natural person may be nominated by the responsible person in written consent for the notification of nanomaterials and shall inform the Commission thereafter. The Commission provides a reference number for the submission of the toxicological profile. If the Commission has concerns regarding the safety of a nanomaterial, the Commission requests the Scientific Committee on Consumer Safety (SCCS) to submit its judgement on the safety of such nanomaterials for use in the relevant categories of cosmetic products and on the rationally predictable exposure conditions. The Commission makes the information public. The SCCS shall provide its opinion within 6 months of the Commission's request. Where the SCCS finds that any necessary data is lacking, the Commission requests the responsible person to provide such data within a clearly mentioned reasonable time, which shall remain non-extendable. The SCCS shall deliver its final opinion within 6 months of submission of additional data. The opinion of the SCCS is then made publicly available.

As per the documents, the Commission is required to submit to the European Parliament and the Council an annual status report in the beginning giving information on evolutions in the use of nanomaterials in cosmetic products within the community, including those used as colourants, UV filters and preservatives in a separate section. The report update was required to summarise mainly the new nanomaterials in new categories of cosmetic products, the number of notifications, the advancement made in developing nano-specific assessment methods and safety assessment guides, and information on international cooperation programs. The Commission is also expected to regularly review the provisions of this Regulation concerning nanomaterials taking into considerations the scientific progress and shall, wherever necessary, propose suitable revisions to those provisions (Commission 2019).

As per labelling requirement on cosmetic products, all ingredients present in the form of nanomaterials shall be clearly indicated in the list of ingredients. The names of such ingredients shall be followed by the word "nano" in brackets. In cosmetic product safety information, specific considerations are required to be given to any possible effects on the safety and toxicological profile due to particle sizes, including nanomaterials.

Colourants, preservatives and UV filters, including nanomaterials, must be authorised. Products containing other nanomaterials not otherwise restricted by the cosmetics regulation are object of a full safety assessment at EU level if the Commission feels any concerns. Carbon black (nano) as colourant and titanium dioxide (nano), tris-biphenyl triazine (nano) and zinc oxide (nano) as UV filters are allowed in cosmetic products with some required characteristics in terms of purity, solubility, particle size, etc. mentioned for all these ingredients (Commission 2019; Dureja et al. 2005).

17.14.3 CDSCO Regulations

In India, CDSCO regulates the cosmetics. CDSCO operates under Directorate General of Health Services, Ministry of Health and Family Welfare, Government of India. It is the National Regulatory Authority (NRA) of India. CDSCO draws its power and functions from the Drugs and Cosmetics Act of 1940 (amended from time to time). The Bureau of Indian Standards (BIS) is the body which issues the standards for ingredient usage in cosmetics.

Under the provisions of Drugs and Cosmetics Act 1940 and rules made thereunder, the manufacture of cosmetics is regulated under a system of inspection and licensing by the State Licensing Authorities appointed by the respective state governments while the import of cosmetics is regulated under a system of registration by the Licensing Authority appointed by the Central Government. Schedule M (II) of Drugs and Cosmetics Act deals with requirements for manufacturing of cosmetics in India. The Drugs Controller General (India) functions as the Licensing Authority who grants the registration certificate and regulates the import of cosmetics in India vide Gazette notification G.S.R. 426(E) under the provisions of Drugs and Cosmetics Act 1940 and rules made thereafter (Guidelines on Registration of Import of cosmetics-CDSCO 2013).

As per Rule 129 of Drugs and Cosmetic Rules, 1945, no cosmetic shall be imported into India unless the product is registered under the rules by the Licensing Authority appointed by the Central Government. Any article falling within the definition of cosmetic is required to be registered along with pack size and manufacturing premises before import into the country. An application for issue of a Registration Certificate for cosmetics intended to be imported into India is made online either by the manufacturer himself/herself or by his/her authorised agent or importer in India. Thus, all the cosmetics which are manufactured and imported must be within the provision of D&C Act (Raj and Chandrul 2016).

17.15 Commercially Available Nanocosmeceuticals

Nanocarriers are available enormously in cosmetic application as listed in Table 17.2 (Maynard 2006).

17.16 Available Patents on Nanocosmeceuticals

The current patents on nanocosmeceuticals are shown in Table 17.3.

17.17 Conclusion

Nanotechnology is becoming increasingly popular in the modern era of cosmetics. In the last decade, the number of nanotechnology-based cosmetic products has increased by nearly 516%. Also, large number of patents based on nanotechnology are in the sphere of personnel care products. This exponential rise in consumption of nanocosmetics has been due to the advantages such as prolongation of residence of

Table 17.2 Marketed products of nanocosmeceuticals

Trade name	Company name	Type of dosage form	Application
Hydro Flash® Broner Daily Face Moisture	Lancome	Nanocapsules	Self-tanning
Renergie Microlift	Lancome	Nanocrystals	Anti-wrinkle
Nano Gold® Energising Cream	Neiman Marcus	Nanoparticles	Anti-ageing, anti-inflammation
Revita Lift® Line	L'Oréal	Niosomes, nanoparticles	Skin tightening, anti-wrinkle
Happylogy® Glowing Skin Essence	Guerlain	Nanoemulsion	Anti-wrinkle
Nanorama™-Nano Gold Mask Pack	Laxon Nanotech Inc.	Nanoparticles	Skin tightening, anti-wrinkle
Royal Jelly	Royalt Jelly	Liposome	Anti-wrinkle
Eye Tender	L'Oréal	Nanosome	Anti-wrinkle
Coside Whitening Mask	Natural Korea	Nanocolloids	Face mask
Elixir Skin-up	Shiseido	Nanoparticles	Make-up foundation
Eye Contour Nanolift	Kara vita	Nanosphere	Anti-wrinkle
Cosile Nano Beauty Soap	Natural Korea	Nanoparticles	Cleanser
Nano Sal TM Moisture Key	Salvonazs	Nanosphere	Moisturiser
Nanosphere Plus	Dermoswiss	Nanosphere	Anti-ageing
Lip Tender	Kara Vita	Nanosphere	Lip moisturiser
Capture	Dior	Liposome	Anti-ageing
Radical Sponge	Vitamins C60 Bioresearch	60 nanoparticles	Skin treatment

Table 17.3 Patents on nanocosmeceuticals

Patent no.	Findings
US9896651B1	The present patent suggests making of antiseptic and fragrance-free soap consisting of deionised water, caustic soda and silver nanoparticles.
WO2011019668A1	The present patent suggests a cosmeceutical composition for administration which is formulated into liposomal delivery system.
DE59300006D1	The present patent suggests formulating aqueous cosmeceutical dermatological preparation containing film-forming oligomer.
FR2219M	The present patent suggests a medicated shampoo composition for treating scalp disorders with colloidal properties.
A61Q19/08	The patent suggests preparation of anti-ageing products.

cosmetic actives, greater stability, protection from degradation of actives, control of release, higher efficacy and better site specificity. However, presently there are safety issues, cost issues and toxicological concerns associated with these nanocosmetics. Reliable and hazard- and safety-revealing evaluations for nanomaterials to study their long-term impact on environment and human health need to be carried out. With the recent technological advancement in the field of nanomaterials and development of newer testing methods, nanocosmetics would definitely provide a beautiful, healthier and safer future.

References

Abbasi E, Aval SF, Akbarzadeh A et al (2014) Dendrimers: synthesis, applications, and properties. Nanoscale Res Lett 9(1):247–256

Abbaszadegan A, Nabavizadeh M, Gholami A et al (2015) Positively charged imidazolium-based ionic liquid-protected silver nanoparticles: a promising disinfectant in root canal treatment. Int Endod J 48(8):790–800

Abbott W, Gopalan S, Marchant GE et al (2006) Cosmeceuticals: an emerging concept. Indian J Pharmacol 37(3):155–159

Afaq F, Katiyar SK (2011) Polyphenols: skin photoprotection and inhibition of photocarcinogenesis. Mini Rev Med Chem 11(14):1200–1215

Aillon KL, Xie Y, El-Gendy N et al (2009) Effects of nanomaterial physicochemical properties on in vivo toxicity. Adv Drug Deliv Rev 61(6):457–466

Aldayel O, Hefne J, Al-Aiyan T (2018) Heavy metals concentration in facial cosmetics. Nat Prod Chem Res 6(1):1–9

Allam KV, Kumar GP (2011) Colourants-the cosmetics for the pharmaceutical dosage forms. Int J Pharm Pharm Sci 3(3):13–21

Allen TM, Cullis PR (2013) Liposomal drug delivery systems: from concept to clinical applications. Adv Drug Deliv Rev 65(1):36–48

Ansell J, Bronaugh R, Carter LK et al (2010) Report of the ICCR Joint Ad Hoc Working Group on Nanotechnology in cosmetic products: criteria and methods of detection, ICCR-4. http://www.nononsensecosmethic.org/wp-content/uploads/2012/12/iccr-4_2010-eng.pdf. Accessed 14 Jul 2010

Antonio JR, Antonio CR, Soares-Cardeal IL et al (2014) Nanotechnology in dermatology. An Bras Dermatol 89(1):126–136

Aranaz I, Acosta N, Elorza B et al (2018) Cosmetics and cosmeceutical applications of chitin, chitosan and their derivatives. Polymers 10(2):213–217

Arul M, Shanmuganathan S (2015) An overview on niosome as carrier in dermal drug delivery. J Pharm Sci Res 7(11):923–927
Attama AA, Momoh MA, Builders PF (2012) Lipid nanoparticulate drug delivery system: a revolution in dosage form design and development. Recent Adv Novel Drug Carrier Syst 5:107–140
Avadhani KS, Manikkath J, Tiwari M et al (2017) Skin delivery of epigallocatechin-3-gallate (EGCG) and hyaluronic acid loaded nano-transfersomes for antioxidant and anti-aging effects in UV radiation induced skin damage. Drug Deliv 24(1):61–74
Ayumi NS, Sahudin S, Hussain Z et al (2019) Polymeric nanoparticles for topical delivery of alpha and beta arbutin: preparation and characterization. Drug Deliv Transl Res 9(2):482–496
Bangale MS, Mitkare SS, Gattani SG et al (2012) Recent nanotechnological aspects in cosmetics and dermatological preparations. Int J Pharm Pharm Sci 4(2):88–97
Baroli B (2010) Penetration of nanoparticles and nanomaterials in the skin: fiction or reality? J Pharm Sci 99(1):21–50
Ben-noun L (2016) Beauty of humans. B.N. Publication House, Israel
Biswas A, Bayer IS, Biris AS et al (2012) Advances in top-down and bottom-up surface nanofabrication: techniques, applications & future prospects. Adv Colloid Interf Sci 170 (1–2):2–27
Borase HP, Patil CD, Salunkhe RB et al (2014) Phytolatex synthesized gold nanoparticles as novel agent to enhance sun protection factor of commercial sunscreens. Int J Cosmet Sci 36 (6):571–578
Boxall ABA, Chaudhry Q, Sinclair C et al (2007) Current and future predicted environmental exposure to nanoparticles. Central Science Laboratory. Report for the Department of Environment Food and Rural Affairs United States Environmental Protection Agency. https://hero.epa.gov/hero/index.cfm/reference/details/reference_id/196111
Brumfiel G (2003) A little knowledge. Nature 424:246–248
Buzea C, Pacheco II, Blandino KR (2007) Nanomaterials and nanoparticles: sources and toxicity. Biointerphases 2(4):MR17–MR71
Carrier WL, Setlow RB, Division B et al (1970) Endonuclease from Micrococcus luteus which has activity toward ultraviolet-irradiated deoxyribo-nucleic acid: purification and properties. J Bacteriol 102(1):178–186
Celia C, Cilurzo F, Trapasso E et al (2012) Ethosomes® and transfersomes® containing linoleic acid: physicochemical and technological features of topical drug delivery carriers for the potential treatment of melasma disorders. Biomed Microdevices 14(1):119–130
Chan HK, Kwok PC (2011) Production methods for nanodrug particles using the bottom-up approach. Adv Drug Deliv Rev 63(6):406–416
Chappell G (2012) Nanomaterials and the EU cosmetics regulation: implications for your company. Global Cosmetic Industry. http://www.gcimagazine.com/business/management/regulation/143553126.html?pa
Chen J, Wei N, Lopez-Garcia M et al (2017) Development and evaluation of resveratrol, vitamin E, and epigallocatechin gallate loaded lipid nanoparticles for skin care applications. Eur J Pharm Biopharm 117:286–291
Choi HS et al (2007a) Renal clearance of quantum dots. Nat Biotechnol 25(10):1165–1170
Choi TY, Kim JH, Ko DH et al (2007b) Zebrafish as a new model for phenotype-based screening of melanogenic regulatory compounds. Pigment Cell Res 20(2):120–127
Crosera M, Bovenzi M, Maina G et al (2009) Nanoparticle dermal absorption and toxicity: a review of the literature. Int Arch Occup Environ Health 82(9):1043–1055
De Jong WH, Delmaar C, Gosens I et al (2016) Description of a nanocosmetics tool for risk assessment, vol 18, pp 1–10
Deng Y, Ediriwickrema A, Yang F et al (2015) A sunblock based on bioadhesive nanoparticles. Nat Mater 14(12):1278–1285
Dhawan A, Taurozzi JS, Pandey AK et al (2006) Stable colloidal dispersions of C60 fullerenes in water: evidence for genotoxicity. Environ Sci Technol 40(23):7394–7401

Dhull K, Tripathy S, Dureja H (2015) Cosmetics: regulatory scenario in USA, EU and India. J Pharm Technol Res Manag 3(2):127–139
Dingley G, Fair J, Glynn J et al (2018) Optical blurring pigment composition suitable for use in cosmetics. US patent 9,968,525, 15 May 2018
Donaldson K, Stone V, Tran CL et al (2004) Nanotoxicology. Occup Environ Med 61(9):727–728
Duarah S, Pujari K, Durai RD (2016) Nanotechnology-based cosmeceuticals: a review. Int J Appl Pharm 8(1):8–12
Dufour EK, Kumaravel T, Nohynek GJ (2006) Clastogenicity, photo-clastogenicity or pseudo-photo-clastogenicity: genotoxic effects of and zinc oxide in the dark, in pre-irradiated or simultaneously irradiated Chinese hamster ovary cells. Mutat Res 607(2):215–224
Duran N, Teixeira Z, Marcato PD (2011) Topical application of nanostructures: solid lipid, polymeric and metallic nanoparticles. In: Beck R, Guterres S, Pohlmann A (eds) Nanocosmetics and nanomedicines. Springer, Berlin, pp 69–99
Dureja H, Kaushik D, Gupta M et al (2005) Cosmeceuticals: an emerging concept. Indian J Pharmacol 37(3):155–159
El-Leithy ES, Makky AM, Khattab AM et al (2018) Optimization of nutraceutical coenzyme Q10 nanoemulsion with improved skin permeability and anti-wrinkle efficiency. Drug Dev Ind Pharm 44(2):316–328
Ernie H (2004) Fullerenes and fish brains: nanomaterials cause oxidative stress. Environ Health Perspect 112(10):A568–A569
European Commission (2019) Document 52018SC0126 title and reference 396, pp 1–25
FDA (2012) FDA regulation of cosmetics and personal care products. CRS report for congress. https://www.everycrsreport.com/files/20120709_R42594_f2c0c94e9b027987b246daa1c2b2ae9defe309c5.pdf
FDA (2013) Guidance for industry. Considering whether an FDA-regulated product involves the application of nanotechnology. http://www.fda.gov/RegulatoryInformation/Guidances/ucm257698.htm
FDA authority over cosmetics: how cosmetics are not FDA-approved, but are FDA-regulated (2018). https://www.fda.gov/cosmetics. Accessed 7 Jan 2018
Ferreira CV, Sartori-Da-Silva MA, Justo GZ (2011) Zebrafish as a suitable model for evaluating nanocosmetics and nanomedicines. In: Beck R, Guterres S, Pohlmann A (eds) Nanocosmetics and nanomedicines. Springer, Berlin, pp 239–251
Fontana MC, Coradini K, Guterres SS et al (2009) Nanoencapsulation as a way to control the release and to increase the photostability of clobetasol propionate: influence of the nanostructured system. J Biomed Nanotechnol 5(3):254–263
Fox ME, Szoka FC, Frechet JM (2009) Soluble polymer carriers for the treatment of cancer: the importance of molecular architecture. Acc Chem Res 42(8):1141–1151
Garg T, Rath G, Goyal AK (2015) Comprehensive review on additives of topical dosage forms for drug delivery. Drug Deliv 22(8):969–987
Gautam A, Singh VR (2011) Dermal exposure of nanoparticles: an understanding. J Cell Tissue Res 11(1):2703–2708
Glowka E, Wosicka-Frackowiak H, Hyla K et al (2014) Polymeric nanoparticles-embedded organogel for roxithromycin delivery to hair follicles. Eur J Pharm Biopharm 88(1):75–84
Griffitt RJ, Hyndman K, Denslow ND et al (2009) Comparison of molecular and histological changes in zebrafish gills exposed to metallic nanoparticles. Toxicol Sci 107(2):404–415
Guery R, Henry B, Martin-blondel G et al (2017) Liposomal amphotericin B in travelers with cutaneous and muco-cutaneous leishmaniasis: not a panacea. PLoS Negl Trop Dis 11(11): e0006094
Guidelines on Registration of Import of Cosmetics—CDSCO (2013). https://cdsco.gov.in/opencms/export/sites/CDSCO_WEB/Pdf-documents/cosmetics/Guidelines_on_Registration_of_Import_of_Cosmetics.pdf. Accessed 2 Jan 2013
Gupta SRN (2018) Applications of gold nano particles in medical research and cosmetics. Int Res J Sci Eng 6(5):181–198

Gupta S, Kesarla R, Omri A (2013) Formulation strategies to improve the bioavailability of poorly absorbed drugs with special emphasis on self-emulsifying systems. Int Sch Res Notices 2013:848043
Gupta S, Kesarla R, Chotai N et al (2017) Systematic approach for the formulation and optimization of solid lipid nanoparticles of efavirenz by high pressure homogenization using design of experiments for brain targeting and enhanced bioavailability. Biomed Res Int 2017:5984014
Gurr JR, Wang AS, Chen CH et al (2005) Ultrafine titanium dioxide particles in the absence of photoactivation can induce oxidative damage to human bronchial epithelial cells. Toxicology 213(1–2):66–73
Hammes C (1997) Cosmeceuticals: the cosmetic-drug borderline, drug discovery approaches for developing cosmeceuticals: advanced skin care and cosmetic products, IBC Library Series. Dermatology 202(4):275–282
Heiden T, Dengler E, Kao W et al (2007) Developmental toxicity of low generation PAMAM dendrimers in zebrafish. Toxicol Appl Pharmacol 225(1):70–79
Hiroyuki T, Takesuke T (2010) Cosmetic containing biocompatible polymer JP2010275249A
Hosseinkhani B, Callewaert C, Vanbeveren N et al (2015) Novel biocompatible nanocapsules for slow release of fragrances on the human skin. New Biotechnol 32(1):40–46
Hu Z, Liao M, Chen Y et al (2012) A novel preparation method for silicone oil nanoemulsions and its application for coating hair with silicone. Int J Nanomedicine 7:5719–5724
Huang X, Robert K, Gann Xu (2005) Peptide-based carbon nanotube hair colourants and their use in hair colourant and cosmetic compositions. US patent 7,452,528, 18 Nov 2008
Huang X Robert KK, Gann Xu (2006) Hair colouring and cosmetic compositions comprising carbon nanotubes. US patent 7,276,088, 2 Oct 2007
Hwang SL, Kim JC (2008) In vivo hair growth promotion effects of cosmetic preparations containing hinokitiol-loaded poly(epsilon-caprolactone) nanocapsules. J Microencapsul 25 (5):351–356
Hyun C, Taik Y, Kyeong K et al (2009) Cosmetic pigment composition containing gold or silver nano-particles. US patent 11/995,847, 22 Jan 2009
Ibrahim KS (2013) Carbon nanotubes-properties and applications: a review. Carbon Lett 14 (3):131–144
Inui S, Aoshima H, Nishiyama A et al (2011) Improvement of acne vulgaris by topical fullerene application: unique impact on skin care. Nanomedicine 7(2):238–241
Jain S, Patel N, Madan P et al (2015) Quality by design approach for formulation, evaluation and statistical optimization of diclofenac-loaded ethosomes via transdermal route. Pharma Dev Technol 20(4):473–489
Jimenez Z, Kim YJ, Mathiyalagan R et al (2018) Assessment of radical scavenging, whitening and moisture retention activities of Panax ginseng berry mediated gold nanoparticles as safe and efficient novel cosmetic material. Artif Cells Nanomed Biotechnol 46(2):333–340
Kabri T, Tehrany E, Belhaj N (2011) Physicochemical characterization of nanoemulsions in cosmetic matrix enriched on omega-3. J Nanobiotechnol 9:41–48
Kaul S, Gulati N, Verma D et al (2018) Role of nanotechnology in cosmeceuticals: a review of recent advances. J Pharm (Cairo) 2018:3420204
Kaur IP, Kapila M, Agrawal R (2007) Role of novel delivery systems in developing topical antioxidants as therapeutics to combat photoageing. Ageing Res Rev 6(4):271–288
Kaushik B, Majumder M (2015) Carbon nanotube: properties and applications. In: Carbon nanotube based VLSI interconnects. Springer, New Delhi, pp 17–37
Khaiat A (2014) Regulations in Asia from China to Japan, Korea, ASEAN. https://asia.incosmetics.com/RXUK/RXUK_InCosmeticsAsia/2014/Documents/AlainKhaiatCosmeticRegulationsFromchinaToJapanKoreaASEAN.pdf?v=635524377936486453
Klaper R, Divino J (2014) Environmental implications of nanotechnology: developing sustainable nanotechnology. Laboratory of Dr. Rebecca Klaper. http://home.freshwater.uwm.edu/klaperlab/environmental-implications-of-nanotechnology

Kovacevic A, Savic S, Vuleta G et al (2011) Polyhydroxy surfactants for the formulation of lipid nanoparticles (SLN and NLC): effects on size, physical stability and particle matrix structure. Int J Pharm 406(1–2):163–172
Kumar S, Randhawa JK (2013) High melting lipid based approach for drug delivery: solid lipid nanoparticles. Mater Sci Eng C Mater Biol Appl 33(4):1842–1852
Lau M, Waag F, Barcikowski S (2017) Direct integration of laser-generated nanoparticles into transparent nail polish: the plasmonic "Goldfinger". Ind Eng Chem Res 56(12):3291–3296
Lena B, Delina L, Ernestm H et al (2009) Comparative photoactivity and antibacterial properties of C60 fullerenes and titanium dioxide nanoparticles. Environ Sci Technol 43(12):4355–4360
Li JJ, Zou L, Hartono D et al (2008) Gold nanoparticles induce oxidative damage in lung fibroblasts in vitro. Adv Mater 20(1):138–142
Liang BA, Hartman KM (1999) It's only skin deep: FDA regulation of skin care cosmetics claims. Cornell J Law Public Policy 8(2):249–280
Logothetidis S (2012) Nanotechnology: principle and applications. In: Nanostructured materials and their applications. Springer, Berlin
Lohani A, Verma A, Joshi H (2018) Nanotechnology-based cosmeceuticals. Int Sch Res Notices 2014:843687
Magnan J, Genard S (2002) Use of hyperbranched polymers and dendrimers comprising a particular group as film-forming agent, film-forming compositions comprising same and use particularly in cosmetics and pharmaceutics. US patent 6,432,423, 13 Aug 2002
Malarkodi C, Rajeshkumar S, Paulkumar K et al (2014) Biosynthesis and antimicrobial activity of semiconductor nanoparticles against oral pathogens. Bioinorg Chem Appl 2014:1
Marchant GE, Sylvester DJ (2006) Transnational models for regulation of nanotechnology. J Law Med Ethics 34(4):714–725
Marty JP, Engelen V (2007) Opinion on safety of nanomaterials in cosmetic products. http://ec.europa.eu/health/ph_risk/committees/04_sccp/docs/sccp_o_099.pdf
Matos BN, Reis TA, Gratieri TGG (2015) Chitosan nanoparticles for targeting and sustaining minoxidil sulphate delivery to hair follicles. Int J Biol Macromol 75:225–229
Matthew C (2009) Nanoparticles from sunscreens damage microbes. Environmental Health News, Scientific American. https://www.scientificamerican.com/article/nanoparticles-in-sunscreen/. Accessed 24 Mar 2009
Maynard A (2006) Nanotechnology: a research strategy for addressing risk, vol 444. Woodrow Wilson International Center for Scholars, pp 267–269
Mehnert W, Mäder K (2012) Solid lipid nanoparticles. Adv Drug Deliv Rev 64(1):83–101
Mei Z, Wu Q, Hu S et al (2005) Triptolide loaded solid lipid nanoparticle-hydrogel for topical application. Drug Dev Ind Pharm 31(2):161–168
Meireles AB, Correa DK, da Silveira JVW et al (2018) Trends in polymeric electrospun fibers and their use as oral biomaterials. Exp Biol Med (Maywood) 243(8):665–676
Michael F, Karin S, Thomas C et al (2000) Cosmetic compositions for hair treatment containing dendrimers or dendrimer conjugates. US patent 6,068,835, 30 May 2000
Millikan LE (2001) Cosmetology, cosmetics, cosmeceuticals: definitions and regulations. Clin Dermatol 19(4):371–374
Mu L, Sprando RL (2010) Application of nanotechnology in cosmetics. Pharm Res 27 (8):1746–1749
Mueller NC, Nowack B (2008) Exposure modeling of engineered nanoparticles in the environment. Environ Sci Technol 42(12):4447–4453
Mukta S, Adam F (2010) Cosmeceuticals in day-to-day clinical practice. J Drugs Dermatol 9:62–69
Nafisi S, Maibach HI (2015) Silica nanoparticles: promising nanoparticles for increasing cosmetic ingredients/drugs efficacy. Cosmetic Toiletries, University of California, San Francisco. https://www.cosmeticsandtoiletries.com/research/chemistry/Silica-Nanoparticles-for-Increased-Cosmetic-Ingredient-Efficacy%2D%2D300987651.html
Nagavarma BVN, Yadav HKS, Ayaz A et al (2012) Different techniques for preparation of polymeric nanoparticles—a review. Asian J Pharm Clin Res 5(3):16–23

Nanda S, Nanda A, Lohan S et al (2016) Nanocosmetics: performance enhancement and safety assurance. In: Grumezescu AM (ed) Nanobiomaterials in galenic formulations and cosmetics. William Andrew, New York, pp 47–67

Nel A, Xia T, Madler L et al (2006) Toxic potential of materials at the nanolevel. Science 311 (5761):622–627

Newgreen DB (2005) Review of the regulation of products at the interface between cosmetics and therapeutic goods. https://www.tga.gov.au/consultation/review-regulation-products-interface-between-cosmetics-and-therapeutic-goods

Nikam S, Chavan M, Sharma PH (2014) Solid lipid nanoparticles: a lipid based drug delivery. Innov Pharm Pharmacother 2(3):365–376

Nohynek GJ, Lademann J, Ribaud C et al (2007) Grey Goo on the skin? Nanotechnology, cosmetic and sunscreen safety. Crit Rev Toxicol 37(3):251–277

Patidar A, Devendra ST, Peeyush K et al (2010) A review on novel lipid based nanocarriers. Int J Pharm Pharm Sci 2(4):30–35

Pentek T, Newenhouse E, O'Brien B et al (2017) Development of a topical resveratrol formulation for commercial applications using dendrimer nanotechnology. Molecules 22(1):137

Peterson RT, Link BA, Dowling JE et al (2000) Small molecule developmental screens reveal the logic and timing of vertebrate development. Proc Natl Acad Sci U S A 97(24):12965–12969

Raj RK, Chandrul KK (2016) Regulatory requirements for cosmetics in relation with regulatory authorities in India against US, Europe, Australia and Asean Countries. Int J Pharma Res Health Sci 4(5):1332–1341

Raj S, Jose S, Sumod US (2012) Nanotechnology in cosmetics: opportunities and challenges. J Pharm Bioallied Sci 4(3):186–193

Rania B, Rainer MV, Muhammad N et al (2007) Medicinal applications of fullerenes. Int J Nanomedicine 2(4):639–649

Rolfe BE, Blakey I, Squires O (2014) Multimodal polymer nanoparticles with combined 19F magnetic resonance and optical detection for tunable, targeted, multimodal imaging in vivo. J Am Chem Soc 136(6):2413–2419

Sanchez-Quiles D, Tovar-Sanchez A (2014) Sunscreens as a source of hydrogen peroxide production in coastal waters. Environ Sci Technol 48(16):9037–9042

Sandoval BM (2009) Perspectives on FDA's regulation of nanotechnology: emerging challenges and potential solutions. Compr Rev Food Sci Food Saf 8(4):375–393

Santos AC, Morais F, Simoes A et al (2019) Nanotechnology for the development of new cosmetic formulations. Expert Opin Drug Deliv 16(4):313–330

Jennifer Sass (2007) Nanotechnology's invisible threat natural resources defense council. https://www.nrdc.org/sites/default/files/nano.pdf

Schaefer H, Redelmeier TE, Lademann J (2011) Skin penetration. In: Johansen AD, Frosch PJ, Lepoittevin JP (eds) Contact dermatitis. Springer, Berlin, pp 215–227

Schaffer T (2007) Lecithin and phospholipids: the optimal choice for natural cosmetics. Euro Cosmet 15(9):18

Seleci DA, Seleci M, Walter J et al (2016) Niosomes as nanoparticular drug carriers: fundamentals and recent applications. J Nanomater 2016:1

Sharma R (2012) Cosmeceuticals and herbal drugs: practical uses. Int J Pharm Sci Res 3(1):59–65

Shetty PK, Venuvanka V, Jagani HV et al (2015) Development and evaluation of sunscreen creams containing morin-encapsulated nanoparticles for enhanced UV radiation protection and antioxidant activity. Int J Nanomedicine 10:6477–6491

Shukla R, Bansal V, Chaudhary M et al (2005) Biocompatibility of gold nanoparticles and their endocytotic fate inside the cellular compartment: a microscopic overview. Langmuir 21 (23):10644–10654

Singh N, Manshian B, Jenkins GJS et al (2009) Nano genotoxicology: the DNA damaging potential of engineered nanomaterials. Biomaterials 30(23–24):3891–3914

So JW, Kim S, Park JS et al (2010) Preparation and evaluation of solid lipid nanoparticles with JSH18 for skin-whitening efficacy. Pharm Dev Technol 15(4):415–420

Soo K (2006) Synthesis of Ag and Ag-TiO2 nanoparticle containing antibiotic and deodorant property
Srinivas K (2016) The current role of nanomaterials in cosmetics. J Chem Pharm Res 8(5):906–914
Sung CM (2013) Compositions and methods for providing ultraviolet radiation protection. US patent 8,481,007, 9 Jul 2013
Terroso T, Kulkamp IC, Jornada DS et al (2009) Development of semi-solid cosmetic formulations containing coenzyme Q10-loaded nanocapsules. Lat Am J Pharm 28(6):819–826
Tournilhac F, Simon P (2001) Cosmetic or dermatological topical compositions comprising dendritic polyesters. US patent 6,287,552, 11 Sept 2001
Usenko CY, Harper SL, Tanguay RL (2007) In vivo evaluation of carbon fullerene toxicity using embryonic zebrafish. Carbon N Y 45(9):1891–1898
Usenko C, Harper SL, Tanguay RL (2008) Fullerene C60 exposure elicits an oxidative, stress response in embryonic zebrafish. Toxicol Appl Pharmacol 229(1):44–55
USFDA (2009) Guidance for industry: colour additive petitions—FDA recommendations for submission of chemical and technological data on colour additives for food, drugs, cosmetics, or medical devices. Center for Food Safety and Applied Nutrition. https://www.fda.gov/regulatory-information/search-fda-guidance-documents/guidance-industry-color-additive-petitions-fda-recommendations-submission-chemical-and-technological
USFDA (2018a) FDA authority over cosmetics: how cosmetics are not FDA-approved, but are FDA-regulated. https://www.fda.gov/cosmetics/cosmetics-laws-regulations/fda-authority-over-cosmetics-how-cosmetics-are-not-fda-approved-are-fda-regulated. Accessed 24 Jul 2018
USFDA (2018b) Is it a cosmetic, a drug, or both? (Or is it soap?). https://www.fda.gov/cosmetics/cosmetics-laws-regulations/it-cosmetic-drug-or-both-or-it-soap. Accessed 2 Aug 2018
Verma P, Pathak K (2010) Therapeutic and cosmeceutical potential of ethosomes: an overview. J Adv Pharm Technol Res 1(3):274–282
Vettor M, Bourgeois S, Fessi H et al (2010) Skin absorption studies of octyl-methoxycinnamate loaded poly(D, L-lactide) nanoparticles: estimation of the UV filter distribution and release behaviour in skin layers. J Microencapsul 27(3):253–262
Watkins R, Wu L, Zhang C et al (2015) Natural product-based nanomedicine: recent advances and issues. Int J Nanomedicine 10:6055–6074
Xia XR, Monteiro-Riviere NA, Riviere JE (2010) Skin penetration and kinetics of pristine fullerenes (C60) topically exposed in industrial organic solvents. Toxicol Appl Pharmacol 242(1):29–37
Xiao L, Takada H, Maeda K et al (2005) Antioxidant effects of water-soluble fullerene derivatives against ultraviolet ray or peroxylipid through their action of scavenging the reactive oxygen species in human skin keratinocytes. Biomed Pharmacother 59(7):351–358
Yilmaz F, Celep G, Tetik G (2016) Nanofibers in cosmetics. Nanofiber Res 2016:127
Zhang Y, Ng W, Feng X et al (2017) Lipid vesicular nanocarrier: quick encapsulation efficiency determination and transcutaneous application. Int J Pharm 516(1–2):225–230
Zhu X, Zhu L, Duan Z et al (2008) Comparative toxicity of several metal oxide nanoparticle aqueous suspensions to zebrafish (Danio rerio) early developmental stage. J Environ Sci Health A Toxic Hazard Subst Environ Eng 43(3):278–284

Nanotechnology in Ocular Drug Delivery 18

Deepa Patel, Drashti Pathak, and Praful Giradkar

Abstract

The nano drug delivery system has already entered into the market for commercialization and so it is dynamically studied for the administration of a variety of drugs for oral applications. The nanocarriers have been proven to have significant contribution in improving therapeutic outcomes and hence novel carriers such as nanovesicles and nanoporous approaches have been established for ocular transport. This chapter outlines basic anatomy and notifies with facts about various delivery means that target distinctive therapeutic spots. The chapter also discusses different obstacles for drug transport and the application of nanocarriers for prolonged retention and improvement in drug targeting. In conclusion, the existing limitations that restrict the quick development of nanotechnology in the ocular field are explained with a prominence on toxicity as well as commercialization of recognized nanocarriers.

Keywords

Nanoformulations · Ocular delivery · Nanovesicles · Site-specific delivery

D. Patel (✉) · D. Pathak
Parul Institute of Pharmacy and Research, Faculty of Pharmacy, Parul University, Vadodara, Gujarat, India
e-mail: pateldeepa18@yahoo.com

P. Giradkar
Formulation and Development, Alembic Research Centre, Vadodara, Gujarat, India
e-mail: praful.giradkar@alembic.co.in

N. Shah (ed.), *Nanocarriers: Drug Delivery System*,
https://doi.org/10.1007/978-981-33-4497-6_18

18.1 Introduction

Topical route is a common and convenient route for ocular administration. Certain natural processes such as eye blinking, lachrymal secretion, and turnover of tears cause fast removal of drugs and other substances from the eye cul-de-sac. These issues keep the ocular delivery a challenging delivery in spite of its easy accessibility for administration. Some novel concepts shall be adopted to get rid of this physical-biological barrier. As a result, such natural barriers may cause improper drug concentration to target tissue. To reach the drugs to the posterior part of the eye is difficult due to eye anatomy; hence there is a requirement of drug delivery which helps to manage posterior-part eye illness, such as diabetic retinopathy, age-related macular degeneration, and optic neuropathy. To deliver the drug at the posterior part such as retina the intravitreal route is a common choice (Tabbara and Ross-Degnan 1986). However, regular administration of drug to retina builds intraocular pressure leading to retinal detachment. Therefore, ocular delivery towards the posterior part of the eye is an important and the most challenging measure that scientists are facing at this moment. Sometimes, drug delivery through systemic route is also carried out via the drug transport across the ocular blood-aqueous and blood-retinal barrier, but this method is moderately difficult (Tabbara and Ross-Degnan 1986; Al-Shamsi et al. 1986).

Nanomedicine is a technology of small particles with particle size typically $\leq$100 nm for treatment of different diseases. Nanomedicine has several advantages over conventional dosage form. Nanomedicine offers sustained, targeted, tissue-specific delivery as well as delivery of hydrophobic drug and large biological molecules with reduced side effect. Apart from nanomedicine liposome is the most widely used drug delivery for eyes. Liposome consists of inner aqueous phase with lipid bilayer which acts as a reservoir for water-soluble and water-insoluble drugs, respectively. Ocular drug delivery with liposome was reported early in 1981 by Smolin et al. (1981).

The idoxuridine loaded into liposomes showed enhanced corneal penetration of idoxuridine compared to free drug for prevention of herpetic keratitis (Duncan and Gaspar 2011). To overcome ocular infection, which can cause permanent blindness if not treated (Farokhzad and Langer 2009), there is a need of dosage form which improves residence time to corneal and consequently enhances the drug absorption by periocular tissue. Improved pre-corneal residence of drug improves bioavailability which helps to reduce the drug dose and frequency of administration.

The effective applications of nanotechnology take place by designing the nano-scale materials or molecules which improves poor water solubility and provides specific site targeting of drug molecules that indirectly prevents accumulation of drugs in healthy tissues.

Nanotechnology is a proficient delivery system for ocular route that may offer extreme residence time to cornea, remove physical barriers, and sustain the drug delivery after topical administration. Novel drug delivery is an evolving arena for ocular administration and significant investigations are going on for the better and effective nanotechnology-centered drug delivery for ocular route (Farokhzad and Langer 2009).

18.1.1 Merits and Demerits

Conventional dosage form such as eye drops either in solution or in suspension form showed poor bioavailability of less than 5%; this poor bioavailability is due to minimum residence time of drug in eye due to drainage of administered dose; this drainage of administrate dose took place due to tear turnover and blinking of eyes. To improve dosage efficacy of eye drop there is a need of frequent addition of eye drop; this can cause side effects like irritation, dryness, and systemic side effect. Tear liquid consists of different components including ions, enzymes, and mucin which act as a protective cover for pathogen and play a crucial role in eye clearance. Tear mucin present in eye has electrostatic charges which form hydrogen bonding with anionic and cationic polymers; hence such polymers as Carbopol® are used to improve the residence time of drug for topical administration of dosage form (Votruba et al. 2010).

Blending of cationic polymers like Eudragit® RL and anionic polymers like Carbopol® with polylactic-co-glycolic acid nanoparticles improves the hydrogen binding interaction with mucin. Cyclosporine A coated with Carbopol-PLGA nanocarriers has negative zeta potential that may express greater tear film concentration in healthy rabbit eyes when comparison is made with plain PLGA particle. Further investigation shows that Eudragit® RL-PLGA nanoparticles with positive potential enhance tear film concentration (C_{max}) and AUC ($AUC_{0-24\ h}$) in healthy rabbit eyes when compared to both non-coated and Carbopol-coated PLGA nanoparticles; this increase in concentration and AUC of positively charged Eudragit-coated nanoparticles due to interaction with negatively charged mucin improves the residence time and increases the cellular uptake of nanoparticles. To improve the covalent bonding of polymers with mucin other approaches like thiolation of polymers are possible, where interaction of polymers takes place with thiol group present in mucin leading to improved interactions by covalent binding. In one study chitosan polymer was thiolated with quaternary ammonium to get quaternary ammonium-chitosan (TCS) conjugates used with sodium alginate nanoparticles (TCS-SA). The thiol group of TCS-SA nanoparticles forms a disulfide bond between thiol group present in mucin, which improves the residence property of nanoparticles on preocular surface. Therefore, surface chemistry of nanoparticles with the use of cationic or anionic polymers can be used to improve the mucoadhesive properties (Votruba et al. 2010; Goldberg et al. 2007; Meisner and Mezei 1995).

18.2 Limitations and Advantages Over Conventional Dosage Forms

Cornea has different proteins like P-glycoproteins (P-gp) and multidrug resistance-associated proteins (MRPs) commonly called efflux protein; this efflux protein pushes the drug out from corneal epithelium cell. Misra et al. have studied the enhancement of corneal drug absorption with reduction of efflux effect after the topical administration (Misra et al. 2009). In one study P-gp and multidrug resistance-associated (MRPs) efflux protein were suppressed by co-administration of steroids (prednisolone) leading to fourfold improvement of the corneal cell uptake of erythromycin compared to plain drug in an in vitro-cultured rabbit primary corneal epithelial cell model for rabbit cornea. Goldberg and Langer have discovered a presence of membrane transport in cornea, conjunctiva, and retina. These transporters performed translocation of xenobiotics and nutrients. Therefore prodrug strategy can improve the ocular drug delivery by improving the absorption properties of poorly permeable parent drug (Goldberg et al. 2007). Antiviral drug acyclovir is poorly aqueous soluble and is a low cornea-permeable parent drug. Its corneal absorption was increased by preparing prodrug with L-aspartate ester. L-Aspartate ester prodrug form of acyclovir acts as a substrate for amino acid transporter resulting in fourfold increase in corneal permeability of acyclovir in the cornea of healthy rabbits. Samad et al. have also showed that acyclovir prodrug with amino acid exhibits threefold increase in aqueous humor concentration compared to plain drug (Samad et al. 2007).

18.2.1 Drug-Eluting Contact Lenses for Sustained Delivery

Nowadays soft contact lens is a prime choice of people; therefore nanoparticle drug formulation may be incorporated in soft contact lenses for their delivery with sustaining action. Drug-loaded contact lenses provide extended action with improved corneal bioavailability. Timolol bioavailability is significantly improved with silicone hydrogel contact lenses for treatment of angle glaucoma in dogs. Prolonged cyclosporine A delivery for the management of dry eyes was improved by addition of vitamin E. Addition of vitamin E into silicone hydrogen contact lenses maintains the therapeutic effect of cyclosporine A for up to 1 month. The imprinted silicone-based hydrogel contact lenses can release small molecules and large molecules like ketotifen fumarate and hyaluronic acid (HA), respectively (Ebrahim et al. 2005; Bochot and Fattal 2012; Smolin et al. 1981; Dharma et al. 1986).

18.2.2 Dendrimer-Based Topical Delivery Systems for Cornea

Dendrimers are ~3–20 nm globular particles having large number of functional groups on the surface with mucoadhesive properties helping to reduce tear washing and dilution for improvement of pre-corneal residence. Some dendrimers show antimicrobial activity and are used as drug carriers and surface modifiers.

In vitro studies of (amido amine) (PAMAM) dendrimers showed interaction of negatively charged ocular mucin with PAMAM dendrimers. Both cationic ($-NH_2$) and neutral (–OH) PAMAM dendrimers showed mucoadhesive properties; these mucoadhesive properties are stronger at pathogenic pH (~5.5). At pH ~5.5, primary amine of $-NH_2$ dendrimers and tertiary amines in the inner cores of both $-NH_2$ and –OH dendrimers are partially protonated and show stronger interaction at target site leading to improvement in residence time of cornea via electrostatic interfaces with the ocular mucins.

PAMAM dendrimers showed significant antibacterial activity which is equivalent to ampicillin, and the antibacterial activity of PAMAM dendrimers may be due to destruction of the bacterial cell wall and opening of the bacterial contents for protein precipitation. In another study amoxicillin interacted with cores of PAMAM dendrimer and cross-linked with polyethylene glycol (PEG) to form a clear hydrogel matrix through disulfide bonds for sustainable injectable drug delivery (Zimmer and Kreuter 1995; Nagarwal et al. 2009). Quinolone molecules are studied as anti-bactericidal agents for ocular applications; quinolone shows low solubility and it has destructive properties to corneal epithelial layers. Cheng et al. reported encapsulation of fluoroquinolones such as nadifloxacin and prulifloxacin within PAMAM dendrimers for improvement of solubility which shows twofold better activity on *E. coli* than free drugs (Cheng et al. 2007).

Dendrimers can be used as topical eye drops which can form gel layer over cornea and provide sustained antimicrobial effect without any side effects like blurred vision, irritation, reduced corneal toxicity, and frequent installation, thereby improving patient compliance. Conditions like trauma, infection, cataracts, and glaucoma infiltration can cause corneal wounds which require sutures to fasten the corneal flap. These sutures can sometimes cause infections, penetrating keratoplasties, corneal scarring, and leaking (Cheng et al. 2008).

In recent years, to avoid problems associated with sutures polymeric corneal glue has been used, which is also called sutureless procedure. These polymeric glues are manufactured with desired physicochemical and biological properties for restoring the integrity of cornea. Cyanoacrylate and fibrin polymeric glues showed beneficial effects over the suture surgery but often cause disadvantages such as stiffness (Agrawal et al. 2012). These hydrogels showed wide applications including wound healing. Use of dendrimers in hydrogel shows high cross-linking in low concentration; this may be due to large surface group of dendrimers (Gratieri et al. 2010). Hydrogel based on photocurable biodendrimer presents secure large central corneal lacerations and fastens allografts in porcine-enucleated eyes. Fascinatingly, the porcine eyes closed with bioadhesive withstand leakage pressure of more than 200 mmHg compared to sutures which show leakage pressure at ~85 mmHg (Ludwig 2008). Sometimes a combination of suture and bioadhesive may be beneficial, as depicted in a porcine allograft model. The allografts with 8 sutures combined with bioadhesive glues showed to withstand 80–85 mmHg leaking pressure which is significantly higher than that with 16 sutures (~45 mmHg). The above study revealed that hydrogel-based dendrimers may be useful to seal the autografts in corneal transplantations; this can lead to potential benefits of reduction of scar formations and graft failure.

18.2.3 Corneal Gene Delivery

Corneal region of eye is a good part for gene therapy because it is simply available, and disconnected with overall circulation and systemic immune system. Gene delivery helps to deliver gene to cornea or adjacent ocular tissues by means of various vector systems. Vectors like cytokines, growth factors, and RNA interference can be utilized for corneal gene transfer. In recent preclinical studies it was found that corneal gene therapy may be beneficial for preventing cornea rejection, neovascularization, and herpetic stromal keratitis (Gipson and Argüeso 2003).

Corneal gene therapy in animal model can be achieved by intrastromal injection of nanoparticles. The vascular endothelial growth factor A through intrastromal injection of PLGA nanoparticles is encapsulated with the plasmid containing a minor hairpin RNA expression cassette for the management of corneal neovascularization. After 5 days of single-dose injection in mice of two micrograms of plasmid eye formulation, significant reduction in corneal protein expression (80%) was observed that is significantly lower compared to control eye without corneal neovascularization. Intrastromally injected nanoparticles retained in stroma and not cleared from the tear liquid lead to long-term gene expression. After 5 weeks, the intrastromal nanoparticle was found with more than threefold improvement than naked plasmid at the same amount (Zambito and Di Colo 2010; du Toit et al. 2011; Mitra 2009).

In gene therapy, co-administration of triamcinolone as steroid helps to improve the therapeutic effect by suppressing the cornea rejection rate. Two-month graft survival rate was more than 92–95% for nonsteroid-administrated group, whereas the graft survival rate was 50%, 30%, and 0% for triamcinolone group (Mitra 2009). Nanoparticles administered by subconjunctival injection were found in the subconjunctival tissue and cornea up to 5 weeks after subconjunctival injection (du Toit et al. 2011). Sustained delivery of therapeutic genes can be achieved by nanoparticle-based gene therapy leading to improvement in cellular uptake and transport to the nucleus.

18.2.4 Bio-distribution of Nanoparticles in the Retina

Bio-distribution in retina is influenced by many factors such as particle size, composition, surface charge, and mode of administration of dosage forms; hence such bio-distribution studies can provide deep understanding of drug bioavailability and its cellular uptake, its duration of action, and drug toxicity. Sakurai et al. studied the ocular bio-distribution of fluorescently labeled polystyrene nanoparticles in rabbit model. Route of administration was intravitreal route. After this administration, large particles (2 μm) were discharged from ocular tissue in less than 6 days. They remained in the vitreous humor cavity near the trabecular meshwork. The 200 nm nanoparticles in inner membrane and vitreous humor cavity were distributed evenly. 2 months postinjection small nanoparticles of 50 nm size were detected indicating crossing of retinal barriers (Sakurai et al. 2001). Size-dependent ocular bio-distribution of gold nanoparticles after I.V. administration was done in mice model (Hariharan et al. 2009; Majumdar et al. 2009). TEM studies were done on two different size gold nanoparticles that were administered intravenously and bio-distribution studies were assessed. 75 ± 5% was found in retinal neurons, 17 ± 6% in endothelial cells, and 8 ± 3% in peri-endothelial glial cells; 20 nm bio-distribution was found in retinal cellular units within 24 h. This nanoformulation did not show any toxicity and structural harm to retina. The absence of particles with 100 nm size in retina represents the impact of size factor on passage through the blood-retinal barriers.

Katragadda et al. have studied fluorescent dye-linked carboxylate-modified polystyrene nanoparticles' bio-distribution in intraocular tissues in Sprague-Dawley rat model. Two types of nanoparticles were used, i.e., 20 and 200 nm. Subconjunctival and trans-scleral administration of 20 nm particles leads to rapid clearance from the site of injection. Rapid clearance may be due to periocular blood or lymphatic circulation (Katragadda et al. 2008). 200 nm nanoparticles were found at the site of administration.

The nanoparticle distribution is affected by surface charge distribution. The aggregation with the vitreous occurs due to adhesion of positively charged nanoparticles with anionic vitreous network. Due to strong interaction of conventional cationic nature of vectors with anionic vitreous this vitreous is considered as a barrier for delivery of such vectors. In order to reach the retina the inclusion of PEG molecule or moiety on the surface of nanoparticles was done (Katragadda et al. 2008; Giannavola et al. 2003).

Block copolymer nanoparticle polyethylene oxide-polyspermine (PEO-PSP) with oligonucleotide was prepared with neutral surface charge. The delivery of nanoparticles of 12 nm range to rat eyes was done intravitreally and these oligonucleotides remained up to 6 days postinjection as detected by confocal microscopy.

For retinal disorders the design and engineering of nanoparticles with suitable surface charge chemistry are of utmost importance; hence steps should be taken to enable these nanoparticles for efficient delivery. Self-assembled amphiphilic nanoparticles were prepared and were administered to healthy rat eyes intravitreally. Due to existence of glycol groups on nanoparticles, the cationic glycol-chitosan nanocarriers and glycol-chitosan/PEI mixed nanocarriers could pass in the vitreal barrier. Several other nanoparticles like anionic HA nanoparticles and HAS nanoparticles could make themselves to retinal pigment epithelium (Jwala et al. 2011).

18.2.5 Topical and Subconjunctival Delivery to the Retina

Drug-loaded nanoparticles and microparticles show sustained release of drugs to the retina after subconjunctival administration. In a study, triamcinolone acetonide (TA)-PLA microparticles revealed sustained administration of drugs for 2 months in both the normal and laser-induced choroidal neovascularization (CNV) Brown Norway rat models whereas no drug levels were detected for TA-PLA nanoparticles (551 nm) or free drug. Lowe et al. discovered subconjunctival administration of degradable hydrogels. *N*-isopropyl acrylamide monomer and a dextran macromer comprising multiple hydrolytically degradable units were utilized to formulate the hydrogels (Jwala et al. 2011). They showed degradability and thermoreactive properties for the sustained release of insulin to the retina. The findings revealed that the hydrogels and their degradation products were not toxic to R28 retinal cells in cell culture (50,000 cells/cm^2) for at least 1 and 7 days, respectively. Thus, subconjunctival hydrogels show promising potential for controlled delivery of drugs to the posterior region of retina.

18.2.6 Intravitreal Delivery Systems

The diffusion or penetration mechanisms are a predominant means for intravitreal administration of the drugs and particles straight towards the vitreous part. Behar-Cohen and coworkers have demonstrated that poly(lactic acid) (PLA) nanoparticles with 310 nm particle size were contained in the retinal pigment epithelium (RPE) cells followed by intravitreal administration to rat model and the holding of nanoformulation in the RPE cells by single intravitreal injection given before 4 months (Jung and Chauhan 2012). Nanoformulations may avoid the rapid clearance and showed improvement in retention in the vitreous and the retina. Steroidal and nonsteroidal drugs are extensively utilized for numerous retinal illnesses due to their anti-inflammatory, antiangiogenic, and neuroprotective characteristics (Kapoor et al. 2009). Any treatment scheme for continued availability of drugs for an extended period of time might support to decrease the frequency of injections and increase drug efficacy by targeted transport (Peng et al. 2012). One dose of systemically delivered D-NAC conjugate gave an improvement in the motor function and significantly enhanced neuronal injury and myelination in the newborn rabbit (Peng and Chauhan 2011).

18.2.7 Subretinal and Systemic Delivery System

For the sake of bypassing the barrier effect from vitreous and inner restraining membrane penetration, injection by subretinal route for gene and drug carrier had been established by the authors. Rod-shaped NPs are able to form through condensation when block copolymer of 30-mer lysine with cysteine group at terminal is conjugated with 10 kDa methoxy-PEG-maleimide (CK30PEG). The size was about

350 ± 5 nm (length ± width) with almost neutral charge. The NPs can directly transfect RPE cells (Ali et al. 2007). The CK30PEG-DNA NPs showed slow retinal degeneration and further activated a structural and functional rescue in retinitis pigmentosa (rds +/−) mouse model through subretinal route. Gene transfer mediated through NPs was able to restore cone function to an almost normal level and improvise the rod function. This was superior in comparison to the subretinal injection by plain DNA; this was confirmed through ERG analysis. These in vivo results promised a good breakthrough for the treatment of various retinal disorders (Ali et al. 2007).

In order to deliver drug to posterior segments of the eyes through BRB, systemic delivery had been explored. The inner BRB retards the entry of drugs from systemic site to the ocular tissue as it contains various tight junctions that form selective barrier. Humphries et al. formulated RNAi-mediated suppression which was a novel approach. This led to an opening of inner BRB which was reversible. The siRNA delivery for targeting claudin-5 in retina made a transient size-selective increase in the permeability of microvessels at paracellular levels. This was achieved without any retinal edema or change in function of retina. Due to creation of the reversible opening, low-molecular-weight drugs were easily delivered to outer layers of retina. However, in case of molecules larger than 1 kDa the delivery was excluded.

To deliver anti-VEGF plasmid surface functionalization was carried out with transferrin, a RGD peptide. These RGD-functionalized PLGA NPs accumulate in the neovascular eye. These enable gene delivery to retinal endothelial cells. After 2 weeks of administration a small CNV area was seen in rats with functionalized NPs in comparison with non-functionalized NPs. Based on histological studies H and E staining-functionalized NPs did not induce cellular filtration. Toxicological studies in ocular tissue can be affected by many factors such as chemistry, size, dose, and time and bio-distribution pattern. In recent times toxicological experiments have been initiated to evaluate toxicity (Ali and Byrne 2009).

Goldberg and coworkers have studied the impact of magnetic nanoparticle size on ocular toxicity in Sprague-Dawley rats measured by IOP, ERG, and histopathology. Significantly the positive impact on the safety of NPs can be assessed by ocular bio-distribution in the presence and absence of disease. Upon intravitreal administration the PAMAM dendrimers were rapidly cleaned in a rat model. The dendrimers were mostly localized to activate glial cells up to 30 days (Goldberg et al. 2007; White et al. 2011; White and Byrne 2010; Mintzer et al. 2009; Wang et al. 2010). Neuroinflammation association with selective localization helps the toxicity profile of dendrimers. The transport of these therapeutic genes by intravitreal injection required lesser amount of drug compared to systemic route that may assist in reducing the harmfulness of injected nanoparticles to other organs.

18.3 Carriers for Ocular Delivery

Surface charge as a factor for ocular delivery

Various nanocarriers have made their way to explore it in the ocular delivery nanocarriers like liposomal formulation, PLGA NPs, gelatin NPs, chitosan NPs, and dendrimers, and micellar formulations, niosomal formulations, and self-emulsification formulations have made their way in ocular delivery (Navath et al. 2011).

In the frontal portion of the human eye, researchers have made notable contributions for enhancing the efficiency of treatments for ocular diseases by improving the drug retention in ocular tissue (Sosa et al. 2008; Cheng et al. 2007). Miki et al. have shown that the surface charge of cornea and conjunctiva is negative and hence cationic NPs will retain on such negatively charged cornea, hence providing more retention of drug in ocular tissue. Smeds et al. demonstrated that the topical administration of positively charged gelatin NPs might prolong the residence time on the negatively charged ocular surface (Smeds et al. 2001).

Size as a factor for ocular delivery

All that is needed for an efficient drug delivery is proper particle size; hence the size of nanocarriers should be small enough to enter ocular barriers. Additionally, the eye-drop preparations comprising gelatin nanoparticles with 180 nm may retain in cornea epithelium cells (Williams and Coster 2010). Certainly, various nanoparticles prepared with synthetic polymers have been discovered for drug transport to the retina through intraocular injection (Williams and Coster 2010; Qazi et al. 2012; Cho et al. 2012). In general, by endocytosis, NPs which are less than 250 nm are usually taken up (Hughes et al. 2005; Duvvuri et al. 2003). Hosoya et al. have stated that NPs can cross via BRB or other ocular barriers (Hosoya and Tachikawa 2009; Hosoya et al. 2011). By exploring several studies further size of NPs with different surface properties can be traced by conjugation of fluorescent dyes with NPs and thereby delivering at the targeted site.

18.3.1 Polymeric Colloidal Nanocarriers for Ocular Drug Delivery

Chronic nature of various ocular diseases and anatomical locations, which is self, unique, and filled with barriers in the eye, gives rise to the requirement of frequent dosing in the drug treatment. Polymeric nanoparticles, which are biodegradable, are able to serve as compatible carriers for reducing the problems of frequent dosing and protection against various proteins and enzymes in the body, resulting in elevated half-life of the drug in the body. Drugs entrapped in the nanoparticles engineered for sustaining and controlling the release at the desired area will lead to reduction in frequent dosing requirement. Size/charge is one of the critical aspects and by modification of the same manipulation the delivery of drug at the desired target region in the eye can be done. Blood-ocular barriers can be easily overcome due to the small size of nanoparticles. In the following section, various applications of biodegradable polymers used to prepare polymeric nanocarriers for ocular gene/drug delivery are summarized.

18.3.1.1 Liposome (Lipid)

Liposomes are one of the important vehicles for drug transport to carry hydrophilic and lipophilic drugs that are bilayer phospholipids of tiny round-shaped spheres like cell membranes. Karn et al. have developed encapsulated liposomes of cyclosporine A (CsA) for the treatment of dry eye syndrome. The study demonstrated topical introduction of CsA-liposomes in eye-drop form to male albino rats who were already induced with DES; the formulation effects were further compared with commercially available CsA emulsion with brand name (Restasis®). The final outcomes specified that a less ocular irritation improved therapeutic efficacy and higher tear amount was achieved with CsA-liposomes (Karn et al. 2014). Sakurai et al. have applied topical delivery to transport drug to the posterior segment of ocular cavity using Annexin A5-associated liposomes (Sakurai et al. 2001). The results of study indicated overcoming of biological barriers such as corneal epithelium in rats/rabbits for delivering to the posterior eye segment. The final concentration obtained after topical administration was 127 ng/g for rat eyes and 18 ng/g for rabbit eyes. Kim et al. have formulated plasmid DNA-encapsulated liposomes through microfluidizer (Kim et al. 2009a). Modified transferrin on the surface of liposome showed high penetration and targeted the RPE. The authors also studied the effects, which were size dependent. Liposomal diameter of less than 80 nm (68 nm and −36 mV) was able to penetrate the RPE layer whereas 100 nm (100 nm and −36 mV) or a bigger diameter was distributed in choroid endothelium. The findings indicated the dependency of size on the effect of liposomes and its distribution in areas of posterior eye segment. Amrite et al. have prepared liposomes with a globule size of 109 nm with a loading efficiency of drug of 94% (Amrite et al. 2008). Latanoprost was delivered by a single injection in subconjunctival part in the superior temporal region of rabbit eye. The drug was released in a sustained manner for about 90 days and had no adverse effects. The single liposomal dose showed superior effect in comparison to single topical latanoprost instillation in eye. The reduction in IOP was about 4.8 ± 1.5 mmHg. So it can be concluded that these nanocarriers loaded with drugs are a potential platform for the treatment of glaucoma (Amrite and Kompella 2005). Zhang et al. have studied and evaluated the effect of tacrolimus which was encapsulated in liposomes and injected through intravitreal route for autoimmune uveoretinitis (EAU) in rats (Kim et al. 2009b). The tacrolimus-encapsulated liposomes were located in the retina's internal limiting membrane and vitreous body. The findings revealed that the liposomes were migrated to the outer nuclear layer from the internal limiting membrane at about 24 h. In about 7–14 days of injection the liposomes reached the retina. Further even after 14 days after the injection, tacrolimus was still detected (the concentration was greater than 50 ng/mL) in the ocular fluid and served the function of reducing the intraocular inflammation without hindering normal retinal function and any rejection of immune system. Avastin™ (bevacizumab) is a large recombinant monoclonal antibody that neutralizes the human vesicular endothelial growth factor and blocks neovascularization. Abrishami and his colleagues formulated multilamellar liposomes loaded with bevacizumab using phospholipid and cholesterol in 1:1 molar ratio for intravitreal delivery. Encapsulation efficiency was about 45% and it was stable. The authors did not furnish the particle size data. The clearance was slower and drug concentration was higher in comparison to antibody solution (Koo et al. 2012).

18.3.1.2 Chitosan Nanoparticles (Polysaccharide Based)

Chitosan is a copolymer, which consists of *N*-acetylglucosamine and glucosamine. Chitosan is formed by deacetylation of chitin of crustacean shells. The obtained product varies with different molecular weights (50–2000 kDa), degree of deacetylation (40–98%), and viscosities.

Chitosan offers advantages such as biodegradability and compatibility with low production cost. Moreover, it is approved by the USFDA as an approved biomaterial. Chitosan is a very good drug carrier for ocular delivery because of its mucoadhesive characteristics and capability to open tight junctions (Peeters et al. 2005).

Various applications of chitosan as nanoparticles had been explored for ocular drug delivery. One of the examples is development of 5-fluorouracil (5-FU)-encapsulated chitosan NPs (CH-DNPs) for ocular transport (Nagarwal et al. 2010; Sanders et al. 2007). The CH-DNP size is about 192 nm with a zeta potential of 42 mV. The administration of these nanoparticles results in the absence of irritation or inflammation in the eyes of rabbit. The concentration of 5-FU was higher in the aqueous humor of eyes treated with CH-DNPs in comparison to free 5FU solution-treated eyes. This higher concentration was obtained due to the mucoadhesive property of chitosan. In findings it was observed that chitosan nanoparticles were used to treat bacterial infection in eyes and to overcome the difficulty of penetrating the barrier of ocular region.

Chitosan can be utilized as a nonviral gene carrier due to its less immunogenicity, high transfection efficacy, and less chances of mutation as in the case of viral vectors (Kadam et al. 2012). Thakur et al. have developed chitosan-DNA nanoparticles in the size range of 98.2 nm with positive charge at 44.1 mV. This study established that the chitosan-DNA NPs could be effectively used as drug carriers for treating corneal diseases (Thakur et al. 2010). Plasmid DNA encapsulation in chitosan NPs did not affect the gene expression capacity. The eyes treated with NPs displayed a substantial quantity of green fluorescent protein (GFP) in the RPE layer compared to saline-treated eyes. There was no effect on retinal function after 30 days of injection and it was further confirmed through electroretinogram. Thus, glycol chitosan NPs ere concluded to be the best suited for gene carrier in the treatment of RPE-associated genetic disease. The linkages between glucosamine and glucosamine, *N*-acetyl-glucosamine and *N*-acetyl-glucosamine, and glucosamine and *N*-acetyl-glucosamine can be degraded by enzyme lysozyme. Chitosan exhibited slow discharge of drug because of degradation by means of several enzymes such as chitin deacetylase and β-*N*-acetyl hexosaminidase (Bourges et al. 2003; Shelke et al. 2011; Bakri et al. 2007).

18.3.1.3 PLGA Nanoparticles

Polylactic-co-glycolic acid (PLGA) is a copolymer of polylactic acid (PLA); it is a FDA-approved biomaterial and is utilized in numerous medical applications such as bone cements/plates/screws and scaffolds. The releasing controlling characteristics can be improved by alteration of PLA/PGA ratio and molecular weights. After degradation of PLGA, non-harmful metabolites such as lactic acid and glycolic acid are produced which are biodegradable in nature. Moreover, PLGA has a good physical strength and hydrophilicity making it a good drug carrier for various intended applications (Gaudreault et al. 2005). By using PLGA-based NPs various advantages had been known for ophthalmic drug delivery. This includes maintenance of slow drug release via polymer degradation, protection against rapid inactivation, high encapsulation efficiency for hydrophobic as well as hydrophilic macromolecules, and targeting site, region, or cells by surface modification (Gupta et al. 2000). Robinson et al. have formulated and evaluated pranoprofen delivery with PLGA NPs in cornea through topical instillation (Robinson et al. 2011). The size of the PF-F-NPs was about 350 nm with a surface charge of −7.41 mV. The encapsulation efficiency was found to be about 80%. To evaluate the cytotoxicity of PF-F-NPs, in vitro cell line study using Y-79 human retinoblastoma was done. The result showed that blank PLGA-NPs were nontoxic to cells and further may be able to lower the cytotoxicity of PF. The effect was compared to commercially available eye-drop formulation and free-form drug solutions in rabbit eye. The corneal permeation was four times higher in comparison to other groups. The nanoparticles have fast onset of action and exhibited sustained residence time in corneal surface, thus reducing ocular edema. These concluded that nanoparticles have great potential for the management of corneal disease associated with chronic inflammation. Surface-modified PLGA NPs of chitosan, polysorbate 80, and glycol chitosan were evaluated for increasing the mucoadhesive property. Unmodified PLGA-NP size was about 224.5 nm with −41.3 mV charge on surface. The size of P80-PLGA NPs was found similar to that of plain PLGA-NPs whereas the size of glycol chitosan and chitosan PLGA-NPs was larger. This was due to adsorption of chitosan molecules on PLGA surface. The surface-charged modified PLGA-NPs of chitosan, glycol chitosan, and P80 were easily able to penetrate the retina of mouse (Iezzi et al. 2012). It was observed that the interaction between PLGA NPs and cell surface to improve penetration can be enhanced by mucoadhesive molecular modification. The results concluded that surface-modified PLGA-NPs are one of the potential carriers for delivery of drugs by topical application (Iezzi et al. 2012).

18.3.1.4 Gelatin Nanoparticles

Gelatin, a biopolymer of natural source, is generally prepared from porcine skin, cow bone, or fish scale. The method used for preparation and purification is acid or alkaline hydrolysis. The triple-helix structure polyampholyte has both cationic and anionic charges. Gelatin NPs showed excellent biocompatibility and have been previously used as a gene drug carrier in ophthalmic application. Gelatin is a FDA-approved biomaterial and since collagen is the major constituent of corneal stroma, the gelatin obtained from it can be used as a potential carrier for improving the bioavailability of drugs (Conley and Naash 2010; Cai et al. 2009, 2010).

Ding et al. have formulated two charged GPs and evaluated its compatibility in the corneal epithelium of human (HCE) cells as well as in rabbit eye. The GPs(+) were prepared with type A gelatin with a size of about 180.6 nm and a charge of 33.4 mV. The GPs(−) were prepared with type B gelatin with a size of about 230.7 nm and a charge of −44.2 mV. The accumulation of intracellular GPs(+)/(−) was confirmed and it was revealed through fluorescence intensity in cell lysates that it was more in GP(+) group than the GP(−) group after 10–60-min cultivation. This phenomenon showed that there is increase in transfection efficiency, thereby increasing bioavailability. Further it was concluded that GP(+) can be used effectively in a concentration of (100 μg/mL, 50 μL) without any irritation. There was no change in IOP and corneal thickness (Ding et al. 2009). There was wide distribution of fluorescence of GP(+) in the corneal region and the drug was retained for longer duration; it was possible due to positively charged GPs(+) that were adsorbed to negatively charged cornea. Boylan et al. have formulated antibacterial GPs loaded with moxifloxacin. They were prepared using type A gelatin (Boylan et al. 2012). GP loaded with moxifloxacin, a type of fluoroquinolone antibiotic, has the size of about 175 nm with a positive charge of 24 mV. In comparison to commercially available antibacterial agent, MoxiGram®, the moxifloxacin-loaded GP effectively delivered drug in New Zealand albino rabbits. The GP showed no irritation and was safe and biocompatible in rabbit cornea. The antibacterial activity also showed good effectiveness against *Staphylococcus aureus* with a zone inhibition diameter of about 13.36 mm at 12 h and 15.46 mm at 24 h. The commercial product (MoxiGram®) showed zone inhibition of 10.49 mm at 12 h versus 12.52 mm at 24 h.

In order to achieve specific blood vessel targeting, surface modification of GPs with arginine-glycine-aspartic acid (RGD) peptide-HA-conjugated complex, named GEH-RGD, was undertaken. The result indicated that GEH-RGD NPs possess a good potential for treating corneal neovascularization through topical administration (Koirala et al. 2011).

18.3.1.5 Niosomes

Niosomes are tiny particles which hold both lipophilic and hydrophilic drugs and have better stability than liposomes. Niosomes have the lowest toxicity due to their nonionic, non-immunogenicity, and biocompatible nature. Liposomes have drawbacks like high cost and fluctuating phospholipid nature.

Discomes are the modified form of niosomes used for ophthalmic drug delivery; discomes are large molecules having a disc-like structure of 10–60 mm, which is derived from niosomes with addition of nonionic surfactant like solulan C24. Discomes have better advantage over niosomes due to their large particle size which will not drain out from eye and will provide the longer corneal residence time leading to improvement of the bioavailability; this may be due to their disc shape nature which fits in the cul-de-sac cavity of eye (Fukuda et al. 2003). For controllable ocular drug delivery of water-soluble drugs discomes are widely used.

18.3.1.6 Microemulsions

Dispersion is made up of oil phase, aqueous phase, surfactant, and co-surfactant at appropriate ratio with colloidal milling to prepare microemulsions. Microemulsions are transparent and translucent liquid having low interfacial tension and droplet size of below 150 nm (Eguchi et al. 2009). Microemulsion contains surfactant and co-surfactant which increase the solubility and improve the permeability and ultimately bioavailability of drugs (Eguchi et al. 2009). Advantage of microemulsions is their ability to increase the solubility of highly lipophilic drugs such as abstention (Gray 2006).

Beilin and coworkers have proved the enhancement of residence time of microemulsions in the eye conjunctival sac by using fluorescent marker in the formulations. Due to oily nature microemulsions show higher residence time in eye leading to improvement of the bioavailability and delay of the residence time with cornea. Ocular bioavailability of indomethacin by topical administration of microemulsions was studied by Muchtar and coworkers; they used poloxamer and lecithin with 0.1% w/v drug in the formulations; the results showed that prepared microemulsions showed increase in the concentration of indomethacin in the cornea.

18.4 Formulation Additives for Ophthalmic Drug Delivery

Better ophthalmic formulations can be prepared by use of proper excipients and vehicles which showed extended retention time in cornea with minimum ocular toxicity and irritation. Formulation additives like viscosity builders, clarifying agent, preservative, buffering agent, and wetting agent are commonly used in formulations. Examples of viscosity builders are HPMC, PVA, PVP, methyl cellulose, poloxamer, carbomer, gellan gum, xanthan gum, and polysaccharides; these polymers show large molecular weight which will not penetrate biological membrane and form three-dimensional structure after contact with water.

Carbomer is commonly used in semisolid and liquid formulations as a suspending agent which increases the viscosity whereas hyaluronic acid is used as a polymer for the formation of biodegradable and biocompatible matrix (Ishibashi et al. 2003). Maximum penetration of formulation through cornea takes place when viscosity is in the range of 20–150 Cps. High amount of viscosity agent forms gel which enables to reduce the dosing frequency and we will be able to develop once-a-day formulations. It has been proved that poloxamer 407 is suitable as a viscosity modifier or carrier for ophthalmic formulation with pilocarpine. The main disadvantage of poloxamer formulation is the blurring of vision, which affects the patient acceptance (Ishibashi et al. 2003; Whitson et al. 2006; Ayaki et al. 2008, 2010). Hydrophilic polymers used in ophthalmic formulation also showed mucoadhesive properties apart from viscosity increment (Whitson et al. 2006). Mucoadhesive properties mainly through noncovalent bonding of polymers with mucin present in eyes increase the contact time in cornea by electrostatic connections. Sodium carboxymethyl cellulose, chitosan, hyaluronic acid, and polyacrylic acid are the examples of mucoadhesive polymers. Cross-linked polyacrylic acid- and carbomer mucoadhesive property-based formulations of timolol maleate, brand name NyoGel (Novartis), and pilocarpine hydrochloride, brand name Pilogel (Alcon Laboratories), are available commercially in the market.

18.4.1 Penetration Enhancer

Penetration enhancer is used in ophthalmic drug delivery to enhance the corneal absorption of drug. Enhancement of drug absorption in cornea is due to modification of corneal epithelium structure. Absorption enhancers like cheating agent (EDTA), preservatives like BKC, surface active agent, and bile salt are most widely used. However, these substances showed a local irritation, which restricts their uses in ocular drug delivery (Ayaki et al. 2008).

Modification of chemical structure—prodrugs

Prodrug is nothing but modification of drug into different salts which helps to increase the corneal absorption. In prodrug there is modification of chemical structure of molecules which gives new improvement to physical properties with high selectivity and low irritation (Ayaki et al.). Different products were developed in recent years for ocular drug delivery including epinephrine, timolol, and pilocarpine (Cholkar et al. 2012). A diester of pivalic acid (dipivefrine) displays 17 times higher absorption through the cornea; this is due to increase in lipophilicity of drug, resulting in smaller dose of dipivefrine application on the eyeball which has similar effect to equivalent epinephrine dose. In conclusion 0.1% dipivefrine gives equal therapeutics effect to 2% epinephrine eye drop.

Cyclodextrin as solubilization

Cyclic oligosaccharides commonly called as cyclodextrin are used to form inclusion complex with drugs to increase the solubility of insoluble drugs without changing the structure. As solubilizing agents, they help to keep insoluble drugs in solution and help to penetrate tissues. In ocular drug delivery cyclodextrin is used in appropriate concentration of less than 15% in eye-drop solutions. Widely used cyclodextrin is 2-hydroxypropyl because these cyclodextrins show very less irritation in eye. Available marketed products which contain cyclodextrin are dexamethasone and pilocarpine which showed increased therapeutics effect compared to conventional routine eye drops.

18.4.2 Antibacterial Agent: Preservatives

Preservative is used in ocular formulation to prevent microbial growth in the formulation in product shelf life. Preservative is mostly used in multidose container because of their multiple uses during administration. Most widely used preservatives in ocular drug delivery system are 50% benzalkonium solution, thimerosal, chlorobutanol, and phenylmercuric nitrate with 0.01–0.2% w/w concentrations. BKC and thimerosal are the common choice for ocular drug delivery because of their broad range of activity against gram-positive and gram-negative bacteria including yeast and fungi. BKC solution is stable over a wide range of pH, temperature, and storage without losing their antibacterial activity. A person wearing soft contact lens daily should avoid the use of BKC solutions because BKC binds to the hydrogel of soft contact lenses.

Alternative to BKC thimerosal is the second choice for ocular drug delivery because it does not bind with hydrogel and can be used by soft contact lens wearer. Thimerosal shows antibacterial and antifungal activity at neutral and alkaline pH but it is bactericidal at lower pH. It is sensitive to light and stable at room temperature. Most of the preservatives show incompatibility with drugs and excipients; hence their choice should be based on compatibility studies. Preservatives should not be used in ocular injection because of their toxicity to eye internal structure when injected.

18.4.3 Oxygen Suppresser: Antioxidant

Oxygen suppresser is known as antioxidants used in ocular drug delivery when active ingredients or any other excipient of formulation is susceptible to oxidation or reduction by free radical mechanism. Most commonly used antioxidants are disodium EDTA sodium metabisulfite and sodium bisulfite. For acid-type formulation the choice is sodium metabisulfite and for neutral pH formulation the choice is sodium bisulfite.

18.4.4 Thickeners: Viscosity Builder

To increase the contact time of formulation in eye and to prolong the action viscosity modifiers are used. There is also increase in bioavailability of drugs due to increase in the residence time of drug in eyes. Choice of viscosity-modifying polymers should be such that it gives up to 50 Cps of viscosity because most ophthalmic products have a viscosity range from 25 to 50 Cps. Viscosity builder should be sterilized by any method before addition because filtration sterilization is difficult in the final product due to presence of viscosity polymers. Widely used viscosity builders include polyvinyl alcohol, polyvinylpyrrolidone, HPMC, methyl cellulose, and hydroxypropyl celluloses. Polyvinyl alcohol and hydroxypropyl methyl cellulose are commonly used in artificial tears and solution for the contact lenses. These polymers are stable at temperature below 110 °C.

18.4.5 Osmotic Agent: Isotonic Agents

Tears present in eye are isotonic which exhibit tonicity equivalent to 0.9% NaCl. Eyes tolerate the tonicity from 0.5% to 1.5%; hence there is a need to adjust the tonicity of formulation in this range. Patient tolerates hypotonic solution compared to hypertonic. Ideal osmolality value is 290 mOsm/L but human eyes tolerate an osmolality from 220 to 550 mOsm/L. Tonicity agents like NaCl, mannitol, dextrose, glucose, and glycerin are most commonly used. Apart from tonicity viscosity builders reduce the pain and provide the smoothness to eyes.

18.4.6 Clarifying Agents

Clarifying agent is used to improve the clarity of the ophthalmic product. Ophthalmic product should be free from particulate matter to reduce the irritation and abrasion of cornea of eye. Filtration of solution through 0.40 μm filter solves the problem without removing the drug but use of clarifying agents like HPMC, Tween 80, and Tween 20 can improve the clarity and viscosity of the product; clarifying agent is most widely used in ophthalmic solution which provides the smoothness properties in eye and helps to reduce the irritation.

18.4.7 Buffering Agents: pH Modifier

Chemical stability of a product is affected by pH which adversely affects potency and formulation effectiveness. Optimum pH is necessary to stabilize the product and selection of buffering agent is based on the chemical and physical nature of active compounds. Drugs which are acidic or neutral in nature will precipitate out in basic solution and effectiveness of the active is reduced. However drugs which are basic in nature precipitate out in acid solution. Precipitation of drug can cause corneal damage and its abrasion. The pH also affects the performance of preservative; in some preservatives effectiveness decreases in extreme low or high pH. As described previously, bacteriostatic nature of thimerosal is observed at neutral or basic pH and bactericidal nature is observed at acidic pH. Ideal pH requirement of ophthalmic solution is between 4 and 7 pH and buffering capacity should be less than 0.05. Most commonly used buffers are acetate buffer, tris buffer, and citric acid and phosphate buffers. Selection of buffer is based on initial development and stability studies of the formulation.

18.4.8 Solvent: Vehicles and Bases

Solvents and vehicles are used to dissolve drugs and excipients to give the final formulation. Vehicles are classified as aqueous and nonaqueous. Selection of vehicles or bases fully depends on the type of dosage prepared and compatibility of drug substances and excipients. Vehicles or bases may contain viscosity agents, preservatives and buffers, electrolytes, clarifying agents, lubricants, and wetting agents.

Aqueous vehicle is used to manufacture the topical solutions and ocular injections, most commonly salt solution or artificial tear solution, are use used as vehicles for topical solutions; available brands differ in their respective content. 5% Dextrose in purified water for injection is used as an aqueous vehicle for vancomycin 25 mg/mL eye-drop solution and survey reported good comfortability of patients. Brands such as polyethylene glycol 400 and propylene glycol component of systane, PVA, and povidone component of murine cannot interchange as generic version. Injection used for ocular drug delivery should contain preservative-free diluents such as isotonic solution 0.9% NaCl, sterile water for injection, and dextrose solution 5% in water; such type of solutions are used to prepare topical solutions.

Nonaqueous vehicles or oleaginous vehicle bases are mostly used to manufacture ophthalmic topical ointments or "oily" solutions. When nonaqueous vehicle is used as bases it should be sterilized before use because ointment bases cannot be filtered at the final stage of sterilization. Ointment which is non-sterile can sterilize by heat called dry heat or autoclave. Non-sterile ointment bases can only be sterilized by dry heat or autoclaves by heat and steam. White petroleum is commonly used as base in ointment than yellow petroleum as yellow petroleum is less purified and more irritant than white petroleum. Corn oil and medium-chain triglyceride are used as topic oily solutions for ophthalmic drug delivery; these are safe and nonirritant and can be sterilized by filtration or dry heat.

18.4.9 Drug Substances

Drugs used to prevent or treat eye infection/disease are called ophthalmic drugs. Active pharmaceutical ingredients for ophthalmic drug delivery are obtained from sterile products, such as non-sterile bulk powders, parenterals, and synthesized chemical or herbal product.

18.5 Ophthalmic Preparation: Compounding

The United States Pharmacopeia Chapter <797> of sterile compounding defines that all ophthalmic preparations either topical or injectable should be sterile. There is a general view in patients about compounders that ophthalmic products are prepared under non-sterile condition or non-GMP environment because once they open the ophthalmic preparation it is no longer sterile after 30 days. A number of literature have reported microbial growth in ophthalmic products which causes eye infection and permanent blindness.

Ophthalmic dosage is often used to treat chronic and acute conditions and it may be needed for quick effect. Ophthalmic dosage form is not considered as a life-threatening treatment. As per the USP Chapter 800 if drug components in compounded ophthalmic preparation are hazardous they must be prepared in negative pressure ISO class seven room with ISO class two biological safety cabinet. Lots of compounded ophthalmic formulations are prepared using sterile ingredients which are available commercially and these formulations are usually topical solutions or ocular injections. Bulk APIs are used to prepare some compounded ophthalmic dosage forms because of manufacturer back orders or unavailability of compounded medication commercially. Bulk API formulation is prepared as single units for individual patients or it is prepared as a batch of multiple units for multiple patients.

At the time of ophthalmic compounding preparation compounders must follow current official United States Pharmacopeia standards; USP has defined the standards including the requirements for compounding personnel; equipment needed for compounding; environment and environmental monitoring during compounding; quality assurance needed for compounding of ophthalmic preparation; packaging handling and labeling of ophthalmic products; and shipment of compounded sterile preparations. Compounders should follow the instructions present in the USP chapter of compounding for preparation of sterile medication; however in addition to USP compounders should follow recently published guidelines that are specific for sterile ophthalmic preparations.

18.5.1 Ophthalmic Preparation: Compounding Practices

Special precaution is needed while calculating the composition of ophthalmic products; for example, in ocular injection which has very concentrated formulation the mathematical calculation used to develop the formula should be double-checked to minimize compounding error. Decimal error will significantly increase or decrease the dose of active ingredients and this may cause serious injury to eyes. To prevent such incidences ingredients should be taken with the smallest size needle and syringe and to improve accuracy in measuring sterile ingredients being withdrawn from vials or bags, use the smallest syringe necessary to measure the desired volume; for example, for dispensing 2 mL solution use 5 mL injection not 10 mL to minimize the error. Filter the liquid from 5 μm filter after immediately withdrawing from ampoules. Sterile powder for injection should also be filtered after reconstitution from 5 μm filter to remove the particulate matter.

Quality control and quality assurance are prime requirements during the preparation of ophthalmic products. Compounder in ophthalmic dosage form must be inspected visually for particulate matter and clarity. Suspension should be checked for presence of any caking and its redispersibility. Ointment should be checked for its smoothness and non-grainy characteristics. The pH should be checked to ensure that the prepared formulation is within the range. The final weights and volume should be calculated and the same should not deviate from 10% of theoretical mean weights or volumes.

Ophthalmic preparation prepared for individual patients does not need to be tested but batch prepared for multiple patients must be tested and should undergo a quarantine period. As per the USP <797> a batch of more than 25 units should undergo sterility testing. Sterility testing is the minimum standard needed for the 25-unit preparations.

Endotoxin testing is not required for the ophthalmic preparation as defined in the USP <797>; however, this endotoxin test must be done for certain ophthalmic products. Endotoxins are very well known for causing savior fever and other brain-related complications when injected into the human bloodstream. Endotoxin contamination of ocular injection is widely reported which causes toxic anterior segment syndrome. This syndrome is nothing but an acute inflammation of the anterior portion of the eye which usually occurs after cataract surgery.

18.6 Novel Ophthalmic Drug Form

18.6.1 Semisolid Forms

18.6.1.1 Sol-to-Gel Systems (In Situ Gels)

Sol-gel systems are semisolid systems which undergo phase transfer from liquid to solid in the presence of external factors like pH, temperature, and electrolyte or enzyme. These are viscous liquid with the ability to transfer from liquid to solid states; this transfer property helps to reduce drug drainage from eyeball surface and helps to enhance drug bioavailability. Sol-gel system is achieved by the use of polymers like gellan gum, carbopol, and cellulose acetate phthalate. This type of research is carried out on drugs including ciprofloxacin hydrochloride, ganciclovir, fluconazole, and pilocarpine.

18.6.1.2 Ophthalmic Ointments

Ophthalmic ointment consists of solid or semisolid hydrocarbon having very low melting point which is close to human body temperature. Ointment is mainly for external uses; after administration to eye ointment melts at body temperature and remains retained for longer time in conjunctival sac, thus helping to enhance drug bioavailability. Ophthalmic ointment shows blurred vision after administration, thus causing patient discomfort, and can be used only at nighttime. Available dosage as ointment is chloramphenicol tube for eye conjunctivitis infection.

18.6.2 Solid Ophthalmic Dosage

18.6.2.1 Drug-Coated Contact Lenses

Drug-coated contact lenses are prepared by adsorbing the drug on the water-soluble surface of lenses from where the drug is released after applying over eyeball for extended period of time. Most widely used polymers to prepare the contact lenses are cross-linked poly-2-hydroxyethyl methacrylate with ethylene glycol dimethyl acrylate. In recent years, lot of research has been carried out on using silicone for preparation of contact lenses. Examples of drugs which are tried as drug-coated contact lens delivery include antibiotics like ciprofloxacin and cyclosporine and steroids like dexamethasone.

18.6.2.2 Ocular Inserts

Ocular inserts are the semisolid or solid dosage form having lot of advantages over the conventional ophthalmic dosage form. Ocular inserts are not susceptible to defense mechanism of eyes like nasolacrimal tear secretion outflow and stay in conjunctival sac for longer period of time and are more stable than conventional ophthalmic dosage form. Ocular insert also shows accurate dosing and extended release of drug in eye sac which reduces dosing frequency and increases patient comfort. Ocular insert does not blur the vision of patients. To develop the ocular insert polymers are widely used including HPMC, methylcellulose and its derivatives, ethylcellulose, povidone, PVA, chitosan, and its derivatives. Polymers undergo solubilization in eye sac and biodegradation. Ocusert (Alza Corporation) is manufactured from copolymer of ethylene and vinyl acetate and is a classic example of ocular insert which contains pilocarpine as model drug for treatment of ocular diseases. Disadvantage of ocular insert is patients' unwillingness due to filling of foreign body in the eye.

18.6.2.3 Soluble Ophthalmic Drug Inserts (SODI)

Acrylamide, polyvinylpyrrolidone, and ethyl acrylate are used to formulate the SODI. These are small oval wafer soluble eye inserts. Tear fluid moistens the SODI after administration in conjunctival sac and then they soften and adhere to eyeball surface. Drug is released from SODI in a controlled manner. Kanamycin, dexamethasone, pilocarpine, atropine, and sulfapyridine drugs are tried in SODI.

18.6.2.4 Minidiscs

Minidiscs look like contact lenses with 4–5 cm diameter; it is convex outside and concave inside to fit the eye surface. Minidisc is formulated using copolymers bis (4-methacryloxy)-butyl poly (dimethylsiloxane) and poly (hydroxyethyl methacrylate). Minidisc is used to manufacture either hydrophilic or hydrophobic drug which is designed to control the hydrophobic and water-soluble drugs. Gentamicin sulfate and sulfisoxazole are the drugs used to prepare the minidisc.

18.6.2.5 Artificial Tear Inserts

Artificial tear inserts are rod-shaped long pellets containing HPC and preservatives. Lacrisert is the brand name available in the market as artificial tear insert used for treatment of dry eye syndrome. Insert absorbs water after inserted in cul-de-sac and forms hydrophilic layer which helps to stabilize the tear film and helps in moistening the cornea.

18.6.2.6 Collagen Shield

Porcine sclera is used to develop collagen shields. Porcine sclera shows similarity with human collagen. Collagen shield is often stored in dry state and should hydrate before application. Collagen shield sometimes causes the patient discomfort and is washed out from eyes.

Research of collagen is carried out on human model and animal model for antibiotics like gentamicin, antiviral drugs, and anti-inflammatory drugs like dexamethasone. Collagen shield shows higher concentration of drug in cornea and aqueous humor compared to contact lenses and eye drops. Now new formulation of collagen has been developed, so-called collasomes, which are a small piece of collagen suspended in 1% of methyl cellulose solution.

18.6.2.7 NODS (New Ophthalmic Drug Delivery System)

It is a film with drug which gets separated from the handle when applied to conjunctival sac and it gets dissolved in the tear fluid and releases the active ingredient called NODS. Smith & Nephew Pharmaceuticals Ltd. patented the new NODS consisting of solidified paper handle and a flag from polyvinyl alcohol. NODS delivers the drug to eyeball and shows increase in the bioavailability of active ingredients. NODS of pilocarpine shows eightfold increase in drug concentration when compared to conventional eye drops. Advantage of NODS is that it does not contain preservative and NODS can be sterilized with gamma rays.

18.6.2.8 Minitablets

Minitablets are the solid, biodegradable dosage form which contains excipients or polymers which form gel after inserted into conjunctival sac. After gel formation minitablets adhere to corneal tissue and increase the residence time and bioavailability of drugs. Advantage of minitablets includes easy application in cul-de-sac cavity and has resistance to tear flow and long contact with cornea, with gradual release of active ingredients.

Minitablets are made up of polymers like hydroxypropyl methylcellulose (HPMC), cellulose, sodium carboxymethyl cellulose, hydroxyethyl cellulose (HEC), and ethyl cellulose acrylates. Minitablets are manufactured by direct compression or indirect method. Indirect method involves preparation of granules and their compression. Active ingredients which are tried for minitablet development are acyclovir, timolol, ciprofloxacin, piroxicam, and gentamicin.

18.7 Methods of Application and Other Ophthalmic Drug Forms

18.7.1 Filter Paper Strips

Filter paper strip is made up of pigments mainly fluorescein or Bengal red. These strips are used to diagnose corneal or conjunctival damages as well as to diagnose microbial infection like herpes simplex viruses in eyeball. Strip is sized approximately 3 mm with 1 mg of sodium fluorescein. Wetting of strip with a drop of saline solution is needed before use.

18.7.2 Sprays

Spray is not a common dosage form and is not a prime choice as ophthalmic drug delivery. Active ingredients which are tried in spray include phenylephrine-tropicamide and phenylephrine-tropicamide-cyclopentolate. Spray should be activated from a distance of 5–10 cm in eye to prevent direct tissue damage by spray force and irritation; spray is mainly used to apply the drug close to eyelid. Spray shows same effectiveness as that of eye drops containing same ingredients. Martini and his associates proved that spray and eye drop showed same miotic effect of pilocarpine hydrochloride when applied in 1% concentration.

18.7.3 Ocular Iontophoresis

Ocular iontophoresis is a noninvasive method to deliver ions to cell or tissue by direct current. Iontophoresis is a fast, safe, and painless therapy of ions by current and is able to attain high concentration in tissue. When iontophoresis is used in pharmacotherapy, positively charged molecules are introduced to tissue from anode and negatively charged from cathode. Drugs which are tried in iontophoresis method are gentamicin, ciprofloxacin, and ketoconazole. Iontophoresis of antibiotics enhances bactericidal activity.

References

Agrawal AK, Das M, Jain S (2012) In situ gel systems as 'smart' carriers for sustained ocular drug delivery. Expert Opin Drug Deliv 9:383–402

Ali M, Byrne ME (2009) Controlled release of high molecular weight hyaluronic acid from molecularly imprinted hydrogel contact lenses. Pharm Res 26:714–726

Ali M, Horikawa S, Venkatesh S et al (2007) Zero-order therapeutic release from imprinted hydrogel contact lenses within in vitro physiological ocular tear flow. J Control Release 124:154–162

Al-Shamsi HN, Dueker DK, Nowilaty SR et al (1986) Neovascular glaucoma at King Khaled eye specialist hospital—etiologic considerations. Middle East Afr J Ophthalmol 16:15–19

Amrite AC, Kompella UB (2005) Size-dependent disposition of nanoparticles and microparticles following subconjunctival administration. J Pharm Pharmacol 57:1555–1563

Amrite AC, Edelhauser HF, Singh SR et al (2008) Effect of circulation on the disposition and ocular tissue distribution of 20 nm nanoparticles after periocular administration. Mol Vis 14:150–160

Ayaki M, Yaguchi S, Iwasawa A et al (2008) Cytotoxicity of ophthalmic solutions with and without preservatives to human corneal endothelial cells, epithelial cells and conjunctival epithelial cells. Clin Exp Ophthalmol 36:553–559

Ayaki M, Iwasawa A, Yaguchi S et al (2010) Preserved and unpreserved 12 anti-allergic ophthalmic solutions and ocular surface toxicity: in vitro assessment in four cultured corneal and conjunctival epithelial cell lines. Biocontrol Sci 15:143–148

Bakri SJ, Snyder MR, Reid JM et al (2007) Pharmacokinetics of intravitreal ranibizumab (Lucentis). Ophthalmology 114:2179–2182

Bochot A, Fattal E (2012) Liposomes for intravitreal drug delivery: a state of the art. J Control Release 161:628–634

Bourges JL, Gautier SE, Delie F et al (2003) Ocular drug delivery targeting the retina and retinal pigment epithelium using polylactide nanoparticles. Invest Ophthalmol Vis Sci 44:3562–3569

Boylan NJ, Suk JS, Lai SK et al (2012) Highly compacted DNA nanoparticles with low MW PEG coatings: in vitro, ex vivo and in vivo evaluation. J Control Release 157:72–79

Cai X, Nash Z, Conley SM et al (2009) A partial structural and functional rescue of a retinitis pigmentosa model with compacted DNA nanoparticles. PLoS One 4:5290

Cai X, Conley SM, Nash Z et al (2010) Gene delivery to mitotic and postmitotic photoreceptors via compacted DNA nanoparticles results in improved phenotype in a mouse model of retinitis pigmentosa. FASEB J 24:1178–1191

Cheng Y, Qu H, Ma M et al (2007) Polyamidoamine (PAMAM) dendrimers as biocompatible carriers of quinolone antimicrobials: an in vitro study. Eur J Med Chem 42:1032–1038

Cheng Y, Xu Z, Ma M et al (2008) Dendrimers as drug carriers: applications in different routes of drug administration. J Pharm Sci 97:123–143

Cho YK, Uehara H, Young JR et al (2012) Flt23k nanoparticles offer additive benefit in graft survival and anti-angiogenic effects when combined with triamcinolone. Invest Ophthalmol Vis Sci 53:2328–2336

Cholkar K, Patel A, Vadlapudi DA et al (2012) Novel nanomicellar formulation approaches for anterior and posterior segment ocular drug delivery. Recent Pat Nanomed 2:82–95

Conley SM, Naash MI (2010) Nanoparticles for retinal gene therapy. Prog Retin Eye Res 29:376–397

Dharma SK, Fishman PH, Peyman GA (1986) A preliminary study of corneal penetration of 125I-labelled idoxuridine liposome. Acta Ophthalmol 64:298–301

Ding XQ, Quiambao AB, Fitzgerald JB et al (2009) Ocular delivery of compacted DNA-nanoparticles does not elicit toxicity in the mouse retina. PLoS One 4:7410

du Toit LC, Pillay V, Choonara YE et al (2011) Ocular drug delivery—a look towards nanobioadhesives. Expert Opin Drug Deliv 8:71–94

Duncan R, Gaspar R (2011) Nanomedicine(s) under the microscope. Mol Pharm 8:2101–2141

Duvvuri S, Majumdar S, Mitra AK (2003) Drug delivery to the retina: challenges and opportunities. Expert Opin Biol Ther 3:45–56

Ebrahim S, Peyman GA, Lee PJ (2005) Applications of liposomes in ophthalmology. Surv Ophthalmol 50:167–182

Eguchi H, Shiota H, Oguro S (2009) The inhibitory effect of vancomycin ointment on the manifestation of MRSA keratitis in rabbits. J Infect Chemother 15:279–283

Farokhzad OC, Langer R (2009) Impact of nanotechnology on drug delivery. ACS Nano 3:16–20

Fukuda M, Hanazome I, Sasaki K (2003) The intraocular dynamics of vancomycin hydrochloride ophthalmic ointment (TN-011) in rabbits. J Infect Chemother 9:93–96

Gaudreault J, Fei D, Rusit J et al (2005) Preclinical pharmacokinetics of Ranibizumab (rhuFabV2) after a single intravitreal administration. Invest Ophthalmol Vis Sci 46:726–733

Giannavola C, Bucolo C, Maltese A et al (2003) Influence of preparation conditions on acyclovir-loaded poly-d,l-lactic acid nanospheres and effect of PEG coating on ocular drug bioavailability. Pharm Res 20:584–590

Gipson IK, Argüeso P (2003) Role of mucins in the function of the corneal and conjunctival epithelia. Int Rev Cytol 231:1–49

Goldberg M, Langer R, Jia X (2007) Nanostructured materials for applications in drug delivery and tissue engineering. J Biomater Sci Polym Ed 18:241–268

Gratieri T, Gelfuso GM, Lopez RFV et al (2010) Current efforts and the potential of nanomedicine in treating fungal keratitis. Expert Rev Ophthalmol 5:365–384

Gray C (2006) Systemic toxicity with topical ophthalmic medications in children. Paediatr Perinat Drug Ther 7:23–29

Gupta SK, Velpandian T, Dhingra N et al (2000) Intravitreal pharmacokinetics of plain and liposome-entrapped fluconazole in rabbit eyes. J Ocul Pharmacol Ther 16:511–518

Hariharan S, Gunda S, Mishra GP et al (2009) Enhanced corneal absorption of erythromycin by modulating P-glycoprotein and MRP mediated efflux with corticosteroids. Pharm Res 26:1270–1282

Hosoya K, Tachikawa M (2009) Inner blood-retinal barrier transporters: role of retinal drug delivery. Pharm Res 26:2055–2065

Hosoya K, Tomi M, Tachikawa M (2011) Strategies for therapy of retinal diseases using systemic drug delivery: relevance of transporters at the blood-retinal barrier. Expert Opin Drug Deliv 8:1571–1587

Hughes PM, Olejnik O, Chang-Lin JE et al (2005) Topical and systemic drug delivery to the posterior segments. Adv Drug Deliv Rev 57:2010–2032

Iezzi R, Guru BR, Glybina IV et al (2012) Dendrimer-based targeted intravitreal therapy for sustained attenuation of neuroinflammation in retinal degeneration. Biomaterials 33:979–988

Ishibashi T, Yokoi N, Kinoshita S (2003) Comparison of the short-term effects on the human corneal surface of topical timolol maleate with and without benzalkonium chloride. J Glaucoma 12:486–490

Jung HJ, Chauhan A (2012) Temperature sensitive contact lenses for triggered ophthalmic drug delivery. Biomaterials 33:2289–2300

Jwala J, Boddu SH, Shah S et al (2011) Ocular sustained release nanoparticles containing stereoisomeric dipeptide prodrugs of acyclovir. J Ocul Pharmacol Ther 27:163–172

Kadam RS, Tyagi P, Edelhauser HF et al (2012) Influence of choroidal neovascularization and biodegradable polymeric particle size on transscleral sustained delivery of triamcinolone acetonide. Int J Pharm 434:140–147

Kapoor Y, Thomas JC, Tan G et al (2009) Surfactant-laden soft contact lenses for extended delivery of ophthalmic drugs. Biomaterials 30:867–878

Karn PR, Do Kim H, Kang H et al (2014) Supercritical fluid-mediated liposomes containing cyclosporin A for the treatment of dry eye syndrome in a rabbit model: comparative study with the conventional cyclosporin A emulsion. Int J Nanomedicine 9:3791

Katragadda S, Gunda S, Hariharan S et al (2008) Ocular pharmacokinetics of acyclovir amino acid ester prodrugs in the anterior chamber: evaluation of their utility in treating ocular HSV infections. Int J Pharm 359:15–24

Kim H, Robinson SB, Csaky KG (2009a) Investigating the movement of intravitreal human serum albumin nanoparticles in the vitreous and retina. Pharm Res 26:329–337

Kim JH, Kim JH, Kim KW et al (2009b) Intravenously administered gold nanoparticles pass through the blood-retinal barrier depending on the particle size, and induce no retinal toxicity. Nanotechnology 20:505101

Koirala A, Makkia RS, Cooper MJ et al (2011) Nanoparticle-mediated gene transfer specific to retinal pigment epithelial cells. Biomaterials 32:9483–9493

Koo H, Moon H, Han H et al (2012) The movement of self-assembled amphiphilic polymeric nanoparticles in the vitreous and retina after intravitreal injection. Biomaterials 33:3485–3493

Ludwig A (2008) The use of mucoadhesive polymers in ocular drug delivery. Adv Drug Deliv Rev 57:1595–1639

Majumdar S, Hingorani T, Srirangam R et al (2009) Transcorneal permeation of L- and D-aspartate ester prodrugs of acyclovir: delineation of passive diffusion versus transporter involvement. Pharm Res 26:1261–1269

Meisner D, Mezei M (1995) Liposome ocular delivery systems. Adv Drug Deliv Rev 16:75–93

Mintzer MA, Dane EL, O'Toole GA (2009) Exploiting dendrimer multivalency to combat emerging and re-emerging infectious diseases. Mol Pharm 9:342–354

Misra GP, Singh RS, Aleman TS et al (2009) Subconjunctivally implantable hydrogels with degradable and thermoresponsive properties for sustained release of insulin to the retina. Biomaterials 30:6541–6547

Mitra AK (2009) Role of transporters in ocular drug delivery system. Pharm Res 26:1192–1196

Nagarwal RC, Kant S, Singh PN et al (2009) Polymeric nanoparticulate system: a potential approach for ocular drug delivery. J Control Release 136:2–13

Navath RS, Menjoge AR, Dai H et al (2011) Injectable PAMAM dendrimer-PEG hydrogels for the treatment of genital infections: formulation and in vitro and in vivo evaluation. Mol Pharm 8:1209–1223

Peeters L, Sanders NN, Braeckmans K et al (2005) Vitreous: a barrier to nonviral ocular gene therapy. Invest Ophthalmol Vis Sci 46:3553–3561

Peng CC, Chauhan A (2011) Extended cyclosporine delivery by silicone-hydrogel contact lenses. J Control Release 154:267–274

Peng CC, Burke MT, Carbia BE et al (2012) Extended drug delivery by contact lenses for glaucoma therapy. J Control Release 162:152–158

Qazi Y, Stagg B, Singh N et al (2012) Nanoparticle-mediated delivery of shRNA.VEGF-A plasmids regresses corneal neovascularization. Invest Ophthalmol Vis Sci 53:2837–2844

Robinson R, Viviano SR, Criscione JM et al (2011) Nanospheres delivering the EGFR TKI AG1478 promote optic nerve regeneration: the role of size for intraocular drug delivery. ACS Nano 5:4392–4400

Sakurai E, Ozeki H, Kunou N et al (2001) Effect of particle size of polymeric nanospheres on intravitreal kinetics. Ophthalmic Res 33:31–36

Samad A, Sultana Y, Aqil M (2007) Liposomal drug delivery systems: an update review. Curr Drug Deliv 4:297–305

Sanders NN, Peeters L, Lentacker I et al (2007) Wanted and unwanted properties of surface PEGylated nucleic acid nanoparticles in ocular gene transfer. J Control Release 122:226–235

Shelke NB, Kadam R, Tyagi P et al (2011) Intravitreal poly(L-lactide) microparticles sustain retinal and choroidal delivery of TG-0054, a hydrophilic drug intended for neovascular diseases. Drug Deliv Transl Res 1:76–90

Smeds KA, Pfister-Serres A, Miki D et al (2001) Photocrosslinkable polysaccharides for in situ hydrogel formation. J Biomed Mater Res 54:115–1121

Smolin G, Okumoto M, Feiler S et al (1981) Idoxuridine-liposome therapy for herpes simplex keratitis. Am J Ophthalmol 91:220–225

Sosa AB, Epstein SP, Asbell PA (2008) Evaluation of toxicity of commercial ophthalmic fluoroquinolone antibiotics as assessed on immortalized corneal and conjunctival epithelial cells. Cornea 27:930–934

Tabbara KF, Ross-Degnan D (1986) Blindness in Saudi Arabia. JAMA 255:3378–3384

Thakur A, Kadam RS, Kompella UB (2010) Influence of drug solubility and lipophilicity on transscleral retinal delivery of six corticosteroids. Drug Metab Dispos 39:771–781

Wang B, Navath RS, Menjoge AR et al (2010) Inhibition of bacterial growth and intra-amniotic infection in a guinea pig model of chorioamnionitis using PAMAM dendrimers. Int J Pharm 395:298–308

White CJ, Byrne ME (2010) Molecularly imprinted therapeutic contact lenses. Expert Opin Drug Deliv 7:765–780

White CJ, McBride MK, Pate KM et al (2011) Extended release of high molecular weight hydroxypropyl methylcellulose from molecularly imprinted, extended wear silicone hydrogel contact lenses. Biomaterials 32:5698–5705

Whitson JT, Ochsner KI, Moster MR et al (2006) The safety and intraocular pressure-lowering efficacy of brimonidine tartrate. 15% preserved with polyquaternium-1. Ophthalmology 113:1333–1339

Williams KA, Coster DJ (2010) Gene therapy for diseases of the cornea—a review. Clin Exp Ophthalmol 38:93–103

Zambito Y, Di Colo G (2010) Thiolated quaternary ammonium-chitosan conjugates for enhanced precorneal retention, transcorneal permeation and intraocular absorption of dexamethasone. Eur J Pharm Biopharm 75:194–199

Zimmer A, Kreuter J (1995) Microspheres and nanoparticles used in ocular delivery systems. Adv Drug Deliv Rev 16:61–73

The manufacturer's authorised representative in the EU is Springer
Nature Customer Service Centre GmbH, Europaplatz 3, 69115 Heidelberg,
Germany. If you have any concerns regarding our products, please
contact ProductSafety@springernature.com

Printed and bound by CPI Group (UK) Ltd, Croydon, CR0 4YY
30/09/2026
02225293-0018